ADVANCES
NITROGEN CYCL
AGRICULTURAL ECOSYSTEMS

Proceedings of the Symposium on Advances in Nitrogen
Cycling in Agricultural Ecosystems held in Brisbane,
Australia, 11–15th May 1987

Edited by J.R. WILSON

Published by
C·A·B International
Wallingford
Oxon OX10 8DE
UK

Tel: Wallingford (0491) 32111
Telex: 847964 (COMAGG G)
Telecom Gold/Dialcom: 84: CAU 001
Fax: (0491) 33508

British Library Cataloguing in Publication Data

Symposium on Advances in Nitrogen Cycling in Agricultural Ecosystems
 (*1987: Brisbane: Qld*)
 Advances in nitrogen cycling in agricultural ecosystems.
 1. Agricultural land. Soils. Nitrogen cycle
 I. Title II. Wilson, J.R.
 631.4'16

 ISBN 0-85198-603-X

Printed in the UK by The Cambrian News Ltd, Aberystwyth

ORGANIZING COMMITTEE

E.F. Henzell

Chief

CSIRO Division of Tropical Crops and Pastures

R.J.K. Myers

Chairman

K.L. Weier

Secretary

G.T. Adams	J.N. Ladd
H.V.A. Bushby	P.G. Saffigna
V.R. Catchpoole	W.M. Strong
R.J. Clements	I. Vallis
J.R. Freney	S.A. Waring
M.P. Hegarty	J.R. Wilson

The committee acknowledges the assistance of Mr E.J. Smith on financial matters, and the help of other administrative and technical staff at the CSIRO, Cunningham Laboratory in the running of the Symposium and the preparation of the Proceedings.

The editor is grateful to Dr J.R. Freney, Dr P.G. Saffigna and Dr I.C. Macrae for assistance with organizing the refereeing of papers, and would like to offer special thanks to Ms K.M. Ward and Miss J.K. Bendixen for their help with preparation of the final manuscript and to Miss M. English for the book index.

CONTENTS

CONTRIBUTORS

J.M. ANDERSON   Department of Biological Sciences, University of Exeter, Exeter, U.K. EX4 4PS.

F.J. BERGERSEN   CSIRO, Division of Plant Industry, P.O. Box 1600, Canberra City, A.C.T., Australia 2601.

A.S. BLACK   School of Agriculture, Riverina–Murray Institute of Higher Education, Wagga Wagga, N.S.W., Australia 2650.

J.M. BREMNER   Department of Agronomy, Iowa State University, Ames, IA U.S.A. 50011.

S. CHRISTENSEN   Institute of Population Biology, Universitets Parken 15, DK–2100 Copenhagen, Denmark.

A.L. COGLE   CSIRO, Division of Tropical Crops and Pastures, Davies Laboratory, P.M.B. Aitkenvale, Qld., Australia 4814.

R.C. DALAL   Queensland Wheat Research Institute, Department of Primary Industries, Toowoomba, Qld., Australia, 4350.

S.K. DE DATTA   IRRI, Agronomy Department, Los Banos, Laguna, P.O. Box 933, Manila, Philippines.

G.D. FARQUHAR   Department of Environmental Biology, Research School of Biological Sciences, Australian National University, Canberra, A.C.T., Australia 2601.

R.C. FOSTER   CSIRO, Division of Soils, Private Bag No. 2, Glen Osmond, S.A., Australia 5064.

J.R. FRENEY   CSIRO, Division of Plant Industry, P.O. Box 1600, Canberra City, A.C.T., Australia 2601.

A.H. GIBSON   CSIRO, Division of Plant Industry, P.O. Box 1600, Canberra City, A.C.T., Australia 2601.

P.M. GROFFMAN   Department of Natural Resources Science, University of Rhode Island, Kingston, R.I., U.S.A. 02881-0804.

D.M. HALSALL   CSIRO, Division of Plant Industry, P.O. Box 1600, Canberra City, A.C.T., Australia 2601.

R.D. HAUCK   National Fertilizer Development Centre, Tennessee Valley Authority, P.O. Box 2040, Muscle Shoals, AL, U.S.A. 35660.

E.F. HENZELL   CSIRO, Division of Tropical Crops and Pastures, 306 Carmody Road, St. Lucia, Queensland, Australia 4067.

D.F. HERRIDGE   New South Wales Department of Agriculture, Agricultural Research Centre, RMB 944, Tamworth, NSW, Australia 2340.

D.S. JENKINSON   Rothamsted Experimental Station, Harpenden, Herts., U.K. AL5 2JQ.

B.T. KANG   International Institute of Tropical Agriculture (IITA), PMB 5320, Ibadan, Nigeria.

J.N. LADD   CSIRO, Division of Soils, Private Bag No. 2, Glen Osmond, S.A., Australia 5064.

S.F. LEDGARD   Ruakura Soil and Plant Research Station, Ministry of Agriculture and Fisheries, Hamilton, New Zealand.

R.L. MC COWN   CSIRO, Division of Tropical Crops and Pastures, Davies Laboratory, P.M.B. Aitkenvale, Qld., Australia 4814.

R.J.K. MYERS   CSIRO, Division of Tropical Crops and Pastures, 306 Carmody Road, St. Lucia, Queensland, Australia 4067.

J.J. NEETESON   Institute of Soil Fertility, P.O. Box 30003, 9750 RA Haren, The Netherlands.

A.P. OCKWELL   ACIAR/CSIRO Dryland Project, P.O. Box 41567, Nairobi, Kenya.

J.S. PATE   Botany Department, University of Western Australia, Nedlands, W.A., Australia, 6009.

E.A. PAUL   Department of Crop and Soil Sciences, Michigan State University, East Lansing, MI, U.S.A. 48824-1325.

M.B. PEOPLES   CSIRO, Division of Plant Industry, P.O. Box 1600, Canberra City, A.C.T., Australia 2601.

R.J. RAISON   CSIRO, Division of Forest Research, P.O. Box 4008, Canberra City, A.C.T., Australia 2601.

T.G. REEVES   Victorian Department of Agriculture and Rural Affairs, P.O. Box 69, Wangaratta, Vic., Australia 3677.

G.P. ROBERTSON   W.K. Kellogg Biological Station, Michigan State University, Hickory Corners, MI, U.S.A. 49060.

P.A. ROGER   IRRI, Soil Microbiology Department, Los Banos, Laguna, P.O. Box 933, Manila, Philippines.

M.M. ROPER   CSIRO, Division of Plant Industry, P.O. Box 1600, Canberra City, A.C.T., Australia 2601.

C.W. ROSE   School of Australian Environmental Studies, Griffith University, Nathan, Brisbane, Qld., Australia 4111.

P.G. SAFFIGNA   Australian School of Environmental Sciences, Griffith University, Nathan, Qld., Australia 4111.

J.O. SKJEMSTAD   CSIRO, Division of Soils, 306 Carmody Road, St. Lucia, Brisbane, Qld., Australia 4067.

C.J. SMITH   Department of Agriculture and Rural Affairs, Institute for Irrigation and Salinity Research, Tatura, Vic., Australia 3616.

K.W. STEELE   Ministry of Agriculture and Fisheries, Invermay Agricultural Centre, Private Bag, Mosgiel, New Zealand.

R.A. STEPHENSON   Maroochy Horticultural Research Station, Queensland Department of Primary Industries, Nambour, Qld., Australia 4560.

J.M. TIEDJE   Department of Crop and Soil Sciences, Michigan State University, East Lansing, MI, U.S.A. 48824-1114.

I. VALLIS   CSIRO, Division of Tropical Crops and Pastures, 306 Carmody Road, St. Lucia, Queensland, Australia 4067.

J.A. VAN VEEN   Research Institute ITAL, P.O. Box 48, 6700 AA Wageningen, The Netherlands.

I. WATANABE   IRRI, Soil Microbiology Department, Los Banos, Laguna, P.O. Box 933, Manila, Philippines.

R.E. WHITE   Department of Soil Science, Massey University, Palmerston North, New Zealand.

J.C. YEOMANS   Department of Agronomy, Iowa State University, Ames, IA, U.S.A. 50011.

PREFACE

This book is the Proceedings of a Symposium held in Brisbane, Australia, in May 1987. The Symposium was the ninth since 1976 of a series of International Symposia and Workshops in which the CSIRO Division of Tropical Crops and Pastures has had an organizing role. Through these Symposia Australia has gained access to new information and expertise helpful to its agriculture, especially in the less-developed subtropical and tropical areas of the north of the continent. Additionally, the reviews of existing knowledge and identification of future research needs contained within these Symposia proceedings have made a valuable contribution to the advancement of science everywhere, not just in Australia.

The subject of this Symposium "Nitrogen Cycling in Agricultural Ecosystems" is of worldwide economic importance. Take wheatgrowing for example, wheat is Australia's most important crop and at the time of the Symposium, in common with the situation in many other exporting countries, was suffering from a worldwide slump in commodity prices. Three of the wheatgrowers' major concerns in these difficult circumstances are directly related to the subject matter of this book:

(1) How to extract maximum efficiency out of the nitrogen fertilizer that they use — worth more than $80M in Australia in 1985. Profitability depends on extracting the largest possible yield increase from that fertilizer.

(2) How to ensure at the same time that as much as possible of the wheat crop earns a premium for high protein content.

(3) Whether to switch to another crop for the moment, such as chickpea or mungbean, and how to manage that legume crop so that it supplies a worthwhile amount of nitrogen to a subsequent wheat crop.

Readers of this book will be able to confirm that there is now a very good understanding of what might happen to the soil and fertilizer nitrogen on its way to creating crop yield and grain quality. The same is true for the nitrogen fixed by pasture or grain legumes. The advances that have been made in this field of science over the last 25 years are truly spectacular.

Nevertheless, there is no cause for complacency. The impressive level of knowledge outlined in this book has not yet been applied to any extent for the benefit of mankind. The fact that the world has avoided catastrophic famine in most developing countries (see Keynote Address by Roland Hauck) is attributable more to a greater total input of fertilizer nitrogen than to any improvement in the efficiency of its use in agriculture.

There are some good reasons for this lack of impact of the new understanding of the nitrogen cycle, and it is not for lack of trying. Nitrogen cycling is an extremely complex phenomenon. Farmers in developed countries, even with computer-aided decision support systems, find it difficult to manage nitrogen efficiently. The problem is even more baffling for third-world farmers. They are so much at the mercy of weather, which governs not only the ability of the crop to use soil nitrogen but also the losses that occur during drying (ammonia volatilization) and wetting (denitrification or leaching). These at present unpredictable weather-driven relationships would be worthy topics for future international deliberation.

E.F. HENZELL

ACKNOWLEDGEMENTS

The Organizing Committee of the Symposium 'Advances in Nitrogen Cycling in Agricultural Ecosystems' gratefully acknowledge financial contributions towards the expenses of the Symposium received from the following:

Ansett Airlines
Australian Development Assistance Bureau
British Council
Incitec Ltd.
Premier's Department, Queensland State Government
Qantas Airways Ltd.
Root Nodule Pty. Ltd.
Utah Foundation
Wheat Research Committee of Queensland

PART I

KEYNOTE ADDRESS

A HUMAN ECOSPHERE PERSPECTIVE OF AGRICULTURAL NITROGEN CYCLING

R.D. Hauck

ABSTRACT

Concepts of nitrogen cycling are expanded to include the cycling of nitrogen throughout world socio-economic systems. Because of its important role in food and fibre production, the extent to which nitrogen is available in these systems contributes to the paradox of overproduction in some parts of the world and inadequate production and supply in others. Expanding the concept of nitrogen cycling into the human ecosphere raises the question: Who are the clients of one's research? This question is discussed in terms of the chronically hungry, farmers, agribusiness, the affluent public, environment, and Self -first, in relation to observations about these clients, then in relation to nitrogen cycle processes.

INTRODUCTION

About $0.5-1.5 \times 10^9$ years ago, ammonia gas dominant in the earth's atmosphere was gradually oxidized to water and dinitrogen. The global nitrogen cycle had begun. Over geologic time between then and now, abiological and biological nitrogen transformation processes have developed and interacted to form the overall global cycle of our current understanding. Through experimentation and observation, we continue to refine this understanding, adding details, quantifying and correcting, identifying and closing knowledge gaps, and viewing the global nitrogen cycle from new perspectives. It is the nitrogen cycle from an agricultural perspective that concerns us here, and more specifically, the cycling of nitrogen in agriculture from the perspective of human needs, desires, and institutions.

THE FOOD/POPULATION EQUATION

"Population, unchecked, increases in a geometric ratio. Subsistence increases only in arithmetical ratio" (Malthus 1798). A century after the grim Malthusian prediction of famine and its consequences, Sir William Crookes echoed the alarm: "... all civilized nations stand in deadly peril of not having enough to eat," concluding that, because of impending shortages of nitrogen, food production might not keep pace with population needs by 1931. "The fixation of atmospheric nitrogen is vital to the progress of civilized humanity" (Crookes 1899).

The simple equation relating world food needs with population became exceedingly more complex within the first quarter of the present century following, and resulting in part from, the discoveries of chemical means of fixing atmospheric dinitrogen on a commercial scale. Within two decades after the first commercial synthesis of ammonia (Curtis 1932), Davis (1939) in *History's Answer to Sir William Crookes* observed that: "The wheat problem of 1931 was one of world surplus, not world shortage, actual or impending." Yet the spectre of famine has been raised time and again. "Ten years from now, parts of the underdeveloped world will be suffering from famine. In fifteen years the famine will be catastrophic and revolutions and social turmoil and economic upheaval will sweep areas of Asia, Africa, and Latin America" (Paddock and Paddock 1967). This dire prediction has not come to pass. Crookes (1899) correctly surmised, "The chemist will step in and postpone the day of famine to so distant a period that we and our sons and grandsons may legitimately live without undue solicitude for the future."

Widespread, catastrophic famine has not occurred. Today, despite minimal increase in land under cultivation and a 50% increase in world population since 1960, per capita food supply has increased about 5%. Through the application of improved technologies, including plant varieties responsive to fertilizer nitrogen, between 1961 and 1980, wheat production tripled in India, and doubled or more than doubled in China, Turkey, and Pakistan. Between 1974 and 1983 world food production increased at an annual rate of 2.2%, almost 0.5% faster than population growth. During this period, developing countries as a whole increased their per capita food production by about 10%, a rate far exceeding that of countries with developed agriculture. But in the 36 least developed countries, the average increase in food production of 2.1% did not keep pace with a population growth rate of 2.6%. Particularly alarming is the current situation in Africa where food production is growing at a rate of 1.3%, but population at a rate of almost 3% per annum.[A] (see footnote next page).

Current trends in the food/population balance forecast an increasing level of extreme malnutrition in some areas of the world (specifically Sub-Sahara Africa and parts of South and West Asia). In the remaining areas, from at least a nutritional point of view, the inhabitants can "live without undue solicitude for the future." But the grandsons of Sir William Crookes' contemporaries are adults today. They and their sons and daughters must resolve the paradox of extreme food/population imbalance —hundreds of millions hungry in some areas of the world while surplus food accumulates in the areas remaining. For how long in the future of an increasingly more complex and interdepedent world of human activities can this paradox be sustained?

Modern agricultural science and technology have demonstrated their power to produce ever-increasing amounts of food and fibre, sufficient, perhaps, even to sustain a human population projected to stabilize at 10-11 billion by the year 2100. But problems of food supply cannot be alleviated through science and technology alone. Neither Malthus nor Crookes, among many, were able to foresee that the food/population equation would become a complex function of humanitarian, socio-economic, environmental, political, and technological variants.

Many conferences have been convened and articles written in recent years concerning the interrelationships among these factors. My intent is not to summarize this information but to raise levels of consciousness about the entire world ecosystem of which agriculture is a part. Because human food supply is indispensably linked to nitrogen supply and use, those interested in the cycling of nitrogen through agricultural systems are asked to keep in mind all aspects, forces, and consequences inherent in the food/population equation.

PHYSICAL AND CONCEPTUAL NITROGEN CYCLING

The human ecosphere

About 60 Mt (million tonnes) of fertilizer nitrogen is now added to world agricultural soils, mainly for the production of cereal grains which yielded 1,660 Mt in 1985/86. An estimated 90 Mt of nitrogen is added annually through biological dinitrogen fixation, of which about 50 Mt is fixed by forage legumes and 40 Mt is fixed during the production of (currently) 145 Mt of legume grain. These amounts of nitrogen are added disproportionately to 11.1% of the earth's total land area, viz., 1,450 million hectares of croplands (1980 estimate). Based on nitrogen consumption trends and projected minimum caloric needs for a human population of 6.16 billion, an annual

---

A Unless otherwise indicated, statistical information presented in this and subsequent sections has been adapted from the following sources: Borlaug, N.E. (1982) – Feeding mankind in the 1980s: The role of international agricultural research. World Bank Tech. Paper No. 1, pp. 57-81. (World Bank: Washington, D.C.); Brown, L. R. (1981) – World food resources and population. The narrowing margin. Population Bull., Vol. 36, no. 3. (Population Reference Bureau, Inc.: Washington, D.C.); Food and Agriculture Organization of the United Nations (1985) – Fertilizer Yearbook, Vol. 34 (FAO: Rome.); Food and Fertilizer Technology Center (1980) – Food Situation and Protein in the Asian and Pacific Region. (FFTC Book Series No. 17: Taipei.); Mellor, J. W. and Johnston, B. F. (1984) – The world food equation: interrelations among development, employment, and food consumption. J. Econ. Lit. Vol. XXII, 531-574; TVA (1980-1986) – Internal statistical information, National Fertilizer Development Center. (Tennessee Valley Authority: Muscle Shoals, Alabama.); World Resource Institute and International Institute for Environment and Development (1986) – World Resources 1986. (Basic Books Inc.: New York.)

consumption of 100 Mt of fertilizer nitrogen is forecast for the year 2000, just over a decade from now.

Most of the increase in nitrogen use will be in developing countries (Harre 1986). Their consumption of nitrogen fertilizers will grow at an expected rate of almost 6% per annum as compared with <1% in developed countries (excluding countries with centrally planned economies). Between the end of World War II and the year 2000, the amount of nitrogen entering the terrasphere each year as a direct result of agricultural production will be increased more than ten-fold. Ever increasing amounts of nitrogen will be cycling through ecosystems in developing countries previously in equilibrium with much lower levels of nitrogen input. Conscious of this fact we now study nitrogen cycling in wetlands and drylands, pastures and forests, and in ley, multiple cropping, and other systems, each ecosystem quantitatively and qualitatively different in the course and extent of nitrogen cycling. How and when this cycling occurs affects the efficiency of nitrogen use. The challenge is to increase this efficiency, thereby increasing the crop production value of nitrogen, increasing farmer profits, conserving nitrogen, energy, and natural resources, and decreasing the risk of adverse effects on the environment which can result from inefficient nitrogen use.

Keeping in mind our usual concepts of nitrogen cycling, let us expand our view of agricultural systems to include nitrogen cycling through the entire human ecosphere. In this context, nitrogen added to land cycles <u>physically</u> within and among plants, soils, and waters, the atmosphere above, animals, and humans, and, <u>conceptually</u> within all of the economic, political, and social institutions that are part of the food/population equation. This expanded view of nitrogen cycling may enable us to better understand the reasons for a particular focus of research and to establish nitrogen research priorities responsive to the needs of a changing agriculture.

Clearly, nitrogen cycling in the biosphere is of interest to investigators representing various scientific disciplines, but these investigators have different objectives, use different approaches, and focus on different levels of integration from the molecular to the global ecosystem. Increasingly, agricultural scientists and those interested in the nutritional and environmental effects of increased agricultural production acknowledge the need to engage in multidisciplinary research. There is need also for increased communication and the sharing of information and ideas between scientists and those who work in nonscientific disciplines. No better opportunity for such information exchange comes to mind than now when thinking people of diverse interests have a common concern about world food supply, food trade, and the environmental effects of technology. Because increased nitrogen use is essential to increased food and fibre production and also is the source of real or perceived environmental problems, investigators of nitrogen cycling, in my opinion, have a special responsibility, as well as opportunity, to think and work in reference to the entire human ecosphere.

Questions

As individuals, few of us have opportunity to make a measurable impact on the entire world ecosystem, but an acute awareness of its interactions stimulates our thinking and may affect our choice and planning of research. As an individual citizen-scientist of the world, useful questions to ask oneself are: What information is being obtained elsewhere which would refine the approach or alter the focus of one's own work? How can one's work be interfaced with that of others working toward the same overall objective, but from a different point of view? Who will use the results of one's research? Who are the clients of one's research? Conferences such as this one, and especially conferences that bring together people with entirely different areas of expertise working on a common problem, provide opportunity for answering these questions.

Who are the clients of one's research? This question may lie imprecisely formulated in the subconscious but needs to be raised to the level of consciousness for rigorous examination in terms of the human ecosphere. Among the possible clients are chronically hungry people, farmers, agribusiness, the affluent public, the environment, and Self. The fact that they are not necessarily mutually exclusive is the basis for variety and priority setting in one's research. Let us examine each client in terms of client needs.

Clients of research

Chronically hungry people

The sophistication of modern agriculture can be measured by the fact that about 3% of the population in the United States, 32% in the Soviet Union, 40% in Latin America, 50% in Asia, and 75% in Africa are directly involved in on-farm crop production.

Three years from now, in 1990, 76% of the world's people will be living in lesser developed countries (LDCs), but 85% of the economic activity and 50% of world grain production will occur in developed countries. The poorest of the LDCs contain most of the chronically hungry of the world. They are hungry because they are poor, not because world food supplies are inadequate. Chronic hunger occurs in countries that lack the infrastructure and resources for intensive food production and lack the purchasing power to compete for available food supplies.

Eventually, in these countries, as resources become available, increases in per-area yields will be achieved with increased use of nitrogen fertilizers, irrigation, and other inputs of intensive agriculture. But now much of the food is produced in these countries by traditional methods on small parcels of land. In the most impoverished countries, new technologies must be introduced slowly with emphasis on the farmer and the farm rather than simply on techniques for increasing production. To increase traditional food supplies and to transform traditional agriculture into modern agriculture will require the help of many well-trained, well-motivated scientists and technologists and an informed extension service working closely with farmers.

Because of the situation described, much nitrogen research specifically directed toward maintaining world food and fibre supplies addresses only incidentally the chronically hungry. But the thinking scientist can find opportunity to work on their behalf without adopting them necessarily as a primary client.

Farmers

Whether they produce food or fibre for barter or for sale, farmers are business people working for profit. They desire maximum yield obtained conveniently at minimum cost, including the cost for nitrogen. Grain and fibre producers in the United States spend about 25% or more of their annual operating funds for nitrogen (see TVA, footnote A). Efficiency of nitrogen use is a primary concern of farmers because inefficient use lowers their profits and increases the risk of adversely affecting environmental quality. Concern for the latter may be for the environment itself or out of fear of impending legislation regulating nitrogen fertilizer use. Most agricultural scientists working with nitrogen in developed countries, and many in LDCs, consciously or not adopt the farmer as their primary client. Most of the technological innovations, e.g., nitrification and urease inhibitors, fluid nitrogen fertilizers, acidified urea formulations, and management practices that minimize nitrogen loss from the soil-plant system, seek to increase the efficiency of uptake and use of applied nitrogen. These innovations make nitrogen use less costly and more convenient, and are directed primarily at farmers in countries with rapidly developing or fully developed modern agriculture. Other farmers benefit from these technologies and practices only when they become affordable under their particular situations of use, and where indigenous skills and infrastructure permit.

Agribusiness

Included in this category of clients are the fertilizer industry, growers of specialty crops, and domestic and international agricultural commodity trade markets. Large tonnages of nitrogen as fertilizer or in foods, feeds, and fibres cycle within agribusiness communities. The flow of nitrogen through agro-economic systems increases continually in magnitude and complexity. Agriculture in developed countries is no longer only a domestic issue because agribusiness is becoming increasingly international. As their income increases, LDCs become important importers of food grown in developed and rapidly developing countries. Thus, world food trade has increased steadily since the 1970's; today about 20% of all cereal grains cross international borders. In spite of this, the paradox remains that chronically hungry nations cannot participate in food trade to the extent needed while overproduction remains a major problem of affluent societies.

A major objective of national agricultural research is to increase the competitive advantage of that nation's farmers in the world agricultural market place. Increasing the efficiency of nitrogen use is an important approach toward achieving this objective. To a lesser but significant extent, some research is conducted mainly for the nitrogen fertilizer industry, e.g., work on product differentiation to give a particular nitrogen formulation a competitive edge. When one accepts the farmer and/or agribusiness as a major client of research, this should be done recognizing that food and fibre production is geared to market incentives that have global implications, some of which are desirable, others, undesirable.

The affluent public

Modern agriculture has been so successful that abundant food supply in affluent countries is taken for granted. Because food production in these countries is geared to market incentives, nations protect food supplies and farm income through price supports, which in turn stimulate still higher production. Because citizens of affluent societies may find unacceptable taxing the public to sustain over-production, the danger is that legislators will be reluctant to fund research that increases production capacity. In matters of environmental concern, including those associated with nitrogen fertilizer use, the courts, legislative bodies, and the public are increasingly involved in decisions previously made only by scientists. The prudent agricultural scientist and administrator must recognize that as a primary client of research, the public is both decision-maker and consumer.

Environment

Concern for conservation of resources and the environment has local, regional, and global implications for nitrogen cycling research. The level and geographical dimensions of concern differ among investigators serving different clients. Because food and fibre production is a compelling concern of investigators in LDCs, they may initially regard the effects of nitrogen use on the environment to be of secondary to little importance. Because a responsible attitude toward nitrogen use is a recognized need in developed agriculture, research on nitrogen use in crop production seeks to provide a basis for an acceptable compromise among the conflicting goals of maximum yield, maximum profit, and zero environmental stress.

Much has been written about the real and/or publicly perceived problems of nitrogen fertilizer use and the quality of surface waters and groundwaters, high nitrate levels in foods and feeds, and the effect of increased levels of fixed nitrogen in the biosphere on ozone depletion in the stratosphere. Legislation has been proposed in Australia, the European Economic Community, and the United States on one or more of the following: animal stocking rates, nitrate contents of drinking waters and certain vegetables, and application of fertilizer nitrogen, animal manures, and sewage sludge to land. Obtaining data that provide a rational basis for decisions about legislation is a major objective of research directed toward the environment and the affluent public as clients.

In LDCs, fertilizer nitrogen use will increase from 14.3 Mt in 1985 to an expected 40 Mt annually by the year 2000. From an environmental perspective, this nitrogen will cycle within two separate but related ecosytems: the agricultural and the urban. Intensive nitrogen use on LDC croplands may eventually result in problems of environmental quality similar to those now being recognized in developed and some rapidly developing countries. Although current nitrogen use in LDCs averages only about 18 kg/ha of cropland, it ranges from a low of 5 kg/ha in Africa to about 200 kg/ha in Korea. As intensive nitrogen use in LDCs becomes increasingly widespread, nitrate enrichment of their drainage waters and groundwaters is highly probable. Nitrate enrichment of drinking water could be a serious health hazard to humans, especially infants, in those LDCs where inadequate sanitation and disease-polluted waters cause chronic intestinal infections that may be fatal when combined with high levels of nitrate.

Nitrogen applied to croplands will flow from fields as food and will concentrate in urban areas. An urban population of 3 billion forecasted for LDCs by the year 2000 will excrete as waste almost 20 Mt of nitrogen per annum. In Bombay, Calcutta, Lagos, Manila, Seoul, and other cities with populations greater than 5 million, the annual nitrogen load in each urban area will exceed 25 million kg, adding further stress to potable water supplies.

Except in most of Africa, intensive use of nitrogen will be coupled with increased use of irrigation, viz., from about 170 million hectares in 1984 to an expected 260 million hectares by 1994 (see World Resources, footnote A). Mismanagement and/or recycling of irrigation waters is resulting in chronically waterlogged soil, denitrification loss, and endemic salinization on an alarming scale.

Solutions to the many problems of food and fibre production and use lie outside the scope of any single discipline, any single area of research, even as broad-based as studies of nitrogen cycling can be. The world is not structured by disciplines; its problems are solved through interdisciplinary effort. As a scientist, one can recognize how isolated areas of nitrogen research relate to the whole of nitrogen cycling through the human ecosphere. Thinking about this ecosphere both as a scientist and as a citizen-scientist inevitably will lead one to ever more relevant activities.

Self

Research can be done for reasons of ego, curiosity, valuing the extension of human thought and knowledge, or for any of the many other reasons that cause a person to work without particular focus on a client other than oneself. (Not included here is serving oneself as we all do to provide for one's physical needs). Sometimes the state of knowledge is such that a client cannot be identified. In that case, one seeks through basic research to extend knowledge to the point where client identity becomes apparent. The extent to which the public should support those who pursue "science for science's sake" is a matter of public judgment, philosophy, and values. In the mainly mission-oriented sciences, such as those serving agriculture, there are fewer opportunities, less public money, and, perhaps, less justification for research pursued primarily for the Self. When public monies stem from public needs, those that are clients of their own research may need to find other than a public source of funding for their research, or serve another client. Collectively, agricultural scientists must give assurance to the public that they are responsive to their public and/or world concerns and needs.

NITROGEN CYCLE PROCESSES

Process/client relationships

Clearly, the research approach to a particular nitrogen cycle process is determined in part by the client to which the research is directed. Although models can be conceived of the total interactions of nitrogen cycling through the human ecosphere, experiments on a particular nitrogen cycle process are made sequentially and often without direct reference to the many processes that simultaneously interact with it. The experimenter usually is aware that these interactions occur but may not be aware of the significance of a particular set of interactions to the different clients of the research. The need for multidisciplinary efforts to improve an understanding of the totality of nitrogen cycling through the human ecosphere is now recognized, but that one can modify one's research to include several clients may not be recognized to the extent needed. The remaining discussion of nitrogen cycling processes will emphasize how research on these processes can be made to serve the needs of more than a single client.

Nitrification: effects, including nitrogen loss

Because of its three major effects on abiological and biological reactions in soils, nitrification has a central role in nitrogen cycle processes. It produces mobile and oxidized forms of nitrogen that can be lost from soil through leaching and runoff, and through reduction to nitrogen gases. It affects the nutrition of both lower and higher plants because of their dissimilar use of the substrate for nitrification, ammonium, and the end product, nitrate. It determines the residence time of an individual nitrogen atom in the terrasphere or hydrosphere before it is released to the atmosphere. Control of nitrification is one approach to increasing the efficiency of plant use of applied nitrogen. This can be done by any one or a combination of several methods, including adding nitrogen at a time and in a manner that permits plants to compete effectively for that nitrogen, use of slow release nitrogen sources, and use of nitrification inhibitors.

Because of the different paths that applied nitrogen can take depending on the extent of its nitrification, control of nitrification may benefit one client and adversely affect another. For example, consider the multiple effects and their possible consequences of adding a nitrification inhibitor with an ammoniacal or ammonium-producing fertilizer early in the growing season:

(i) If the nitrification inhibitor is effective, much of the available nitrogen supply for seedlings and young crop plants will be relatively immobile.

(ii) The ratio of ammonium to nitrate will change at a slower rate than if inhibitor were absent.

(iii) Nitrification of ammonium derived from soil organic matter may be repressed, the extent depending on the amount of nitrification inhibitor used with the applied nitrogen and the manner of fertilizer application, i.e., banded or broadcast.

(iv) There may be less denitrification and movement of nitrogen as nitrate early in the growing season.

(v) There may be more heterotrophic immobilization of fertilizer-derived ammonium early in the season.

(vi) The slower rate of nitrification may adversely affect growth of ammonium-sensitive plants such as the Solanaceae, or may benefit the growth of certain cereal grains, e.g., wheat, that may respond to higher levels of available ammonium later in the growing season.

(vii) After harvest, applied nitrogen that has become immobilized, subsequently mineralizes and nitrifies; the nitrate that is formed during this reaction sequence may benefit growth of an immediately succeeding crop, or in the absence of such a catch crop, may leach or denitrify.

In the field evaluation of nitrification inhibitors, a first step is to delineate crop management systems in which loss of nitrogen as a result of nitrification can be prevented through use of an inhibitor or where adjusting the ratio of ammonium to nitrate may benefit crop growth. The next step is to find the best way to manipulate the fertilizer-crop management system so as to maximize the benefits of using nitrification inhibitors and minimize their possible adverse effects. The potential clients of such research are the farmer, agribusiness, the affluent public, and the environment. However, one can reasonably assume that situations may arise where use of a nitrification inhibitor may increase crop yields while at the same time increasing rather than decreasing the potential for nitrogen loss resulting from late-season nitrate formation. In that case the farmer and agribusiness experience a short-term benefit while the affluent public and the environment are disadvantaged. Two questions in this regard yet to be answered using long-term field experiments with ^{15}N are:

(i) How does immobilization of applied nitrogen during the growing season affect its subsequent mineralization, nitrification, and loss?

(ii) Does repeated use of a nitrification inhibitor affect soil nitrifying capacity?

Are the chronically hungry clients of this research? At the present state of knowledge of nitrification inhibitor use, the farmer must risk paying for a technology that in many circumstances could provide only marginal benefit; this militates against nitrification inhibitor use by the poorer farmers. In rapidly developing countries farmers may profitably use such inhibitors to improve the efficiency of nitrogen fertilizer use, e.g., for direct-seeded, flooded rice under good water management. In this system use of a nitrification inhibitor may prevent nitrate formation during the pre-flood period, thereby minimizing denitrification loss after application of permanent floodwater. Where maintenance of floodwater is entirely dependent on rainfall and the paddy is subject to periodic wetting and drying, nitrification is more difficult to control than where floodwater is maintained by irrigation. By directing research toward solving the problems of nitrification/denitrification in paddy systems with uncertain water supplies, an investigator in a developed country serves the needs of the chronically hungry through research that provides information that is useful to scientists working in those LDCs where nitrogen use for rice is extremely inefficient.

Nitrogen may be lost as nitrous oxide (N_2O) formed during nitrification (Bremner and Blackmer 1978). Because only a fraction of one percent of applied nitrogen usually is evolved as a result of this process, the relatively small amount of nitrogen lost is of little practical importance to farmers. However, from the perspective of the global nitrogen cycle, the cumulative annual emission of N_2O into the atmosphere from this process may be significant. Assuming conservatively that an average of 5 kg of nitrate-nitrogen is formed annually from each hectare of the earth's croplands, permanent pasture, and forests, and assuming that 0.05% of the nitrogen that is nitrified is evolved as N_2O during the reaction, then the annual total emission of nitrogen as N_2O would exceed 20 million kg.

Another possible avenue of nitrogen loss during nitrification involves the accumulation and reaction of nitrite in soils to form nitrogen gases (Nelson 1982). The practical significance of such nitrite reactions is unknown. The client of research in this area is as yet unidentified but will become apparent when sufficient quantitative information has been obtained.

Ammonia volatilization and denitrification

Studies of denitrification, and studies of ammonia (NH_3) volatilization have both local and global implications. On a local scale either process may contribute to the inefficiency of nitrogen use, affecting mainly the farmer. Studies on the local scale are directed toward improved understanding of the processes so that they can eventually be manipulated for farmer benefit. Decreasing denitrification loss is best approached through control of nitrification. Decreasing nitrogen loss by NH_3 volatilization focuses mainly on decreasing NH_3 loss from urea or animal manures added to the soil surface or from urea added to rice paddy floodwaters. Improved management practices such as soil incorporation or deep placement of urea can benefit both poor and affluent farmers. More sophisticated approaches such as use of urease inhibitors or acid-forming urea formulations are more applicable to affluent farmers and the fertilizer industry than to poor farmers in LDCs.

Models of NH_3 fluxes estimate that > 1,100 Mt of nitrogen as NH_3/NH_4^+ could circulate annually between the earth and its atmosphere (Hauck 1984). Although most estimates of NH_3 transfer from and into the atmosphere are much lower (and estimates vary considerably), they all suggest that NH_3 is being evolved in very large quantities into the atmosphere from a source or sources not generally being measured. Direct and indirect evidence from greenhouse and field experiments indicates that plants may be a major source of atmospheric NH_3, not only from senescing tissues but also from actively growing leaves (Wetselaar and Farquhar 1980). The NH_3 subcycle of the global nitrogen cycle affects acid deposition from the atmosphere to the land and waters and the possible transfer of nitrogen between ecosystems. Research on the possible loss of NH_3 from the phyllosphere, especially from crop plants, has both the farmer and the global environment as clients.

Quantification of denitrification loss and NH_3 volatilization on a global scale remains imprecise. Identifying and measuring the contributions of all sources of and sinks for N_2O (a major product of denitrification) in the troposphere are research activities of potential interest to all humans because of the observed relationship between tropospheric N_2O content and the regulation of stratospheric ozone. However, whether increased levels of fixed nitrogen in the biosphere and increased N_2O formation will result in progressive depletion of ozone has not been established. Should such a sequence of reactions pose a major health hazard to humans, then research on denitrification will have an urgency beyond that of improving the efficiency of nitrogen use and may well focus on changing the ratio of N_2O/N_2 that is evolved during denitrification. Other aspects of denitrification that may need to be considered in environmental studies include the horizontal transport of dissolved N_2O to surface waters (Bowden and Bormann 1986), possible leaching of dissolved N_2O, and the occurrence of denitrification in groundwaters by Thiobacillus denitrificans using ferrous sulfide as a reductant (Strebel et al. 1985).

Biological nitrogen fixation

Biological N_2 fixation as a key to world food production is an idea accepted by most investigators of nitrogen cycle processes, but much research on this is far removed from helping the chronically hungry as a client. Moreover, how enhancing biological N_2 fixation and increasing food production in LDCs are related merits examination.

During the period 1979-1984 when cereal production in the developing world increased 37%, total production of pulses increased by only 6.8%. Most of these legumes were grown for direct human consumption. During the same period, soybean production in the developing countries increased 93.7%, mainly because of the high cash value of this legume. The increased production of cereal grains for carbohydrate production has not been matched worldwide by an increase in production of plant protein. In LDCs, total calories per capita derived from vegetables and meat rose 114 and 129%, respectively. Total meat consumption during 1979–1984 increased by more than 25%, being related directly to increase in wealth and standard of living. To some extent the rate of increase in production of food for human consumption has been lessened by increased production of animal feed and forage. Regardless of whether the increased production of plant protein for human consumption would be more desirable from an economic or environmental point of view, the trend toward more meat consumption by an increasing proportion of the world human population should be considered in planning long-term strategies for biological N_2 fixation research.

The relatively high cost of nitrogen from fertilizers versus biologically fixed nitrogen has been cited as a major reason for enhancing the use of biological N_2 fixation in LDCs. The actual price of fertilizer nitrogen to a farmer varies considerably; it is determined by the costs of production, storage, handling, and distribution, and, in many LDCs, by the amount of government subsidy. The cost of the starting material for most nitrogen fertilizers, anhydrous ammonia, when indexed for inflation, decreased during the period 1945–1985, with the exception of certain periods (e.g., 1974–1975, years of unusually high oil prices). The price of urea, which accounts for about 80% of all nitrogen fertilizer used in Asia, has paralleled that of ammonia. However, in poorer countries where fertilizers are least available, their cost relative to other agricultural inputs remains high. Although replacing at least part of the expected increase in nitrogen fertilizer use with nitrogen obtained through biological N_2 fixation remains a desirable goal, this should be considered in the context of the totality of human needs, including the immediate need for carbohydrate and protein, as well as future food consumption patterns.

An enormous gap has developed between research on the fundamental aspects of biological N_2 fixation and the practical application of basic discoveries. This gap is widening because of lack of focus of this basic research on crop production. Among the most signal contributions that basic biological N_2 fixation research could make to crop production, both in LDCs and in developed countries, would be to develop legume–rhizobial systems that will efficiently fix atmospheric N_2 in the presence of high levels of fixed nitrogen. Concomitant with this basic research can be the investigation of green manures and other substances that may release nitrogen at a rate that benefits legume growth but does not interfere with nodule formation and N_2 fixation activity. Some biological N_2 fixation systems have evolved where the N_2 fixing component of the system is physically separated from the medium from which the host plant obtains part of its nitrogen needs, e.g., the Azolla–Anabaena symbiosis, or the legume Sesbania rostrata which is nodulated on both stems and roots.

The high energy requirement for fixation is probably the greatest limitation of the biological N_2 system. More emphasis is needed on improving the transfer of photosynthate from the leaf to the nodule without the accumulation of undesirable metabolites. Because of the comparative ease with which sophisticated instrumentation and powerful genetic engineering tools can be applied to microbiological research, or because of the large volume of work that can be completed in a relatively short period of time, perhaps because of the elegance of the research per se, regardless of the reasons, our knowledge of the physiology and molecular biology of the microbial symbiont is far greater than our understanding of the host plant. Until major multidisciplinary efforts emphasize strategic research that advances basic findings closer to applications in productive ecosystems, no client of basic biological N_2 fixation research, other than Self, will be well served in the foreseeable future.

Nitrogen balance

Conceptual and quantitative understandings of nitrogen cycling are described in models of nitrogen balance. They describe inputs to and outputs from ecosystems and consider internal nitrogen flows and fluxes among nitrogen storage pools. The residence time of nitrogen in any given compartment of the system may vary from seconds to eons. The spatial scale varies from the microsite to global dimensions. Increased understanding of nitrogen cycling is of interest not only to investigators mainly concerned with food and fibre production but also to a variety of biological and physical scientists who study the effects of nitrogen in different environments and who pursue short-term and long-term objectives relevant to local, regional, and world concerns. There is increasing need for investigators of nitrogen cycling to think on all levels of integration, from the microcosm to the macrocosm, to enable them to accurately and effectively use all available and appropriate information to serve different clients.

A nitrogen balance study that characterizes the flooded rice system, or one that provides a long-term measure of nitrogen fertility in a given crop management system, can be so constructed as to provide data of use in different ways to all clients of research under discussion here. Every investigator understands that single nitrogen cycle processes do not exist unrelated to other processes, that nitrogen cycling in agricultural systems is a subset of global nitrogen cycling. Whenever possible, this understanding should be translated into activities that become part of the multidisciplinary, multiobjective efforts that are needed to sustain the human ecosphere.

Methodology and technology

Methodology

Increased understanding and more accurate quantification of nitrogen cycle processes rapidly follow advancements in methodology. Included among important advances are the development of techniques for the now rather routine use of stable nitrogen tracers; automated instruments for the rapid, precise, and accurate determination of nitrogen isotope-ratios on microgram quantities of nitrogen; the acetylene reduction and acetylene blockage techniques for determinations (with awareness of their limitations) of nitrogenase activity and denitrification loss, respectively; automated instrumentation for gas analysis by combustion and/or chromatography and ion analyses by colorimetry and high performance chromatography; methods for assaying nitrate reductase, nitrate permease, and other enzymes that permit rapid survey of plants for traits important to efficient use of nitrogen; and computer processing of large amounts of interrelated nitrogen cycling information.

Use of these advances leads to the refinement of information needed to generate new ideas for improved management of plants, soils, and agricultural inputs and to better understanding of the implications of changing levels and patterns of nitrogen use.

All clients of agricultural research usually are directly or indirectly benefited by use of these methodologies except in those not too infrequent cases where use of the methodology itself becomes the objective and Self-interest the client of the research. Regardless of the major client of one's research, awareness of other client needs may stimulate developments that serve other clients, e.g. a micro-diffusion technique or a semi-micro distillation apparatus that is equally suitable for use under substandard, as well as ideal laboratory conditions.

Biotechnology

Some view cellular and subcellular biotechnology of the 1980s as fostering revolutionary breakthroughs in agriculture; others view it merely as the most recent major research tool for the progress of biological research. From either perspective, modern biotechnology differs from other research innovations in the speed and magnitude of its expected productivity gains. Major gains probably will be in animal and seed production and in alleviating environmental stress in crop production. Direct effects of biotechnology on nitrogen cycling are likely to be minimal. Molecular biology has led to rapid progress in the identification of nif genes and how they are organized in the Rhizobium-legume symbiosis, but this information has yet to result in improved N_2 fixation systems for use by farmers. With non-legumes, research on actinorhizal plants coupled with biotechnology has resulted in large-scale replication of Frankia-infected Alnus clones, which in extensive plantings could significantly increase the nitrogen economy of marginal lands. Because nitrogen cycling is inherent in any agricultural system, it is necessarily affected by biotechnological innovations introduced into that system. But the effects of such innovations will be small in the foreseeable future.

UNIFIED PERSPECTIVES

Concepts of nitrogen cycling become increasingly complex in response to need and purpose. As knowledge increases, so does awareness of the unknown and our understanding of why this knowledge gap should be filled. At some point in the acquisition of knowledge, it is appropriate to question whether emphasis should be shifted from adding detail and refinement to a concept to expanding the concept into new and specific areas of application. Expansion, not towards increasing complexity, but towards understanding why and for whom research is conducted, is the focus of the nitrogen cycling concept presented here. Not discussed, but kept in mind, were the interactions of microflora and microfauna; nitrogen cycling within the plant and between the plant and its environment; nitrogen cycling among diverse ecosystems; or the entire spectrum of subcycles on one level of integration interacting with those immediately above and below. Research from fundamentally basic to immediately applicable is needed in all aspects of nitrogen cycling. In expanding nitrogen cycling in agricultural ecosystems to include its cycling through the human ecosphere, the intent is not to pass judgment on the value or direction of any research endeavor. Such judgments are left to individual investigators of nitrogen cycling who are invited to consider the purpose and clients of their research.

The often repeated calls for multidisciplinary research attest to the recognition of its need. On the level of the individual scientist, multidisciplinary research begins with an awareness and appreciation of multiple interactions. Studies of nitrogen cycling further such understanding. Aware as we are of nitrogen's vital role in food and fibre production and environmental effects, as individual scientists, we are seemingly helpless to solve problems of society that are inextricably intertwined with our technologies. As citizen–scientists, we are obligated to try, and collectively we can make an impact. The greatest advances in nitrogen cycling research in agricultural ecosystems will develop from this unified effort.

REFERENCES

Bowden, W.B., and Bormann, F.H. (1986). Transport and loss of nitrous oxide in soil water after forest clear cutting. Science (Washington, D.C.) 233, 867–9.

Bremner, J.M., and Blackmer, A.M. (1978). Nitrous oxide emission from soils during nitrification of fertilizer nitrogen. Science (Washington, D.C.) 199, 295–6.

Crookes, W. (1899). The wheat problem. Presidential address to Brit Assoc. Advancement Sci., London, 7 Sept., 1898. Rep. 68th Meeting, pp. 2–38.

Curtis, H.A. (1932). A history of nitrogen fixation processes. In 'Fixed Nitrogen'. (Ed. H.A. Curtis) pp. 77–89. (Amer. Chem. Soc. Monograph Ser., The Chemical Catalog Co., Inc.: New York).

Davis, J.S. (1939). The spectre of dearth of food: History's answer to Sir William Crookes. In 'On Agricultural Policy 1926–1939'. (Ed. J.S. Davis) pp. 3–23. (Research Institute: Stanford Univ., CA.)

Harre, E.A. (1986). Outlook for fertilizer chemicals. Presented at World Chemical Congress, Newport Beach, CA, 10 Sept., 1986. (Nat. Fert. Dev. Center: Muscle Shoals, AL.)

Hauck, R.D. (1984). Atmospheric nitrogen chemistry, nitrification, denitrification, and their interrelationships. In 'The Handbook of Environmental chemistry'. (Ed. O. Hutzinger) Vol. 1, Part C, pp. 105–125. (Springer-Verlag: Berlin.)

Malthus, T.R. (1798). Essay on the principles of population. (Published anonymously, 1798: London.)

Nelson, D.W. (1982). Gaseous losses of nitrogen other than through denitrification. In 'Nitrogen in Agricultural Soils' (Ed. F.J. Stevenson) pp. 327–63. (Amer. Soc. Agron.: Madison, WI.)

Paddock, W., and Paddock, P. (1967). 'Famine-1975! America's Decision: Who Will Survive ' p.8. (Little, Brown, and Co.: Boston.)

Strebel, O., Boettcher, J., and Koelle, W. (1985). Use of input-output balances of solutes in ground water in a recharge area for evaluation and prediction of groundwater quality problems. Z. Dtsch. Geol. Gels. (J. Ger. Geol. Soc.) 136(2), 533–41.

Wetselaar, R., and Farquhar, G. D. (1980). Nitrogen losses from tops of plants. Adv. Agron. 33, 263–302.

PART II

NITROGEN TRANSFORMATION PROCESSES

ROLE OF THE CROP PLANT IN CYCLING OF NITROGEN

J.S. Pate and G.D. Farquhar

ABSTRACT

This review examines the principal events and processes involved in the uptake, assimilation, and partitioning of nitrogen by crop species, and the possible avenues of exchange of nitrogen between plant and environment as the crop develops and matures. Nitrogen status of the crop is examined in relation to plant age, growth and nitrogen availability, and case studies are presented illustrating contrasting patterns of nitrate utilization, involving reduction predominantly in root or shoot. An empirical model of nitrogen partitioning in white lupin (Lupinus albus) is discussed, and experimental studies on the cycling of nitrogen through roots of other species evaluated. Problems associated with the measurement of gaseous or non-gaseous losses of N from above or below ground parts of the growing crop are examined within the context of the overall nitrogen economy of the crop and its impact on the ecosystem. Nitrogen transfer to fruits is examined in terms of physiological repercussions within the maturing plant, harvest index for nitrogen, and potential value of unharvested crop residues.

INTRODUCTION

The typical cropping programme for arable soils involves sowing of seed at the commencement of a growing season, usually after several months of fallow associated with unfavourable conditions for plant growth. Mineralization before or shortly after planting causes peak values for indigenous soil ammonium and nitrate, so that an emerging crop with initially small demands for N is usually oversupplied with N. There then follows an exponentially increasing phase of growth up to and possibly beyond flowering, during which soil mineral N levels fall and additional N inputs as fertilizer are likely to be necessary to maintain maximum crop growth rate. Finally, as reproduction of the crop gets under way, further uptake of N from the soil diminishes and extensive internal mobilization of N occurs from vegetative parts to fruits and seeds until the crop is harvested at maturity. Within this complex progression of changing supply and demand lies the key to understanding how the growing crop exchanges N with its parent ecosystem, while at the same time providing a challenge to agronomists for developing N fertilizer practices optimizing crop growth while minimizing environmental pollution.

PLANT GROWTH AND NITROGEN STATUS

The first set of questions which one might pose concerns the basic N composition of crops; viz. what concentrations of N occur in total crop biomass and individual organs, into what forms is this N incorporated, and what criteria should one use when attempting to relate N status to growth and final productivity?

In contrast to carbon, which is readily stored as sugar, starch and other oligosaccharides, N is rarely present in plant tissues grossly in excess of current usage. There are some exceptions, notably certain nitrophiles which may accumulate nitrate up to 1–2% of their dry weight (McKee 1962). In other species, soluble N usually fails to account for more than 10–20% of the total N of the vegetative organs of a plant, and much of this may indeed represent solutes in transport rather than true cytoplasmic reserves (Wallace and Pate 1967; Pate et al. 1981). Insoluble forms of N other than those in cytoplasmic proteins or protein of metabolically-important organelles such as chloroplasts and mitochondria, are normally not present in large amount in vegetative parts of the plant. Concentrated reserves of N do however occur in the highly specialized protein bodies of seeds and occasionally in the underground perennating organs of certain geophytes (Dixon et al. 1983). These general constraints, coupled to the fact that the single largest N resource of the vegetative plant is likely to be committed to photosynthetically-operative proteins, reduce greatly the capacity of species to accommodate N above the basic requirements for growth.

It follows from the above that crop species tend not to differ greatly between one another in N concentration in dry matter of vegetative organs of similar kind and age. However, data for a broad range of crops (Gladstones and Loneragan 1975; Pate and Layzell 1981), suggest a higher range of values for legumes than for cereals and other non legumes, and a tendency for shoots to be consistently higher in N (range for mid season harvests 1.9–5.1 %N in dry matter) than roots (1.2–2.8 %N). These differences, and the even wider differences in protein content between seeds of different species, have great impact upon the manner in which a species deploys its N during growth.

Within each species N content of each plant part varies predictably with age. A recent case study of narrow-leaved lupin (Lupinus augustifolius) illustrates this point (Fig. 1), each vegetative organ declining progressively in N concentration with season. The data of Fig. 1 relate to plants 80% dependent on N_2 fixation for their N supply and therefore tightly regulated in respect of growth and N supply.

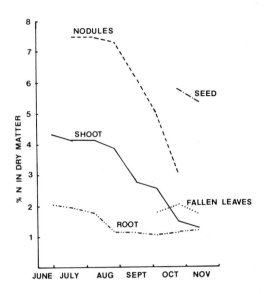

Fig. 1: Nitrogen concentrations in dry matter of organs of a field-grown crop of narrow-leaved lupin (<u>Lupinus</u> <u>angustifolius</u> L. cv Illyarrie), Geraldton, W. Australia, 1986. Crop duration was 181 days (J.S. Pate, M. Unkovich, J. Hamblin and C.A. Atkins, <u>unpublished data</u>).

In non-legumes under varying levels of N supply, N concentrations in dry matter increase predictably from a minimum value (N min) representing the N level just sustaining survival, through to an upper limit (N max) defining both the upper limit of uptake and the maximum crop growth rate (Angus and Moncur 1985). In wheat, both N max and N min decline predictably with growth (Fig. 2), essentially reflecting an increasing relative amount of older and structurally supporting tissues of low N content in the biomass of both root and shoot. Based on these concepts Angus and Moncur (1985) propose a simple model for growth response to N, defining the relative N concentration (RNC) at a particular stage of plant growth in terms of N max, N min, and total mass of nitrogen (N) as a proportion of total biomass (W):-

$$RNC = (N/W - N\ min)/(N\ max - N\ min)$$

The reader is referred to Agren (1985) and Greenwood <u>et</u> <u>al</u> (1986) for further evaluations of the quantitative relationships between growth rate and N content. These are applicable to other species as well as to wheat.

UPTAKE AND ASSIMILATION OF NITRATE

Leaving aside certain exceptionally acid or poorly-aerated soils in which nitrifying bacteria are relatively inactive, cultivated soils have most of

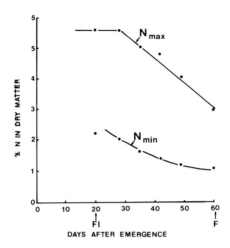

Fig. 2: Critical levels for nitrogen in whole plant dry matter of solution culture-grown wheat (<u>Triticum</u> <u>sativum</u> L. cv. Gabo). N max defines the upper limit of uptake and the N level for maximum growth rate, N min is the concentration just sustaining survival. FI – floral initiation, F–flowering (data redrawn from Angus and Moncur 1985).

their available N in the form of NO_3^- due to the rapid nitrification of ammonium N released from decomposing organic matter or ammoniacal fertilizer. Much of the following discussion accordingly centres upon patterns of uptake, storage and reduction of NO_3^- in plants, and how availability of NO_3^- as opposed to NH_4^+ affects the composition and productivity of a crop, and hence its overall impact upon N cycling within the parent ecosystem.

Most crop plants can absorb NO_3^- from the soil environment rapidly and effectively, to the extent that their tissues and xylem sap may often be many times more concentrated in the ion than in the outer medium. Both NO_3^- and NH_4^+ are absorbed by inducible, energetically-dependent uptake mechanisms (Pate and Atkins 1983; Smirnoff and Stewart 1985; Lewis 1986). Whereas NH_4^+ is toxic and hence must be assimilated into organic compounds immediately upon absorption by the root, NO_3^- can enter into storage pools of root and shoot, or alternatively be reduced at or close to the site of uptake by an inducible nitrate reducing system. The latter involves a cytoplasmically-located reductase coupled to a nitrate-reducing complex located in plastids. In non-photosynthetic tissues, NADH derived from glycolysis, mitochondrial dehydrogenases, or the pentose phosphate pathway provides reductant (Abrol <u>et al</u>. 1983), while in the light NO_3^- may be assimilated at essentially no cost using surplus photosynthetically-generated reductant (Smirnoff and Stewart 1985). The energetics of the process are recently discussed by Pate and Layzell (1987).

26

The kinetics of NO_3^- uptake, storage and radial transfer to the root xylem are highly complex and dependent on NO_3^- level in the medium as well as on the current storage status of the root (Pate and Atkins 1983). Activity of the root nitrate reductase relative to uptake of further NO_3^- will obviously affect the extent to which a species passes free NO_3^- to its shoot, and hence, with increasing spillover of the ion, the extent to which NO_3^- is available for storage in the shoot. Some species (pineapple, paw paw, beet and other members of the Chenopodiaceae, certain Solanaceae (e.g. Capsicum), Boraginaceae and Asteraceae (e.g. Xanthium, lettuce)) are regarded as classic NO_3^- storing species, often with relatively small pools of soluble organic N and therefore dependent on current assimilation of recently absorbed or stored NO_3^- for their growth (Pate 1973, 1983; Smirnoff and Stewart 1985). In these species, high tissue levels of NO_3^- do not necessarily signify supraoptimal supply in the root environment. By contrast in other species (e.g. Lupinus, see Atkins et al. 1979) reduction of NO_3^- is so effective that little or no NO_3^- accumulates under optimal supply and soluble N currently in excess of the demands of growth accumulates principally in the form of amides (asparagine and glutamine).

Differences in composition of plants feeding on NO_3^- or NH_4^+ derive primarily from the above-mentioned variations in assimilation. A good example is provided by Scaife et al. (1986) who exposed lettuce to a range of N treatments given as $Ca(NO_3)_2$, NH_4NO_3 or $(NH_4)_2SO_4$, the latter with nitrapyrin in the potting mix to inhibit nitrification. The plants exhibited essentially similar growth responses as N dose increased to an optimum, but at final harvest up to 200 fold greater levels of mid-rib sap NO_3^- were present in the $Ca(NO_3)_2$-fed plants (1840 mg NO_3/L) than in the $(NH_4)_2SO_4$ treatment (8 mg NO_3/L). As expected the NH_4NO_3-fed plants showed intermediate values for NO_3^- storage. As a second illustration, one may quote data for water-cultured barley seedlings provided with N as NO_3^-, $NO_3^- + NH_4^+$ (1:1), or NH_4^+ (Lewis et al. 1982). Xylem sap analysis of the NO_3^- -fed plants showed 82% of the N exported from the root as NO_3^- and only 6% as the principal compound glutamine, versus only trace amounts (5%) in NO_3^- and 53% as glutamine in the NH_4^+ treatment. Again the NH_4NO_3-fed plants responded in an intermediate fashion.

These results have clear implications for the development of fertilizer practices aimed at reducing NO_3^- levels in plant tissue destined for human consumption, as well as for minimizing groundwater nitrate pollution, say through the use of ammonium-based fertilizers coupled with nitrification inhibitors. Successful switching of plants to NH_4^+ as opposed to NO_3^- nutrition requires that the crop adapts successfully without symptoms of ammonia toxicity. According to literature reviewed by Lewis (1986) most agriculturally-important species have this capacity.

The classic methods used for estimating how much of the NO_3^- reducing activity of a plant takes place in its shoot compared with its root have been to measure the ratio of reduced organic N to NO_3-N in the xylem stream moving

27

from root to shoot (Pate 1973, 1980; Smirnoff and Stewart 1985), and to determine the relative proportions of a plant's total in vitro or in vivo nitrate reductase activity (NRA) recoverable in above-and below-ground plant parts (Sutherland et al. 1985; Gojan et al. 1986). Both forms of assessment need to be evaluated critically in relation to a number of possible shortcomings or sources of error.

Xylem sap may be collected from decapitated plants bleeding naturally under root pressure, from root systems enclosed under pressure or, by applying mild vacuum to segments of stem tissue. The relative proportion of total xylem sap N in organic form probably overestimates root assimilation of NO_3^- since some of the reduced N in xylem is likely to be cycling through the root from the shoot or being released following N turnover in root tissue. With vacuum extracted stem tracheal sap there is always the possibility of xylem contents being augmented in or depleted of either NO_3 or organic N due to lateral exchanges between the xylem stream and bordering vascular parenchyma (Pate 1980). Similarly, bleeding sap exuding slowly under root pressure may contain a proportionately greater amount of reduced N than in the xylem fluid of a normally transpiring plant. Because, with faster rates of water flux in the intact plant, NO_3^- is likely to enter to a larger extent by apoplastic pathways which circumvent the root's NO_3^- assimilating system (Pate and Layzell 1987).

Where assessments of NO_3 assimilation are based on in vivo assays of NO_2 accumulation, the use of dark anaerobic conditions to inhibit NO_2 reduction is also likely to affect adversely the activity and turnover of nitrate reductase (Hunter 1985). The choice has to be made between basing NRA on NO_2 generated from endogenous NO_3 or from a supposedly saturating amount of applied NO_3 (Hunter 1985). Neither approach is fully satisfactory. The main problem with in vitro assays of nitrate reductase is the likelihood of the enzyme being incompletely extracted or inadequately protected against inactivation during assay (Oaks and Hirel 1985). Assays also need to be conducted over several diurnal cycles and a range of nutritional conditions before the reductase potential of a species can be properly compared with that of other species.

Recent studies have combined $^{15}NO_3$ feeding experiments with assays of NRA and nitrogenous solutes of root bleeding sap (e.g. Breteler et al. 1980; Morgan et al. 1985; Gojan et al. 1986). As shown by Rufty and Volk (1986), a full time course study of exchange of the applied ^{15}N with the unlabelled pools of NO_3^- and reduced forms of N in xylem sap and root and shoot tissue is required before meaningful interpretations can be made. Even then, serious complications arise if large amounts of already existing N are cycling between root and shoot when an experiment is conducted.

Bearing in mind the above limitations, one must conclude that presently available experimental procedures provide only a general indication of whether a species normally reduces NO_3^- principally in its root or shoot. Indeed, even where a species apparently exhibits extreme behaviour in its pattern of utilization of NO_3^- data must be interpreted with caution. For example the extremely low levels of NO_3^- but massive levels of reduced N in xylem of woody plants (Bollard 1960; Pate 1980; Bray 1983; Smirnoff and Stewart 1985) have been widely interpreted as showing that reduction of NO_3^- must be occurring principally, if not exclusively, in roots. However, woody species under horticulture (Pate 1980), or in natural ecosystems following periods of intense nitrification, may show appreciable amounts of NO_3^- in their xylem and NRA in their leafy shoots (Stewart and Orebamjo 1983; Smirnoff et al. 1984). Some species (e.g. certain Gymnospermae, Ericaceae and Proteaceae) nevertheless show consistently negligible levels of leaf nitrate reductase even under cultivated conditions in which NO_3^- is freely available (Smirnoff et al. 1984).

Flora indigenous to acid soils deficient in total N, waterlogged soils, or soils in which N becomes available almost entirely through mycorrhizal activity, may normally encounter little or no NO_3^-. When cultivated these same speices may still exhibit a distinct preference for NH_4-N (e.g. cranberry (Vaccinium, see Pate 1980 and Greidanus et al. 1972) and display a marked intolerance of NO_3-N (e.g. certain Acacia spp. Hansen and Pate 1987).

Dealing with the other extreme of behaviour, it has been concluded (Pate 1973, 1980) that genera such as Xanthium (Wallace and Pate 1967; Keltjens et al. 1986), Gossypium and Datura (Lewis 1986), which regularly show 90% or more of their xylem sap N as NO_3^-, must be reducing NO_3^- predominantly in their leaves. But even here it is possible that a major proportion of the plant's xylem-borne NO_3^- might be circulating within the plant and being drawn off and assimilated only as required by either root or shoot. This interpretation requires that NO_3^- is freely mobile in phloem as well as xylem of the species in question, a situation shown not to be the case for most species whose phloem sap has been examined (Pate 1980; Pate et al. 1984b).

CYCLING OF NITROGEN WITHIN THE VEGETATIVE PLANT

The preceding sections have shown that considerable variation occurs between and within plant species in patterns of uptake, assimilation and distribution of N between organs of different type and age. Inherent within such variation must be an ability to partition N internally in a highly characteristic fashion, and to modulate such behaviour in relation to nutritional circumstances. Embodied in these processes are capacities to exchange N continuously between root and shoot, to cycle N through mature leaves, and to redistribute N progressively from old vegetative parts to growing regions and reproductive structures. One must understand how these transfer processes are implemented and regulated before being able to appreciate how the N metabolism of a particular species adapts to differing availability of N.

The legume, Lupinus albus, has featured prominently in studies of the modelling of N translocation because of the ease with which its phloem sap can be collected (Pate et al. 1979; Layzell et al. 1981). Each empirically-based model uses C/N ratios of xylem and phloem sap, photosynthetic gains and respiratory losses of C by plant parts, and net changes in C and N of dry matter of these same parts over specific 7 – 10 day intervals of growth. Studies so far have been mostly restricted to symbiotically-dependent plants grown in the absence of inorganic N. The fixed N exported from nodules moves initially upwards to the shoot in the xylem, so that roots distal to nodules can gain N solely by phloem translocation from the shoot (Oghoghorie and Pate 1972; Pate 1973). Furthermore, since it is possible to collect xylem sap from both the top of the root and from roots below nodules, one can readily distinguish between newly fixed N exported from the nodules and N cycling through or released from already existing pools of N in the root (see Pate 1984; Pate et al. 1984a).

The most sophisticated model of N flow so far constructed for Lupinus albus (Fig. 3) depicts N partitioning amongst nodules, roots, developing inflorescence and developing lateral upper shoots, and four age groups (I to IV) of leaflet and stem and petiole segments on the main stem. In the one week period of study 35 mg N was fixed. This N together with a small amount of N mobilized from lower leaves ended up mainly (44%) in the upper two strata of leaflets, 18% in the inflorescence plus lateral apices, 14% in nodules, 13% in the root and 11% in main stem and young laterals.

The N partitioning process is seen to be implemented primarily by the longitudinally-oriented long distance transport pathways of the plant, namely upward flow in xylem, and multidirectional outward transport from leaves in phloem (Fig. 3). The driving forces of these complementary flow processes are, respectively, transpirational loss of water and the mass flow export of photosynthetically-produced sugar. Interacting in ordered fashion with these two transport processes are five distinct radially-directed short distance transport activities (see Fig. 3). These are (a) xylem to phloem cycling through leaves, (b) xylem to phloem transfer in the upper stem, (c) cycling of shoot-derived N from root back to shoot by phloem to xylem transfer, (d) self nourishment of nodules by direct incorporation of newly fixed N, and (e) xylem to xylem transfer in nodal tissue of lower regions of the stem. In the example cited (Fig. 3), processes (a) to (e) represent the equivalents respectively of 39, 16, 15, 9 and 5% of the total N increment of the plant during the period of study.

Each of these radial transport components is vital to the nourishment of at least one part of the system. Firstly, apical growing parts of the shoot receive through process (b) a considerable amount of N additional to that acquired by direct transpirational attraction of xylem sap or as translocate from nurse leaves. Secondly, the top leaves and apical regions of the shoot receive a distinct additional component of transpirationally-attracted N through process (e), though this will obviously deprive by an equal amount the transpirational rewards for N to lower mature leaves. Thirdly, the root

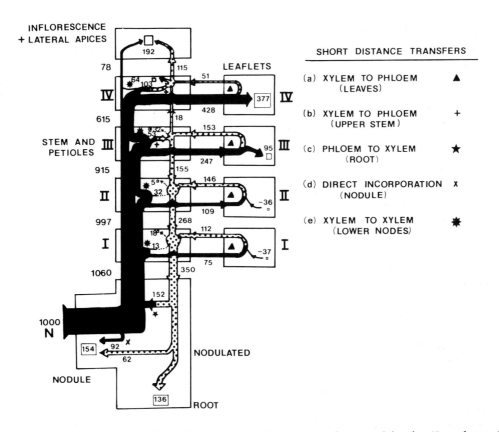

Fig. 3: Empirical model of N flow and utilization for symbiotic N_2- dependent plants of white lupin (Lupinus albus L. cv Ultra) for the week immediately following anthesis (51-58 days after sowing). I-IV refer to four successive strata of leaflets, and stem and petiole segments on the main shoot. N flow is depicted on a proportional basis in terms of fixation of 1000 units by weight of N_2. (Modified from Pate 1986).

derives by process (a) a major fraction of the N which has cycled through lower mature leaves. Fourthly, the cycling of N from root back to shoot by process (c) provides a means whereby any N currently in excess of a root's requirements for growth can return to the shoot, thereby further differentially enriching the shoot with N. Finally, as outlined by Atkins et al. (1984), the capacity of nodules to incorporate N directly into their bacterial tissues (process d) is crucial to at least the early stages of nodule development. This is because the N content per unit dry matter of mature nodules is 3-5 times greater than that of the root making it impossible for nodules to attract sufficient N for growth solely by import of relatively N-poor phloem translocate from the shoot.

In the absence of information on phloem composition it becomes **extremely** difficult, if not impossible, to make reliable assessments of even the major transport exchanges for N within a plant, let alone piece together a complete flow profile for N. Additionally, when considering non N_2-fixing plants feeding on NO_3^-, uncertainties remain regarding the relative activities of alternative sites of N assimilation, and whether or not unreduced NO_3^- is mobile in phloem. Age-related effects and supra optimal supplies of N may also influence the conclusions drawn for a species, especially if young seedlings are used in which massive amounts of N from cotyledon catabolism may transiently predominate within a plant's transport pathways (see Peoples et al. 1986).

Notwithstanding these problems, a number of authors have attempted to quantify cycling of N between shoot and root of various crop species. The experimental strategies used have included (1) $^{15}NO_3$ feeding of foliage and subsequent examination of xylem sap for ^{15}N (Oghoghorie and Pate 1972), (2) studies of nitrate reductase activity, xylem sap composition and N transfer in split root studies involving application of $^{15}NO_3$ to one half of the root system (Keltjens et al. 1986), (3) a similar split root study also including examination of ^{15}N in xylem sap from the unlabelled half of the root system (Cooper et al. 1986a), and (4) courses of ^{15}N labelling of the inorganic and organic fractions of xylem sap after application of $^{15}NO_3$ to water-cultured seedlings (Rufty and Volk 1986) or older plants under field conditions (Cooper et al. 1986b).

Taken collectively, the above studies indicate considerable variation between and within species in relation to age and nutrition. For example, in a study by Keltjens et al. (1986), white lupin, maize and cocklebur (Xanthium strumarium) plants were estimated to reduce 80, 50 and 6% respectively of their currently-absorbed NO_3^- in roots, and although this provided more than enough N for root growth in the first two species, both still showed a significant return of N to the root from the shoot via the phloem. Somewhat surprisingly, cocklebur, whilst showing the lowest proportional NRA in its roots, translocated less N from shoot to root than did lupin or maize. Nonetheless this shoot-derived N was still judged to be a quantitatively important nutritional supplement to the cocklebur root.

Studies (Cooper et al. 1986a,b) on winter wheat also provided conclusive proof of extensive cycling of N through roots, especially in young seedlings in which 60% of the amino N of xylem was regarded to be circulating as opposed to coming from assimilation of incoming NO_3^-. Furthermore, older plants under field conditions showed only 30% of their xylem ^{15}N in reduced form shortly after applying $^{15}NO_3$, suggesting that N cycling through roots remained an important active process in reducing NO_3^-. Conversely, with soybean seedlings, Rufty and Volk (1986) concluded that NRA of the root was contributing only 30-42% of the reduced-N found in xylem exudate.

Translocation of N in seedling wheat plants has been examined in two further studies. In the first (Simpson et al. 1982), upward flow of N in xylem was measured as the product of transpiration rate and N concentration in the transpiration stream, and found to be grossly in excess of the shoot's requirements for growth. A significant fraction of the currently mobile N was therefore concluded to have been cycling between root and shoot. In the second study (Lambers et al. 1982) a split-root technique was employed; one half of the root supplied with all nutrients, the other with N-free nutrient solution. Growth of the N-deprived half of the root system was then assumed to have come entirely from N cycled via the shoot, so that by measuring respiration and C and N increments of these roots the C:N ratio of the phloem stream feeding them could be estimated.

An important function for phloem translocation from shoot to root is to provide N for deeply penetrating roots exploring soil profiles deficient in N. This is analogous to the situation for distal roots of N_2-dependent legumes, and in either situation the large sink capacity of the N-starved roots would be likely to minimize return flow N to the shoot via xylem. Conversely, where ample N is available to depth in the soil and where current uptake and assimilation of N by the root easily meets its demand for N, a much larger proportion of a plant's N resource is likely to be circulating in the vascular system. Indeed, under luxury uptake the cycling component might comprise most of the reduced N leaving the root system. In this situation it would be clearly unacceptable to use ratios of inorganic (NO_3)N: reduced amino N in xylem sap to indicate proportional activity of a root in reducing incoming NO_3^-.

One might regard this circulating fraction of the plant's N resource as an eminently labile and hence potentially leachable component, even if it represents only a few percent of the plant's total N (white lupin, Pate et al. 1981), or only a day or so of growth requirement for N (wheat, Cooper et al. 1986a). Furthermore, the circulating process may play a fundamental role in regulating further uptake of N, possibly by negative feedback control of uptake, and may also be viewed as a principal means whereby N mobilized during senescence of lower leaves may be routed back to become available to growing parts higher up the shoot. This pathway is clearly depicted as of some significance at flowering of white lupin (Fig. 3), and may assume even greater prominence during reproduction as further uptake of N wanes and reserves of N become progressively monopolized by developing seeds.

TRANSFER OF NITROGEN TO DEVELOPING FRUITS AND SEEDS

Crops differ enormously in the timing of their flowering and fruit maturation relative to the assimilation of C and N which takes place over the growth cycle. For instance, Pate and Minchin (1980) show for grain legumes that from 6 to 37% of the total net CO_2 assimilation and from 11 to 50% of the total N_2 fixation taking place over the growth cycle of a genotype or species can be accomplished during vegetative growth up to flowering. These substantial variations between species carry special significance in terms of N, since, as shown for the same grain legumes by Pate and Flinn (1973) and

33

Peoples et al. (1983), some 60–85% of the plant's total N resource is exploited indiscriminately for fruit filling. Nitrogen assimilated in early vegetative growth will thus pass through repeated cycles of turnover and successive transfers between different age groups and classes of organs before finally moving to the reproductive organs. By contrast only a few percent of the C fixed before flowering is mobilized to seeds (Pate and Flinn 1973; Pate et al. 1980), so that photosynthesis during late pod fill is paramount for seed development.

Based on the above analysis, one can envisage two radically different physiological programmes for providing fruits of a seed crop species with N. The first would simply be for the plant to create a sufficiently large reserve of N by the end of flowering to meet demands during fruiting. With no energy requirement for further N assimilation there would be the greatest possible proportional allocation of post anthesis photosynthate to seed-fill. This programme is deemed appropriate for determinate, large-seeded, quick maturing grain legumes (e.g. mung bean, cowpea, snap bean), since it is clearly more expeditious for a sudden large peak demand for N to be met by mobilization of reserves than by slower continued acquisition of new N from soil or N_2-fixing nodules. However, as pointed out for seed crops generally by Sinclair and de Wit (1976), internal retrieval of N is a potentially self destructive process since the largest single reservoir of N is in chloroplast protein, and once this has been signficantly eroded, photosynthetic performance will be inevitably affected. This disadvantage could be partly offset were the N reserves of organs other than leaves to be used first, thereby retaining photosynthetic capacity until the last possible moment. There is little evidence of such a strategy in crop species (e.g. Pate and Flinn 1973; Peoples et al. 1983), although certain grain legumes (e.g. broad bean) contain proportionally greater amounts of mobilizable N in stems, petioles, roots and nodules than do others such as cowpea and chickpea (Pate and Minchin 1980).

The second, alternative programme is for N reserves to continue accumulating throughout flowering and pod fill. It would be especially effective for slow maturing indeterminate cultivars with small N capital accumulated at the time of flowering (e.g. lupin, broad bean and certain genotypes of soybean). The drawbacks are that generation of further N capital requires retention of photosynthetic capacity sufficient to fuel both seed filling and additional assimilation of N, and that associated costs are incurred in maintenance of tissue integrity in stem and root.

A sudden leaf senescence programme carries the obvious limitation of there simply not being sufficient total N in vegetative organs to fill the full complement of fruits. This would apply especially to seed crops with high levels of seed protein (e.g. lupin and soybean, 35–50% protein by dry weight), since these would have to tap three or more times the catchment of leaf N in filling each unit weight of seed than, say, a cereal with only 10–15% seed protein. Conversely, the alternative programme of near full retention of photosynthetic surfaces, requires a protracted period of time for assimilating the necessary N, during which the crop is likely to become critically short of water or subject to heat stress.

NON–GASEOUS LOSSES OF NITROGEN FROM A CROP DURING GROWTH

One most important aspect of the seasonal dynamics of N for a crop is the extent to which net losses of N occur to soil or atmosphere at times before plant maturity. One way of obtaining such information is to draw up for different stages in crop growth careful balance sheets for N in above- and below-ground living biomass or as shed plant parts, such as flowers, aborted fruits and dead shoots or foliage. By recording also changes in N content and concentration in each part of the shoot and in the root material from different horizons of the soil, one obtains some indication of how N is currently being partitioned, and which plant parts are losing N to the rest of the plant or possibly to the surrounding environment.

It should be emphasized that this approach gives no indication of the extent to which N is being exchanged additional to that suggested from the net balance sheet, nor, indeed, whether aerial atmosphere or rhizosphere is the ultimate recipient of any of the released N. Ideally, parallel direct measurements should be made of gaseous losses of N from above ground parts of the crop, and of N released from root biomass in turnover of old root material. These forms of measurement particularly those associated with root sampling, are notoriously difficult to accomplish in the field.

We will use studies with narrow-leaved lupin (Lupinus angustifolius) to aid further discussion. Fig. 4 shows the seasonal changes in organ contents of N of a field-grown crop. Peak biomass N was recorded in both root and shoot at 125 days after sowing, i.e. 30 days after the commencement of flowering, and 14 days after the first significant recoveries of N as dead leaves and aborted fruits in litter traps under the canopy. For the remaining 56 days until full maturity living biomass N was reduced by 35% (231 mg N/plant), the below ground loss (55 mg N/plant) amounting to approximately one third of that from shoots (176 mg N). A further deficit of 76 mg N was accounted for as fallen litter, leaving a surprisingly large unresolved loss of N (100 mg N/plant) from shoot biomass during reproductive development of the crop. The most logical avenues for such a loss would be leaching from shoots by rain, consumption by insects, gaseous losses of N to atmosphere, and downward translocation to roots and thence a loss to the soil. According to the model of N flow in lupin (Fig. 3) downward flow of N to roots is a significant element in N partitioning so long as N is still accumulating rapidly in the plant. Whether such translocation continues to achieve prominence during late crop growth is debatable, although the relatively low C:N ratios of phloem sap collected from the stem base of old plants (Layzell et al. 1981) would indicate that, with root metabolism still attracting a significant share of a plant's net photosynthate, downward flow of N to roots would remain signficant. However, the N flow profile for lupin (Fig. 3) allows one to argue equally convincingly that losses of N from root and nodule

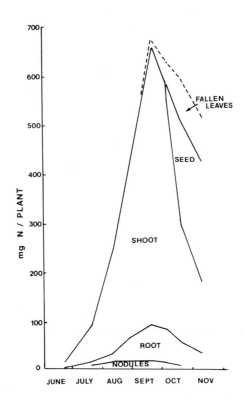

Fig. 4: Distribution of total N in organs of a field-grown crop of narrow-leaved lupin (<u>Lupinus</u> <u>angustifolius</u> L. cv. Illyarrie), Geraldton, W. Australia, 1986. Data plotted cumulatively, crop duration 181 days. N concentrations in organs of the same crop are shown in Fig. 1 (J.S. Pate, M. Unkovich, J. Hamblin and C.A. Atkins, <u>unpublished data</u>).

biomass might have been coupled to upward transport of N to the shoot <u>via</u> xylem.

The above study on lupin was carried out on deep sand of extremely low N content (0.01 - 0.04% N), without added N fertilizer. Plants were shown by the ^{15}N natural abundance technique to be obtaining 80% of their N from symbiotic activity. Nitrogenase activity ceased within 14 days after peak biomass N had been achieved, so one can assume that any subsequent losses of N from shoot and root were unlikely to have been signficantly ameliorated by further intake of N from soil or atmosphere.

A parallel glasshouse study of symbiotically dependent lupin grown in quartz sand provided further evidence of considerable net loss of total plant

N from the time of peak biomass until plant maturity (Table 1). The loss appeared to have occurred to almost as great an extent from the roots as from shoots. Soluble N in leachates from roots was assayed following each watering of the pots with distilled water or —N culture solution. Residual N left in the rooting medium was also determined after recovery of all sievable root biomass at final harvest. The leachates plus residual N accounted for all of the recorded loss in root biomass plus 60% of the otherwise unaccounted for deficit in aboveground biomass N (Table 1). While dangerous to extrapolate from these results to the field, the data (Table 1) suggest that the bulk of the N released from a maturing plant may well return to the rooting medium, and that much of this may be incorporated into insoluble, organically—bound material formed following decay of roots. This essentially non—leachable form of N in the soil, may well comprise a major long term asset to subsequent cereal crops.

Table 1: Nitrogen balance sheet for pot—grown symbiotically—dependent plants of narrow—leaved lupin (Lupinus angustifolius cv. Yandee) (J.S. Pate, unpublished data).

Item	Total N
	mg N/plant
Whole plant N at time of maximum N content (119 days after sowing)	479
Losses of N from dry matter from time of (plant) maximum N content until maturity (178 days)	
(a) loss from shoot	59
(b) loss from root	51
Total N recovered in leachates of roots over growth cycle	31
Residual N left in rooting medium (after root removed)	55

1. Plants grown in pots of organic matter—free quartz sand, supplied with an effective Rhizobium at sowing, and supplied throughout growth with N—free culture solution.

The use of N_2—fixing leguminous crops in markedly N deficient soils or in N—free media has decided practical advantages for modelling N turnover during growth, but since N inputs are in this case likely to be tightly coupled to growth, the data may bear little relevance as to how crops might behave under heavy N fertilization. Juding from the wide range of total and soluble N levels in single crop species across a range of availability levels of N, one may imagine similar variation in the extent of 'leakiness' of N. Compounded with this, species may well differ basically from one another in their proneness to lose N from shoot or root, whether as gaseous or solute losses from living biomass, or as unretrieved N following organ senescence. Broadly based information is simply not available in this regard. One also clearly requires much more information at species level, with respect to the effect of N source especially comparisons between applied soluble N <u>versus</u> a slowly-mineralizing bank of indigenous organic N.

GASEOUS LOSSES OF PLANT NITROGEN DURING CROP GROWTH

Information is also needed concerning the fluxes of N—containing gases between shoots and the air. Wetselaar and Farquhar (1980) reviewed data on crop N contents where large declines in N from above—ground parts were reported (up to 75 kg N/ha in 10 weeks). They noted that such losses, if in the gaseous form, would require an efflux of 18 nmol N/m^2 leaf surface/s from a crop with LAI of 5. It is important to realise that the review by Wetselaar and Farquhar was of <u>losses</u>, so that reports of continually increasing, or stable crop N contents, were not included. To that extent, the above figures may overestimate the average situation. On the other hand, gaseous and other forms of loss pass unnoticed in crop N records if the rate of loss is less than the rate of uptake of N by the crop from the soil. Studies with N isotopes are now also revealing N losses from the system considered (e.g. Witty 1983, clear in peas but also apparent in other species), but the majority of cases involve more classical techniques and typically report a decline in above ground N between anthesis and harvest (e.g. Cox <u>et al</u>. 1985 in some but not all lines of wheat examined). Wetselar and Farquahar (1980) concluded that losses were greatest from plants with high N contents, suggested that these may occur when plant N exceeds 1/NHI times the capacity of the reproductive sinks, and that loss in gaseous form directly from the tops may be an important mechanism. (NHI is the nitrogen harvest index, the ratio between the N content of the reproductive organs and the total N content of the tops).

The history of research into gaseous uptake and loss of N—containing compounds by leaves has been reviewed by Wetselaar and Farquhar (1980) and Farquhar <u>et al</u>. (1983). We consider here more recent data relating to plants and volatile nitrogeneous compounds, whether in the oxidised form, the reduced form, or as N_2.

Losses as dinitrogen

The detection of N_2 loss is most difficult because of its preponderance in the atmosphere. However, Radmer and Ollinger (1982) have demonstrated the evolution of N_2 from isolated chloroplasts treated with hydroxylamine (NH_2OH), and that this was caused by NH_2OH competing with water for oxidation by photosystem II. <u>In vivo</u>, NH_2OH may be produced via nitrite reductase. The potential danger of NO_2^- accumulation should be recognised. Nitrite can also act as a substrate for N_2 volatilization in other reactions, with for example, reduced NAD(P)H, oximes and ascorbic acid.

Losses in the oxidised form

The above reactions of NO_2^- also evolve the gaseous oxides of N (NO_x = NO, N_2O and NO_2). Considerable amounts of NO_x are evolved when herbicides are applied which block NO_2^- reduction (Klepper 1979). Harper (1981) found considerable evolution of NO_x from the in vivo (+NO_3^-) assay system for nitrate reductase of soybean leaves. He found evolution to be enhanced under anaerobic conditions, but detected no losses under ambient conditions when tissue NO_2^- content was low and similar to the content under field conditions. This conclusion, and a similar one by Klepper (1979), should be treated with caution since techniques capable of measuring the fluxes they report (> 200 $nmol/m^2/s$) may not be sufficiently sensitive to detect those of interest here (< 20 $nmol/m^2/s$) (Farquhar et al. 1983). Mulvaney and Hageman (1984) have since shown that the compound evolved in Harper's assay was acetaldehydeoxime (CH_3 CH:NOH), with smaller amounts of nitrous oxide (N_2O).

Mulvaney and Hageman summarized their unpublished data which suggested considerable gaseous losses from soybeans, mostly as oxidised N compounds. The rates were 12.3 or 18 $nmol/m^2/s$ (assuming leaf area of 300 cm^2 per 24 day-old plant), depending on whether or not a trap containing KOH and $KMnO_4$ (which would trap acidic compounds or those that can be oxidised to acidic ones) was included. However, the flow of air through the chamber containing four seedlings (presumably c. 1200 cm^2) was so small (18L/h) that we calculate that the CO_2 concentration would have dropped to the compensation point, allowing a net CO_2 assimilation rate of less than 0.2 $\mu mol/m^2/s$ (c.f. healthy leaf = 20 $\mu mol/m^2/s$). The air in the chamber would thus have become saturated with water vapour, thereby reducing transpiration rate to an unnaturally low value. Interpretation of the N-loss data is thus difficult.

Weiland and Omholt (1985) used much greater ratios of flow rates to area but still had problems with condensation. When net loss from the plants was observed 70% of the 'volatile' N was in the water used to rinse the bag which formed the leaf 'chamber'. These authors observed losses of NO_x from field plants of Zea mays ranging from −1.6 to +1.6, with a mean of 0.2 $nmol/m^2/s$.

Losses in the reduced form

Potential candidates for N losses in the reduced form are HCN, and, more likely, amines and NH_3. Farquhar et al. (1980) showed that the rate of uptake of NH_3, J, by a leaf is given by $J = g (P_a - \gamma)/P$, where g is the conductance in series of the stomata and the boundary layer to the diffusion of NH_3, P is the absolute pressure, P_a the partial pressure of NH_3 in the air and γ is the NH_3 compensation point. The conductance to diffusion of NH_3 is 0.92 times that for water vapour. The NH_3 compensation point increases with temperature, being typically a few nbar for healthy leaves. The ambient partial pressures of NH_3 are of a similar magnitude. Thus plants can either take up or release small amounts of NH_3, depending on whether or not P_a exceeds γ. Since the review by Farquhar et al. (1983), Mulvaney and Hageman (1982) have reported losses of reduced N from soybean shoots averaging $\underline{c}$. 7 nmol/m^2/s (again we assume a leaf area of 300 cm^2/plant). However we urge caution in accepting these results because of the extremely low rates of CO_2 assimilation and of transpiration allowed in the experiments.

The greatest flux involving NH_3 is that of the photorespiratory C oxidation pathway, so one would suppose that factors which increase photorespiratory flux might tend to increase the tissue concentration of NH_4^+. These include increased partial pressure of O_2, decreased partial pressure of CO_2 (as imposed by Mulvaney and Hageman 1982), and increased temperature. Increase of temperature might also increase the rate of NH_3 assimilation, so its effect cannot be predicted. However, the partial pressure of NH_3 (e.g. in the intercellular air spaces) in equilibrium with a certain concentration of NH_4^+ (e.g. in the apoplast) increases with temperature, in a similar manner to the compensation point (Farquhar $\underline{et}$ $\underline{al}$. 1980). Weiland and Stutte (1985) showed increases of rate of loss of reduced N with increase of O_2 pressure, from 1.9, through 2.5 to 3.4 nmol/m^2/s, as O_2 concentration increased from 1, through 20 to 40%. [The fluxes were incorrectly labelled as per cm^2, and should have been per dm^2, as for other fluxes in the same table (R.T. Weiland, personal communication)]. Weiland and Stutte (1985) reported that rates of loss of reduced N from sorghum (a C_4 species with diminished photorespiration) were insensitive to O_2 pressure, averaging 3.9 nmol/m^2/s at ambient O_2 pressure. Foster and Stutte (1986) report losses of reduced plus oxidised N equivalent to 4 nmol/m^2/s from soybean under the same conditions in which the comparable Weiland and Stutte (1985) results were 3.3 nmol/m^2/s. Foster and Stutte (1986) showed that spraying soybean leaves with inhibitors of either of two key enzymes involved in the reassimilation of NH_3 released during photorespiration, viz. glutamate synthase and glutamine synthetase, increased the rate of N loss to about 18 nmol/m^2/s. This rate and those reported by Weiland and Stutte are much smaller than the rates of photorespiratory release and refixation of NH_3 (typically $\underline{c}$. 4000 nmol/m^2/s, Farquhar $\underline{et}$ $\underline{al}$. 1980).

The rates of loss of reduced N from soybean leaves in 1% O_2 reported by Weiland and Stutte (1985) ranged from 64 to 92% of the rates of loss at 20% O_2, again indicating to us that the photorespiratory NH_3 cycle is effectively isolated from the intercellular air spaces of leaves. This isolation is probably achieved by two lines of defence. Firstly, by glutamine synthetase whose affinity for NH_3 is great (Km (NH_4^+) = $\underline{c}$. 20 μM), and secondly by an ammonium transporter in the plasmalemma; evidence for its presence in Phaseolus vulgaris (C_3 species) was published by Raven and Farquhar (1981). The observations that the NH_3 compensation points of some C_3 and C_4 species are comparable (Farquhar et al. 1980) and that losses of reduced N from C_4 species are sometimes larger than those from C_3 species (Weiland and Stutte 1985; Foster and Stutte 1986) support the view that C_3 species have largely overcome the potential damage of NH_3 loss from the photorespiratory cycle.

Weiland and Omholt (1985) examined NH_3 exchange between Zea mays and the atmosphere in the field, and observed rates of evolution ranging from 0.02 to 2.5, with a mean of 0.9 nmol/m^2/s, similar to fluxes observed previously from this species in the laboratory (Farquhar et al. 1979; Weiland and Stutte 1979; Farquhar et al. 1980).

Total rates of gaseous N loss

Summarizing the previous sections on data which have been presented since 1982, most recent evidence suggests that gaseous losses in vivo are generally < c. 5 nmol/m^2/s (e.g. Weiland and Stutte 1985; Foster and Stutte 1986 [except Palmer amaranth, a C_4 species, with losses of 12 nmol/m^2/s at 20°C]). In what we regard as the most careful field study, Weiland and Omholt (1985) concluded that the total rate of loss from maize was c. 1 nmol/m^2/s.

We conclude that the losses of NH_3 from the photorespiratory pathway are generally small, as are the net losses of NH_3 from most tissue before flowering. These conclusions are consistent with past (Denmead et al. 1976) and recent micrometeorological measurements of the very small NH_3 fluxes above crops (O.T. Denmead and R. Wetselaar, unpublished data).

Whereas many atmospheric scientists have regarded vegetation as passive platforms for deposition and possible absorption of NH_3 and other nitrogenous gases and particulates, Stutte and co-workers have tended to present vegetation as a considerable source of volatile N. From the middle ground, it has been suggested previously (Farquhar et al. 1979) that vegetation tends to pull the atmosphere NH_3 content (up or down) toward the level inside the airspaces of leaves. NH_3 fluxes between vegetation and the atmosphere are probably more important in maintaining the Gaia [the conserved properties of the geosphere-biosphere complex (Lovelock 1979)] than in causing crop losses of N.

Nevertheless, in the crop context, we are cautious about other nitrogenous compounds, especially in the period between anthesis and maturity, and about the effects of high temperature if uptake of N into sinks is slow. Rapid progress in the assessment of gaseous losses from shoots will require the development of sensitive detectors of NO_x and other compounds (Fahey et al. 1985), hopefully with sufficient precision and robustness to enable reliable micrometeorological measurements to be made over crops.

NITROGEN AT CROP HARVEST

The N still in above ground biomass of the crop is partitioned upon harvesting into seed and unrecovered straw or stover. The amounts of N in these two fractions are relatively easy to determine, thus enabling one to compute a crop's nitrogen harvest index (NHI) (Canvin 1976), i.e. the proportion of the final above-ground N recovered as grain. Although widely used as a measure of N use efficiency (Austin et al. 1977; Halloran 1981; Rattunde and Frey 1986), NHI fails to take into account N returned to the soil as root material, and any losses of N from plant biomass prior to harvest. For example, for lupin (Fig. 4) NHI is 0.63, but if N harvested is expressed as the ratio of seed N to total N recoverable in shoot, root and shed leaves at final harvest, the value becomes 0.48. If based on a comparison between final seed N and peak biomass N (Fig. 4), the apparent index of crop efficiency is reduced to only 0.37. This last expression, though tedious to estimate, provides the most meaningful evaluation of the crop's performance in cycling of N within the context of its ecosystem. Ideally, all three forms of expression of NHI should be computed. For example, the difference between 0.48 and 0.37 in the above example would express 'leakiness' of N before final harvest, the difference between 0.63 and 0.48 the prominence of leaf litter and root residues in returning N to the cropping system, and the conventionally expressed NHI value of 0.63 a measure of how efficiently N is finally mobilized from non-deciduous parts of the shoot to the seeds.

The above approach to N partitioning has special relevance to a grain legume such as the narrow leaf lupin in which it is a common experience for crops achieving near equal peak biomass in early fruiting to range almost three-fold in harvested yield of seed between sites and seasons (J. Hamblin, personal communication). Much of this difference stems from relative success in fruit set, and comprehensive inventories of pre-harvest losses and residue returns of N under such variable reproductive loads would provide an instructive commentary on the nature and extent of residual benefit to subsequent crops. One important point is that leaf litter from a heavily fruiting crop is likely to be less rich in N than in a crop reproducing poorly due to less efficient remobilization of N from the latter crop prior to leaf fall. Patterns of breakdown of organic residues of differing C:N ratio are known to be very different.

Field data for other crop species relate almost exclusively to NHI data based on above ground final harvest. Considerable variability in NHI exists between and within species, e.g. 0.25-0.51% for Avena sterilis compared with 0.42-0.67 for A. sativa (Fawcett and Frey 1982), and the somewhat higher range of values for both spring wheat (0.57 - 0.74, Loffler and Busch 1982) and durum wheat (0.57 - 0.86, Desai and Bhatia 1978). Highly repeatable differences between cultivars of a species have also been recorded across a range of final yields and soil N availabilities as shown recently by Rattunde and Frey (1986) for oats (Avena sativa).

CONCLUDING COMMENT

The methodologies currently used in quantifying the processes involved in the uptake, assimilation and partitioning of N by crop species are clearly inadequate, giving only a general indication of how plants respond to N source and availability. Comparisons between species are possible only with respect to a few simple measurements such as critical N concentrations in plant tissues for maximum growth and harvest indices for N. Even then, it is often impossible to judge whether apparent differences in behaviour relate to growth conditions or exclusively to genotype. As this review has indicated, much of the information available is essentially descriptive and relates to only a few facets of plant response to N. Above all we remain woefully ignorant of how N uptake, assimilation, storage and retranslocation are regulated, and how these activities interact in turn with plant growth and productivity.

REFERENCES

Abrol, Y.P., Sawhney, S.K., and Naik, M.S. (1983). Light and dark assimilation of nitrate in plants. Plant Cell Environ. 6, 595–9.

Agren, G.I. (1985). Theory of growth of plants derived from the nitrogen productivity concept. Physiol. Plant 64, 17–28.

Angus, J.F., and Moncur, M.W. (1985). Models of growth and development of wheat in relation to plant nitrogen. Aust. J. Agric. Res., 36, 537–44.

Atkins, C.A., Pate, J.S., and Layzell, D.B. (1979). Assimilation and transport of nitrogen in non-nodulated (NO_3-grown) Lupinus albus L. Plant Physiol. 64, 1078–82.

Atkins, C.A., Shelp, B.J., Kuo, J., Peoples, M.B., and Pate, J.S. (1984). Nitrogen nutrition and the development and senescence of nodules of cowpea seedlings. Planta. 162, 316–26.

Austin, R.B., Ford, J.A., Edrich, J.A., and Blackwell, R.D. (1977). The nitrogen economy of winter wheat. J. Agric. Sci. 88, 349–57.

Bollard, E.G. (1960). Transport in the xylem. Ann. Rev. Plant Physiol. 11, 141–66.

Bray, C.M. (1983). Nitrogen metabolism in plants. In 'Plants'. (Longman: London).

Breteler, H., and Ten Cate, C.H.H. (1980). Fate of nitrate during initial nitrate utilization by nitrogen-depleted dwarf bean. Physiol. Plant 48, 292–6.

Canvin, D.T. (1976). Interrelationships between carbohydrate and nitrogen metabolism. In 'Genetic Improvement of Seed Proteins'. pp. 172–95 (National Academy of Sciences: Washington, D.C.).

Cooper, H.D., Clarkson, D.T., Johnston, M.G., Whiteway, J.N., and Loughman, B.C. (1986a). Cycling of amino-nitrogen between shoots and roots in wheat seedlings. Plant Soil. 91, 319–22.

Cooper, H.D., Clarkson, D.T., Ponting, H.E., and Loughman, B.C. (1986b). Nitrogen assimilation in field-grown winter wheat: Direct measurements of nitrate reduction in roots using ^{15}N. Plant Soil 91, 397–400.

Cox, M.C., Qualset, C.O., and Rains, D.W. (1985). Genetic variation for nitrogen assimilation and translocation in wheat. II. Nitrogen assimilation in relation to grain yield and protein. Crop Sci. 25, 435–40.

Denmead, O.T., Freney, J.R., and Simpson, J.R. (1976). A closed ammonia cycle within a plant canopy. Soil Biol. Biochem. 8, 161–4.

Desai, R.M., and Bhatia, C.R. (1978). Nitrogen uptake and nitrogen harvest index in durum wheat cultivars varying in their grain protein concentrations. Euphytica 27, 561–6.

Dixon, K.W., Kuo, J., and Pate, J.S. (1983). Storage reserves of the seed-like, aestivating organs of geophytes inhabiting granite outcrops in south-western Australia. Aust. J. Bot. 31, 85–103.

Fahey, D.W., Eubank, C.S., Hubler, G., and Fehsenfeld, F.C. (1985). Evaluation of a catalytic reduction technique for the measurement of total reactive odd-nitrogen NO_y in the atmosphere. J. Atmos. Chem. 3, 435–68.

Farquhar, G.D., Wetselaar, R., and Firth, P.M. (1979). Ammonia volatilization from senescing leaves of maize. Science, 203, 1257–8.

Farquhar, G.D., Wetselaar, R., and Weir, B. (1980). On the gaseous exchange of ammonia between leaves and the environment: determination of the ammonia compensation point. Plant Physiol. 66, 710–4.

Farquhar, G.D., Wetselaar, R. and Weir, B. (1983). Gaseous nitrogen losses from plants. In 'Gaseous Loss of Nitrogen from Plant-Soil Systems'. (Eds J.R. Freney and J.R. Simpson). pp. 159–80. (Martinus Nijhoff/Dr W. Junk: The Hague.)

Fawcett, J.A., and Frey, K.J. (1982). Nitrogen harvest index variation in Avena sativa and A. sterilis. Proc. Iowa Acad. Sci. 89, 155–9.

Foster, E.F., and Stutte, C.A. (1986). Glutamine synthetase activity and foliar nitrogen volatilization in response to temperature and inhibitor chemicals. Ann. Bot. 57, 305–7.

Gladstones, J.S., and Loneragan, J.F. (1975). Nitrogen in temperate crop and pasture plants. Aust. J. Agric. Res. 26, 103–12.

Gojan, G.A., Passama, L., and Robin, P. (1986). Root contribution to nitrate reduction in barley seedlings (Hordeum vulgare L.). Plant Soil 91, 339–42.

Greenwood, D.J., Neeteson, J.J., and Draycott, A. (1986). Quantitative relationships for the dependence of growth rate of arable crops on their nitrogen content, dry weight and aerial environment. Plant Soil 91, 281–301.

Greidanus, T., Paterson, L., Schrader, L.A., and Dana, M.N. (1972). Essentiality of ammonium for cranberry nutrition. J. Am. Soc. Hortic. Sci. 97, 272–7.

Halloran, G.M. (1981). Cultivar differences in nitrogen translocation in wheat. Aust. J. Agric. Res. 32, 535–44.

Hansen, A.P., and Pate, J.S. (1987). Evaluation of the ^{15}N natural abundance method and xylem sap analysis for assessing N_2 fixation of understorey legumes in jarrah (Eucalyptus marginata) forest in S.W. Australia. J. Exp. Bot. (in press).

Harper, J.E. (1981). Evolution of nitrogen oxide(s) during in vivo nitrate reductase assay of soybean leaves. Plant Physiol. 68, 1488–93.

Hunter, W.J. (1985). Soyabean stem in vivo nitrate reductase activity. Ann Bot. 55, 759–61.

Keltjens, W.J., Nieuwenhuis, J.W., and Nelemans, J.A. (1986). Nitrogen retranslocation in plants of maize, lupin and cocklebur. Plant Soil 91, 323–7.

Klepper, L. (1979). Nitric oxide (NO) and nitrogen dioxide (NO_2) emissions from herbicide treated soybean plants. Atmos. Environ. 13, 537–42.

Lambers, H., Simpson, R.J., Beilharz, V.C., and Dalling, M.J. (1982). Growth and translocation of C and N in wheat (Triticum aestivum) grown with a split root system. Physiol. Plant 56, 421–9.

Layzell, D.B., Pate, J.S., Atkins, C.A., and Canvin, D.T. (1981). Partitioning of carbon and nitrogen and the nutrition of root and shoot apex in a nodulated legume. Plant Physiol. 67, 30–6.

Lewis, O.A.M. (1986). 'Plants and Nitrogen'. (Edward Arnold: London.)

Lewis, O.A.M., James, D.M., and Hewitt, E.J. (1982). Nitrogen assimilation in barley (Hordeum vulgare L. cv. Mazurka) in response to nitrate and ammonium nutrition. Ann. Bot. 49, 39–49.

Loffler, C.M., and Busch, R.H. (1982). Selection for grain protein, grain yield, and nitrogen partitioning efficiency in hard red spring wheat. Crop Sci. 22, 591–5.

Lovelock, J.E. (1979). 'Gaia'. (Oxford Uni. Press: Oxford).

McKee, H.S. (1962). 'Nitrogen Metabolism in Plants'. (Clarendon Press: Oxford).

Morgan, M.A., Jackson, W.A., and Volk, R.J. (1985). Uptake and assimilation of nitrate by corn roots during and after induction of the nitrate uptake system. J. Exp. Bot. 36, 859–69.

Mulvaney, C.S., and Hageman, R.H. (1982). Evolution of N compounds from soybean seedlings under ambient conditions. Agron. Abstracts, p. 105.

Mulvaney, C.S., and Hageman, R.H. (1984). Acetaldehyde oxime, a product formed during the in vivo nitrate reductase assay of soybean leaves. Plant Physiol. 76, 118–24.

Oaks, A., and Hirel, B. (1985). Nitrogen metabolism in roots. Ann. Rev. Plant Physiol. 36, 345–65.

Oghoghorie, C.G.O., and Pate, J.S. (1972). Exploration of the nitrogen transport system of a nodulated legume using ^{15}N. Planta 104, 35–49.

Pate, J.S. (1973). Uptake, assimilation and transport of nitrogen compounds by plants. Soil Biol. Biochem. 5, 109–19.

Pate, J.S. (1980). Transport and partitioning of nitrogenous solutes. Ann. Rev. Plant Physiol. 31, 313–40.

Pate, J.S. (1983). Patterns of nitrogen metabolism in higher plants and their ecological significance. In 'Nitrogen as an Ecological Factor'. (Eds J.A. Lee, S. McNeill and I.H. Rorison). pp. 225–55. (Blackwell Scientific Publications: Oxford).

Pate, J.S. (1984). Nitrogen metabolism of lupins. Proc. 3rd Int. Lupin Congress, La Rochelle, France 1984, pp. 290–309.

Pate, J.S. (1986). Xylem-to-phloem transfer – vital component of the nitrogen-partitioning system of a nodulated legume. In 'Phloem Transport'. (Eds J. Cronshaw. W.J. Lucas and R.T. Giaquinta), pp. 445–62. (A.R. Liss, Inc.: New York).

Pate, J.S., and Atkins, C.A. (1983). Nitrogen uptake, transport and utilization. In 'Nitrogen Fixation'. Volume 3: Legumes. (Ed. W.J. Broughton), pp. 245–98. (Oxford University Press: U.S.A.).

Pate, J.S., Atkins, C.A., Hamel, K., McNeil, D.L., and Layzell, D.B. (1979). Transport of organic solutes in phloem and xylem of a nodulated legume. Plant Physiol. 63, 1082–8.

Pate, J.S., Atkins, C.A., Herridge, D.F., and Layzell, D.B. (1981). Synthesis, storage and utilization of amino compounds in white lupin (Lupinus albus L.). Plant Physiol. 67, 37–42.

Pate, J.S., Atkins, C.A., Layzell, D.B., and Shelp, B.J. (1984a). Effects of N_2 deficiency on transport and partitioning of C and N in a nodulated legume. Plant Physiol. 76, 59–64.

Pate, J.S., Atkins, C.A., and Perry, M.W. (1980). Significance of photosynthate produced at different stages of growth as carbon source for fruit filling and seed reserve accumulation in Lupinus angustifolius L. Aust. J. Plant Physiol. 7, 283–97.

Pate, J.S., and Flinn, A.M. (1973). Carbon and nitrogen transfer from vegetative organs to ripening seeds of field pea (Pisum arvense L.). J. Exp. Bot. 24, 1090–9.

Pate, J.S., and Layzell, D.B. (1981). Carbon and nitrogen partitioning in the whole plant — a thesis based on empirical modelling. In 'Nitrogen and Carbon Metabolism'. Chapter 4. (Ed. J.D. Bewley), pp. 94–134. (Martinus Nijhoff/Junk: The Hague).

Pate, J.S., and Layzell, D.B. (1987). Energetics and biological costs of nitrogen assimilation. In 'The Biochemistry of Plants: A Comprehensive Treatise'. Vol. 13, Intermediary Nitrogen Metabolism. (Ed. B.J. Miflin), in press. (Academic Press: Florida, USA).

Pate, J.S., and Minchin, F.R. (1980). Comparative studies of carbon and nitrogen nutrition of selected grain legumes. In 'Advances in Legume Science'. (Eds R.J. Summerfield and A.H. Bunting), pp. 105–14. (Royal Botanic Gardens: Kew).

Pate, J.S., Peoples, M.B., and Atkins, C.A. (1984b). Spontaneous phloem bleeding from cryopunctured fruits of a ureide–producing legume. Plant Physiol. 74, 499–505.

Peoples, M.B., Pate, J.S., and Atkins, C.A. (1983). Mobilization of nitrogen in fruiting plants of a cultivar of cowpea. J. Exp. Bot. 34, 563–78.

Peoples, M.B., Pate, J.S., Atkins, C.A., and Bergersen, F.J. (1986). Nitrogen nutrition and xylem sap composition of peanut (Arachis hypogaea L. cv. Virginia bunch). Plant Physiol. 82, 846–52.

Radmer, R., and Ollinger, D. (1982). Nitrogen and oxygen evolution by hydroxylamine–treated chloroplasts. FEBS Letters 144, 162–6.

Rattunde, H.F., and Frey, K.J. (1986). Nitrogen harvest index in oat. Crop Sci. 26, 606–10.

Raven, J.A. and Farquhar, G.D. (1981). Methlammonium transport in Phaseolus vulgaris leaf slices. Plant Physiol. 67, 859–63.

Rufty, T.W. Jr., and Volk, R.J. (1986). Alterations in enrichment of NO_3^- and reduced–N in xylem exudate during and after extended plant exposure to $^{15}NO_3^-$. Plant Soil 91, 329–32.

Scaife, A., Ferreira, M.E.S., and Turner, M.K. (1986). Effect of nitrogen form on the growth and nitrate concentration of lettuce. Plant Soil 94, 3–16.

Simpson, R.J., Lambers, H., and Dalling, M.J. (1982). Translocation of nitrogen in a vegetative wheatplant (Triticum aestivum). Physiol. Plant. 56, 11–17.

Sinclair, T.R., and de Wit, C.T. (1976). Analysis of the carbon and nitrogen limitations to soybean yield. Agron. J. 68, 319–24.

Smirnoff, N., and Stewart, G.R. (1985). Nitrate assimilation and translocation by higher plants: Comparative physiology and ecological consequences. Physiol. Plant 64, 133–40.

Smirnoff, N., Todd, P., and Stewart, G.R. (1984). The occurrence of nitrate reduction in the leaves of woody plants. Ann. Bot. 54, 363–74.

Stewart, G.R., and Orebamjo, T.O. (1983). Studies on nitrate utilization by the dominant species of regrowth vegetation of tropical West Africa: A Nigerian example. In 'Nitrogen as an Ecological Factor'. (Eds J.A. Lee, S. McNeill and I.H. Rorison). pp. 167–88. (Blackwell: Oxford).

Sutherland, J.M., Andrews, M., McInroy, S., and Sprent, J.I. (1985). The distribution of nitrate assimilation between root and shoot in Vicia faba L. Ann. Bot. 56, 259–63.

Wallace, W., and Pate, J.S. (1967). Nitrate assimilation in higher plants with special reference to the cocklebur (Xanthium pennsylvanicum Wallr.). Ann. Bot. 31, 213–28.

Weiland, R.T., and Omholt, T.E. (1985). Method for measuring nitrogen gas exchange from plant foliage. Crop Sci. 25, 359–61.

Weiland, R.T., and Stutte, C.A. (1979). Pyro–chemiluminescent differentiation of oxidized and reduced N form evolved from plant foliage. Crop Sci. 19, 545–7.

Weiland, R.T., and Stutte, C.A. (1985). Oxygen influence on foliar nitrogen loss from soyabean and sorghum plants. Ann. Bot. 55, 279–82.

Wetselaar, R.L., and Farquhar, G.D. (1980). Losses of nitrogen from the tops of plants. Adv. Agron. 33, 263–302.

Witty, J.F. (1983). Estimating N_2–fixation in the field using ^{15}N-labelled fertilizer: some problems and solutions. Soil Biol. Biochem. 15, 631–9.

SYMBIOTIC NITROGEN FIXATION

D.F. Herridge and F.J. Bergersen

ABSTRACT

Values for the proportions (P) of plant nitrogen derived from fixation of atmosphere N_2 and for yields and N contents are reviewed for some tropical legume crops. Actual and theoretical levels of N_2 fixation by the crops are compared. We conclude that, under the best conditions of experimental agriculture, and with some legumes, N_2 fixation can supply most of the N requirements of these crops. However, the potential levels of N_2 fixation are seldom realized in commercial agriculture. Substantial improvements to N_2 fixation will be achieved only when both N yields and P are improved. Strategies for such improvements are examined in terms of crop and soil management, inoculation and plant breeding and selection. Benefits of legumes for succeeding crops sometimes surpass those arising directly from N_2 fixation; reasons for this are examined and we conclude that "sparing" of soil N by nodulated crops is an important component.

INTRODUCTION

Despite a century of study, the role of N_2 fixation in the cycling of N in agricultural systems remains undefined. Legumes are confronted with two sources of N for growth, viz. mineral N in the soil and N fixed in root nodules. Total crop N can be estimated simply but biologically-fixed N_2 can be estimated only when total N is partitioned between these two sources. Advances in methodology in recent years have allowed estimates of seasonal fixation to be made for most of the common species. Unfortunately, these estimates are site, season, and in some cases method, specific, making it difficult to develop general principles which adequately describe the process. There is a diversity of opinion on the level of fixation by the average legume crop. Burns and Hardy (1975) calculated fixation to be c. 140 kg N/ha/year. Within three years this figure was halved (Burris 1978). LaRue and Patterson (1981) averaged reported levels of symbiotic dependence [P = 100(crop N from N_2/ total crop N)] for soybean, and combined the value obtained (50%) with average commercial yields to estimate fixation at 75 kg N/ha/year, a figure similar to the one offered by Burris. Other authors have proposed more optimistic estimates, eg. Rennie (1985).

Efficient management of legumes in agriculture depends upon accurate assessment of N_2 fixation in the field. This knowledge not only provides an insight into the N economy of the legume but adds to our understanding of the general N cycle. Using the information gained, strategies can be developed to solve problems involving N in agricultural and natural systems. Solutions will come in the form of cropping and tillage systems to enhance N_2 fixation, improved legume genotypes, improvements in methods of inoculation, etc. Not only will data from individual experiments be used to solve problems, but they will be used also to develop principles and, eventually, functional models. Fixation should then be able to be predicted or estimated, even by farmers, given relevant information on paddock history, climatic data, plant species and pre-crop soil nutrient tests.

This paper reviews N_2 fixation by tropical crop legumes by first examining the possibility that the energy costs of fixation reduce potential yield. We then examine published estimates of fixation and use these data to calculate the levels of fixation achieved by the "average" commercial crop. We consider the role of the legume in the maintenance of soil N fertility and finally describe strategies to increase the proportions and amounts of N_2 fixed. Tropical forage legumes are not included in this review; for accounts of their N_2 fixation, readers are referred to Henzell and Vallis (1977), Vallis et al. (1977), and Vallis and Gardner (1985).

IS N_2 FIXATION MORE ENERGY DEMANDING THAN NITRATE ASSIMILATION?

Havelka et al. (1982) argued persuasively that the supply of energy (photosynthate) is a primary regulator of N_2 fixation activity. Evidence is drawn from the many experiments involving source-sink manipulations, although alternative interpretations such as treatment effects on growth rates and on the export and demand for N are possible (Mahon 1983). There is little doubt, however, that considerable energy is required by the nodulated legume to support fixation which in the nitrate-fed plant would be targeted to produce more photosynthetic capacity (Mahon and Child 1979; Schubert and Ryle 1980). A cost of fixation of 6.5 g C/g N (Mahon 1979; Ryle et al. 1979a) translates into a theoretical loss of 15 to 20 kg dry matter for every kg N fixed (LaRue and Patterson 1981). Indoor gas exchange studies suggest that nitrate-fed plants produce 13 to 28% more dry matter per unit of photosynthate than do plants dependent upon fixation (Silsbury 1977; Pate et al. 1979; Ryle et al. 1979b). In the field, differences in growth between fixing and nitrate-supplied plants are not always apparent although several reports suggest variable (0 to 28%) reductions in whole plant and seed yield associated with 20 to 30% increases in P (see Mahon 1983). In one report, a 29% increase in P co-incided with a reduction of only 2% in plant N and equivalent seed N.

Our data (D.F. Herridge and J. Brockwell, unpublished data) comparing soybean cv. Bragg grown on a low N soil, either heavily fertilized with N (300 kg/ha) or essentially dependent on N_2 fixation, illustrate the same finding (Table 1). At shoot yields of 8.0 to 8.5 t/ha and seed yields of 3.0 to 3.2 t/ha, there were only small differences in dry matter and N yields between nodule–dependent and fertilizer–N–dependent plants. The very small reduction in the shoot yield is in contrast to the expected dry matter reduction of 2.0 to 3.0 t/ha calculated from the theoretical dry matter loss (15 to 20 kg dm/kg N fixed) associated with the measured fixation of $\underline{c}$. 150 kg N/ha.

Table 1: Shoot and seed dry matter and N yields of soybean cv. Bragg dependent on N_2 fixation relative to the yields from uninoculated, heavily N–fertilized plots at Breeza, north–western NSW (D.F. Herridge and J. Brockwell, unpublished data).

		Relative yields			
Nodule mass[1]	Ureide index[2]	Shoot d.m.	Shoot N	Seed d.m.	Seed N
mg/plant	*%*	*%*	*%*	*%*	*%*
334	55 (16)	92	90	93	98

1. At 70 days after planting; 2. Ureide method, mean of 4 samplings between days 70 and 109, N–fertilized value in brackets.

We must conclude that the extra cost of N_2 fixation <u>can</u> be safely carried by most field–grown crops with little loss of production, and that generally factors other than the supply of energy for N_2 fixation limit legume growth. Possibly, the extra energy cost may become a burden at the upper levels of productivity. In this context, soybean seed yields in excess of 4.2 t/ha (viz. up to 7.2 t/ha) were only achieved by Troedson et al. (1984a) in the saturated soil culture system with the addition of fertilizer–N. At the more average yields in conventionally–irrigated plots of 3.5 to 4.2 t/ha, yield responses to N fertilizer were not evident.

We have calculated the upper limits of N productivity for the tropical crop legumes by combining maximum yield data for the various species with seed protein levels and estimated harvest indices for N (Table 2). The values range from $\underline{c}$. 200 kg N/ha for the two species of mung bean to 700 kg N/ha for soybean. In the following section we review published estimates of N_2 fixation for these crops and consider factors other than fixation–related energy costs which might be responsible for the differences between these estimates and the values in Table 2.

Table 2: Theoretical estimates for upper limits of total crop N based upon maximum seed yield data

Species	Maximum yield	Seed protein[1]	Seed N[2]	Harvest index	Estimated crop N	Reference (yield data)
	t/ha	*%*	*kg/ha*		*kg/ha*	
soybean	8.5	39	580	0.80	730	3
pigeonpea	8.0	22	280	0.65	430	4
groundnut	9.6	28	310	0.80	390	5
lima bean	7.0	21	240	0.65	370	5
common bean	5.0	25	196	0.65	300	5
cowpea	4.0	25	160	0.65	250	6
black gram	4.0	23	150	0.65	230	6
green gram	3.0	25	120	0.65	180	6

1. From Norton <u>et al</u>. 1985; groundnut value, M.B. Peoples, <u>personal communication</u>; 2. Soybean protein=Nx5.71; remainder (protein) =Nx6.25; groundnut in shell 3.25% N; 3. Troedson <u>et al</u>. 1984b; 4. Wallis <u>et al</u>. 1983; 5. Summerfield and Roberts 1985; 6. R. Lawn, <u>personal communication</u>

ESTIMATES OF N_2 FIXATION

Most reviews of the agronomy of symbiotic N_2 fixation (e.g. Nutman 1971; LaRue and Patterson 1981; Herridge 1982b; Graham and Chatel 1983; Rennie 1985) tabulate values for fixation over the whole season and this review will be no exception. Methods such as the acetylene reduction assay and measurement of total plant N, used to provide many of the values in the early reviews, have since come under question and data from them have not been included here. Nor have we considered reports where growth and fixation in pots and lysimeters were extrapolated to an area basis or where soil was amended with high C:N material to enhance <u>P</u>. We restrict ourselves to reports using ^{15}N techniques (isotope dilution, A–value, natural abundance), N difference and N balance methods and the ureide technique. In many reports total crop N, and therefore total N fixed, was underestimated because below–ground organs and shed foliage were not included. These sampling problems should not have given a biased estimated of <u>P</u> if ^{15}N techniques were used.

Published estimates of N_2 fixation are presented in Table 3. The list for soybean is not exhaustive, examples were chosen to cover the range of quantities and proportions of N fixed. Estimates of N_2 fixation by common bean are also readily available but for other widely–grown tropical species they are almost non–existent. If we assume the data for soybean and common bean are reasonably representative of all the crop legumes, a number of observations can be made.

Table 3: Estimates of seed yield, N productivity and the proportion (P) and amount of N obtained from symbiotic fixation for commonly-grown tropical crop legumes.

Species	Seed yield	Total crop N	N$_2$ fixed P	N$_2$ fixed Amount	Method	Reference
	t/ha	*kg/ha*	*%*	*kg/ha*		
soybean	0.50–3.31	41–355	0–71	0–236	ureide	1
	1.38–2.76	100–187	13–40	14– 75	N diff.	2
	n.a.	150–334	14–62	33–151	^{15}N	3
	2.02–2.70	147–182	26–38	38– 70	^{15}N	4
	2.16–3.96	346–406	34–67	139–234	^{15}N	5
	1.84–2.89	246–273	42–78	116–192	^{15}N	6
	2.54–4.55	260–450	25–78	73–288	ureide	7
	2.40–2.80	260–400	71–80	185–320	N diff.	8
	3.41–4.49	368–387	71–80	263–311	N diff.	9
	n.a.	143–216	84–87	125–188	^{15}N	10
common bean	1.80–3.47	112–184	16–52	20– 92	N bal.	11
	n.a.	136–191	54–72	100–104	^{15}N	12
	n.a.	100–169	38–68	40–125	^{15}N	3
	n.a.	59–238	0–73	0–121	^{15}N	13
cowpea	n.a.	76–134	61–76	47– 97	^{15}N	10
	n.a.	n.a.	n.a.	60–120	N bal.	14
groundnut	3.06–3.77	236–382	40–58	94–222	N diff.	15
pigeonpea	n.a.	78–134	17–51	13– 69	N diff.	16

n.a. – not available; 1. Herridge and Holland 1987; 2. Weber 1966; 3. Rennie 1984; 4. Armager et al. 1979. 5. Bergersen et al. 1985/Brockwell et al. 1985; 6. Vasilas and Ham 1984; 7. Herridge 1982a; 8. Nelson and Weaver 1980; 9. Bezdicek et al. 1978; 10. Eaglesham et al. 1982. 11. Westerman et al. 1981; 12. Rennie and Kemp 1984; 13. Rennie and Kemp 1983; 14. Wetselaar 1967; 15. Ratner et al. 1979; 16. Rao et al. 1981.

The N fixed by soybean and common bean in the field varies enormously, viz. from 0 to 320 kg N/ha (Table 3). Cowpea and groundnut show less of a range, with the maximum for groundnut approaching that for soybean whilst cowpea and pigeonpea had similar N$_2$ fixation to common bean. We found no reliable reports on N$_2$ fixation by green gram or black gram. A number of reports gave N$_2$ fixation estimates by soybean of >300 kg N/ha (Bezdicek et al. 1978; Nelson and Weaver 1980; Troedson et al. 1984a (not shown in Table 3)). Values for P varied enormously also; viz. from 0 to 87%, 0 to 73%, 61 to 76%,

40 to 58%, and 17 to 51% for soybean, common bean, cowpea, groundnut and pigeonpea respectively.

Notwithstanding the efforts to provide these values, we must basically agree with LaRue and Patterson (1981) that "...There is not a single legume crop for which we have valid estimates of the N fixed in agriculture...". By agriculture, we assume they meant commercial, rather than experimental, agriculture. All but one of the reports in Table 3 relate to experimental plantings which can be quite atypical of commercial crops, especially in terms of soil nutrient status and plant husbandry. The values for total plant N in Table 3 are probably substantially larger than are found in the average commercial crop. On the other hand, the experimental P values might be lower than for commercial crops because of the generally higher N status of soils on experimental farms.

Accepting these qualifications, these data are combined with those in Table 2 to estimate N_2 fixation by an average commercial crop. We also compare these estimates for each of the species with a calculated maximum level of fixation. Because of limited data, we shall restrict the preliminary calculations of maximum and average P to soybean and common bean and extrapolate the results to the other species. The highest estimate of P for soybean was 87%, which we consider a realistic figure given that nitrate is likely to be available in moderate quantities in all soils capable of supporting high yields. The average P was 51%, a figure almost identical with that proposed by Rennie et al. (1982) for soybean based upon isotope dilution studies in six different countries. An average P of 47% was calculated for common bean; we must, however, treat this value with caution. Although P for common bean can be as high as 73% in experimental fields, almost all of the commercial plantings of common and lima beans are fertilized with N and are not inoculated (Freire 1982; Lyman et al. 1985; Weiser et al. 1985). Crops that depend upon fixation therefore appear to be the exception (Westermann et al. 1981). A more realistic P for the average commercial bean crop would be c. 10%.

When the highest-recorded and average seed yields for the various species are combined with the highest-recorded and average levels of fixation, it becomes possible to compare theoretical upper limits of fixation with more realistic "average" levels and to partition the difference between the two according to factors associated with either growth or P (Table 4). The upper limits of N_2 fixation ranged from 160 for green grams to 635 kg N/ha for soybean, and the average levels were from 7 for common bean to 150 kg N/ha for soybean. For pigeonpea, groundnut, cowpea and the two species of mung bean, it is clear from columns 5 and 6 in Table 4 that the differences between the theoretical upper limits of fixation and the calculated levels achieved by the average crop are associated principally with low yield. Increasing P from average (0.50) to maximum (0.87) at the same level of yield results, on average, in an additional 19 kg N/ha fixed. For common and lima bean however, increasing P from average (0.10) to maximum (0.73) results in an extra 60 kg

Table 4: Calculations of realistic maximum– and global average–levels of N_2 fixation by the crop legumes, and gains resulting from improvements in yield and in the proportion (P) of crop N derived from symbiotic fixation.

Species	Theoretical maximum crop N[1]	Average seed yield[2]	Calculated N_2 fixation			
			av.yld & av.P[3]	av.yld & max.P[4]	max.yld & av.P	max.yld & max.P
			kg/ha			
soybean	730	1750	150	260	365	635
pigeonpea	430	680	37	64	215	370
groundnut	390	1000	40	70	195	340
lima bean	370	1200	12	88	37	270
common bean	300	600	7	50	30	220
cowpea	250	240	15	26	125	220
black gram	230	350	20	35	115	200
green gram	180	380	20	35	90	160

1. From column 6, Table 2; 2. From Summerfield and Roberts 1985; 3. Calculated as average seed N x (1/N harvest index) x average P (0.10 for common bean, lima bean; 0.50 for remainder); N harvest indices were half the values used in Table 2; 4. Maximum P of 0.73 for common and lima beans; 0.87 for remainder.

N/ha fixed at the average yield level (mean of the two species). With soybean, an extra 110 kg N/ha is fixed when P is increased from 0.5 to 0.87. Increasing yield whilst maintaining P at average levels results in increases of only 23 and 25 kg fixed N/ha for lima and common bean in contrast to average increases of 140 kg N/ha for the remainder. Although the absolute values in the last four columns relied on a number of disputable assumptions, it is clear that the theoretical upper levels of fixation will be achieved only when crop yield and P are increased concomitantly.

DOES SYMBIOTIC N_2 FIXATION INCREASE THE SIZE OF THE SOIL N POOL?

Benefits of tropical crop legumes to subsequent cereal crops are consistent and measureable. The contribution of the legume is often quantified as the amount of fertilizer-N required in the monocrop to match production by the rotation, and/or by the yield improvement in the absence of fertilizer-N. Responses in grain yield lie mainly in the 1 to 3 t/ha range, representing 50 to 100% increases over yields in the monoculture (Table 5). These responses are equivalent to those from applications of 50 to 100 kg N/ha. The benefits of the legume are thus obviously large, but we are again faced with the problem of direct extrapolation to commercial planting because productivity in experimental plots is almost always substantially greater than in the average commercial crop.

Table 5: Cereal grain yield responses to previous legume crops relative to the monocrop sequence

Crop sequence	Increase in yield	Relative increase	Fertilizer-N equivalent	Reference
	t/ha	*%*	*kg/ha*	
pigeonpea–sorghum	1.06	354	40	2
cowpea–corn	0.64	31	60	3
black gram–sorghum	3.68	79	68	4
green gram–sorghum	2.82	61	68	4
soybean–corn[1]	1.25–3.70	35–113	90–168	5
soybean–sorghum	2.56	86	135	5
soybean–corn	n.a.	49	>84	6
soybean–sorghum[1]	1.94	42	76	7

1. Data averaged over treatments involving the same species; 2. Rao et al. 1983; 3. Mughogho et al. 1982; 4. Doughton and MacKenzie 1984; 5. Welch 1985; 6. Baldock et al. 1981; 7. Clegg 1982.

The apparent enrichment of the soil–N pools by the legumes (column 4, Table 5) is in sharp contrast to the expected levels of residual N resulting from fixation. When the quantities of N involved in plant growth, in fixation and in the seed are calculated, we find that levels of fixation are often insufficient to offset the N removed with the harvested seed, resulting in net losses of N from the system (column 5, Table 6). At the upper levels of productivity by pigeonpea, lima and common bean, cowpea and black and green gram, a $P < 65\%$ will result in a net loss of N (see N harvest index values, Table 2). At the average yield of these crops, the N harvest indices would be substantially lower and therefore the P required to at least cover the N requirements of the seed would be proportionally less. The higher harvest indices proposed in Table 2 for soybean and groundnut suggest that higher P would be required by these species before they could contribute to the soil–N pool.

If the benefits of the crop legumes in rotations cannot be explained in terms of residual fixed N, then what are the sources of the benefits? Most certainly a number of factors operate, the relative importance of each dictated by site, season and crop sequence. Improvements in soil structure following legumes (Hearne 1986) and phytotoxic and allelopathic effects of different crop residues (Sanford and Hairston 1984) can result in extra yield from the rotation. Legumes break cycles of pests and diseases of cereals. Control of these can account for all of the legume effect (Dyke and Slope 1978) or part of the effect (Reeves et al. 1984). Nodulation and N_2 fixation interfere with the legume's capacity to use mineral N (Wych and Rains 1978). Whether this is due to nodulation–induced retardation of root growth, in turn

Table 6: Nitrogen balances for several legume crops

Species	Crop N	N fixed	Seed N	N balance	Reference
			kg/ha		
soybean[1]	100	13	80	− 67	3
	187	75	159	− 84	3
soybean[1]	224	90	206[2]	− 116	4
soybean[1]	332	173	191	− 18	5
soybean[1]	242	172	169	+ 3	6
soybean[1]	164	92	63[2]	+ 29	7
	315	195	128[2]	+ 67	7
soybean[1]	378	287	257[2]	+ 30	8
green gram	172	112	89	+ 23	9

1. Data averaged over sites, years and treatments where appropriate; 2. Assumed seed contained 6.5% protein-N; 3. Weber 1966; 4. Ham 1978; 5. Herridge 1982a; 6. Thurlow and Hiltbold 1985; 7. Hunt et al 1985; 8. Bezdicek et al. 1978; 9. Chapman and Myers 1987.

reducing sites for absorption of mineral-N (Rigaud 1981), or to a more direct effect on nitrate uptake/metabolism remains unclear (Wych and Rains 1978). Thus, post-harvest levels of nitrate are often higher in soils after legumes than after non-fixing crops (Table 7). Differences will be expanded in time because of the generally faster mineralization of N from the lower C:N residues of legumes relative to non-legumes (Henzell and Vallis 1977; Sanford and Hairston 1984). Thus, the legume benefit may be due to "nitrate sparing" rather than to enrichment of the soil-N pool (Doughton and MacKenzie 1984; D.F. Herridge and J.F. Holland, unpublished data). Another source of potential benefit to soil N may be the N-rich legume root and nodule material which is shed during growth. Measurement of this contribution has proved difficult although recent studies based on ^{15}N analysis of soil in the root zone of the legume indicate that it may be substantial (Poth et al. 1986).

STRATEGIES TO ENHANCE N$_2$ FIXATION

We have considered the amounts of N fixed by the crop legumes and have argued that reduced fixation activity is associated with low plant yield and reduced P. Yield of the crop legumes is influenced by agronomic, climatic and genetic factors. It is not our intention to consider all factors here and readers are referred to the reviews of Gibson (1977) and Graham and Chatel (1983). What has not been considered to the same extent are the modifications to the plant and bacterial genes, to agronomic practice, and to the environment, which result in increases in P, and from which may arise increases in N fixed; these may be independent of effects on yield. Most

Table 7: Extra soil nitrate after growth of nodulated crop legumes compared with cereals and unnodulated legumes

Crops	Extra soil nitrate (kg/ha) post-harvest[1]	pre-sowing[2]	Reference
soybean, sorghum	+ 23	+ 62	Herridge & Holland, unpub.data
green gram, sorghum	+ 26	+ 57	Doughton & MacKenzie 1984
black gram, sorghum	+ 38	+ 68	Doughton & MacKenzie 1984
soybean, unnod. soybean	+ 22	–	Herridge et al. 1984

1. Immediately after harvest of crop;   2. Just before sowing next crop

certainly, much of the variation in P is related to variation in soil nitrate. High soil nitrate usually results in low levels of P and vice versa. We have shown a strong inverse relationship between the amount (kg/ha) of soil nitrate-N to a depth of 1.2 metres at planting (x) and P (Herridge 1982a; Herridge et al. 1984; D.F. Herridge, unpublished data):

$$P = 82.7 - 0.184x \quad (r = 0.88)$$

The levels of soil nitrate-N involved in the derivation of this expression ranged from 83 to 400 kg N/ha. An improved version would include estimates of P where soil nitrate levels were close to zero at planting. Further improvement will accrue when the effect of crop size on P is taken into account. Progress towards enhancing fixation will be achieved by reducing the legume sensitivity to soil nitrate, by interfering with uptake and utilization of nitrate or by reducing the amount of nitrate available to the plant.

Crop and Soil Management

Both yield and P can be influenced by crop and soil management; we examine two important practices.

Tillage

Cultivation accelerates the oxidation of organic matter in soils (Doran 1980) and generally results in higher nitrate-N in the profile (e.g. Thomas et al. 1973; Dowdell et al. 1983). Cultivation may also decrease the rates of denitrification (Doran 1980; Rice and Smith 1982), immobilization (Rice and Smith 1982) and leaching (Thomas et al. 1973) of nitrate compared to untilled soils. For cereals under no-tillage, additional fertilizer-N may be required to supplement the reduced soil nitrate-N, but for legumes, the lower nitrate-N should result in enhanced fixation activity.

In cropping systems research involving four species of crop legume grown at three sites in north-western NSW, soil nitrate levels were reduced in the no-tillage relative to the cultivated plots. At sites A (high fertility) and C (low fertility), fixation by soybean was effectively suppressed in all treatments due to a combination of severe moisture stress restricting nodulation and crop growth (both sites) and high levels of soil nitrate (A site only). At the moderate fertility (B) site, the absence of tillage resulted in enhanced nodulation, a higher value for P (52% for the cultivated plots versus 66% for the no-tillage plots), and a positive N balance (Table 8).

Table 8: Effect of tillage on soil nitrate, nodulation and N measurements of crop growth and fixation by field-grown Forrest soybean

Tillage	Soil nitrate[1]	Nodule mass	Crop N[2]	Seed N	N fixed[3]	N balance[4]
	kg/ha	*mg/plant*			*kg/ha*	
Cultivated	214	69	252	161	132	−29
No-tillage	185	219	355	182	236	+54

1. At sowing, 0 to 1.2 m depth;   2. Including roots;   3. Ureide method (Herridge 1984);   4. N fixed minus seed N

The use of no-tillage techniques resulted also in increased plant growth and accumulation of N in seven of the eleven crops (Fig. 1). As we saw in Table 4, enhancement of fixation by the crop legumes will be linked to both improvements in plant productivity and P. In the 1984 season, when fixation activity by all species was substantial, the higher amount of N fixed by Forrest in the no-tillage plots (+79%) resulted from a 27% increase in P coupled to a 41% improvement in plant growth (Table 8).

Cropping intensity and sequence

The quantity of soil nitrate available to a legume crop can be influenced by the recent use of the soil. Irrigated soybeans grown immediately after an oat crop fixed 244 kg N/ha compared with 143 kg N/ha fixed by soybeans growing in previously fallowed soil; the proportions of fixed N in seed were 68 and 33% and net balances were +39 and −44 kg N/ha, respectively (Bergersen et al. 1985). At another irrigated site, winter cereal cropping removed 130 kg N/ha and left 19 μg/g (0-30 cm) of extractable mineral N at soybean planting, compared with 38 μg/g in winter-fallowed soil. Both treatments produced similar soybean yields (3.5 and 3.3 t/ha respectively) but P was 57% in winter-cropped and only 7% in winter-fallowed soils (Bergersen 1987). Thus, with proper choice of cereals and crop legumes, cropping systems can be managed for improved N_2 fixation. In a different approach, Rerkasem et al.

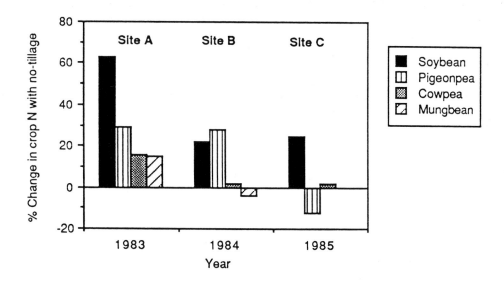

Fig. 1: Relative changes in total N uptake by four crop legumes under no-tilled compared to tilled conditions. Soybean data for 1983, 1984 and 1985 are the means of 1,2 and 3 varieties, respectively (D.F. Herridge and J.F. Holland, <u>unpublished</u> <u>data</u>).

(1985) found that intercropping of maize and ricebean (<u>Vigna</u> <u>umbellata</u>) increased P for the legume, due to competition for soil N by the maize. The symbiotic dependence of the monocrop legume was much less. The net result was a higher total N yield of the intercrop (maize plus legume) relative to the combined N yield of maize and ricebean monocrops.

Inoculation

Legume inoculation is a long-established and successful practice, especially with particular crops in the more technically-advanced countries. Vincent (1965) and others (e.g. Allen and Allen 1958) have argued that it is a desirable practice in most agricultural soils throughout the world. Date (1977), however, cautioned that the need to inoculate was not universal and should be carefully determined for each individual situation before investing in inoculant production and use. Indeed, most responses to inoculation in tropical soils appear confined to crops such as soybean and groundnut (Ayanaba 1977), both of which have specific <u>Rhizobium</u> requirements (Halliday 1985). Such assessments are not usually made on the basis of measurements of N_2 fixation, but rather following failure to produce discernible increases in nodulation and/or yield. We have noted from our own experiments that responses to inoculation, in terms of P (and sometimes in plant N), may be found in the absence of dry matter yield responses. Not infrequently, such responses occur as a consequence of high rates of application of inoculants.

In the experiment reported in Table 9, both yield and _P_ were increased through inoculation at low soil nitrate; at high soil nitrate, only _P_ was affected. In both situations, total N fixed was increased. Bergersen (1987) reported similar results at another site.

Table 9: Effect of inoculation on nodulation, N_2 fixation and production of dry matter and N in whole plants and seed of Bragg soybean sown into low and high nitrate soils (D.F. Herridge and J. Brockwell, <u>unpublished</u> <u>data</u>)

Treatment	Nodule mass[1]	Ureide index[2]	Shoot[3] Dry matter	N	Seed yield
	mg/plant	_%_	_t/ha_	_kg/ha_	_t/ha_
Low nitrate soil					
Uninoculated	0	23	4.9	66	1.7
Normal inoculation	72	44	6.8	168	3.3
High inoculation	334	55	7.0	195	3.0
High nitrate soil					
Uninoculated	0	16	8.9	205	3.2
Normal inoculation	4	16	8.0	196	2.9
High inoculation	50	27	8.5	213	3.3

1. Sampled 70 days after planting; 2. Mean value of 4 samplings between days 70 and 109; 3. Sampled at day 124

As a farming practice, inoculation remains the exception rather than the rule (Vincent 1982). Legume inoculants require exacting technology for production, distribution and use (Date and Roughley 1977). In Australia and the USA, legume inoculation has played a fundamental role in the development of the livestock, soybean and other industries. Less use has been made of inoculants in developing countries. In Latin America, only two countries use inoculants to any extent and even in Brazil, the largest producer of the seed legumes, common beans are fertilized with N rather than inoculated (Freire 1982).

Where populations of infective rhizobia exist in high numbers in soils, they present a formidable barrier to the successful exploitation of superior strains of <u>Rhizobium</u> used as inoculants (Devine 1984). In the USA, large populations of the soybean <u>Rhizobium</u> have developed with cropping so that now less than 10% of nodules on soybean are formed by the inoculant and yield responses to inoculation are rare (Devine 1984; Halliday 1985). In response, research programs in several laboratories (e.g. Devine 1984) aim to produce cultivars of the host that bypass the resident rhizobia in the soil to be nodulated by better, selected inoculant strains. This strategy assumes that fixation is limited by the effectiveness of the resident rhizobia. Although true in some situations (Jenkins et al. 1954; Ham et al. 1971) it is likely to be an isolated rather than a widespread phenomenon.

Plant breeding and selection

The challenge to improve the N_2 fixation capacity of the legumes through selection and breeding is complex because there are two components to consider: the host plant and the rhizobia. Early research focussed on selecting more effective Rhizobium strains. Recent research has looked to the host to provide the variation in fixation capacity. Two strategies appear to offer success: selection of plant genotypes with enhanced capacity for fixation under low N conditions (symbiotic vigour) and selection of genotypes with greater ability to nodulate and fix N in the presence of soil nitrate (nitrate tolerance).

Enhanced capacity for N_2 fixation

Of the commercial crop legumes, the common bean is regarded as the weakest at fixing N and is supplied with fertilizer-N in most cases; hence, increasing its fixation capacity has enjoyed considerable attention during the past decade. High levels of fixation were found to be associated with late maturity and climbing habit (Graham and Rosas 1977; Rennie and Kemp 1983). This implied a simple relationship between leaf area duration and N_2 fixation (see also Wynne et al. 1982) although Graham and Halliday (1977) suggested a link between carbohydrate distribution within the plant and fixation.

A breeding programme by Bliss and co-workers at the University of Wisconsin has produced new genotypes of common bean with increased levels of N_2 fixation. Selected hybrid lines have displayed between three and seven fold increases in total N fixed and two to four fold increases in P, relative to the commercial parent Sanilac (Table 10, Experiment 1). For all of the hybrids, however, fixation capacity was substantially less than the capacity of Puebla 152, the high-fixing donor. In a second experiment, the superiority of hybrid line 24-21 was obvious. The line displayed higher rates of growth and N accumulation (+36%) and more total growth (+79%) than the commercial parent whilst retaining the short season and determinate characteristics. The enhanced growth and fixation capacity of line 24-17 were linked to the undesirable traits of late maturity and indeterminateness. Line 24-55, although agronomically-acceptable, displayed only marginal increases in growth and fixation, relative to Sanilac.

Nitrate tolerance

Crop legumes utilize substantial amounts of soil nitrate during growth (Table 3, and Harper and Gison 1984). Development of symbioses where P is maintained at near maximum levels in the presence of high soil nitrate could provide the biggest single advance in the improvement of N_2 fixation by legumes.

Table 10: Summary of data from two experiments from a breeding programme to increase N_2 fixation by the common bean. (Attewell and Bliss 1985)

Parent or line	Experiment 1			Experiment 2		
	N fixation		Seed yield	Time to maturity	Determinate habit	Total N
	Amount	P				
	mg/plant		*g/plant*	*days*		*mg/plant*
Sanilac	76	12	18	85	yes	591
24–17	583	48	31	110	no	1068
24–21	216	25	19	91	yes	1045
24–55	192	22	23	94	yes	668
Puebla	852	57	38	120	no	1429

It is possible to select genotypes of soybean that are partially insensitive to nitrate (Table 11, and Betts and Herridge 1984). All lines had similar shoot yields suggesting that the higher levels of fixation by the nitrate-tolerant Korean lines reduced their use of soil N. Post-harvest measurements of soil nitrate confirmed this; up to 34 kg/ha of additional N was recovered from the Korean plots immediately after seed harvest relative to the Bragg plots. Seed yield of the Korean lines was c. 30% less than the average yield of the two commercial lines. Correlation matrices among the indices of nodulation and N_2 fixation and plant growth and seed yield revealed independence between the symbiotic- and yield-related characters. Therefore it should be possible to use these nitrate-tolerant lines as high-fixing donor parents in a breeding programme with selection for both seed yield and fixation capacity.

CONCLUDING REMARKS

The level of fixation in the "average" legume crop is only a fraction of the potential, and substantial improvements in fixation are possible through breeding and crop and soil management. Legume crops can be grown in soils of low N-fertility without the addition of fertilizer-N. When the amount of crop N derived from fixation exceeds that harvested with the seed, legume cropping also represents a means of stabilizing, and in some cases improving, the N-fertility of the soil. Costings of the energy requirements of N_2- and nitrate-dependent plants suggest a higher energy cost for the former. However, we should also take account of the costs associated with the provision of adequate levels of nitrate-N; in fact, the energy required to manufacture and transport fertilizer-N are considerable and may well eliminate the difference between the two biological systems.

Fixed-N enters the soil-N pool as stable organic-N and is resistant to short-term losses *via* leaching and denitrification. Excessive use of highly

Table 11: Measurements of nodulation and N_2 fixation by, and growth and yield of, nitrate tolerant and commercial genotypes of soybean in a high nitrate field soil. (Herridge and Betts (1985) and D.F. Herridge and J.H. Betts, unpublished data)

Genotype	Nodule/plant		N fix. index[1]	Shoot yield	Seed yield	Residual soil nitrate[2]
	Weight	Number				
	mg		*%*	*g/plant*	*t/ha*	*kg/ha*
Nitrate tolerant lines						
Korean 466	376	34.5	36	45.9	1.56	64
" 468	254	16.8	27	43.3	1.74	79
" 469	176	19.5	30	41.6	1.43	76
Commercial cultivars						
Bragg	24	2.0	11	39.7	2.23	45
Davis	40	1.3	15	48.5	2.17	n.a.

1. Ureide method (Herridge 1984);   2. To a depth of 1.2 m

soluble N fertilizers in Europe and the USA during the past 30 years has created ecological problems in ground water supplies and even in marine environments. There is an urgent need to rationalize the use of N fertilizers both from the point of cost and of pollution. There is also an urgent need to increase the uptake of N by cereal and oilseed crops in many countries of the world, especially in the third world. Nitrogen fixation can provide an answer. The challenge is to increase the levels of fixation achieved by commercial crops in the field to a point at or very near to their potential.

REFERENCES

Allen, E.K., and Allen, O.N. (1958). Biological aspects of symbiotic nitrogen fixation. In 'Encyclopedia of Plant Physiology'. (Ed. W. Ruhland.) pp. 48–118. (Springer–Verlag: Berlin.)

Amarger, N., Mariotti, A., Mariotti, F., Durr, J.C., Bourguignon, C., and Lagacherie, B. (1979). Estimate of symbiotically fixed nitrogen in field grown soybeans using variations in ^{15}N natural abundance. Plant Soil 52, 269–80.

Attewell, J., and Bliss, F.A. (1985). Host plant characteristics of common bean lines selected using indirect measures of N_2 fixation. In 'Nitrogen Fixation Research Progress'. (Eds H.J. Evans, P.J. Bottomley and W.E. Newton.) pp. 3–9. (Martinus Nijhoff: Boston.)

Ayanaba, A. (1977). Toward better use of inoculants in the humid tropics. In 'Biological Nitrogen Fixation in Farming Systems of the Tropics'. (Eds A.N. Ayanaba and P.J. Dart.) pp. 181–7. (John Wiley & Sons: Chichester, UK.)

Baldock, J.O., Higgs, R.L., Paulson, W.H., Jackobs, J.A., and Shrader, W.D. (1981). Legume and mineral N effects on crop yields in several crop sequences in the upper Mississippi Valley. Agron. J. 73, 885–90.

Bergersen, F.J. (1987). Concluding remarks. In 'A Century of Nitrogen Fixation Research: Present Status and Future Prospects'. (Eds F.J. Bergersen and J.R. Postgate) In Press (Phil. Trans. Roy. Soc. Lond.).

Bergersen, F.J., Turner, G.L., Chase, D.L., Gault, R.R., and Brockwell, J. (1985). The natural abundance of ^{15}N in an irrigated soybean crop and its use for the calculation of nitrogen fixation. Aust. J. Agric. Res. 36, 411–23.

Betts, J.H., and Herridge, D.F. (1984). Effects of nitrate supply on nodulation and nitrogen fixation by soybean. In 'Seventh Australian Legume Nodulation Conference'. (Eds I.R. Kennedy and L. Copeland.) Aust. Inst. Agric. Sci. Occ. Publ. No. 12, pp 65–6.

Bezdicek, D.F., Evans, D.W., Abede, B., and Witters, R.E. (1978). Evaluation of peat and granular inoculum (Rhizobium japonicum) for soybean yield and nitrogen fixation under irrigation. Agron. J. 70, 865–8.

Brockwell, J., Gault, R.R., Chase, D.L., Turner, G.L., and Bergersen, F.J. (1985). Establishment and expression of soybean symbiosis in a soil previously free of Rhizobium japonicum. Aust. J. Agric. Res. 36, 397–409.

Burns, R.C., and Hardy, R.W.F. (1975). 'Nitrogen Fixation in Bacteria and Higher Plants'. (Springer–Verlag: Berlin.)

Burris, R.H. (1978). The global nitrogen budget—science or seance? In 'Nitrogen Fixation Volume 1: Free Living Systems and Chemical Models'. (Eds W.E. Newton and W.H. Orme–Johnson.) pp. 7–16. (University Park Press: Baltimore, USA.)

Chapman, A.L., and Myers, R.J.K. (1987). Nitrogen contributed by grain legumes in rotation with rice on the Cununurra soils of the Ord Irrigation Area, Western Australia. Aust. J. Exp. Agric. 27, 155–63.

Clegg, M.D. (1982). Effect of soybean on yield and nitrogen response of subsequent sorghum crops in eastern Nebraska. Fld Crops Res. 5, 233–9.

Date, R.A. (1977). The development and use of legume inoculants. In 'Biological Nitrogen Fixation in Farming Systems of the Tropics'. (Eds A.N. Ayanaba and P.J. Dart.) pp. 169–80. (John Wiley & Sons: Chichester, UK.)

Date, R.A. and Roughley, R.J. (1977). Preparation of legume seed inoculants. In 'A Treatise on Dinitrogen Fixation IV. Agronomy and Ecology'. (Eds R.W.F. Hardy and A.H. Gibson.) pp. 243–75. (John Wiley & Sons: New York.)

Devine, T.E. (1984). Genetics and breeding of nitrogen fixation. In 'Biological Nitrogen Fixation'. (Ed. M. Alexander.) pp. 127–54. (Plenum Publ. Corp., USA.)

Doran, J.W. (1980). Soil microbial and biochemical changes associated with reduced tillage. Soil Sci. Soc. Am. J. 44, 765–71.

Doughton, J.A., and MacKenzie, J. (1984). Comparative effects of black and green gram (mung beans) and grain sorghum on soil mineral nitrogen and subsequent grain sorghum yields on the eastern Darling Downs. Aust. J. Exp. Agric. Anim. Husb. 24, 244–9.

Dowdell, R.J., Crees, R., and Cannell, R.Q. (1983). A field study of contrasting methods of cultivation on soil nitrate content during autumn, winter and spring. J. Soil Sci. 34, 367–79.

Dyke, G.V., and Slope, D.B. (1978). Effects of previous legume and oat crops on grain yield and take–all in spring barley. J. Agric. Sci (Camb.). 91, 443–51.

Eaglesham, A.R.J., Ayanaba, A., Ranga Rao, V., and Eskew, D.L. (1982). Mineral N effects on cowpea and soybean crops in a Nigerian soil. II. Amounts of N fixed and accrued to the soil. Plant Soil 68, 183–92.

Freire, J.R. (1982). Research into the Rhizobium/Leguminosae symbiosis in Latin America. Plant Soil 67, 227–39.

Gibson, A.H. (1977). The influence of the environment and managerial practices on the legume–Rhizobium symbiosis. In 'A Treatise on Dinitrogen Fixation IV. Agronomy and Ecology'. (Eds R.W.F. Hardy and A.H. Gibson.) pp. 393–450. (John Wiley & Sons: New York.)

Graham, P.H., and Chatel, D.L. (1983). Agronomy. In 'Nitrogen Fixation Volume 3: Legumes'. (Ed. W.J. Broughton) pp. 56–98. (Oxford University Press: Oxford.)

Graham, P.H., and Halliday, J. (1977). Inoculation and nitrogen fixation in the genus Phaseolus. In 'Exploiting the Legume–Rhizobium Symbiosis in Tropical Agriculture'. (Eds J.M. Vincent, A.S. Whitney and J. Bose.) pp. 313–34 (Univ. Hawaii College Tropic. Agric. Misc. Publ. 145.)

Graham, P.H., and Rosas, J.C. (1977). Growth and development of indeterminant bush and climbing cultivars of Phaseolus vulgaris L. inoculated with Rhizobium. J. Agric. Sci.(Camb.) 88, 503–8.

Halliday, J. (1985). Biological nitrogen fixation in tropical agriculture. In 'Nitrogen Fixation Research Progress'. (Eds H.J. Evans, P.J. Bottomley and W.E. Newton.) pp. 675–81. (Martinus Nijhoff: Boston.)

Ham, G.E. (1978). Use of ^{15}N in evaluating symbiotic N_2 fixation of field–grown soybeans. In 'Isotopes in Biological Dinitrogen Fixation'. pp. 151–61. (IAEA: Vienna.)

Ham, G.E., Cardwell, V.B., and Johnson, H.W. (1971). Evaluation of Rhizobium japonicum inoculants in soils containing naturalized populations of rhizobia. Agron. J. 63, 301–3.

Harper, J.E., and Gibson, A.H. (1984). Differential nodulation tolerance to nitrate among legume species. Crop Sci. 24, 797–801.

Havelka, U.D., Boyle, M.G., and Hardy, R.W.F. (1982). Biological nitrogen fixation. In 'Nitrogen in Agricultural Soils'. (Ed. F.J. Stevenson.) pp. 365–422. (American Society of Agronomy: Madison, USA.)

Hearne, A.B. (1986). Effect of preceding crop on the nitrogen requirements of irrigated cotton (Gossypium Hirsutum L.) on a vertisol. Fld Crops Res. 13, 159–75.

Henzell, E.F., and Vallis, I. (1977). Transfer of nitrogen between legumes and other crops. In 'Biological Nitrogen Fixation in Farming Systems of the Tropics'. (Eds A.N. Ayanaba and P.J. Dart.) pp. 73–88 (John Wiley & Sons: Chichester, UK.)

Herridge, D.F. (1982a). Use of the ureide technique to describe the nitrogen economy of field-grown soybeans. Plant Physiol. 70, 7–11.

Herridge, D.F. (1982b). A whole-system approach to quantifying biological nitrogen fixation by legumes and associated gains and losses of nitrogen in agricultural systems. In 'Biological Nitrogen Fixation for Tropical Agriculture'. (Eds P.H. Graham and S.C. Harris.) pp. 593–608 (CIAT: Colombia.)

Herridge, D.G. (1984). Effects of nitrate and plant development on the abundance of nitrogenous solutes in root-bleeding and vacuum-extracted exudates of soybean. Crop Sci. 24, 173–9.

Herridge, D.F., and Betts, J.H. (1985). Nitrate tolerance in soybean; variation between genotypes. In 'Nitrogen Fixation Research Progress'. (Eds H.J. Evans, P.J. Bottomley and W.E. Newton.) p. 32. (Martinus Nijhoff: Boston.)

Herridge, D.F., and Holland, J.F. (1987). Effects of tillage on plant available nitrogen and N_2 fixation by soybean. In. 'Nitrogen Cycling in Agricultural Systems of Temperate Australia'. (Eds P.E. Bacon, J. Evans, R.R. Storrier, and A.C. Taylor.), pp. 390–6. (Aust. Soc. Soil Sci. Inc.: Wagga Wagga.)

Herridge, D.F., Roughley, R.J., and Brockwell, J. (1984). Effects of rhizobia and soil nitrate on the establishment and functioning of the soybean symbiosis in the field. Aust. J. Agric. Res. 35, 149–61.

Hunt, P.G., Matheny, T.A., and Wollum, II, A.G. (1985). Rhizobium japonicum nodular occupancy, nitrogen accumulation, and yield for determinant soybean under conservation and conventional tillage. Agron. J. 77, 579–84.

Jenkins, H.V., Vincent, J.M., and Waters, L.M. (1954). The root-nodule bacteria as factors in clover establishment in the red basaltic soils of the Lismore district, New South Wales. III. Field inoculation trials. Aust. J. Agric. Res. 5, 77–89.

LaRue, T.A., and Patterson, T.G. (1981). How much nitrogen do legumes fix? Adv. Agron. 34, 15–38.

Lyman, J.M., Baudoin, J.P., and Hidalgo, R. (1985). Lima Bean (Phaseolus lunatus L.). In 'Grain Legume Crops'. (Eds R.J. Summerfield and E.H. Roberts.) pp. 477–519. (Collins: London.)

Mahon, J.D. (1979). Environmental and genotypic effects on the respiration associated with symbiotic nitrogen fixation in peas. Plant Physiol. 63, 892–7.

Mahon, J.D. (1983). Energy relationships. In 'Nitrogen Fixation Volume 3: Legumes'. (Ed. W.J. Broughton) pp. 299–325. (Oxford University Press: Oxford.)

Mahon, J.D., and Child, J.J. (1979). Growth response of inoculated peas (Pisum sativum) to combined nitrogen. Can. J. Bot. 57, 1687–93.

Mughogho, S.K., Awai, J., Lowendorf, H.S., and Lathwell, D.J. (1982). The effect of fertilizer nitrogen and Rhizobium inoculation on the yield of cowpeas and subsequent crops of maize. In 'Biological Nitrogen Fixation for Tropical Agriculture'. (Eds P.H. Graham and S.C. Harris.) pp. 297–301. (CIAT: Colombia.)

Nelson, A.N., and Weaver, R.W. (1980). Seasonal nitrogen accumulation and fixation by soybeans grown at different densities. Agron. J. 72, 613–6.

Norton, G., Bliss, F.A., and Bressoni, R. (1985). Biochemical and nutritional attributes of grain legumes. In 'Grain Legume Crops'. (Eds R.J. Summerfield and E.H. Roberts.) pp. 73–114. (Collins: London.)

Nutman, P.S. (1971). Perspectives in biological nitrogen fixation. Sci. Prog., Oxf. 59, 55–74.

Pate, J.S., Layzell, D.B., and Atkins, C.A. (1979). Economy of carbon and nitrogen in a nodulated and nonnodulated (NO_3-grown) legume. Plant Physiol. 64, 1083–8.

Poth, M., La Favre, J.S., and Focht, D.D. (1986). Quantification by direct ^{15}N dilution of fixed N_2 incorporation into soil by Cajanus cajan (pigeon pea). Soil Biol. Biochem. 18, 125–7.

Rao, Kumar, J.V.D.K., Dart, P.J., Matsumoto, T., and Day, J.M. (1981). Nitrogen fixation by pigeonpea. Proc. Int. Workshop on Pigeonpeas. Vol. 1., pp. 190–9. (ICRISAT: Hyderabad.)

Rao, Kumar, J.V.D.K., Dart, P.J., and Sastry, P.V.S.S. (1983). Residual effect of pigeonpea (Cajanus cajan) on yield and nitrogen response of maize. Exptl. Agric. 19, 131–41.

Ratner, E.I., Lobel, R., Feldhay, H., and Hartzook, A. (1979). Some characteristics of symbiotic nitrogen fixation, yield, protein and oil accumulation in irrigated peanuts (Arachis hypogaea L.). Plant Soil 51, 373–86.

Reeves, T.G., Ellington, A., and Brooke, H.D. (1984). Effects on lupin-wheat rotations on soil fertility, crop disease and crop yields. Aust. J. Exp. Agric. Anim. Husb. 24, 595–600.

Rennie, R.J. (1984). Comparisons of N balance and ^{15}N isotope dilution to quantify N_2 fixation in field-grown legumes. Agron. J. 76, 785–90.

Rennie, R.J. (1985). Nitrogen fixation in agriculture in temperate regions. In 'Nitrogen Fixation Research Progress'. (Eds H.J. Evans, P.J. Bottomley and W.E. Newton.) pp. 659–65. (Martinus Nijhoff: Boston.)

Rennie, R.J., Dubetz, S., Bole, J.B., and Muendel, H.H. (1982). Dinitrogen fixation measured by ^{15}N isotope dilution in two Canadian soybean cultivars. Agron. J. 74, 725–30.

Rennie, R.J., and Kemp, G.A. (1983). N_2-fixation in field beans quantified by ^{15}N isotope dilution. I. Effect of strains of Rhizobium phaseoli. Agron. J. 75, 640–4.

Rennie, R.J., and Kemp, G.A. (1984). ^{15}N-determined time course for N_2 fixation in two cultivars of field bean. Agron. J. 76, 146–54.

Rerkasem, B., Rerkasem, K., and Bergersen, F.J. (1985). Yield and nitrogen fixation advantage in corn–ricebean intercrop. Proc. 10th Nth. Amer. Rhiz. Conf., p. 72.

Rice, C.W., and Smith, M.S. (1982). Denitrification in no–tilled and ploughed soils. Soil Sci. Soc. Am. J. 46, 1168–73.

Rigaud, J. (1981). Comparison of the efficiency of nitrate and nitrogen fixation in crop yield. In 'Nitrogen and Carbon Metabolism'. (Ed. J.D. Bewley.) pp. 17–48. (Martinus Nijhoff: The Hague.)

Ryle, G.J.A., Powell, C.E., and Gordon, A.J. (1979a). The respiratory costs of nitrogen fixation in soyabean, cowpea, and white clover. I. Nitrogen fixation and the respiration of the nodulated root. J. Exp. Bot. 30, 135–44.

Ryle, G.J.A., Powell, C.E., and Gordon, A.J. (1979b). The respiratory costs of nitrogen fixation in soyabean, cowpea, and white clover. II. Comparisons of the cost of nitrogen fixation and the utilization of combined nitrogen. J. Exp. Bot. 30, 145–53.

Sanford, J.O., and Hairston, J.E. (1984). Effects of N fertilization on yield, growth, and extraction of water by wheat following soybeans and grain sorghum. Agron. J. 76, 623–7.

Schubert, K.R., and Ryle, G.J.A. (1980). The energy requirements for nitrogen fixation in nodulated legumes. In 'Advances in Legume Science'. (Eds R.J. Summerfield and A.H. Bunting) pp. 85–96. (Royal Botanic Gardens: Kew.)

Silsbury, J.H. (1977). Energy requirements for symbiotic nitrogen fixation. Nature. 267, 149–50.

Summerfield, R.J., and Roberts, E.H. (1985). 'Grain Legume Crops'. (Collins: London.)

Thomas, G.W., Blevins, R.L., Phillips, R.E., and McMahon, M.A. (1973). Effect of a killed sod mulch on nitrate movements and corn yield. Agron. J. 63, 736–9.

Thurlow, D.L., and Hiltbold, A.E. (1985). Dinitrogen fixation by soybeans in Alabama. Agron. J. 77, 432–6.

Troedson, R.J., Lawn, R.J., Byth, D.E., and Wilson, G.L. (1984a). Nitrogen fixation by soybeans in saturated soil. In 'Seventh Australian Legume Nodulation Conference'. (Eds I.R. Kennedy and L. Copeland.) Aust. Inst. Agric. Sci. Occ. Publ. No. 12, pp. 17–8.

Troedson, R.J., Lawn, R.J., and Byth, D.E. (1984b). Water management for peak soybean production in clay soils. In 'Root Zone Limitations to Crop Production on Clay Soils'. (Eds W.A. Muirhead and E. Humphreys.) pp. 257–65. (CSIRO: Melbourne.)

Vallis, I., and Gardener, C.J. (1985). Effect of pasture age on the efficiency of nitrogen fixation by 10 accessions of Stylosanthes spp. Aust. J. Exp. Agric. 25, 70–5.

Vallis, I., Henzell, E.F., and Evans, T.R. (1977). Uptake of soil nitrogen by legumes in mixed swards. Aust. J. Agric. Res. 28, 413–25.

Vasilas, B.L., and Ham, G.E. (1984). Nitrogen fixation in soybeans: an evaluation of measurement techniques. Agron. J. 76, 759–64.

Vincent, J.M. (1965). Environmental factors in the fixation of nitrogen by the legume. In 'Soil Nitrogen'. (Eds W.V. Bartholomew and F.E. Clark.) pp. 384–435. (American Soc. Agron: Madison.)

Vincent, J.M. (1982). Role, needs and potential of the nodulated legume. In 'Nitrogen Fixation in Legumes'. (Ed. J.M. Vincent.) pp. 263–84. (Academic Press: Australia.)

Wallis, E.S., Byth, D.E., Whiteman, P.C., and Saxena, K.B. (1983). Adaption of pigeonpea (Cajanus cajan) to mechanical culture. Proc. Aust. Plant Breed. Conf., Adelaide 1983, pp. 142–5.

Weber, C.R., (1966). Nodulating and nonnodulating soybean isolines: II. response to applied nitrogen and modified soil conditions. Agron. J. 58, 46–9.

Weiser, G.C., Grafton, K.F., and Berryhill, D.L. (1985). Nodulation of dry beans by commercial and indigenous strains of Rhizobium phaseoli. Agron. J. 77, 856–9.

Welch, L.F. (1985). Rotational benefits to soybeans and following crops. In 'World Soybean Research Conference III: Proceedings'. (Ed. R. Shibles.) pp. 1054–60 (Westview Press: Boulder.)

Westermann, D.T., Kleinkopft, G.E., Porter, L.K., and Leggett, G.E. (1981). Nitrogen sources for bean production. Agron. J. 73, 660–4.

Wetselaar, R. (1967). Estimation of nitrogen fixation by four legumes in a dry monsoonal area of north–western Australia. Aust. J. Exp. Agric. Anim. Husb. 7, 518–22.

Wych, R.D., and Rains, D.W., (1978). Simultaneous measurements of nitrogen fixation estimated by the acetylene–ethylene assay and nitrate absorption by soybeans. Plant Physiol. 62, 442–8.

Wynne, J.C., Ball, S.T., Elkan, G.H., Isleib, T.G., and Schneeweis, T.J. (1982). Host-plant factors affecting nitrogen fixation of the peanut. In 'Biological Nitrogen Fixation for Tropical Agriculture'. (Eds P.H. Graham and S.C. Harris.) pp. 67-75. (CIAT: Colombia.)

NITROGEN FIXATION NOT ASSOCIATED WITH LEGUMES

A.H. Gibson, M.M. Roper and D.M. Halsall

ABSTRACT

Research in recent years has demonstrated that significant levels of N_2 fixation occur naturally in many agricultural and natural ecosystems. Foremost in this research has been the study of rhizosphere associations, or associative symbiosis, usually involving Azospirillum spp. and graminaceous plants. Although the initial optimism created by this work has not been fulfilled, a better, and improving, understanding of these associations is developing. Studies involving free-living bacteria in soil, with or without organic matter amendment, have also revealed higher levels of N_2 fixation than hitherto considered. Despite these advances, knowledge of the ecology of free-living diazotrophs in the field is sparse and requires close attention if management practices, or inoculation practices, are to be developed to fully utilise the potential of their N_2-fixing activities.

INTRODUCTION

During the past 15 years, there has been a rapid and widespread increase in N_2 fixation research, ranging from fully symbiotic associations, through facultative associations, to free-living forms that may be heterotrophic, chemolithotrophic or phototrophic. Additional N_2-fixing organisms (diazotrophs) have been found, e.g. Azospirillum spp., some Pseudomonas spp., Campylobacter spp., but in most instances research has been directed towards understanding, improving and evaluating N_2 fixation by recognised diazotrophs (e.g. Klebsiella, Azotobacter) or diazotrophic associations (e.g. Azolla-Anabaena, actinorhizal nodules) (Gibson and Jordan 1983). Apart from the mid-70's oil crisis and its effect on N-fertilizer prices, two major scientific developments have facilitated these studies. The acetylene reduction assay, which measures nitrogenase activity by determining the rate at which acetylene is reduced to ethylene (Hardy et al. 1968) has enabled both laboratory and field studies that would otherwise be very time-consuming, expensive, or even impossible. The other development has been the use of molecular biology techniques to provide an understanding of the genetic control and regulation of nitrogenase in a wide range of diazotrophs. Although the application of molecular biology has yet to achieve significant impact in promoting N_2 fixation in the agricultural systems, the accumulating research findings provide great potential to develop novel bacteria, and associations, that could be highly beneficial to global agriculture.

In this paper, we review the common diazotrophs and their associations, examine their physiology in relation to known and possible habitats, and consider their contribution to various agricultural systems. Emphasis is given to those bacteria primarily located in tropical regions although it should be appreciated that current research also involves temperate and arctic ecosystems. Finally we consider the more significant issues to be addressed in attempting to improve N_2 fixation by bacteria and their associations, excluding legumes, in agricultural systems. Detailed consideration of N_2 fixation associated with rice culture is provided by Watanabe et al. (in these Proceedings).

DIAZOTROPHS AND THEIR ASSOCIATIONS

Nitrogen fixation is widely distributed within the procaryotes, but has not been confirmed in any eucaryotes, even when the responsible genes have been transferred from procaryotes (Zamir et al. 1981). Nitrogenase is a two component enzyme system comprising a Mo—Fe protein and a smaller Fe-protein. In Klebsiella pneumoniae, at least 17 genes are involved in the regulation and synthesis of nitrogenase, and based on their known functions a similar number can be expected in other diazotrophs. The responsible genes may be clustered, as in Klebsiella (Robson et al. 1983), located at various sites around the chromosome, as in Azotobacter (Bishop and Brill 1977) or even located on a plasmid, as in fast-growing Rhizobium spp. (see Evans et al. 1985). Apart from rhizobia, little is known of the genetics of symbiosis in those bacteria forming associations, or in their respective hosts.

The major physiological requirement for N_2 fixation is a low concentration of O_2 at the sites of enzyme synthesis and function. Synthesis of nitrogenase is repressed at high O_2 concentrations, and both components, but especially the Fe—protein and the electron carriers to nitrogenase, are inactivated by O_2 (Robson and Postgate 1980). Various protective mechanisms have evolved to provide diazotrophs with a degree of tolerance to O_2, enabling some to function under normal aerobic conditions. Combined N, particularly NH_4^+, represses nitrogenase synthesis and function at relatively low concentration. In some bacteria, the inhibitory effect of nitrate is a consequence of denitrification producing nitrite ions that inhibit the enzyme system. Energy for growth and N_2 fixation is usually obtained from the uptake and metabolism of simple sugars, organic acids and sugar alcohols but some bacteria can utilize xylans (Halsall et al. 1985) and in one case, they can utilise cellulose (Waterbury et al. 1983). Significant contributions are also made by phototrophs, such as the cyanobacteria, and by chemotrophs that derive their energy for CO_2 fixation from the utilisation of reducible sulphur compounds, hydrogen or methane.

In Table 1, the genera of bacteria known to include free-living diazotrophs are listed according to their tolerance of oxygen under N_2-fixing conditions. For many genera, not all species, or strains within a species, are capable of N_2 fixation.

Table 1: Representative diazotrophs, grouped according to the sensitivity of N_2 fixation to oxygen.

Classification	Taxonomic Group
HETEROTROPHS	
Aerobes	Azotobacter, Azomonas, Derxia, Beijerinckia
Microaerophiles	Azospirillum, Herbaspirillum, Thiobacillus, Pseudomonas, Xanthobacter, Bradyrhizobium, Campylobacter and various Methylcoccaceae.
Facultative Anaerobes	Klebsiella, Citrobacter, Enterobacter, Erwinia, Bacillus
Strict Anaerobes	Clostridium, Desulfovibrio, Desulfotomaculum
PHOTOTROPHS	
Aerobes	Heterocystous cyanobacteria (Anabaena, Nostoc, Calothrix)
Microaerophiles	Non-heterocystous cyanobacteria (Plectonema, Lyngbya, Oscillatoria, Gloeothece)
Facultative Anaerobes	Rhodospirillum, Rhodopseudomonas
Strict Anaerobes	Chromatium, Thiocystis

Aerobes

The two principal groups are the cyanobacteria, formerly known as blue green algae, and the Family Azotobacteraceae, although the latest edition of Bergey (Holt 1984) indicates that Beijerinckia and Derxia should be excluded from this Family. The ability of Azotobacter to fix N_2 under aerobic conditions in the laboratory has been attributed to various mechanisms which serve to protect nitrogenase from O_2-mediated inactivation. Foremost is respiratory protection, by which mechanism the bacteria raise their rate of respiration to lower the O_2 level within the cells (Dalton and Postgate 1969), albeit at a loss of efficiency of N_2 fixation. The activity of uptake hydrogenase (H_2 oxidising) enzymes will also lower the O_2 level in the cells, and theoretically at least, improve the efficiency of N_2 fixation by recycling the H_2 evolved during N_2 fixation (Robson and Postgate 1980). Conformational changes, involving the association of the two nitrogenase proteins with an FeS protein similar in some regards to a ferredoxin (Robson and Postgate 1980), are also involved. Beijerinckia and Derxia are less tolerant of O_2 than Azotobacter but are still regarded as aerobic diazotrophs. The belief that protection from O_2 was provided by the gummy polysaccharides secreted by these bacteria has been questioned because non-gummy mutants of Derxia gummosa are unchanged in their O_2 tolerance (Pedrosa et al. 1980).

The cyanobacteria show various forms of protection of nitrogenase from O_2. Many of the filamentous forms are characterised by thick-walled heterocysts within which the nitrogenase is located. The O_2-evolving Photosystem II is absent from the heterocysts (Cossar et al. 1985), while further protection is provided by high activity of the oxidative pentose phosphate pathway and the presence of an uptake hydrogenase (Stewart 1980). The filamentous, non-heterocystous forms such as Oscillatoria and Spirulina are also aerobic N_2-fixers; they appear to derive protection of nitrogenase through a switch off/on mechanism that reversibly converts the active enzyme to an O_2-insensitive form, as well as having basically high rates of nitrogenase synthesis. The unicellular forms, such as Gloeothece and Gloeocapsa are also aerobic diazotrophs but N_2 fixation is restricted mainly to the dark periods; in the light, nitrogenase activity is supported by respiration rather than by photosynthesis (Maryan et al. 1986).

Microaerophiles

This group lacks the protective mechanism of Azotobacter and requires microaerobic conditions for growth dependent on N_2 fixation; e.g. in semi-solid media, they form a pellicle below the surface and this provides protection by retarding O_2 diffusion to the nitrogenase-active cells. Azospirillum spp. are commonly found in association with the roots of cereals and C_4 grasses (Patriquin et al. 1983) and with straw residues (Roper and Halsall 1986). Xanthobacter was created to include hydrogen-oxidising diazotrophs such as X. flavus (formerly Mycobacterium flavum) and X. autotrophicus (formerly Corynebacterium autotrophicum). A new addition to the diazotrophic genera is Pseudomonas, following isolation of N_2-fixing isolates from grass and cereal roots (Barraquio et al. 1983; Haahtela et al. 1983; Krotzky and Werner 1987) and from a forest soil (Chan et al. 1986). One, at least, of these strains can utilize simple phenolics as an energy source (Chan et al. 1986).

Chemoautotrophs in this group include Thiobacillus ferro-oxidans, which obtain energy for CO_2 fixation by oxidising dissolved ferrous ions to ferric, and the Methylcoccaceae, which are able to use methane but not all of which are diazotrophs. Within this group, acetylene inhibits the methane-oxidising

69

enzyme system and hence, nitrogenase activity and growth; methanol can be used as an energy source for the acetylene reduction assay (Postgate 1978). Some strains of Bradyrhizobium can fix N_2 under microaerobic conditions (Pagan et al. 1975).

Facultative anaerobes

This group is capable of aerobic growth when supplied with combined N but requires anaerobic conditions, or conditions of very low O_2 concentration, for N_2 fixation. Typical of the group are the diazotrophic members of the Enterobacteriaceae (Klebsiella, Enterobacter, Citrobacter, and Erwinia). Studies with Klebsiella pneumoniae have provided much of the basic knowledge on the genetics of N_2 fixation including its regulation (Cannon et al. 1985), although studies now extend to other genera (see Evans et al. 1985). Only two Bacillus species (B. macerans and B. polymyxa) are known as diazotrophs.

Rhodopseudomonas and Rhodospirillum are two phototrophs in this group, and in Rhodospirillum, there is a necessity for a Mn^{2+}-dependent enzymic activation of the Fe-protein before nitrogenase can function. Although diazotrophy is widespread in these bacteria, it is not universal; nor is it necessarily light dependent.

Strict anaerobes

Clostridium pasteurianum, and some strains of C. acetobutylicum are diazotrophs typical of this group. They are frequently found in silage and compost (Harper and Lynch 1984). Desulfotomaculum, a genus of chemotrophs, are spore-formers that use reducible sulphur compounds as electron acceptors. Although they are common inhabitants in the soil and water in geothermal regions, the thermophilic species are not diazotrophs (Postgate 1981). Other strict anaerobic diazotrophs are Desulfovibrio (found primarily in polluted water and marine environments) and phototrophic types which include Chromatium and Thiocystis.

Symbiotic diazotrophs

Most widespread in this group are the cyanobacteria, notably species of the heterocystous, filamentous Nostoc. The associations range from lichens (Millbank 1974) through mosses, ferns, and cycads, to the Angiosperm Gunnera spp. in which the cyanobacteria invade the stem glands at the base of the leaves. In the last association, the organisms are intracellular, but in the other associations, they remain outside the host cells, albeit inside the plant tissue (cycads, Azolla), or in close association with fungi, as in some lichens. A high proportion of the cyanobacterial cells (up to 40%) transform into heterocysts. Phycobiliproteins in the endophyte can be involved with photosynthesis, but in some associations the endophyte is completely, or partially, dependent on the host for fixed carbon.

The most intensively studied association in recent years involves the fern, Azolla spp. and Anabaena azollae. This fern is common in rice paddies and bodies of slow-moving water in both the tropics and temperate regions. The fern can double in biomass in 2.5 days under ideal conditions and has value as a green manure crop or as livestock feed. Temperatures above 35°C can be detrimental and in the field, phosphorus can be a major limiting element. A more complete description of this, and other cyanobacterial associations, has been provided by Peters et al. (1986).

The other major diazotrophic symbiont is the actinomycete, Frankia sp., which nodulates approximately 160 species from 10 phylogenetically diverse orders. Most of these species are shrubs or trees in such genera as Casuarina, Allocasuarina, Purshia, Elaeagnus, Myrica, and Comptonia. Some species, especially within the Casuarina group, have great potential in reafforestation, in agro-forestry, and in reclamation and colonisation of denuded or poor land. Although the nodules bear a superficial resemblance to legume nodules, internally they are characterised by a central vascular system and a cortex packed with cells of the endophyte (Torrey 1978). In all species examined, except the Casuarinas, nitrogenase activity is confined to thick-walled vesicles developing at the tips of hyphal branches (Torrey 1985). The endophyte has been isolated in pure culture (Torrey 1985) but it is very slow-growing and inoculant production in quantity will be difficult. Host x strain specificity exists with regards to nodule formation and the effectiveness of N_2 fixation (Reddell and Bowen 1985; Simon et al. 1985).

Parasponia spp., of the Family Ulmaceae, form nodules with Bradyrhizobium sp. These woody shrubs are found in New Guinea and Indonesia where they are cultivated as shade trees in plantations. The nodules resemble actinorhizal nodules (Trinick 1979). Five species of the Family Zygophyllaceae reportedly nodulate with Bradyrhizobium sp. but little is known about the biology of these associations (Athar and Mamood 1981).

ASSOCIATIVE NITROGEN FIXATION

Associative N_2 fixation has received much attention since the observations of Döbereiner that Azotobacter paspali fixed N_2 in the rhizosphere of Paspalum notatum (Döbereiner et al. 1972) and that Azospirillum spp. (previously classified as Spirillum spp.) were common inhabitants of the root zone of a wide range of tropical grasses, including Digitaria decumbens, Pennisetum purpureum, Brachiaria mutica, Panicum maximum, and Zea mays (von Bulow and Döbereiner 1975; Döbereiner and Day 1976). Since 1976, more than 300 papers have been published describing N_2 fixation in the root zone (within the roots or in the rhizosphere) or the isolation of diazotrophs from roots. Many of the papers concern tropical grasses or cereals such as wheat, sorghum, millet, maize and rice, (see van Berkum and Bohlool 1980; Weier 1980; Vose 1983; Plant and Soil, vol. 90, 1986) but others concern aquatic plants such as Spartina alternifolia, Scirpus olneyi (bulrush - van Berkum and Sloger 1981), Zostera marina (Capone and Budin 1982),

71

Avicennia marina (mangrove – Hicks and Silvester 1985) and *Typha latifolia* (Biesboer 1984), or dicotyledonous species such as *Ipomoea batatas* (sweet potato – Hill et al. 1983) and *Brassica juncea* (Saha et al. 1985). Where the principal diazotrophs have been isolated, *Azospirillum* spp. are often predominant, but *Azotobacter*, *Beijerinckia*, *Klebsiella*, *Bacillus* and *Pseudomonas* species have also been found in significant numbers in particular circumstances (see Kosslak and Bohlool 1983). Host specificity towards certain diazotrophs has been described, such as *Azospirillum lipoferum* associating with the roots of C_4 grasses and *A. brasilense* with C_3 plants such as wheat and rice (Baldani and Döbereiner 1980), *Azotobacter paspali* with *Paspalum notatum* cv. *batatais*, but not cv. *pensacola* (Boddey et al. 1983) and *Klebsiella pneumoniae* with *Poa pratensis* cv. Park (Lee et al. 1986). Inheritance of the ability to promote associative N_2 fixation has been demonstrated in maize (von Bulow and Döbereiner 1975; Ela et al. 1982), wheat (Larson and Neal 1978), rice (Iyama et al. 1983) and pearl millet (Bouton et al. 1985).

Whether true symbiotic associations develop within roots has not been determined to everyone's satisfaction. A number of groups have isolated diazotrophs from thoroughly surface sterilised (?) roots (van Berkum and Bohlool 1980), and microscopy has revealed bacteria with strong tetrazolium-reducing activity in the cortex (Döbereiner and Day 1976) and protoxylem vessels (Patriquin et al. 1983). However it has not been resolved whether these are casual infections occupying dead cells or whether specific diazotrophs are infecting the roots and establishing a true symbiosis. There are several reports of enhanced vesicular arbuscular mycorrhizal infection in the presence of root-associated diazotrophs (Pacovsky et al. 1985; Subba Rao et al. 1985).

Another form of associative N_2 fixation involves epiphylls or leaf surface diazotrophs. Ruinen (1974, 1975) isolated a number of diazotrophs from tropical plants in south-east Asia, Africa and South America, and Bentley and Carpenter (1980) have measured nitrogenase activity on leaves of numerous species in the jungles of Costa Rica. Sengupta et al. (1982) found that more than half the Angiosperms in eastern India had diazotrophs in the phyllosphere. Inoculation of wheat and rice with 22 of 161 isolates enhanced yield by more than 50%. Murty (1983, 1984) observed nitrogenase activity on leaves of several crop plants, including rice, sugarcane and sorghum, with *Beijerinckia* as the predominant organism, while Roskoski (1980) found low levels of nitrogenase activity on the leaves of *Coffea arabica*.

COLOGICAL REQUIREMENTS

Two major limitations on the development of a particular strain of microorganism within a mixed population are energy supply and the ability to compete with other components of the microflora. Energy supply may be limiting in an absolute sense, or for particular microorganisms, the carbon substrate present may not be available due to the organisms' inability to metabolise the complex compounds. Competition is primarily concerned with tolerance or resistance to antibiotics, toxins, phage and bacteriocins produced by other microorganisms but also involves the ability to use limiting substrates. For many diazotrophs, two further important limitations are the level of available N in the soil, as this affects the growth of other microorganisms as well as inhibiting N_2 fixation, and the degree of aeration, or O_2 level, of the environment.

Energy supply

For diazotrophs in the rhizosphere, energy supply may be less limiting than it is outside this zone, although the competition for substrates may be greater. Studies involving root exudation, and the effect of photosynthesis, have given variable results, especially in trying to predict host line or bacterial strain responses to inoculation. Working with Pennisetum americanum, Bouton et al. (1985) observed higher numbers of Azospirillum brasilense on the roots of plants derived from "high nitrogenase activity" parents, relative to those derived from "low" parents. However differences in nitrogenase activity were only evident at low O_2 levels. Similar results were recorded by Kipe-Nolt et al. (1985) using Sorghum bicolor and strains of Azospirillum and Enterobacter. In another study with this species, inoculation with Azotobacter or Azospirillum caused an increase in the exudation of carbon compounds (Lee and Gaskins 1982) as also found with P. americanum (Bouton et al. 1985). However, Jain and Rennie (1986) concluded that the spermosphere model, based on the use of exudates from germinating seeds, is not a practical technique for screening potential host x strain combinations of wheat and Azospirillum for nitrogenase activity. Attempts to show a direct utilisation of photosynthate by rhizosphere populations by examining diurnal variation in nitrogenase activity have had variable success (e.g. Döbereiner et al. 1972, cf. Döbereiner and Day, 1976 working with Paspalum notatum, and Balandreau and Dommergues 1971 cf., Watanabe and Cabrera 1979 working with Oryza sativa). Smith et al. (1984) also report variable effects of shading on nitrogenase activity in the root zone of Pennisetum americanum. A rapid increase in nitrogenase activity was observed when photosynthesis of Spartina alternifolia was stimulated by increasing the light intensity or raising the CO_2 level (Whiting et al. 1986).

Apart from the direct effect on Azolla, free-living cyanobacteria and such photosynthetic diazotrophs as Rhodopseudomonas, light influences crop growth and, in some instances, associative N_2 fixation. In grasslands, N_2 fixation at the soil surface has been attributed to cyanobacteria (Paul et al. 1971), while Day et al. (1975) considered that the activity of cyanobacteria at the soil surface made a very significant contribution to the N economy of the Broadbalk experiment at Rothamsted. For free-living heterotrophs outside the rhizosphere, plant residues are the major source of energy. As early as 1888, Barthelot observed an increase in the N content of unsterilised soils, especially during the summer, an effect he believed due to the cooperation of microorganisms, while Gautier and Drouin (1888) found a N gain in soil following the addition of organic matter. Jensen (1941) and Jensen and Swaby (1941) observed that Azotobacter and Clostridum butyricum, were able to fix N in cellulose-culture with a range of aerobic cellulolytic bacteria and fungi. In the field, the incorporation of plant residues may stimulate nitrogenase activity (Abd-el-Malek 1971; Yoneyama et al. 1977; Charyulu et al. 1981; Roper 1983; Ladha et al. 1986). These residues contain cellulose, hemi-celluloses, lignins and small amounts of protein and soluble carbohydrates (Harper and Lynch 1981). A wide variety of bacteria and fungi are able to break down these residues to produce disaccharides, hexoses and dicarboxylic acids (Cheshire 1977) which are readily used by other members of the soil microflora. Diazotrophs are thought to rely on simple organic substrates for their energy supply, but some at least can utilise more complex substrates. For example, Azospirillum spp. can utilise xylan, a principal component of hemi-cellulose (Halsall et al. 1985) and recent results indicate that other diazotrophs have similar capabilities (D.M. Halsall and A.H. Gibson, unpublished data). To the present time, the only cellulolytic diazotroph is a bacterium found in shipworms (Waterbury et al. 1983). The actinorhizal symbiont, Frankia sp., and some strains of Bradyrhizobium sp., fix N_2 in culture (Pagan et al. 1975; Torrey et al. 1981), but little is known of their N_2-fixing activity in soil, nor of their use of complex carbon sources.

Competition

Competition between diazotrophs and other soil microorganisms, and the survival of introduced diazotrophs, has received less study than has been warranted by the various reports of the failure to establish inoculants, e.g. Azospirillum (Smith et al. 1984; Gaskins et al. 1984/5), Klebsiella and Enterobacter (Wright and Weaver 1982) and Azotobacter (Hegazi et al. 1979). An exception was the report of establishment of several Azospirillum brasilense strains in the rhizosphere and in roots of wheat, and of A. lipoferum strains with sorghum (Baldani et al. 1986). Improved establishment of inoculant Azospirillum on wheat roots was achieved by the inclusion of antibacterial and antifungal agents to which the inoculant was tolerant (Bashan 1986). Actinomycetes of the genera Nocardia and Streptomyces may inhibit diazotrophs of various genera (Zuberer and Roth 1982; Kulinska and Drozdowicz 1983). Phages capable of attacking Azospirillum have been described but they do not appear to be of major significance (Germida 1986).

Soil environment

Soil moisture

In addition to affecting microbial activity, soil moisture affects gas exchange and the level of O_2 in the soil environment. In the laboratory, high moisture contents (from 50% field capacity to waterlogging) are needed for significant nitrogenase activity (Rice et al. 1967; Hegazi et al. 1979). However under field conditions, where microaerobic and anaerobic microsites are relatively undisturbed, nitrogenase activity has been observed at less than 50% field capacity, albeit at low levels. The response is dependent on the soil type (Roper 1985). Rice et al. (1967) suggested that a combination of aerobic and anaerobic or microaerobic conditions was required for plant residue breakdown and N_2 fixation respectively, and this is supported by further laboratory studies (Veal and Lynch 1984; Roper 1985). Associative N_2 fixation is greater at higher moisture levels than in drier soils (Weier et al. 1981; Souto and Döbereiner 1985). In a field study involving Panicum maximum, Weier et al. (1981) found a significant positive correlation between moisture level and nitrogenase activity (r = 0.59), as did Roper (1983) in straw amended soil. In both cases, the activity of many cores was improved after watering to field capacity.

Soil temperature

Temperature influences the activity of both the cellulolytic and other microorganisms involved in residue degradation as well as N_2 fixation by diazotrophs. Pal et al. (1975) showed that residue decomposition increased from 7 to 37°C, while Roper (1985) determined that straw breakdown occurred, but not nitrogenase activity, up to 50°C. Using two glucose-amended soils, Roper (1985) observed different temperature-response curves, and suggested that there were differences in the soil populations of diazotrophs (see Roper and Halsall 1986) or in the temperature adaptation of the two populations. Azolla have temperature maxima for nitrogenase activity below that of free-living cyanobacteria, and in China, when temperatures rise above 35°C, the Azolla dies, sinks to the bottom of the paddies, and is replaced by cyanobacteria. Associative N_2 fixation is also temperature responsive, and in the study with Panicum maximum, Weier et al. (1981) found a significant negative correlation (r =-0.78) between nitrogenase activity and temperature over the range 14-28°C. In a straw amended soil, there was a significant positive correlation between nitrogenase activity and temperature between 19 and 30°C (Roper 1983).

Mineral nutrients

Where plant residues have a high C:N ratio, nitrogen supply may limit residue decomposition (Stotzky and Norman 1961; Bhardwaj and Novak 1978) but the addition of inorganic N may suppress nitrogenase activity (Knowles and Denike 1974). Although mineral N, as NH_4^+ and NO_3^-, represses nitrogenase activity in culture, little is known of the sensitivity of the various diazotrophs under soil conditions. Sulphur and phosphorus additions may enhance the decomposition of cereal straws under nutrient deficient conditions (Stotzky and Norman 1961; Stewart et al. 1966) but little is known of the effect of minerals on nitrogenase activity in the field. Working with grazed Brachiaria decumbens, Miranda et al. (1985) showed a beneficial effect of Mo supplementation on the numbers of Azospirillum spp. in the rhizosphere and on the N yield (29%) of the grass. The recent observation in Azotobacter of a second nitrogenase system, containing vanadium instead of molybdenum, that (i) develops under conditions of molybdenum deficiency and (ii) has much less ability to reduce acetylene than the Mo–nitrogenase (Joerger et al. 1986; Robson et al. 1986), raises interesting possibilities for future research.

Soil properties

Clay content and pH have a major influence on microbial activity in soil, including plant residue decomposition and nitrogenase activity. Whereas soil bacteria function in residue decomposition at pH 6.5–8.5 (Parr and Papendick 1978), fungi tend to be more active at lower pH. In culture, diazotrophs tend to favour pH 7, although Azotobacter is more tolerant to high pH levels and Beijerinckia tolerates pH down to 5.0–5.5 (Alexander 1961). The highly reactive nature of clay colloids can provide protection from desiccation (Bushby and Marshall 1977) and from predation and parasitism (Roper and Marshall 1978). Clays also form complexes with organic matter (Mortland 1970) and this can reduce the rate of organic matter decomposition (Ladd et al. 1981). Little is known of the effect of clay on N_2 fixation although Macura and Pavel (1959) observed higher levels of nitrogenase activity by Azotobacter when the liquid culture medium was amended with montmorillonite clay. In sand culture systems, montmorillonite increased activity 10–fold (M.M. Roper, unpublished data). In an examination of 20 soils, there was a broad positive correlation between the number of diazotrophs and soil clay content (D.M. Halsall and A.H. Gibson, unpublished data).

METHODS FOR MEASUREMENT OF NITROGEN FIXATION

There are three major approaches to the measurement of N_2 fixation, viz. direct methods involving ^{15}N, indirect methods such as the acetylene reduction assay, and N balance which involves measurement of all N inputs, outputs and residuals. The methods are detailed in Bergersen (1980) and Elkan (1987), while Knowles (1981), Vose (1983), Watanabe and Roger (1984) and Chalk (1985) have critically evaluated the different approaches.

^{15}N–based Techniques

Direct method

$^{15}N_2$ is incorporated into the atmosphere in and above the soil, and after an appropriate incubation period, total N and ^{15}N determinations on component parts are made. Nitrogen fixed is calculated, using the known level of ^{15}N in the atmosphere and correcting for natural abundance of ^{15}N in that atmosphere and in untreated samples (c. 0.3663%). With systems involving soil, the major problem lies in determining the concentration of $^{15}N_2$ in the atmosphere at the sites of N_2 fixation. This concentration is likely to vary in space and time due to differential diffusion rates unless action is taken to circulate the gas (e.g., Eskew et al. 1981; Watanabe and Ventura 1982) or to evacuate a closed system before introducing the labelled $^{15}N_2$ (e.g. Morris et al. 1985). No ready solution to these problems has been devised for in situ field assays.

^{15}N-isotope dilution

This method utilises ^{15}N-labelled NO_3^-, NH_4^+, urea or an organic N-source. Comparison is made between the isotopic ^{15}N concentration in the N-fraction of the treated sample with that in a similar system known not to be fixing N_2. The extent to which the ^{15}N concentration is lower in the treated samples provides an estimate of N_2 fixed. This approach has been used on grass-diazotroph associations (Boddey et al. 1983; Boddey and Victoria 1986) and on associations involving rice (Ventura and Watanabe 1983) and sugarcane (de Freitas et al. 1984). The major difficulty lies in achieving non-fixing control plants with similar physiology and growth habit to the test plants (Knowles 1981; Boddey et al. 1983). The level of supplied N should be low in order to avoid both inhibition of N_2 fixation and promotion of excessive plant growth. This technique is too insensitive to measure N_2 fixation by free-living heterotrophs in natural environments.

^{15}N-abundance methods

These rely on organic and inorganic N fractions in soil having a higher level of ^{15}N than atmospheric N_2, and on the high sensitivity and accuracy of the new generation of mass spectrometers (Ledgard and Peoples, in these Proceedings). The necessary non-fixing control system may be difficult to achieve for both associative fixation and free-living N_2-fixing systems in the field. There is also concern about different levels of natural abundance at different depths in the soil and over time. Despite these problems, the technique has been used successfully to determine N_2 fixation in sugarcane associations (Vose 1983).

Acetylene reduction assay

This assay is based on the ability of nitrogenase to reduce acetylene to ethylene which is determined by gas chromatography. Various assay procedures have been described (Turner and Gibson 1980) and the advantages and disadvantages of the assay reviewed by Knowles (1981). Central to the use of the assay systems is the determination of a $C_2H_2:N_2$ ratio appropriate to the system. The theoretical ratio is 4:1 when allowance is made for the normal production of 1 H_2 (at least) for each N_2 reduced; in the presence of C_2H_2, all electrons are utilised in C_2H_2 reduction, without any H_2 production. Determined ratios for various systems range from 1.3:1 to 8.1:1 (Hardy et al. 1973) and hence calibration of any system is essential if the estimate of N_2 fixation is to have credibility. Problems in determining the "N_2 fixed" component of the ratio, be it by ^{15}N techniques or total N increment, should not be ignored. Acetylene is 65x more soluble in water than N_2 and slow equilibration of $^{15}N_2$ may cause under-estimation of N_2 fixation in short assays. In some natural systems, C_2H_4 is produced continuously from various biological processes, and is immediately oxidized by other microorganisms. Acetylene inhibits C_2H_4 oxidation (de Bont 1976) but not C_2H_4 production (Nohrstedt 1975/6, 1983; Witty 1979) so that C_2H_2 reduction may be overestimated. Various methods are available to determine endogenous C_2H_4 production during assays (Nohrstedt 1983, 1984).

Acetylene inhibits N_2 fixation and results should be interpreted with caution where a population of diazotrophs is developing during the assay (Halsall and Gibson 1985). Questions have been raised relating to the effect of C_2H_2 on O_2 exchange in legume nodules as it affects C_2H_2 reduction (Minchin et al. 1986); similar questions are relevant to the use of the assay with actinorhizal nodules.

The C_2H_2 reduction assay may be applied to enclosed and semi-enclosed systems, but rarely to open systems. The efficacy of semi-enclosed systems is dependent on little or no loss of gas, often achieved by wetting the soil around the container (Roper 1983). A tracer gas such as propylene or propane, with similar solubility and diffusive properties as C_2H_4, is incorporated into the assay vessel to act as a reference standard. Diffusion rates of C_2H_2 into the soil and of C_2H_4 into the atmosphere vary and estimates should not commence until a steady C_2H_4 production rate is established.

The assay was used to show the first high levels of nitrogenase activity in the rhizosphere of various tropical grasses and maize (von Bulow and Döbereiner 1975; Döbereiner and Day 1976). These assays used washed, excised roots in bottles flushed with N_2 and supplemented with 2-3% O_2 (in an attempt to simulate the original conditions and allow the systems to "repair" following exposure to ambient O_2) before adding C_2H_2 16 hours later. Subsequently, it was considered that the resulting high activity was due to an unnatural build-up of diazotrophs during the "repair" period (Tjepkema and van Berkum 1977; van Berkum and Bohlool 1980). Hence the apparent high levels of nitrogenase activity recorded in earlier studies should be regarded with due caution; in the current literature, authors often stress that nitrogenase activity was observed from the commencement of the assay. The approach adopted by Weier (1980) in maintaining field cores under the conditions applicable at the time of collection is to be commended. In any closed or semi-enclosed system, maintenance of ambient O_2 levels in the atmosphere above the soil is essential.

Nitrogen increment or N-balance

This approach involves the determination of total N before and after the imposition of a treatment, and the measurement of all inputs and outputs. High background N in many soils, and large sample variability, requires a considerable increase in total N to determine N_2 fixed with any accuracy. In a field system, any N increase is a net change as there will be losses due to denitrification and leaching; similarly where plants are growing in the soil, N-uptake by the roots at depth can influence the assessment of N_2 fixed unless suitable controls without associative or free living N_2 fixation can be established.

In conclusion, there are still severe problems in measuring N_2 fixation, especially under field conditions. The systems and approaches described have their disadvantages as well as advantages. In recent years, techniques for determining N_2 fixation by nodulated legumes, and by laboratory-based free-living diazotrophs and associations have had success (Herridge and Bergersen, in these Proceedings). Further developments are required for the accurate determination of N_2 fixation by these latter diazotrophic systems in the field.

ESTIMATES OF NITROGEN FIXED

The determination of the contribution of biological N_2 fixation to soil and plant N is exceedingly difficult, and the assessment of the values obtained is confounded by the very different methods of determination and variation in the expression of the results. Representative estimates of N_2 fixation from studies with grasses and cereal crops, with or without inoculation, range from negligible to 103 kg N/ha/yr (Table 2), with the latter arising in N-balance studies with a double-cropping rice system over 14 years (App et al. 1984). Rather higher values from N-balance studies have been reviewed by Dart (1986) who cautioned the need for extreme care in the interpretation of these field assessments. With grasses in tropical and subtropical regions, the natural levels of N_2 fixation appear to range from 20 to 30 kg N ha/yr, with little evidence for any benefits from inoculation with Azospirillum or other diazotrophs. Of concern is that many estimates based on C_2H_2 reduction, with both grasses and cereals, use the traditional 3:1 $C_2H_2:N_2$ ratio, despite the fact that the theoretical ratio is at least 4:1 and, in practice, the appropriate ratio may be greater than, or less than, this value.

The situation with crops is more difficult to assess. Initially, inoculation of wheat, maize and sorghum with Azospirillum was reported to give yield increases up to 100% (e.g. Kapulnik et al. 1981). Subsequent research has indicated that the N-yield increase, in some instances at least, is probably associated with phytohormonal effects on root growth (e.g. Kapulnik et al. 1985a,b) or other causes (Boddey et al. 1986). In one case, up to 32% of plant N was considered to be derived from recent N_2 fixation although there was no increase in total plant N (Rennie et al. 1983). The inclusion of straw and other plant residues promoted N_2 fixation following inoculation of cereals (e.g. Hegazi et al. 1983; Ladha et al. 1986), an effect consistent with field studies on free-living N_2 fixation in fallow soil amended with residues (Roper 1983, 1985). Nitrogen fixation in leaf litter mulch under sugar cane has been estimated as high as 155 g N/ha/day (Patriquin 1982).

In the U.K., there has been a concerted program to study the enrichment of straw residues through N_2 fixation, with the aim of developing N-enriched composts (Lynch and Harper 1985).

Table 2: Representative examples of N_2 fixation benefit in pasture grasses and cereal crops.

Species	N increase	Method	Reference
GRASSES			
Cynodon, Paspalum, Brachiaria, Andropogon	up to 33 kgN/ha/yr	C_2H_2	Weaver et al. 1980
Panicum	26.3 gN/ha/day	C_2H_2	Weier et al. 1981
Brachiaria spp.	30–45 kgN/ha/yr	^{15}N(O.M.)	Boddey & Victoria 1986
Paspalum	21 kgN/ha/yr	^{15}N, C_2H_2	Boddey et al. 1983
Pennisetum, Sorghum	9–39% increase	C_2H_2,N yld	Smith et al. 1984
Leptochloa (Pak.)	Significant	C_2H_2	Zafar et al. 1986
39 grasses (temp. Australia)	< 1 kgN/ha/yr	C_2H_2	Thompson et al. 1984
Tussock grasses (N.Z.)	0–3 kgN/ha/yr	C_2H_2	Line & Loutit 1973
CROPS			
Maize	Negligible	C_2H_2	Albrecht et al. 1981
Maize	c.40 kgN/ha/yr	N balance	deFreitas et al. 1982
Maize	51 kgN/ha/crop	C_2H_2,N yld	Hegazi et al. 1983
Wheat	Negligible	C_2H_2	Kapulnik et al. 1985b
Wheat	Negligible	N yld	Millet & Feldman 1984
Wheat, barley	Negligible	C_2H_2, ^{15}N	Lethbridge & Davidson 1983
Wheat	No N incr.: up to 32% Ndfa[1]	^{15}N	Rennie et al. 1983
Wheat	Signif.; not N_2 fixation	^{15}N	Boddey et al. 1986
Wheat (leaf inoc.)	70–80% incr.	N yld	Pati & Chandra 1981
Sugar cane	17 kgN/ha/yr	^{15}N	Vose 1983
Rice (2 crops/yr)	79, 103 kgN/ha/yr	N balance	App et al. 1984

1. Nitrogen derived from the atmosphere.

Measurements of N_2 fixation in marine environments (e.g. Capone and Budin 1982), in wood litter (Roskoski 1981), by epiphylls associated with coffee plants (Roskoski 1980), in leaf litter (Jones and Bangs 1985), in tree barks (Yatazawa et al. 1983), in tree boles (Silvester et al. 1982), and in association with duckweed mats (Zuberer 1982) all show small to modest levels of N_2 fixation that undoubtedly contribute to the N-economy of their respective ecosystems.

THE FUTURE

The inability to measure accurately the amount of N_2 fixed remains a major limitation to future research involving free-living diazotrophs, including those associated with growing plants. Broadly-based N-balance studies, and those dependent on C_2H_2 reduction, do little more than indicate that N_2 fixation has occurred. Isotope dilution experiments can be valuable but caution is required in their application and interpretation (Boddey and Victoria 1986), a point that is also relevant in attempting to determine $C_2H_2:N_2$ ratios for the simpler acetylene reduction assay. Natural abundance techniques have some promise in certain areas of research, but caution must be exercised due to different levels of the isotope at different depths and in different fractions. Without consolidation of measurement techniques, assessment of the short and long term benefits of research into free-living (including associative) N_2 fixation will be difficult.

Clarification of the role of diazotrophs such as Azospirillum and Azotobacter in promoting growth through phytohormonal effects (a form of N-mining?) and/or through N_2 fixation, is required. Both undoubtedly play a role, maybe even together, but current results make it difficult to assess the benefits of inoculation, especially as the techniques for determining N_2 fixation are often inadequate.

Azospirillum, and perhaps other diazotrophs, undoubtedly enter the roots of various graminaceous species, often exhibiting a degree of host x strain specificity. Is this a true symbiosis? If so, can it be promoted, or, through the application of molecular biology techniques, can it be developed into a more recognizable and profitable association?

The failure to achieve consistent establishment of inoculated diazotrophs (e.g. Schank and Smith 1984) points to our poor understanding of the ecology of these organisms. In the laboratory, on defined media and under defined conditions, much is known of their response to temperature, pH, combined nitrogen and oxygen. However, few attempts have been made to understand the importance of these factors in the field. Even less is known of their interactions with other microorganisms, or of their use of natural substrates, alone or in combination with bacteria and fungi able to breakdown complex residues. Without this information, inoculation must be a "hit-or-miss" affair. Knowledge of the eco-physiology of these bacteria is essential in any attempt to manage naturally-occurring diazotrophs in any system. The difficulties in obtaining pure cultures of some diazotrophs (e.g. Balandreau 1983; Vose 1983; Lindberg and Granhall 1984) suggests that many bacteria, adapted to certain environments, are not being recovered and recognised through the application of conventional techniques for isolation. Similarly, the presence in <u>Azotobacter</u> of a Va-nitrogenase that reduces acetylene to ethylene poorly, raises the possibility that other bacteria, not currently recognised as diazotrophs, may be contributors to the N-economy of soil.

CONCLUDING COMMENTS

Research over the last 15 years indicates that significant biological N_2 fixation is occurring in most natural ecosystems and in crop situations, a fact that had not been properly appreciated in the past. Future research should aim to capitalise on these systems through better management and through the introduction of bacteria better able to fix N_2 under the conditions existing or imposed. These goals will only be achieved through a better understanding of the microorganisms involved and their interaction with plants or plant residues. The prospects of developing N_2-fixing plants, without bacterial involvement, appear remote at this stage, but we should be aware that modern molecular biology techniques offer an opportunity to better utilise existing systems, provided of course it is known what the goals are and what is needed to achieve them.

REFERENCES

Abd-El-Malek, Y. (1971). Free-living nitrogen-fixing bacteria in Egyptian soils and their possible contribution to soil fertility. <u>Plant Soil, Special Volume</u>, 423-42.

Albrecht, S.L., Okon, Y., Lonnquist, J., and Burris, R.H. (1981). Nitrogen fixation by Corn-Azospirillum associations in a temperate climate. <u>Crop Sci.</u> 21, 301-6.

Alexander, M. (1961). 'Introduction to Soil Microbiology'. (John Wiley and Sons: New York.)

App, A., Santiago, T., Daez, C., Menguito, C., Ventura, W., Tirol, A., Po, J., Watanabe, I., de Datta, S.K., and Roger, P. (1984). Estimation of the nitrogen balance for irrigated rice and their contribution to phototrophic nitrogen fixation. <u>Field Crop Res.</u> 9, 17-27.

Athar, M. and Mamood, A. (1981). Extension of Rhizobium host range to Zygophyllaceae. <u>In</u> 'Current Perspectives in Nitrogen Fixation'. (Eds A.H. Gibson and W.E. Newton) p. 481. (Aust. Acad. Sci.: Canberra.)

Balandreau, J. (1983). Microbiology of the association. <u>Can. J. Microbiol.</u> 29, 851-9.

Balandreau, J., and Dommergues, Y. (1971). Mesure <u>in situ</u> de l'activitie nitrogenasique. <u>C.R. Acad. Sci., Paris</u> 273, 2020-3.

Baldani, V.L.D., Alvarez, M.A. de B., Baldani, J.I., and Dobereiner, J. (1986). Establishment of inoculated Azospirillum spp. in the rhizosphere and in the roots of field grown wheat and sorghum. <u>Plant Soil</u> 90, 35-46.

Baldani, V.L.D. and Dobereiner, J. (1980). Host plant specificity in the infection of cereals with Azospirillum spp. <u>Soil Biol. Biochem.</u> 12, 433-9.

Barraquio, W.L., Ladha, J.K., and Watanabe, I. (1983). Isolation and identification of N_2-fixing Pseudomonas associated with wetland rice. <u>Can. J. Microbiol.</u> 29, 867-73.

Bashan, Y. (1986). Enhancement of wheat root colonization and plant development of Azospirillum brasilense Cd. following temporary depression of rhizosphere microflora. <u>Appl. Environ. Microbiol.</u> 51, 1067-71.

Bentley, B.L., and Carpenter, E.J. (1980). Effects of desiccation and rehydration on nitrogen fixation by epiphylls in a tropical rainforest. <u>Microb. Ecol.</u> 6, 109-13.

Bergersen, F.J. (1980). 'Methods for Evaluating Biolgical Nitrogen Fixation'. (Wiley: Chichester.)

Bhardwaj, K.K.R., and Novak, B. (1978). Effect of moisture and nitrogen levels on the decomposition of wheat straw in soil. <u>Zbl. Bakt. II Abt.</u> 133, 477-82.

Biesboer, D.D. (1984). Seasonal variation in nitrogen fixation, associated microbial populations, and carbohydrates in roots and rhizomes of Typha latifolia (Typhaceae). Can. J. Bot. 62, 1965–67.

Bishop, P.E. and Brill, W.J. (1977). Genetic analyses of Azotobacter vinelandii mutant strains unable to fix nitrogen. J. Bacteriol. 130, 954–6.

Boddey, R.M., Baldani, V.L.D., Baldani, J.I., and Dobereiner, J. (1986). Effect of inoculation of Azospirillum spp. on nitrogen accumulation by field grown wheat. Plant Soil 95, 109–21.

Boddey, R.M., Chalk, P.M., Victoria, R.L., Matsui, E., and Dobereiner, J. (1983). The use of ^{15}N isotope dilution technique to estimate the contribution of associated biological nitrogen fixation to the nitrogen nutrition of Paspalum notatum cv. batatais. Can. J. Microbiol. 29, 1036–45 (also see Soil Biol. Biochem. 15, 25–32.)

Boddey, R.M., and Victoria, R.L. (1986). Estimation of biological nitrogen fixation associated with Brachiaria and Paspalum grasses using ^{15}N labelled organic matter and fertilizer. Plant Soil 90, 265–92.

Bouton, J.H., Albrecht, S.L., and Zuberer, D.A. (1985). Screening and selection of pearl millet for root associated bacterial nitrogen fixation. Field Crops Res. 11, 131–40.

Bushby, H.V.A., and Marshall, K.C. (1977). Water status of rhizobia in relation to their susceptibility to desiccation and to their protection by montmorillonite. J. Gen. Microbiol. 99, 19–27.

Cannon, F., Beynon, J., Buchanan-Woolaston, V., Burghoff, R., Cannon, M., Kwiatkowski, R., Lauer, G., and Rubin, R. (1985). Progress in understanding organization and expression of Nif genes in Klebsiella. In 'Nitrogen Fixation Research Progress'. (Eds H.J. Evans, P.J. Bottomley and W.E. Newton). pp. 453–60. (Martinus Nijhoff: Dordrecht.)

Capone, D.G., and Budin, J.M. (1982). Nitrogen fixation associated with rinsed roots and rhizomes of Eelgrass, Zostera marina. Plant Physiol. 70, 1601–4.

Chalk, P.M. (1985). Estimation of N_2 fixation by isotope dilution: an appraisal of techniques involving ^{15}N enrichment and their application. Soil Biol. Biochem. 17, 389–410.

Chan, Y.-K., Wheatcroft, R., and Watson, R.J. (1986). Physiological and genetic characterization of a diazotrophic Pseudomonad. J. Gen. Microbiol. 132, 2277–85.

Charyulu, P.B.B.N., Nayuk, D.N., and Rao, V.R. (1981). $^{15}N_2$ incorporation by rhizosphere soil. Plant Soil 59, 399–405.

Cheshire, M.V. (1977). Origins and stability of soil polysaccharides. J. Soil Sci. 28, 1–10.

Cossar, J.D., Rowell, P., Darling, A.J., Murray, S., Codd, G.A., and Stewart, W.D.P. (1985). Localization of ribulose, 1,5-bisphosphate carboxylase/oxygenase in the N_2-fixing cyanobacterium Anabaena cylindrica. FEMS Microb. Lett. 28, 65–8.

Dalton, H., and Postgate, J.R. (1969). Effect of oxygen on growth of Azotobacter chroococcum in batch and continuous cultures. J. Gen. Microb. 54, 463–73.

Dart, P.J. (1986). Nitrogen fixation associated with non-legumes in agriculture. Plant Soil 90, 303–34.

Day, J.M., Harris, D., Dart, P.J., and van Berkum, P. (1975). The Broadbalk experiment. In 'Nitrogen Fixation by Free-living Microorganisms'.(Ed. W.D.P. Stewart). pp. 71–84. (Cambridge Univ. Press: Cambridge).

deBont, J.A.M. (1976). Bacterial degradation of ethylene and acetylene reduction test. Can. J. Microbiol. 22, 1060–62.

deFreitas, J.L.M., da Rocha, R.E.M., Pereira, P.A.A., and Dobereiner, J. (1982). Organic matter and inoculation with Azospirillum on nitrogen incorporation by field grown maize. Pesq. Agropec. Bras. 17, 1423–32.

deFreitas, J.R., Victoria, R.L., Ruschel, A.P., and Vose, P.B. (1984). Estimation of N_2-fixation by sugarcane, Saccharum sp., and soybean, Glycine max, grown in soil with ^{15}N-labelled organic matter. Plant Soil 82, 257–61.

Dobereiner, J., and Day, J.M. (1976). Associative symbiosis in tropical grasses: characterization of microorganisms and dinitrogen-fixing sites. In 'First International Symposium N_2 Fixation'. (Eds W.E. Newton and C.J. Nyman). pp. 518–38. (Washington State Univ. Press: Pullman).

Dobereiner, J., Day, J.M., and Dart, P.J. (1972). Nitrogenase activity and oxygen sensitivity of the Paspalum notatum – Azotobacter paspali association. J. Gen. Microbiol. 71, 103–6.

Ela, S.W., Anderson, M.A., and Brill, W.J. (1982). Screening and selection of maize to enhance associative bacterial nitrogen fixation. Plant Physiol. 70, 1564–7.

Elkan, G.H. (1987). 'Symbiotic Nitrogen Fixation Technology'. (Marcel Dekker: New York).

Eskew, D.L., Eaglesham, A.R.J., and App, A.A. (1981). Heterotrophic $^{15}N_2$ fixation and distribution of newly fixed nitrogen in a rice-flooded soil system. Plant Physiol. 68, 48–52.

Evans, H.J., Bottomley, P.J., and Newton, W.E. (1985). 'Nitrogen Fixation Research Progress'. (Martinus Nijhoff: Dordrecht).

Gaskins, M.H., Albrecht, S.L., and Hubbell, D.H. (1984/5). Rhizosphere bacteria and their use to increase plant productivity: a review. Agric. Ecosys. Environ. 12, 99–116.

Gautier, A., and Drouin, R. (1888). Compt. Rend. (Paris) 106: 754, 863, 1098, 1174, 1232 and 113: 820.

Germida, J.J. (1986). Population dynamics of Azospirillum brasilense and its bacteriophage in soil. Plant Soil 90, 117–28.

Gibson, A.H., and Jordan, D.C. (1983). Ecophysiology of nitrogen–fixing systems. In 'Encyclopedia of Plant Physiology. Physiological Plant Ecology.' Vol. 12C. (Eds O.L. Lange et al.). pp. 301–90. (Springer Verlag: Berlin).

Haahtela, K., Helander, I., Nurmiaho–Lassila, E.–L., and Sundman, V. (1983). Morphological and physiological characteristics and lipopolysaccharide composition of N_2–fixing (C_2H_2–reducing) root associated Pseudomonas sp. Can. J. Microb. 29, 874–80.

Halsall, D.M., and Gibson, A.H. (1985). Cellulose decomposition and associated nitrogen fixation by mixed cultures of Cellulomonas gelida and Azospirillum species or Bacillus macerans. Appl. Environ. Microbiol. 50, 1021–6.

Halsall, D.M., Turner, G.L., and Gibson, A.H. (1985). Straw and xylan utilization by pure cultures of nitrogen–fixing Azospirillum spp. Appl. Environ. Microbiol. 49, 423–8.

Hardy, R.W.F., Burns, R.C., and Holsten, R.D. (1973). Applications of the acetylene–ethylene assay for measurement of nitrogen fixation. Soil Biol. Biochem. 5, 47–81.

Hardy, R.W.F., Holsten, R.D., Jackson, E.K., and Burns, R.C. (1968). The acetylene–ethylene assay for N_2 fixation: laboratory and field evaluation. Plant Physiol. 43, 1185–207.

Harper, S.H.T., and Lynch, J.M. (1981). The chemical components and decomposition of wheat straw leaves, internodes and nodes. J. Sci. Food Agric. 32, 1057–62.

Harper, S.H.T., and Lynch, J.M. (1984). Nitrogen fixation by cellulolytic communities at aerobic–anaerobic interfaces in straw. J. Appl. Bacteriol. 57, 131–7.

Hegazi, N.A., Monib, M., Amer, H.A., and Shokr, E.–S. (1983). Response of maize plants to inoculation with azospirilla and (or) straw amendment in Egypt. Can. J. Microbiol. 29, 888–94.

Hegazi, N.A., Vlassak, K., and Monib, M. (1979). Effect of amendments, moisture, and temperature on acetylene reduction in Nile delta soils. Plant Soil 51, 27–37.

Hicks, B.J., and Silvester, W.B. (1985). Nitrogen fixation associated with the New Zealand mangrove (Avicennia marina). Appl. Environ. Microbiol. 49, 955–9.

Hill, W.A., Bacon–Hill, P., Crossman, S.M., and Stevens, C. (1983). Characterization of N_2–fixing bacteria associated with sweet potato roots. Can. J. Microbiol. 29, 860–2.

Holt, J.G. (1984). 'Bergey's Manual of Systematic Bacteriology', Vol. 1. (Williams and Wilkins: Baltimore.)

Iyama, S., Sano, Y., and Fujii, T. (1983). Diallel analysis of nitrogen fixation in the rhizosphere of rice. Plant Sci. Lett. 30, 129–35.

Jain, D.K., and Rennie, R.J. (1986). Use of spermosphere model for the screening of wheat cultivars and N_2–fixing bacteria for N_2 fixation. Can. J. Microbiol. 32, 285–8.

Jensen, H.L. (1941) Nitrogen fixation and straw decomposition by soil microorganisms. III. Clostridium butyricum in association with aerobic cellulose decomposers. Proc. Linn. Soc. N.S.W. 66, 239–49.

Jensen, H.L., and Swaby, R.J. (1941). Nitrogen fixation and straw decomposition by soil microorganisms. II. The association between Azotobacter and facultative–aerobic cellulose decomposers. Proc. Linn. Soc. N.S.W. 66, 89–106.

Jones, K., and Bangs, D. (1985). Nitrogen fixation by free–living heterotrophic bacteria in an oak forest: the effect of liming. Soil Biol. Biochem. 17, 705–9.

Joerger, R.D., Premakumar, R., and Bishop, P.E. (1986). Tn5–induced mutants of Azotobacter vinelandii affected in nitrogen fixation under Mo–deficienit and Mo–sufficient conditions. J. Bact. 168, 673–82.

Kapulnik, Y., Okon, Y., and Henis, Y. (1985a). Changes in root morphology of wheat caused by Azospirillum inoculation. Can. J. Microbiol. 31, 881–7.

Kapulnik, Y., Feldman, M., Okon, Y., and Henis, Y (1985b). Contribution of nitrogen fixed by Azospirillum to the N nutrition of spring wheat in Israel. Soil Biol. Biochem. 4, 509–15.

Kapulnik, Y., Sarig, S., Nur, I., Okon, Y., Kigel, J., and Henis, Y. (1981). Yield increases in summer cereal crops in Israeli fields inoculated with Azospirillum. Exp. Agric. 17, 179–87.

Kipe–Nolt, J.A., Avalakki, U.K., and Dart, P.J. (1985). Root exudation of sorghum and utilization of exudates by nitrogen–fixing bacteria. Soil Biol. Biochem. 17, 859–63.

Knowles, R. (1981). The measurement of nitrogen fixation. In 'Current Perspectives in Nitrogen Fixation'. (Eds A.H. Gibson and W.E. Newton) pp. 327–33 (Aust. Acad. Sci.: Canberra).

Knowles, R., and Denike, D. (1974). Effect of ammonium–, nitrite– and nitrate–nitrogen on anaerobic nitrogenase activity in soil. Soil Biol. Biochem. 6, 353–8.

Kosslak, R.M., and Bohlool, B.B. (1983). Prevalence of Azospirillum spp. in the rhizosphere of tropical plants. Can. J. Microbiol. 29, 649–52.

Krotzky, A., and Werner, D. (1987). Nitrogen fixation in Pseudomonas stutzeri. Arch. Microb. 147, 48–57.

Kulinska, D., and Drozdowicz, A. (1983). Occurrence of microorganisms antagonistic to Azospirillum spp. Zbl. Mikrob. 138, 585–94.

Ladd, J.N., Oades, J.M., and Amato, M. (1981). Microbial biomass formed from ^{14}C, ^{15}N-labelled plant material decomposing in soils in the field. Soil Biol. Biochem. 13, 119–26.

Ladha, J.K., Tirol, A.C., Garoy, M.L.G., Caldo, G., Ventura, W., and Watanabe, I. (1986). Plant-associated N_2 fixation (C_2H_2 reduction) by five rice varieties, and relationship with plant growth characters as affected by straw incorporation. Soil Sci. Plant Nutr. 32, 91–106.

Larson, R.I., and Neal, J.L. (1978). Selective colonisation of the rhizosphere of wheat by nitrogen-fixing bacteria. Ecol. Bull., Stockholm 26, 331–42.

Lee, K.J., and Gaskins, M.H. (1982). Increased root exudation of ^{14}C-compounds by sorghum seedlings inoculated with nitrogen-fixing bacteria. Plant Soil 69, 391–9.

Lee, K.-K., Shearman, R.C., and Klucas, R.V. (1986). Nitrogen fixation (acetylene reduction) by lines composing 'Park' Kentucky bluegrass. Can. J. Microbiol. 32, 348–52.

Lethbridge, G., and Davidson, M.S. (1983). Root-associated nitrogen-fixing bacteria and their role in the nitrogen nutrition of wheat estimated by ^{15}N isotope dilution. Soil Biol. Biochem. 15, 365–74.

Lindberg, T., and Granhall, U. (1984). Isolation and characterization of dinitrogen-fixing bacteria from the rhizosphere of temperate cereals and forage grasses. Appl. Environ. Microbiol. 48, 683–9.

Line, M.A., and Loutit, M.W. (1973). Studies on nonsymbiotic nitrogen fixation in New Zealand tussock-grassland soils. N.Z. J. Agric. Res. 16, 87–95.

Lynch, J.M., and Harper, S.H.T. (1985). The microbial upgrading of straw for agricultural use. Phil. Trans. Roy. Soc. Lond. B310, 221–6.

Macura, J., and Pavel, L. (1959). The influence of montmorillonite on nitrogen fixation by Azotobacter. Folia Microbiologia 4, 82–90.

Maryan, P.S., Eady, R.R., Chaplin, A.E., and Gallon, J.R. (1986). Nitrogen fixation by Gloeothece sp. PCC6909: respiration and not photosynthesis supports nitrogenase activity in the light. J. Gen. Microbiol. 132, 789–96.

Millbank, J.W. (1974). Associations with blue-green algae. In 'The Biology of Nitrogen Fixation'. (Ed. A. Quispel). pp. 238–64 (North Holland: Amsterdam).

Millet, E., and Feldman, M. (1984). Yield response of a common spring wheat cultivar to inoculation with Azospirillum brasilense at various levels of nitrogen fertilization. Plant Soil 80, 255–9.

Minchin, F.R., Sheehy, J.E., and Witty, J.F. (1986). Further errors in the acetylene reduction assay: effects of plant disturbance. J. Exp. Bot. 37, 1581–91.

Miranda, C.H.B., Seiffert, N.F., and Dobereiner, J. (1985). Effect of molybdenum on numbers of Azospirillum on Brachiaria decumbens production. Pesq. Agropec. Bras. 20, 509–13.

Morris, D.R., Zuberer, D.A., and Weaver, R.W. (1985). Nitrogen fixation by intact grass-soil cores using $^{15}N_2$ and acetylene reduction. Soil Biol. Biochem. 17, 87–91.

Mortland, M.M. (1970). Clay-organic complexes and interactions. Adv. Agron. 22, 75–117.

Murty, M.G. (1983). Nitrogen fixation (acetylene reduction) in the phyllosphere of some economically important plants. Plant Soil 73, 151–3.

Murty, M.G. (1984). Nitrogenase activity (acetylene reduction), associated with leaves of different cultivars of paddy (Oryza sativa L.). Plant Soil 77, 253–61.

Nohrstedt, H.-O. (1975/6). Decomposition of ethylene in soils and its relevance to the measurement of nitrogenase activity with the acetylene reduction method. Grundforbättring 27, 171–8.

Nohrstedt, H.-O. (1983). Natural formation of ethylene in forest soils and methods to correct results given by the acetylene-reduction assay. Soil Biol. Biochem. 15, 281–6.

Nohrstedt, H.-O. (1984). Carbon monoxide as an inhibitor of N_2ase activity (C_2H_2) in control measurements of endogenous formation of ethylene by forest soils. Soil Biol. Biochem. 16, 19–22.

Pacovsky, R.S., Fuller, G., and Paul, E.A. (1985). Influence of soil on the interaction between mycorrhizae and Azospirillum in sorghum. Soil Biol. Biochem. 17, 525–31.

Pagan, J.D., Child, J.J., Scowcroft, W.R., and Gibson, A.H. (1975). Nitrogen fixation by Rhizobium cultured in a defined medium. Nature 256, 406–7.

Pal, D., Broadbent, F.E., and Mikkelsen, D.S. (1975). Influence of temperature on the kinetics of rice straw decomposition in soils. Soil Sci. 120, 442–9.

Parr, J.F., and Papendick, R.I. (1978). Factors affecting the decomposition of crop residues by microorganisms. In 'Crop Residue Management Systems'. pp. 101–29. (Amer. Soc. Agron.: Madison).

Pati, B.R., and Chandra, A.K. (1981). Effect of spraying nitrogen–fixing phyllosphere bacterial associations on wheat plants. Plant Soil 61, 419–27.

Patriquin, D.G. (1982). Nitrogen fixation in sugar cane litter. Biol. Agric. Hort. 1, 39–64.

Patriquin, D.G., Döbereiner, J., and Jain, D.K. (1983). Site and processes of association between diazotrophs and grasses. Can. J. Microbiol. 29, 900–15.

Paul, E.A., Myers, R.J.K., and Rice, W.A. (1971). Nitrogen fixation in grassland and associated cultivated ecosystems. Plant Soil Special Volume, 495–507.

Pedrosa, F.O., Döbereiner, J., and Yates, M.G. (1980). Hydrogen–dependent growth and autotrophic carbon dioxide fixation in Derxia. J. Gen. Microb. 119, 547–51.

Peters, G.A., Toia, R.F., Calvert, H.E., and Marsh, B.H. (1986). Lichens to Gunnera – with emphasis on Azolla. Plant Soil 90, 17–34.

Postgate, J.R. (1978). 'Nitrogen Fixation'. (Edward Arnold: London.)

Postgate, J.R. (1981). Microbiology of the free–living nitrogen–fixing bacteria, excluding cyanobacteria. In 'Current Perspectives in Nitrogen Fixation'. (Eds A.H. Gibson and W.E. Newton). pp. 217–28. (Aust. Acad. Science: Canberra.)

Reddell, P., and Bowen, G.D. (1985). Frankia source affects growth, nodulation and nitrogen fixation in Casuarina species. New Phytol. 100, 115–22.

Rennie, R.J., deFreitas, J.R., Ruschel, A.P., and Vose, P.V. (1983). ^{15}N isotope dilution to quantify dinitrogen (N_2) fixation associated with Canadian and Brazilian wheat. Can. J. Bot. 61, 1667–71.

Rice, W.A., Paul, E.A., and Wetter, L.R. (1967). The role of anaerobiosis in asymbiotic nitrogen fixation. Can. J. Microbiol. 13, 829–36.

Robson, R.L., Eady, R.R., Richardson, T.H., Miller, R.W., Hawkins, M., and Postgate, J.R. (1986). The alternative nitrogenase of Azotobacter chroococcum is a vanadium enzyme. Nature 322, 388–90.

Robson, R.L., Kennedy, C., and Postgate, J.R. (1983). Progress in comparative genetics of nitrogen fixation. Can. J. Microbiol. 29, 954–67.

Robson, R.L., and Postgate, J.R. (1980). Oxygen and hydrogen in biological nitrogen fixation. Ann. Rev. Microbiol. 34, 183–207.

Roper, M.M. (1983). Field measurements of nitrogenase activity in soils amended with wheat straw. Aust. J. Agric. Res. 34, 725–39.

Roper, M.M. (1985). Straw decomposition and nitrogenase activity (C_2H_2 reduction): effects of soil moisture and temperature. Soil Biol. Biochem. 17, 65–71.

Roper, M.M., and Halsall, D.M. (1986). Use of products of straw decomposition by N_2–fixing (C_2H_2–reducing) populations of bacteria in three soils from wheat–growing areas. Aust. J. Agric. Res. 37, 1–9.

Roper, M.M., and Marshall, K.C. (1978). Effects of a clay mineral on microbial predation and parasitism in Escherichia coli. Microb. Ecol. 4, 279–89.

Roskoski, J.P. (1980). N_2 fixation (C_2H_2 reduction) by epiphylls on coffee, Coffea arabica. Microb. Ecol. 6, 349–55.

Roskoski, J.P. (1981). Comparative C_2H_2 reduction and $^{15}N_2$ fixation in deciduous wood litter. Soil Biol. Biochem. 13, 83–5.

Ruinen, J. (1974). Nitrogen fixation in the phyllosphere. In 'The Biology of Nitrogen Fixation'. (Ed. A. Quispel). pp. 121–67. (North Holland: Amsterdam.)

Ruinen, J. (1975). Nitrogen fixation in the phyllosphere. In 'Nitrogen Fixation by Free–living Microorganisms'. (Ed. W.D.P. Stewart). pp. 85–100. (Cambridge Univ. Press: Cambridge.)

Saha, K.C., Sannigrahi, S., and Mandal, L.N. (1985). Effect of inoculation of Azospirillum lipoferum on nitrogen fixation in rhizosphere soil, their association with root, yield and nitrogen uptake by mustard (Brassica juncea). Plant Soil 87, 273–280.

Schank, S.C., and Smith, R.L. (1984). Status and evaluation of associative grass bacteria N_2–fixing systems in Florida. Proc. Soil Crop Sci. Soc. Florida 43, 120–3.

Sen Gupta, B., Nandi, A.S., and Sen, S.P. (1982). Utility of phyllosphere N_2–fixing microorganisms in the improvement of crop growth. I. Rice. Plant Soil 68, 55–67.

Silvester, W.B., Sollins, P., Verhoeven, T., and Cline, S.P. (1982). Nitrogen fixation and acetylene reduction in decaying conifer boles. Can. J. For. Res. 12, 646–52.

Simon, L., Stein, A., Cote, S., and Lalonde, M. (1985). Performance in in vitro propagated Alnus glutinosa clones inoculated with Frankiae. Plant Soil 87, 125–33.

Smith, R.L., Schank, S.C., Milam, J.R., and Baltensperger, A.A. (1984). Responses of Sorghum and Pennisetum species to the N_2–fixing bacterium, Azospirillum brasilense. Appl. Environ. Microbiol. 47, 1331–6.

Souto, S.M., and Dobereiner, J. (1985). Season variation in dinitrogen fixation and assimilation of nitrate in tropical forage grasses. Pesq. Agropec. Bras. 20, 319–34.

Stewart, B.A., Porter, L.K., and Viets, F.G. (1966). Effect of sulfur content of straws on rates of decomposition and plant growth. Soil Sci. Soc. Amer. Proc. 30, 355–8.

Stewart, W.D.P. (1980). Some aspects of structure and function in N_2-fixing cyanobacteria. Ann. Rev. Microbiol. 34, 497–536.

Stotzky, G., and Norman, A.G. (1961). Factors limiting microbial activities in soil. Arch. Mikrobiol. 40, 341–69.

Subba Rao, N.S., Tilak, K.V.B.R., and Singh, C.S. (1985). Effect of combined inoculation of Azospirillum brasilense and vesicular–arbuscular mycorrhiza on pearl millet (Pennisetum americanum). Plant Soil 84, 283–6.

Thompson, J.A., Gemell, L.G., Roughley, R.J., and Evans, J. (1984). Nitrogenase activity associated with pasture grasses in northern New South Wales. Soil Biol. Biochem. 16, 217–22.

Tjepkema, J.D., and van Berkum, P. (1977). Acetylene reduction by soil cores of maize and sorghum in Brazil. Appl. Environ. Microbiol. 33, 626–9.

Torrey, J.G. (1978). Nitrogen fixation by actinomycete nodulated angiosperms. BioScience 28, 586–92.

Torrey, J.G. (1985). The site of nitrogenase in Frankia in free–living culture and in symbiosis. In 'Nitrogen Fixation Research Progress'. (Eds H.J. Evans, P.J. Bottomley, and W.E. Newton). pp. 293–299. (Martinus Nijhoff: Dordrecht.)

Torrey, J.G., Tjepkema, J.D., Turner, G.L., Bergersen, F.J., and Gibson, A.H. (1981). Dinitrogen fixation by cultures of Frankia sp. CPI1 demonstrated by $^{15}N_2$ incorporation. Plant Physiol. 68, 983–4.

Trinick, M.J. (1979). Structure of nitrogen–fixing nodules formed by Rhizobium on roots of Parasponia andersonii. Can. J. Microbiol. 25, 565–78.

Turner, G.L., and Gibson, A.H. (1980). Measurement of nitrogen fixation by indirect means. In 'Methods for Evaluating Biological Nitrogen Fixation'. (Ed. F.J. Bergersen). pp. 111–38. (J. Wiley and Son: Chichester.)

van Berkum, P., and Bohlool, B.B. (1980). Evaluation of nitrogen fixation by bacteria in association with roots of tropical grasses. Microb. Rev. 44, 491–517.

van Berkum, P., and Sloger, C. (1981). Comparing time course profiles of immediate acetylene reduction by grasses and legumes. Appl. Environ. Microbiol. 41, 184–89.

Veal, D.A., and Lynch, J.M. (1984). Associative cellulolysis and dinitrogen fixation by co–cultures of Trichoderma harzianum and Clostridium butyricum. Nature 310, 695–7.

Ventura, W., and Watanabe, I. (1983). ^{15}N dilution technique of assessing the contribution of nitrogen fixation to the rice plant. Soil Sci. Pl. Nutr. 29, 123–31.

von Bulow, J.F.W., and Dobereiner, J. (1975). Potential for nitrogen fixation in maize genotypes in Brazil. Proc. Nat. Acad. Sci., U.S.A. 72, 2389–93.

Vose, P.B. (1983). Developments in nonlegume N_2-fixing systems. Can. J. Microbiol. 29, 837–50.

Watanabe, I., and Cabrera, D.R. (1979). Nitrogen fixation associated with the rice plant grown in water culture. Appl. Environ. Microbiol. 37, 373–8.

Watanabe, I., and Roger, P.A. (1984). Nitrogen fixation in wetland rice field. In 'Current Development in Biological Nitrogen Fixation'. (Ed. N.S. Subba Rao). pp. 237–76. (Oxford and IBH Publishing: New Delhi).

Watanabe, I., and Ventura, W. (1982). Nitrogen fixation by blue–green algae associated with deep water rice. Curr. Sci. 51, 462–5.

Waterbury, J.B., Calloway, C.B., and Turner, R.D. (1983). A cellulolytic nitrogen–fixing bacterium cultured from the gland of Deshayes in shipworms. Science 221, 1401–3.

Weaver, R.W., Wright, S.F., Varanaka, M.W., Smith, O.E., and Holt, E.C. (1980). Dinitrogen fixation (C_2H_2) by established forage grasses in Texas. Agron. J. 72, 965–8.

Weier, K.L. (1980). Nitrogen fixation associated with grasses. Trop. Grassl. 14, 194–201.

Weier, K.L., MacRae, I.C. and Whittle, J. (1981). Seasonal variation in the nitrogenase activity of Panicum maximum var. trichoglume pasture and identification of associated bacteria. Plant Soil 63, 189–97.

Whiting, G.J., Gandy, E.L., and Yoch, D.C. (1986). Tight coupling of root associated nitrogen fixation and plant photosynthesis in the salt marsh grass Spartina alternifolia and carbon dioxide enhancement of nitrogenase activity. Appl. Environ. Microbiol. 52, 108–13.

Witty, J.F. (1979). Acetylene reduction assay can overestimate nitrogen fixation in soil. Soil Biol. Biochem. 11, 209–10.

Wright, S.F., and Weaver, R.W. (1982). Inoculation of forage grasses with N_2-fixing Enterobacteriaceae. Plant Soil 65, 414–9.

Yatazawa, M., Hambali, G.G., and Uchino, F. (1983). Nitrogen–fixing activity in warty lenticellate tree barks. Soil Sci. Pl. Nutr. 29, 285–94.

Yoneyama, T., Lee, K.-K., and Yoshida, T. (1977). Decomposition of rice residues in tropical soils. IV. The effect of rice straw on nitrogen fixation by heterotrophic bacteria in some Philippine soils. Soil Sci. Pl. Nutr. 23, 287–95.

Zafar, Y., Ashraf, M. and Malik, K.A. (1986). Nitrogen fixation associated with roots of Kallar grass (Leptochloa fusca). Plant Soil 90, 93–106.

Zamir, A., Maina, C.V., Fink, G.R., and Szalay, A.A. (1981). Stable chromosomal integration of the entire nitrogen fixation gene cluster from Klebsiella pneumoniae in yeast. Proc. Nat. Acad. Sci. U.S.A. 78, 3496–500.

Zuberer, D.A., (1982). Nitrogen fixation (acetylene reduction) associated with duckweed (Lemnaceae) mats. Appl. Environ. Microbiol. 43, 823–8.

Zuberer, D.A., and Roth, M. (1982). In vitro inhibition of non–symbiotic nitrogen–fixing bacteria by rhizosphere actinomycetes associated with grasses. Can. J. Microbiol. 28, 705–9.

THE ROLE OF SOIL FAUNA IN AGRICULTURAL SYSTEMS

J.M. Anderson

ABSTRACT

The role of soil fauna in nitrogen cycling can not simply be quantified as metabolic transfers and tissue turnover because they have a wide range of indirect effects on the soil environments for microbial populations and processes, and plant root growth and nutrient uptake. Furthermore, the hydrology of soils, and hence soil erosion and nutrient leaching from the system, are also affected by the burrowing and feeding activities of soil fauna; notably by termites and earthworms.

The scale at which the animals affect soil processes and their overall contribution to ecosystem-level processes is related to the body size and population densities of the animals. Size is of particular significance in determining whether animals directly affect microbial populations through feeding activities or indirectly influence the environments in which they operate by modifying soil physicochemical properties.

The agronomic significance of soil fauna (other than pest species) depends upon the intensity of cropping systems because grazing, fire, tillage, inorganic fertilizers and pesticides generally reduce the species complement and population densities of soil fauna, and override their contribution to soil processes. In minimum-tillage and pasture systems their effects become more evident but are sometimes difficult to separate from improvements in soil properties as a consequence of management practices.

It is concluded that in intensively cultivated systems the potential value of soil fauna in crop production may be manifested in microfaunal/ microbial activities in the rhizosphere and with decreasing intensity of management the role of macrofauna on soil structure is more significant. But the main constraint to constructing possible scenarios for manipulating soil fauna to

*improve crop production is a lack of understanding of the links
between gross soil processes and the structure of soil organism
communities.*

INTRODUCTION

The soil fauna includes a wide range of taxonomic groups from protozoa
and microarthropods to moles and larger vertebrates which live and feed in
soil. It is axiomatic, therefore, that they will variously affect soil
physical, chemical and microbiological processes at some scale and to some
degree. There is a vast literature on the role of invertebrates in soils. A
wide spectrum of vertebrate and invertebrate effects is considered by Hole
(1981), Zlotin and Khodashova (1980), Anderson et al. (1981, 1984) and Fitter
et al. (1985), and the roles of earthworms in soils are covered by Ghilarov
(1982), Satchell (1983) and Lee (1985). Despite extensive research in this
field there is little quantitative information on the role of soil animals in
mediating nutrient fluxes at the ecosystem level or on the agronomic
importance of soil invertebrates other than earthworms and pest species.

The question addressed here is whether significant quantitative effects
on crop production occur as a consequence of the presence, absence or temporal
change in the activities of invertebrate populations in soils and their
interactions with microorganisms involved in N cycling. Plant growth is taken
as the ultimate criterion for the importance of these soil invertebrate
effects because roots ramify through the soil matrix at a density necessary to
optimise nutrient uptake (Robinson and Rorison 1983; Bowen 1984). Nitrogen
mineralization in soil microsites is therefore integrated by plant uptake over
spatial and temporal scales which may not be detectable by bulk soil
measurements (Anderson 1987). Plant growth can also be indirectly influenced
by invertebrates through changes in soil structure and the physical
environment of roots, effects on mycorrhizas and root pathogens in the
rhizosphere and the production of plant growth substances. These, and other
effects of invertebrates in soils, are undoubtedly real phenomena but at what
frequency or intensity must they occur for crop production to be affected?
Similar considerations apply to processes affecting N losses through nitrate
leaching, denitrification or removal of organic N from the system (e.g. by
termites).

Soil invertebrates range in size from amoebae to giant earthworms of more
than a metre in length. Even within groups the size variation may be extreme;
such as differences in mass of 50,000 between the smallest and largest
earthworms (Lee 1985). Population densities of macro-, meso- and micro-fauna
also exhibit wide variation from a few individuals to many millions/m^2 (Fig.
1) with generation times ranging from a year to a few hours respectively.
Hence the invertebrate contribution to N fluxes in soils is the sum of
activities of a great diversity of organisms and their interactions over a
wide range of spatial and temporal scales, and involving both direct and
indirect effects of their activities which affect plant growth.

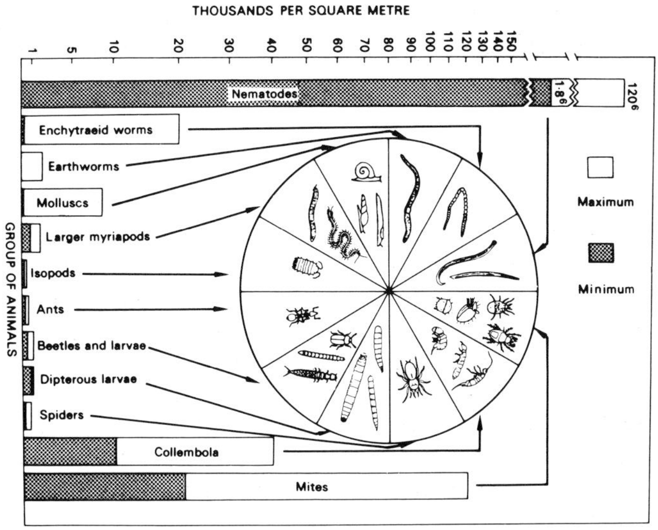

Fig. 1: Major groups of soil invertebrates and range of population densities typical of a cool, temperate grassland in Europe. (From Kevan 1965).

This paper considers invertebrate populations and activities in the context of the soil environment (which is both a constraint on, and an expression of, their activities), effects on soil processes including activities which indirectly influence N cycling, and the effects of agricultural practices on soil invertebrate populations.

FUNCTIONAL ATTRIBUTES OF SOIL FAUNA AND THEIR INTERACTIONS WITH MICROORGANISMS

Soil organisms can be classified by width into microflora and micro–, meso– and macro–fauna (Fig. 2). The definition of these categories is arbitrary since some groups have overlapping size distributions and some of the larger invertebrates could be classified as mesofauna in early developmental stages. Most Classes or higher taxa of soil organisms include saprotrophs (organisms utilizing dead organic matter), necrotrophs (organisms killing their food resource), biotrophs (exploiting living organisms), and non-specialised organisms which exploit more than one of these trophic modes. These, and finer, trophic distinctions have been defined down to species level for well–studied groups such as protozoa (Clarholm 1984), nematodes (Freckman 1982), mites and collembola (Petersen and Luxton 1982; Visser 1985) and earthworms (Lee 1985). None the less, there are a whole variety of properties related to size and mass which characterise the functional attributes of decomposer communities (Cousins 1980): mass is negatively correlated with the intrinsic reproductive rate (r_m) and positively correlated with generation times of organisms (Heron 1972): production and metabolic rates of soil invertebrates are also a positive function of mass (Petersen and Luxton 1982).

91

In addition, size and mass determine the extent to which the activities of the animals are constrained by soil structure or modify their habitat through burrowing and feeding activities (Anderson 1987).

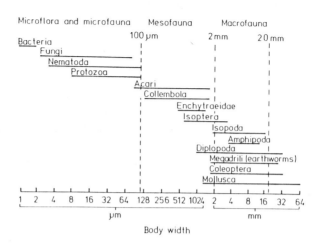

Fig. 2: Classification of soil organisms by thallus or body width. (After Swift et al. 1979).

Microfauna

The microfauna (mainly nematodes and protozoa) occupy water-filled soil pores and water films in the soil at similar scales to fungi and bacteria (root-feeding nematodes are not considered here). Microbial generation times in soil are in the order of hours, days or even longer according to carbon availability (Chapman and Gray 1985; Clarholm 1985) with higher rates in the rhizosphere than bulk soil as a consequence of root exudates (Bowen and Theodorou 1979).

Nematodes and protozoa feeding on bacteria and fungi utilize higher quality food resources in microbial protoplasm than the plant structural materials exploited by saprotrophic microorganisms; consequently their generation times can be faster than those of their microbial prey which are limited by carbon and nutrient availability from higher plant materials (Ingham et al. 1985). Bacterial- and fungal- feeding nematodes and protozoa are thus able to respond rapidly to microbial growth, exploit local food resources and then encyst until the next pulse of activity (Clarholm 1984).

The distribution of microbial cells within soil pores of different sizes, and the continuity of water in capillary spaces connecting pores, impose constraints on the capability of these microfauna to exploit their microbial food resources (Elliott et al. 1980; Anderson 1987). Darbyshire et al. (1985)

measured pore-size distribution in a clay soil under arable cultivation and recorded a modal pore diameter frequency of $10 \mu m$ diameter to a depth of 25cm. Mobility of a major proportion of nematode populations is limited by pores below c. 20 μm with optimum conditions of pore and particle sizes at least an order of magnitude larger (Wallace 1962) but some flagellates and naked amoebae can penetrate pores as small as 2 μm diameter (Clarholm 1984).

There is some evidence that predation by protozoa can decrease bacterial population densities both in the rhizosphere of crop plants (Darbyshire and Greaves 1967) and in bulk soil. Protozoa can consume 10^3 to 10^5 bacterial cells per division (Clarholm 1985) and nematodes consume in the order of 2 to 72×10^5 bacteria/nematode/day (Ingham et al. 1985). Stout and Heal (1967) calculated that, in a temperate arable soil, protozoa could consume between 150 and 900g bacteria/m^2/y equivalent to between 15 and 85 times the standing crop of bacteria. Reciprocal changes in bacteria and protozoa numbers have been detected in a semi-arid grassland (Ingham et al. 1986a) and in a comparative study of tilled and no-till arable systems (Elliott et al. 1984). Schnurer et al (1986) suggest that to invoke a causal relationship between these trends it is necessary to sample soils at intervals of a few hours. Using daily enumerations Cutler et al. (1922) were able to demonstrate an inverse relationship between bacteria and amoebae in 86% of samples during the course of a year. Recent laboratory studies show both increases and decreases in bacterial and fungal densities as a consequence of feeding by nematodes and protozoa (Ingham et al. 1985) but the extent to which these interactions are density dependent and, in the strict sense, regulate microbial populations has not been demonstrated.

Mesofauna

Collembola and mites are the most abundant mesofauna in temperate grasslands and arable soils where they mainly inhabit surface litter and air-filled macropores down to c. 100 μm diameter. Many species are fungal-feeders and laboratory studies have shown that they have the potential to affect the distribution, growth and activities of saprotrophic fungi (Seastedt 1984; Visser 1985), adversely affect plant growth by grazing on vesicular-arbuscular mycorrhiza (Warnock et al. 1982), and reduce the inoculum potential of root-infecting pathogens such as Pythium, Fusarium and Rhizoctonia on seedlings. Ulber (1983) found that the proportion of sugarbeet seedlings lethally infected by Pythium ultimum was reduced from 92% to 9% in sterile soils inoculated with the collembola Folsomia firmentaria. However, another collembola, Onychiurus armatus, is one of a group of soil invertebrates which can cause economic damage to sugarbeet crops by feeding on the roots of seedlings causing death by damage or through enhancing susceptibility to pathogens (Brown 1985). Attempts to quantify these types of counterbalancing effects influencing the growth of field crops have proved intractable for two main reasons. Firstly, it is not yet possible to relate the plant response to the amount of the total microbial biomass, or root mass, influenced by invertebrate feeding. Secondly, the experimental removal of key species in the community using biocides may result in changes in the activities of other organisms in the community.

It has been established for some time that collembola populations may increase following applications of DDT, aldrin and some carbamate and organophosphorus pesticides as a consequence of reduced predation by mesostigmatid mites which are more susceptible to these agrochemicals than collembola (Edwards and Lofty 1969). More recently, complex indirect effects of biocides have been shown by Santos et al. (1981) in a desert system where the elimination of nematophagous tydaeid mites in litter resulted in an increase in populations of nematodes feeding on baccteria and yeast, and a decrease in litter decomposition rates as a consequence of reduced microbial activity. Similar studies by Ingham et al (1986b) using a number of selective biocides on the total soil biota of a semi-arid grassland produced complex compensatory changes in the soil biota which are difficult to interpret. The complexity and variability of these responses bring into question whether current biocides can be used to unravel the functional links in more complex communities where more species and trophic links are operating.

As yet there is little evidence that mesofauna feeding activities significantly affect the distribution and abundance of bacteria and fungi in grassland and arable soils.

Macrofauna

The soil macrofauna involved in litter decomposition (millipedes, isopods, earthworms, termites, etc.) are large enough to modify soil structure and the distribution of organic matter on and within the soil through their feeding and burrowing activities. The earthworms and termites are the best documented groups in terms of these ecological roles and their effects on soils.

Earthworms

Earthworm casting rates range from 75 to 250 t/ha/y in temperate soils (Edwards and Lofty 1972) to 1200 t/ha/y in a tropical savanna where geophagous species were estimated to ingest 5-36 times their body weight per day (Lavelle 1978). Even higher casting rates for tropical earthworms, up to 2600 t/ha/y, are reported by Edwards and Lofty (1972) but the extrapolation of short-term studies to an annual estimate is often the basis of these high values.

Litter-feeding species have low assimilation efficiencies (in the order of 10%) and it has been calculated in a number of studies that moderate population densities of earthworms turnover the annual carbon inputs to the system several times a year (Lee 1983). This intimate mixing of organic and mineral material has been found in pot experiments to increase stabilized, clay-bound carbon by up to 17%, depending on soil type (Shaw and Pawluk 1986), though field evidence of this phenomenon is lacking. Many field studies have, however, reported that earthworm-worked soils generally have higher porosity, increased field water-holding capacity, 2 to 10 times higher water infiltration rates, more water-stable aggregates, and increased availability of plant nutrients than soils where worms are scarce or absent (Lee 1985).

But the causal relationship between these properties and earthworm activities should not be assumed because of the covariance between earthworm populations and other plant and soil parameters considered earlier. Nordstrom and Rundgren (1973) could find no significant correlation between soil porosity and earthworm biomass in 20 forest and pasture sites in Sweden and concluded that the turnover of roots had a major effect on pore space. On the other hand close relationships between burrow densities and infiltration rates have been shown in a number of temperate and tropical studies. Ehlers (1975), for example, found that infiltration rates, which were five times higher in a zero-till cornfield than normally tilled plots, were related to earthworm burrow densities. Aina (1984) recorded infiltration rates of 82 L/h and 6 L/h respectively for an old bush-fallow and a cleared site cultivated under cassava for 12 years. In the bush-fallow, earthworm channels (1 to 10 mm diameter with a modal frequency of 3 to 5 mm) averaged $220/m^2$ compared with the arable plot where channels (mainly in a 1 to 3 mm diameter range) averaged $36/m^2$.

The clearest demonstration of the agronomic importance of these effects is where exotic species of earthworms have been introduced to soils where earthworms are absent or have different feeding activities, or where earthworms are eliminated by agrochemicals.

Earthworm introductions have been carried out in a number of geographic regions (Ghilarov and Perel 1984). In New Zealand production of some pastures increased initially by more than 70% when introduced earthworms (Lumbricus rubellus and Aporrectodea caliginosa) mobilized nutrients accumulated in grass litter mats (Stockdill 1982). The nutrient flush subsequently declined but yield enhancement remained at c. 25-30% as a consequence of these introductions (Syers and Springett 1984).

Springett (1985) documented the further effects of introducing Aporrectodea longa, which is a deep-burrowing earthworm, into pastures where populations of surface active species (A. caliginosa, A. trapezoides and Lumbricus rubellus) were already established. Prior to the introduction of A. longa, the top 5-7 cm of the soil was well structured with no turf mat but roots only penetrated to the depth of this surface horizon and were susceptible to rotting when the soils were water-logged over winter. After 18 months the patches where A. longa had been introduced doubled total porosity below 10 cm depth, infiltration rates also doubled, rooting depth increased and there was a small increase in pasture production. Similar effects on soil properties followed the introduction of mixed populations of surface-active and deep-burrowing lumbricids into pastures on reclaimed polder soils in the Netherlands (Hoogerkamp et al. 1983). After 8 years the inoculation sites could be seen in aerial photographs as patches of enhanced sward growth, infiltration increased to 120 to 140 times that of control plots and gleying of the profile was reduced.

Edwards and Lofty (1980) used a fumigant (DD) to eliminate earthworms from soil used for continuous cultivation of cereals for 6 years using direct-drilling. Some soils were inoculated with deep-burrowing earthworm species (A. longa and L. terrestris) or shallow-burrowing species (A. caliginosa and A. chlorotica), with unfumigated soils and fumigated soils without added worms as controls. Barley production was similar for the different treatments but root mass and rooting depth were increased in the worm treatments. Earlier experiments by Edwards and Lofty (1978) had shown increased barley shoot and root development in large, undisturbed soil cores to which mixed populations of soil arthropods had been added. Very large increases in cereal yields have been attributed to the activities of earthworm populations by Atlavinyte and colleagues in the USSR (Lee 1985) but details of field studies are not available and the basis of the improved fertility cannot be established.

Dung beetles

The burial of dung by beetles is analogous to the role of earthworms in producing a surface water sink at the site of deposition, relocating the dung in an environment more conducive to microbial decomposition and reducing the period of contaminated herbage avoidance by stock. McKinney and Morley (1975) calculated, however, that at 2 cows/ha in Australia about 4% of the pasture would be covered by dung and hence this represents a maximum ameliorative effect of dung beetles assuming that they were the sole agents of dung removal. Similarly, the zone of enhanced herbage growth at the site of dung burial was less than 8–15% of the total area giving an enhanced herbage production of 1–3%; and agronomically insignificant contribution to pastures receiving regular applications of fertilizer.

Termites

Termites affect soil properties and processes through four main types of activity (summarised from Wood 1987).
1. Physical modification of soil profiles through the construction of mounds, sheeting, foraging runways and infilling of large food masses.
2. Changes in soil texture involving the selection of clay fractions from sub-soil for constructions.
3. Changes in the nature and distribution of organic matter and plant nutrients through litter brought into the nest, which is more or less completely catabolised (digestion and microbial activity) in situ and the composition of termite-worked soil which has higher CEC and exchangeable bases than the surrounding soil.
4. Construction of subterranean galleries which affect drainage and moisture regimes in the soil.

96

The quantitative significance of these processes can only be illustrated here. In some African savannas up to 10t/ha/y of soil is brought to the surface by termites (mostly Macrotermitinae) which can remove 60% of wood and grass litter production (Collins 1983). Wielemaker (1984) reports that even on intensively cultivated, 20% slopes in Kenya intense termite activities virtually eliminated surface runoff and erosion because of the stability and porosity of the soil. In many regions of Africa and India, surface mounds, which can represent considerable concentrations of available plant nutrients in the system, are often used for soil amelioration by farmers. Lal (1987) cites examples of cultural practices in Tanzania where mounds, containing as much as 3.5% $CaCO_3$, contributed the equivalent of 11 t/ha of lime, but in Zaire mounds on clay soil had lower fertility than surrounding soil and were an impediment to mechanised farm operations. The agronomic significance of termites is clearly determined by a fine balance between their beneficial effects on soil conservation and fertility, and their pest status which can vary from year to year according to climatic conditions.

Having established the many ways and spatiotemporal scales at which soil invertebrates affect soil processes their net contributions to soil N fluxes will be considered next.

EFFECTS OF INVERTEBRATES ON NITROGEN FLUXES

Soil invertebrates contribute both directly and indirectly to N fluxes. The direct contributions are the mobilization of N through trophic transfers in food webs from trophic level <u>n</u> to level <u>n+1</u> (feeding, excretion) and turnover of tissue production. The indirect effects involve feedbacks to the controls over the activities of organisms of lower trophic levels, e.g. predators altering prey/microbial interactions and the activities of meso- and macro-fauna changing the physicochemical conditions of microbial habitats.

Direct effects

Estimates of N fluxes through the soil biota can be made for systems where sufficient data are available on soil invertebrate and microbial populations. Information on feeding rates, assimilation, excretion and other physiological constants are available in the literature and can be applied according to defined, functional groups of organisms. There are few detailed studies of N fluxes in grassland or arable systems using this approach.

System budgets

Hunt <u>et al</u>. (1987) investigated the soil N transformations in a shortgrass prairie in the USA, which will be functionally similar to many semi-arid rangelands. They used a food-web model which involved 8 trophic levels and 12 faunal groups with resource inputs separated into resistant and labile fractions to represent the widely divergent decomposition rates of plant and soil organic matter fractions. Groups excluded from the model, because of their low biomass on this site, included ciliates, earthworms, termites and insect larvae. Bacteria were estimated to mineralise the most N ($4.5g/m^2/y$ followed by the invertebrates ($2.9g/m^2/y$) and fungi ($0.3g/m^2/y$). In this system the fauna mediated about 38% of the 77 kg/ha annual flux of mineral N with bacterial-feeding amoebae and nematodes accounting for over 84% of the N mineralized by the fauna.

At the other extreme, detailed N budgets have been drawn up for four cropping systems within the Swedish Ecology of Arable Lands Programme (Rosswall and Paustian 1984): a lucerne ley without fertilizer, a meadow fescue ley receiving 200 kg N/ha/y, barley receiving 120 kg N/ha/y and barley without fertilizer N. Sampled fauna were divided into trophic groups and metabolic constants, and transfer efficiencies between trophic levels, were then applied from the literature. The N budget for barley without fertilizer is shown in Fig. 3. It was concluded that the soil fauna were important in releasing mineral N through consumption of microbial biomass with a low C/N ratio. Only the protozoa and nematode faunas were estimated to excrete significant amounts of ammonium. The direct contribution of the fauna to N

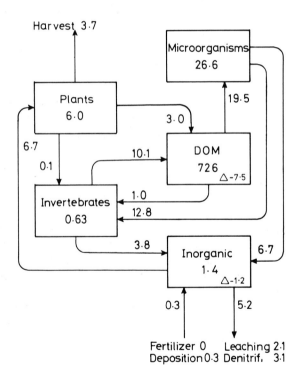

Fig. 3: Summarized N budget for an unfertilized annual barley crop in Sweden. Compartment values are in g/m^2 and represent peak standing crop for plants and mean annual values for all others. Delta symbols indicate net changes, otherwise compartments are assumed to be in steady state. All flows expressed as $g/m^2/y$. DOM = Dead organic matter. (After Rosswall and Paustian 1984).

98

mineralization was highest in the leys (c. 42-50 kg/ha/y) and lowest in the barley plots (c. 38-40 kg/ha/y). However the proportion of total mineralization attributed to the fauna was 31-36% in the barley plots compared with 19-20% in the leys. Earthworm populations in the leys, with a biomass of 50 kg d.w./ha, were considered to have an insignificant contribution to N fluxes but this probably underestimated their roles in the system.

Nitrogen mobilization by earthworms

Lee (1983) calculated that an earthworm biomass of 5 g d.w./m^2 (50 kg/ha), represents a turnover of about 1 gN/m^2/y, assuming a production to biomass quotient of 2-5 and tissue comprising 10-12.5% N. With excretion contributing a further 1.8 gN/m^2/y the total flux through worm tissues is equivalent to 28 kgN/ha/y. For the most densely populated New Zealand pastures with a worm biomass of 450 kg d.w./ha the total could be in the order of 275 kgN/ha. The total N flux mediated by the worms will be much higher because a major proportion of the ingested N is not assimilated. Syers et al. (1979) found that surface casting in a New Zealand pasture, by a (undefined) population of L. rubellus and A. caliginosa, amounted to 33 t/ha/y and contained 73% (132 kg N) of the total N content of litter removed by the worms. Of the total N in the casts only 3.5 kg/ha was in mineralized forms, a smaller mineral N pool than in control litter during the study period, and they concluded that casting was not an agronomically significant N flux pathway in these pastures.

Andersen (1983) investigated N turnover by earthworms in arable plots in the Netherlands receiving inputs of 120 kgN/ha/y as compound fertilizer plus varying treatments of up to 400 t/ha/y of slurry or farmyard manure. Earthworm biomass was 30 g(fresh wt.)/m^2 or less in all plots (Table 1), equivalent to about 5 g d.w./m^2, assuming 16% dry matter (Lakhani and Satchell 1970). Production estimates, derived using a number of different methods, resulted in P/B quotients of 2-4. The results, summarised in Table 1, show that the worms consumed 3-4 t/ha/y of organic matter and turned over 9.5-13.3 kg N/ha/y. While this theoretically represents only 5-6% of the N flux to plants in Andersen's plots, if the figures for the control plots receiving inorganic fertilizer are extrapolated to the Swedish study considered above then a biomass of 50 g/m^2 would mediate a flux of about 95 kg N/ha/y; or about twice the estimate of Lee (1983). Such extrapolation is not really justified but it does illustrate the difficulty of determining N turnover through populations of accessible invertebrates such as worms let alone the problems of estimating populations, processes and N fluxes for mesofauna and microfauna.

Termites and N fluxes

The pathways and processes of N fluxes mediated by termites are complex and will not be considered here (see Wood and Sands 1977; Collins 1983). Termite populations do not turnover N in situ in the same way as earthworms

Table 1: Earthworm populations; feeding activities and N turnover in arable plots in the Netherlands receiving annual applications of compound fertilizer (control), farmyard manure (FYM) or slurry (SL). (Data from Andersen 1983).

	Control	100t/ha FYM	100t/ha SL
Application (kg/ha)			
inorganic N	120	147	266
organic N		412	187
Plant uptake of N (kg/ha)	189	211	230
N turnover by worms (kg/ha)	9.5	13.3	12.0
N flux via worms as % of plant uptake	5	6	5
Worm density (no./m^2)	123	219	237
Worm biomass (g f.w./m^2)	23.6	29.4	26.8
Consumption of organic matter (kg/ha/y)	2 950	4 040	3 700

but transport nutrients to their nests, to varying extents, from which N is mainly returned to the system through predation of workers and alates. Termite biomass generally ranges 0.2–5.0 g d.w./m^2, sometimes as high as 20g/m^2, and N returned to the system through biomass turnover is normally 1–25 kg/ha (Lee 1983). Collins (1983) has calculated for <u>Macrotermes</u> in Kenya and Nigeria that the annual turnover of N is <u>c</u>. 7.6 and 75 kg N/ha/y respectively. These estimates indicate the potential significance of litter-feeding termites in savannas where N turnover between plants and soil is 20–40 kg/ha/y under dry regimes and up to 80 kg N/ha/y with increasing rainfall and herbivory (Bate 1981),

It is important to recognise that the N fluxes calculated in these budgets represent turnover by the biota and not necessarily N mineralization. Excretory ammonium–N is immediately available to plants but N organic excretory products (urea, uric acid and amino acids) and dead tissues must be reutilized and recycled many times for complete ammonification. Hence the mineral N flux to plants mediated by fauna will usually be much less than N turnover through earthworm populations. These estimates of the role of fauna in soil N fluxes are therefore based on necessarily simplistic models of N transformations from tissue N to excretory ammonium with no quantification of the complexities of N recycling in the biota. It is also assumed that the fauna have no indirect effects on microbial N flux pathways. An alternative approach to investigating animal–mediated N fluxes to plants is to set up exclusion experiments with animals which are assumed to have predominantly direct effects on N mineralization. Only nematodes and protozoa, which have fast turnover rates for living and dead tissues, feed on fungi and bacteria, and do not affect soil structure, meet these specifications. However, such defined experiments can only be set up in the laboratory.

Microfauna mediated N uptake by plants

Enhanced N uptake by the grass Bouteloua gracilis occurred in the presence of bacteriophagous protozoa (Elliott et al. 1979; Ingham et al. 1985) or with fungal-feeding nematodes (Ingham et al. 1985). Also, Clarholm (1984) found wheat plants grown in autoclaved (but unleached) arable soil had 60% higher mass and N content when inoculated with protozoa and a natural complement of soil bacteria than when inoculated with bacteria alone (Table 2). A lower shoot/root ratio was found in the presence of protozoa suggesting an improved nutrient supply to the roots. The mechanism put forward by Clarholm (1985) to account for the enhanced N mineralization is that N immobilized in the rhizosphere, as a consequence of bacterial stimulation by root exudates, is mobilized as excretory ammonium by protozoa. It is debatable, however, that this phenomenon is significant in mediating a significant proportion of the N uptake by arable crops since these require low rooting densities to efficiently exploit a mobile ion such as nitrate which is the main form of N taken up under field conditions (Robinson and Rorison 1983).

Table 2: Dry weights and N contents of wheat plants grown for 6 weeks in sterilized soil reinoculated with bacteria alone or bacteria plus protozoa. Values are expressed as mg per experimental chamber containing 3 plants. (From Clarholm 1984).

Component		Bacteria alone	Bacteria + Protozoa
Dry weight	shoots	116	213
	roots	65	113
	total plant	181	326
N content	shoots	3.92	4.39
	roots	0.93	1.41
	total plant	4.85	5.80
N content	planted seeds	3.24	3.24
N mineralized from soil (plant N – seed N)		1.61	2.56

Indirect affects

Indirect effects of invertebrates on N fluxes from dead organic matter to root uptake by plants include: the removal of litter N pools from the system; influences on litter decomposition rates by transport to more or less favourable environments for decomposition; alteration of the conditions for root growth and activity; and rhizosphere effects on pathogens and mycorrhizas. Springett and Syers (1979) also mention 'plant growth substances' in earthworm casts; a hypothesis put forward by earlier workers to account for aberrant root growth patterns in grass seedlings. The present discussion is restricted to indirect effects on microbial processes of N mineralization and N availability to plants.

Indirect effects are difficult to demonstrate and even more difficult to quantify, particularly for micro-fauna, except by the removal or addition of key animals to the system and monitoring the response. Salawu and Estey (1979), for example, carried out pot experiments in which soybeans were grown in sterile soil alone, in soil inoculated with spores of Glomus (a VA mycorrhizal fungus), and Glomus spores plus fungal-feeding nematodes (Aphelenchus avenae). After 45 days the top-growth of plants treated with Glomus alone was up to 40% higher than for treatments with Glomus and nematodes; but there was only a 12.9% difference between control plants without the mycorrhiza and treatments with nematodes and mycorrhiza. It was also found that the simultaneous addition of both nematode and VA fungus resulted in the formation of only 3 nodules/plant compared with 24 nodules/plant for the treatment with Glomus alone; a difference attributed to the indirect affects of the nematodes on P uptake by the soybean plants.

It has proved difficult to carry out comparable manipulations of microfauna in the field. Ingham et al (1986b) used pulse teatments of selective biocides in a semi-arid grassland to reduce populations of bacteria (streptomycin), fungi including VA mycorrhizas (captan and PCNB), nematodes (carbofuran) and microarthropods (cygon), and monitored responses for 7 months after biocide application. Overall, groups were reduced rather than eliminated by the biocides and responses were complex. But the results are of particular interest since they show that the community did not return to the original composition within 7 months but, after an initial flush of mineral N, no further changes in mineral N fluxes were observed because of compensatory flux pathways operating within the food web.

The effects of macrofauna on microbial populations are best documented for base-rich soils which are intensively worked by earthworms. Bhatnager (1975) estimated that 40% of aerobic, non-symbiotic N_2-fixing bacteria, 13% of anaerobic N_2-fixers and 16% of denitrifying bacteria in the total soil volume were located in a narrow zone of a few millimetres depth around earthworm burrows in a French pasture. The casts contained fewer fungal propagules and denitrifying bacteria, higher counts of total bacteria and more hemicellulolytic, amylolytic and nitrifying bacteria than unworked soil (Loquet et al. 1977).

When casts are first voided mineral N predominates as ammonium but nitrification progresses rapidly at field temperatures under which the worms are active (Syers et al. 1979). The conditions for microbial growth in the casts are largely determined by selective feeding by the worms on plant materials and soil organic matter, and will therefore be determined by the ecology of different species. Ingestion of high quality food will produce casts with higher mineral N and available carbohydrate than the surrounding soils (Syers et al. 1979). As a consequence of enhanced heterotrophic microbial activity in the casts oxygen is depleted leading to the development of anaerobic conditions in which denitrification can take place (Svensson et al. 1986). Casts of L. terrestris fed on lucerne (Medicago sativa) were shown by Svensson et al. (1986) to consistently maintain higher rates of nitrous oxide production (using the acetylene inhibition assay) than soil controls. The maximum rates of denitrification in the casts, approaching 0.2 mgN/g soil, cannot be extrapolated from these laboratory studies to the field but these workers suggest that the phenomenon may contribute to the high spatial variability of denitrification rates observed in field studies. This is supported by a preliminary results (P. Elliott, D. Knight and J.M. Anderson, unpublished data) for a permanent pasture in the UK where mean ($\pm$SE) denitrification rates in earthworm casts, during a spring period of low rainfall, were 0.380 $\pm$ 0.24 mgN/g soil compared with 0.03 $\pm$ 0.27 mgN/g from paired soil cores. It is of interest in this respect that large-earthworm casts are very similar in size and texture to the soil aggregates which Tiedje et al (1984) have shown to form anaerobic cores and constitute sites of denitrification during nitrate flushes in a well-drained silty loam under continuous arable cultivation.

Surface runoff represents another pathway of N loss which is influenced by termites and earthworms. Sharpley et al (1979) treated permanent pasture with carbaryl and found that the elimination of earthworms doubled surface runoff. This was attributed to a three fold reduction in infiltration rate in the absence of surface casting by earthworms which maintain the openings of the burrows to the soil surface. The decreased infiltration in the absence of the worms resulted in greater net losses of dissolved inorganic N (13.8 kg/ha/y) in the treated plots than in the untreated grassland with casts present (2.2 kgN/ha/y).

Soil fauna clearly have a whole spectrum of effects which affect soil fertility and plant growth in agricultural systems. It is appropriate, therefore, to consider the effects of agricultural management practices on their populations.

INVERTEBRATE POPULATIONS AND MANAGEMENT PRACTICES

The two main determinants of species diversity and population densities of soil invertebrates excluding climate, are the quality and quantity of detritus inputs to the system (Heal and Dighton 1985), and the stability and complexity of soil and litter habitats (Anderson 1978). The two factors interact through the regulatory role of resource quality on litter decomposition rates (Swift et al. 1979; Heal and Dighton 1985). As resource

103

quality increases the accumulated litter and cellular soil organic matter pools decrease as a consequence of faster decomposition. Hence microhabitat complexity and diversity of associated species are also likely to decline. Conversely, populations of larger soil invertebrates and their contribution to total fauna biomass tend to increase as some function of increased N turnover in natural systems (Heal and Dighton 1985).

Earthworms and other macrofauna consume soil and organic matter forming a microhabitat for smaller invertebrates, and Yeates (1981) has suggested that their feeding activities may constitute an incidental but density-dependent source of mortality for soil nematodes. In natural systems these patterns of community structure are to some extent predictable because of the integration of plant production, litter quality, soil organic matter pools and N cycling in relation to parent soil types and climate. But in agricultural systems factors affecting the soil invertebrate community (Fig. 4) may be imposed independently or in combinations which have interactive effects which are poorly understood.

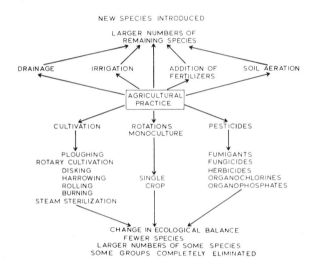

Fig. 4: Ecological effects of agricultural practice on soil fauna populations. (From Edwards and Lofty 1969).

Pesticides

Pesticide effects are particularly complex because of differences in exposure and sensitivity of non-target species (Edwards and Lofty 1969; Smith et al. 1980; and other reviews in Dindal 1980), changes in soil conditions influencing the inactivation or degradation of the pesticide (McColl 1984) and trophic interactions in the community (Parker et al. 1984; Ingham et al. 1986b) which are considered below. Herbicides also give complex effects by providing a pulsed input of plant material to the soil which, for example, can produce a short term increase in earthworm populations but a long term decline through sward removal (Edwards and Brown 1982). The organophosphorus and carbamate pesticides have transient effects on nematodes and microarthropods but populations of earthworms, such as L. terrestris, may take longer to recover because of lower reproductive rates and colonising abilities than smaller invertebrates. Standard spraying regimes of the fungicide benomyl can significantly reduce earthworm numbers and activity (Stringer and Wright 1973; Keogh and Whitehead 1975) and some insecticides are also extremely toxic to earthworms. After three years spraying pasture in the UK with the insecticide phorate (3.3 kg/ha/month) earthworm populations were eliminated with consequent changes in soil physical properties (Clements 1978). In comparison with the untreated plots, infiltration rates and hydraulic conductivity were reduced by c. 93% and bulk density increased by up to 17%; none-the-less, herbage yield increased 40% as a consequence of controlling root-feeding beetle larvae.

Perfect et al. (1979) found that spraying cowpea with DDT over a 4-year period doubled grain yield in a Nigerian bush-fallow rotation with no fertilizer inputs. But the decline in yield over the cropping period, as is usual in these systems, was faster in the treated plots and was particularly marked in a subsequent maize crop. The effect was not attributable to greater nutrient depletion in the more heavily cropped plots. It was concluded that differences in soil fertility arose through supression of earthworm effects on soil properties at concentrations which had little effect on population densities (Perfect et al. 1979). Edwards and Thompson (1973) concluded that in temperate, cultivated soils the overall effects of tillage have greater long-term impact on soil invertebrate communities than modern pesticides.

Other management practices

In temperate grasslands, old leys and permanent pastures have the highest population densities of most invertebrate groups. Microclimatic conditions and soil surface microhabitats associated with grass cover, litter and the root mats favour microarthropods such as mites and collembola which can have population densities as high as $300000/m^2$ (Curry 1969). Stocking densities of cattle or sheep influence the quantity, quality (dung rather than litter) and distribution of resource inputs to the soil, herbage cover and soil porosity through trampling. Hutchinson and King (1980) found that sheep stocking density (10, 20 and 30 animals/ha) affected the abundance and biomass of all invertebrates. But while numbers of small animals (mites, collembola, enchyraeid worms, nematodes, etc.) were inversely related to sheep density many of the larger invertebrates, such as earthworms and scarabeid larvae, reached peak abundance under the intermediate stocking level.

Organic manures (such as slurry and farm-yard manure), and inorganic fertilizers increase soil fauna populations through increased pasture growth and higher residue quality. Cotton and Curry (1980) recorded worm populations of 400–500/m^2 (100–200g f.w./m^2) in grassland fertilized with slurry while applications of superphosphate with lime in the clover–rich pastures of New Zealand have resulted in very large lumbricid populations of more than 1000/m^2 (c. 300 g f.w./m^2) (Lee 1985). Comparable worm population densities of 800/m^2 have been recorded in northern England for a hay meadow dressed with fertilizer and farmyard manure to a total of 118, 42 and 111 kg/ha/y N, P and K respectively. Soil pH and manure applications were positively correlated with earthworm populations in this study but enchytraeid populations, which are highest in acid soils, were negatively correlated with earthworm biomass.

Drainage of soils under permanent, fertilized grassland (Gilbey 1985) or forage crops (Carter et al. 1982) increased earthworm populations but interpretation is complicated by improved soil environmental conditions and plant production.

When grassland is ploughed and maintained under arable cultivation the number of invertebrate species and population densities are usually drastically reduced. The changed soil environment and disturbance selects for small species with high reproductive rates and against larger invertebrates, such as earthworms, with a life cycle of more than a year (Edwards and Lofty 1969). The restructured community may include higher proportions of insect pests (Edwards and Lofty 1969). Andren and Lagerlof (1980) observed major differences in the composition of microarthropod communities in a 5 year–old ley compared to the first and second barley/ley rotations. Collembola were dominant (c. 60000/m^2) in the maintained ley but reduced to a similar density to mites in the first crop rotation and to only 10000/m^2 in the second rotation where small, fungal–feeding prostigmatid mites (Tarsonemoidae) were most abundant (c. 35000/m^2). Contrary to the observation by Ferris (1982) that plant parasitic nematodes, and to some extent bacterial feeders (Sohlenius and Bostrom 1986) show a proportional increase in arable systems, Andren and Lagerlof (1980) found that all trophic groups of nematodes were reduced by c. 50% from 8 x 10^6/m^2 in the ley to about 2 x 10^6/m^2 in the second crop rotation.

Low (1972) compared earthworm populations in old grassland with grassland cultivated for three years and with fields tilled for 25 years on the same soil series. After three year's cultivation the worm populations were half those in the grassland and reduced to about 14% over longer periods of cultivation.

In the short term, shallow tillage is much less damaging to worm populations than deep ploughing; particularly for deep–burrowing species (Edwards 1983). But in the long term the loss of surface litter and reduced organic matter content of soils is mainly responsible for the decline in worm populations in tilled soils (Lee 1985). Minimum tillage practices, through reduced soil disturbance and increased weed and crop residue inputs, generally show higher populations of nematodes (Stinner and Crossley 1982), microarthropods (Edwards and Lofty 1969) and earthworms (Lee 1985) than conventionally tilled systems. Edwards and Lofty (1982) recorded a thirty-fold increase in worm populations after eight years of direct drilling fields which had previously been ploughed.

DISCUSSION

Natural grassland and forest ecosystems develop towards a state of dynamic equilibrium between organic N inputs to the soil and mineral N uptake by roots. Once this state is reached plant communities with similar net primary production can exhibit significant differences in the size and turnover of different nutrient pools within the system (Cole and Rapp 1981) and in the structure and functioning of soil organism communities. In functional terms, therefore, the various animal and microbial processes will equilibrate with the availability of N in the system and hence the various pathways involved in N mineralization though faunal effects may not be of direct consequence for plant production under 'steady state' conditions. Similarly, water infiltration rates are usually high, and soil erosion and surface-water runoff usually low, under undisturbed vegetation cover associated with faunal communities which are very different in structure and functioning.

If, however, some component of the system is changed, such as quality or quantity of soil organic matter and nutrient inputs, climate or soil parameters (including the introduction or removal of soil fauna), then the whole functioning of the community may shift to a new equilibrium (Anderson 1987). During the transitional phase soil fauna may act as driving variables determining both the rate of change and the new equilibrium state. Once the new state is reached it maintains some degree of resilience to small and frequent, or severe and infrequent perturbations. For example, Anderson et al. (1985) established a series of lysimeters in an oak woodland with and without a small biomass of soil macrofauna (about 3.7 g d.w./m^2 of millipedes, woodlice and surface-living earthworms) and with and without tree roots to act as water and nutrient sinks. After a year the humus form and the mineral N fluxes were different in the four treatments with enhanced total mineral N losses from the macrofauna treatments of c. 12 kg/ha over the summer/autumn period and reduced mineral N losses over winter/spring. But when the soil fauna were killed by a single, large application of carbofuran there were negligible changes in the N losses from experimental soils over a period of months indicating that the functional attributes of the system were the consequence of indirect effects of fauna on soil properties and processes. Similarly, the introduction of exotic earthworms into New Zealand pastures caused a large initial increase in herbage production, associated with the mobilization of nutrients from accumulated litter. Thereafter, pasture production settled down to an improved yield as a consequence of changes in soil conditions for root growth.

When continuous vegetation cover is replaced by intensive arable cropping the dynamic balance is disturbed and production occurs in cohorts of crops sustained by pulsed inputs of fertilizers. Tillage, pesticides and fertilizers to maintain production in this non-equilibrium state not only override faunal processes but also further reduce and obscure their contribution to soil fertility. But when the system employs residue conservation, organic manures and/or minimum-till cultivation, the soil biological processes are more evident. Under these conditions faunal processes could have an impact on plant production if we knew how to manage them to achieve specific effects within the life span of the crop plant. The role of earthworms is better understood in this respect than most other groups; although we lack detail of their interactions with soil bacteria and fungi to appreciate their manipulative potential beyond their physical effects on soils. Far less is known about invertebrate/microbial interactions in the rhizosphere.

Whipps and Lynch (1986) have reviewed the influence of the rhizosphere on crop productivity and detail the microbial responses to changes in organic exudates from crop roots following changes in the plant environment (fertilizers, pesticides, pathogens, moisture stress, etc.). At the present time the implications of these responses for the activity of root pathogens, mycorrhizas, growth-promoting rhizobacteria, rhizobia and associative N_2-fixers is poorly understood but may have considerable agronomic potential. It is already established, however, that biological control of microbial pathogens and some root-feeding nematodes can be achieved by competitive displacement, induced resistance, predation or parasites (Baker and Cook 1974). But little is known about faunal effects which may enhance or inhibit these responses though the studies reviewed earlier show that the feeding activities of nematodes and protozoa can affect microbial populations at this scale and particularly in the rhizosphere.

In conclusion, the roles of soil fauna in N cycles are difficult to quantify because in addition to their direct contributions to N fluxes there are a wide range of indirect effects on the soil as an environment for microorganisms and plant roots. Consequently, treatments which enhance or reduce soil invertebrate activities, particularly those of macrofauna, may not be interpretable in terms of simple cause/effect relationships. It would appear that there are soils and management circumstances where these net effects on soil processes may be important. The current interest on less intensive, more organically based agriculture in Europe and N. America, and on improving and sustaining low-input agriculture in the tropics (Swift 1985), may result in stimulating the critical interdisciplinary approaches which are still needed to resolve the roles of fauna in soil processes.

REFERENCES

Aina, P.O. (1984). Contribution of earthworms to porosity and water infiltration in a tropical soil under forest and long-term cultivation. Pedobiologia 26, 131-6.

Andersen, N.C. (1983). Nitrogen turnover by earthworms in arable plots treated with farmyard manure and slurry. In 'Earthworm Ecology'. (Ed. J.E. Sachell). pp. 139-50. (Chapman & Hall: London.)

Anderson, J.M. (1978). Inter- and intra-habitat relationships between woodland Cryptostigmata species diversity and the diversity of soil and litter microhabitats. Oecologia 32, 341–8.

Anderson, J.M. (1987). Spatiotemporal effects of invertebrates on soil processes. Biol. Fertil. Soils (in press).

Anderson, J.M., Huish, S.A., Ineson, P., Leonard, M.A. and Splatt, P.R. (1985). Interactions of invertebrates, micro-organisms and tree roots in nitrogen and mineral element fluxes in deciduous woodland soils. In 'Ecological Interactions in Soil'. (Eds A.H. Fitter, D. Atkinson, D.J. Read and M.J. Usher). pp. 377–92. (Blackwell Scientific Publications: Oxford.)

Anderson, J.M., Rayner, A.D.M. and Walton, D.W.H. (1984). 'Invertebrate-microbial Interactions'. (Cambridge University Press: Cambridge.)

Anderson, R.V., Coleman, D.C., and Cole, C.V. (1981). Effects of saprotrophic grazing on net mineralization. In 'Terrestrial Nitrogen Cycles'. (Eds F.E. Clarke and T. Rosswall) pp. 201–15. (Swedish Natural Science Research Council: Stockholm.)

Andren, O., and Lagerlof, J. (1980). The abundance of soil animals (micro-arthropoda, enchytraeidae, nematoda) in a crop rotation dominated by ley and in a rotation with varied crops. In 'Soil Biology as Related to Land-use Practices'. (Ed. D. Dindal). pp. 274–9. (EPA: Washington.)

Baker, K.F., and Cook, R.J. (1974). 'Biological Control of Plant Pathogens'. (Freeman: San Francisco.)

Bate, G.C. (1981). Nitrogen cycling in savanna ecosystems. In 'Terrestrial Nitrogen Cycles'. (Eds F.E. Clark and T. Rosswall). pp. 463–75. (Swedish Natural Science Research Council: Stockholm.)

Bhatnagar, T. (1975). Lombriciens et humification: un aspect nouveau de l'incorporation microbienne d'azote induite par les vers de terre. In 'Biodegradation et Humification'. (Eds G. Kilbertus, O. Reisinger, A. Mourney and J.P. Cansella da Fonseca). pp. 157–68. (Pierron: Sarraeguemines.)

Bowen, G.D. (1984). Tree roots and use of soil nutrients. In 'Nutrition of Plantation Forests'. (Eds G.D. Bowen and E.K.S. Nambiar). pp. 147–9. (Academic Press: London.)

Bowen, G.D., and Theodorou, C. (1979). Interactions between bacteria and ectomycorrhizal fungi. Soil Biol. Biochem. 11, 119–26.

Brown, R.A. (1985). Effects of some root-grazing arthropods on the growth of sugar beet. In 'Ecological Interactions in Soil'. (Eds A.H. Fitter, D. Atkinson, D.J. Read and M.J. Usher). pp. 285–95. (Blackwell Scientific Publications: Oxford.)

Carter, A., Heinonen, J., and de Vries, J. (1982). Earthworms and water movement. Pedobiologia 23, 395–7.

Chapman, S.J., and Gray, T.R.G. (1985). Importance of cryptic growth, yield factors and maintenance energy in models of microbial growth in soils. Soil Biol. Biochem. 18, 1–4.

Clarholm, M. (1984). Heterotrophic, free-living protozoa: neglected micro-organisms with an important task in regulating bacterial populations. In 'Current Perspectives in Microbial Ecology'. (Eds M.J. Klug and C.A. Reddy.) pp. 321–6. (American Society of Microbiology: Washington.)

Clarholm, M. (1985). Possible roles for roots, bacteria, protozoa and fungi in supplying nitrogen to plants. In 'Ecological Interactions in Soil'. (Eds A.H. Fitter, D. Atkinson, D.J. Read and M.J. Usher). pp. 355–65. (Blackwell Scientific Publications: Oxford.)

Clements, R.O. (1978). The benefit and some long-term effects of controlling invertebrates in a perennial rye grass field. Sci. Proc. Roy. Dubl. Soc. Ser. A. 6, 335–41.

Cole, D.W., and Rapp, M. (1981). Element cycling in forest ecosystems. In 'Dynamic Properties of Forest Ecosystems'. (Ed. D.E. Reichle). pp. 341–409. (Cambridge University Press: Cambridge.)

Collins, N.M. (1983). The utilization of nitrogen resources by termites (Isoptera). In 'Nitrogen as an Ecological Factor'. (Eds J.A. Lee, S. McNeill and I.H. Rorison), pp. 381–412. (Blackwell Scientific Publications: Oxford.)

Cotton, D.C.F., and Curry, J.P. (1980). The response of earthworm populations (Oligochaeta, Lumbricidae) in grassland managed for silage production. Pedobiologia 20, 181–8.

Cousins, S.H. (1980). A trophic continuum derived from plant structure, animal size and a detritus cascade. J. Theor. Biol. 82, 607–18.

Curry, J.P. (1969). The qualitative and quantitative composition of the fauna of an old grassland site at Celbridge, Co. Kildare. Soil Biol. Biochem. 1, 219–27.

Cutler, D.W., Crump, L.M., and Sandon, H. (1922). A quantitative investigation of the bacterial and protozoan fauna. Phil. Trans. Roy. Soc. Ser. B. 211, 317–47.

Darbyshire, J.F., and Greaves, M.P. (1967). Protozoa and bacteria in the rhizosphere of Sinapis alba L. and Lolium perenne L. Can. J. Microb. 13, 1057–68.

Darbyshire, J.F., Robertson, L., and Mackie, L.A. (1985). A comparison of two methods of estimating the soil pore network available to protozoa. Soil Biol. Biochem. 17, 619–24.

Dindal, D.L. (1980). 'Soil Biology as Related to Land Use Practices'. (Envir. Prot. Agency: Washington.)

Edwards, C.A. (1983). Earthworm ecology in cultivated soils. In 'Earthworm Ecology'. (Ed. J.E. Satchell). pp. 123–38. (Chapman and Hall: London.)

Edwards, C.A., and Lofty, J.R. (1969). The influence of agricultural practice on soil microarthropod populations'. In 'The Soil Ecosystem'. (Ed. J.G. Sheals). pp. 237–47. (Systematics Association: London.)

Edwards, C.A., and Lofty, J.R. (1972). 'Biology of Earthworms'. (Chapman and Hall: London.)

Edwards, C.A., and Lofty, J.R. (1978). The influence of arthropods and earthworms upon root growth of direct drilled cereals. J. Appl. Ecol. 15, 789–95.

Edwards, C.A., and Lofty, J.R. (1980). Effects of direct drilling and minimal cultivation on earthworm populations in agricultural soils. J. Appl. Ecol. 19, 723–34.

Edwards, C.A., and Thompson, A.R. (1973). Pesticides and soil fauna. Resid. Rev 45, 1–80.

Edwards, P.J., and Brown, S.M. (1982). Use of grassland plots to study the effect of pesticide on earthworms. Pedobiologia 24, 145–50.

Ehlers, W. (1975). Observations of earthworm channels and infiltration on tilled and untilled loess soil. Soil Sci. 119, 242–9.

Elliott, E.T., Anderson, R.U., Coleman, D.C., and Cole, C.V. (1980). Habitable pore space and microbial trophic interactions. Oikos 35, 327–35.

Elliott, E.T., Coleman, D.C., and Cole, C.V. (1979). The influence of amoebae on the uptake of nitrogen by plants in gnotobiotic soil. In 'The Soil–Root Interface'. (Eds J.L. Harley and R. Scott–Russell). pp. 221–9. (Academic Press: New York.)

Elliott, E.T., Horton, K., Moore, J.C., Coleman, D.C., and Cole, C.V. (1984). Mineralization dynamics in fallow dryland wheat plots, Colorado. Plant Soil 76, 149–55.

Ferris, H. (1982). The role of nematodes as primary consumers. In 'Nematodes in Soil Ecosystems'. (Ed. D.W. Freckman). pp. 3–13. (University of Texas Press: Austin.)

Fitter, A.H., Atkinson, D., Read, D.J., and Usher, M.B. (1985). 'Ecological Interactions in Soil'. (Blackwell Scientific Publications: Oxford.)

Freckman, D. (1982). 'Nematodes in Soil Ecosystems'. (University of Texas Press: Austin.)

Ghilarov, M.S. (1982). Proceedings of the Vth International Colloquium on Terrestrial Oligochaeta. Pedobiologia 23, 3/4, 173–319.

Ghilarov, M.S., and Perel, T.S. (1984). Transfer of earthworms (Lumbricidae, Oligochaeta) for soil amelioration in the USSR. Pedobiologia 27, 107–13.

Gilbey, J. (1985). The effects of grassland improvement on soil faunal populations. Brit. Grassl. Soc. Occ. Symp. 20, 94–6.

Heal, O.W., and Dighton, J. (1985). Resource quality and trophic structure in the soil system. In 'Ecological Interactions in Soil'. (Eds A.H. Fitter, D. Atkinson, D.J. Read and M.J. Usher). pp. 339–54. (Blackwell Scientific Publications: Oxford.)

Heron, A.C. (1972). Population ecology of a colonizing species: the pelagic tunicate Thalia democratica. Oecologia 10, 269–312.

Hole, F.D., (1981). Effects of animals on soil. Geoderma 25, 75–112.

Hoogerkamp, M., Rogaar, H., and Eijsackers, H.J.P. (1983). Effect of earthworms in grassland on recently reclaimed polder soils in the Netherlands. In 'Earthworm Ecology'. (Ed. J. Satchell), pp. 85–105. (Chapman and Hall: London.)

Hunt, H.W., Coleman, D.C., Ingham, E.R., Ingham, R.E., Elliott, E.T., Moore, J.C., Rose, S.L., Reid, C.P.P., and Morley, C.R. (1987). The detrital food web in a shortgrass prairie. Biol. Fertil. Soils 3, 57–68.

Hutchinson, K.L., and King, K.L. (1980). The effects of sheep stocking level on invertebrate abundance, biomass and energy utilization in a temperate, sown grassland. J. Appl. Ecol. 17, 369–87.

Ingham, E.R., Trofymow, J.A., Ames, R.N., Hunt, H.W., Morley, C.R., Moore, J.C., and Coleman, D.C. (1986a). Trophic interactions and nitrogen cycling in a semi–arid grassland soil. I. Seasonal dynamics of the natural populations, their interactions and effects on nitrogen cycling. J. Appl. Ecol. 23, 597–614.

Ingham, E.R., Trofymow, J.A., Ames, R.N., Hunt, H.W., Morley, C.R., Moore, J.C., and Coleman, D.C. (1986b). Trophic interactions and nitrogen cycling in a semi–arid grassland soil. II. System responses to removal of different groups of soil microbes or fauna. J. Appl. Ecol. 23, 615–30.

Ingham, R.E., Trofymow, J.A., Ingham, E.R., and Coleman, D.C. (1985). Interactions of bacteria, fungi and their nematode grazers: effects on nutrient cycling and plant growth. Ecol. Monogr. 55, 119–40.

Keogh, R.G., and Whitehead, P.H. (1975). Observations on some effects of pasture spraying with benomyl and carbendazim on earthworm activity and litter removal from pasture. N.Z. J. Exp. Agric. 3, 103–4.

Kevan, D.K.McE. (1965). The soil fauna — its nature and biology. In 'Ecology of Soil—Borne Pathogens'. (Eds K.F. Baker and W.C. Snyder). pp. 33—51. (University of California Press: San Francisco.)

Lakhani, K.H., and Satchell, J.E. (1970). Production by Lumbricus terrestris L. J. Anim. Ecol. 39, 473—92.

Lal, R. (1987). 'Tropical Ecology and Physical Edaphology'. (John Wiley and Sons: Chichester.)

Lavelle, P. (1978). Les vers de terre de la savane Lamto (Cote d'Ivoire): peuplements, populations et fonctions dans l'ecosysteme. Publ. Lab. Zool. E.N.S. 12, 1—301.

Lee, K.E. (1983). The influence of earthworms and termites on soil nitrogen cycling. In 'New Trends in Soil Biology'. (Eds P. Lebrun, H.M. Andre, A. de Medts, C. Gregoire—Wibo and G. Wauthy), pp. 35—48. (Dieu—Brichart: Louvain—la—Neuve.)

Lee, K.E. (1985). 'Earthworms: Their Ecology and Relationships with Land Use'. (Academic Press: Sydney.)

Loquet, M., Bhatnagar, T., Bouche, M.B., and Rouelle, J. (1977). Essai d'estimation de l'influence ecologique des lombriciens sur les microrganismes. Pedobiologia 17, 400—17.

Low, A.J. (1972). The effect of cultivation on the structure and other physical characteristics of grassland and arable soils. J. Soil Sci. 23, 363—80.

McColl, H.P. (1984). Nematicides and field populations of enchytraeids and earthworms. Soil Biol. Biochem. 16, 139—44.

McKinney, G.T., and Morley, F.H.W. (1975). The agronomic role of introduced dung beetles in grazing systems. J. Appl. Ecol. 12, 831—7.

Nordstrom, S., and Rundgren, S. (1973). Associations of lumbricids in southern Sweden. Pedobiol. 14, 1—27.

Parker, L.W., Santos, P.F., Phillips, J., and Whitford, W.G. (1984). Carbon and nitrogen dynamics during the decomposition of litter and roots of a Chihuahuan desert annual, Lepidium lasiocarpum. Ecol. Monogr. 54, 339—60.

Perfect, T.J., Cook, A.G., Critchley, B.R., Critchley, U., Davies, A.L., Swift, M.J., Russell— Smith, A., and Yeadon, R. (1979). The effect of DDT on the productivity of a cultivated forest soil in the sub—humid tropics. J. Appl. Ecol. 16, 705—19.

Petersen, H., and Luxton, M. (1982). A comparative analysis of soil fauna populations and their role in decomposition processes. Oikos 39, 287—388.

Robinson, D., and Rorison, I.H. (1983). Relationships between root morphology and nitrogen availability in a recent theoretical model describing nitrogen uptake from soil. Plant Cell. Envir. 6, 641—7.

Rosswall, T., and Paustian, K. (1984). Cycling of nitrogen in modern agricultural systems. Plant Soil 76, 3—21.

Salawu, E.O., and Estey, R.H. (1979). Observations of the relationships between a vesicular arbuscular fungus, a fungivorous nematode and the growth of soybeans. Phytoprotection 60, 99— 102.

Santos, P.F., Phillips, J., and Whitford, W.G. (1981). The role of mites and nematodes in early stages of buried litter decomposition in a desert. Ecology 62, 664—9.

Satchell, J.E. (1983). 'Earthworm Ecology: from Darwin to Vermiculture'. (Academic Press: London.)

Schnurer, J., Clarholm, M., and Rosswall, T. (1986). Fungi, bacteria and protozoa in soil from four arable cropping systems. Biol. Fert. Soils 2, 119—26.

Seastedt, T.R. (1984). The role of microarthropods in decomposition and mineralization processes. Ann. Rev. Entomol. 29, 25—46.

Sharpley, A.N., Syers, J.K., and Springett, J.A. (1979). Effect of surface—casting earthworms on the transport of phosphorus and nitrogen in surface runoff from a pasture. Soil Biol. Biochem. 11, 459—62.

Shaw, C., and Pawluk, S. (1986). Faecal microbiology of Octolasion tyrtaeum, Aporrectodea turgida and Lumbricus terrestris and its relation to the carbon budgets of three artificial soils. Pedobiologia 29, 377—89.

Smith, T.D., Keith, D., Kevan, McE., and Hill, S.B. (1980). Effects of six biocides on non—target soil mesoarthropods from pasture on Ste. Rosalie clay loam, St. Clet, Quebec. In 'Soil Biology as Related to Land—Use Practices'. (Ed. D. Dindal). pp. 57—70. (EPA: Washington.)

Sohlenius, B., and Bostrom, S. (1986). Short—term dynamics of nematode communities in arable soil: influence of nitrogen fertilization in barley crops. Pedobiologia 29, 183—91.

Springett, J.A. (1985). Effects of introducing Allolobophora longa Ude on root distribution and some soil properties in New Zealand pastures. In 'Ecological Interactions in Soil'. (Eds A.H. Fitter, D. Atkinson, D.J. Read and M.B. Usher). pp. 339—405. (Blackwell Scientific Publications: Oxford.)

Springett, J.A., and Syers, J.K. (1979). The effects of earthworm casts on ryegrass seedlings. In 'Proceedings on the 2nd Australasian Conference on Grassland Invertebrate Ecology'. (Eds T.K. Crossby and R.P. Pottinger). pp. 44—7. (Government Printer: Wellington.)

Stinner, B.R., and Crossley, C.A. (1982). Nematodes in no-tillage agro-ecosystems. In 'Nematodes in Soil Ecosystems'. (Ed. D.W. Freckman). pp. 14–28. (University of Texas Press: Austin.)

Stockdill, S.M.J. (1982). Effect of introduced earthworms on the productivity of New Zealand pastures. Pedobiologia 24, 29–35.

Stout, J.D., and Heal, O.W. (1967). Protozoa. In: 'Soil Biology'. (Eds A. Burges and F. Raw). pp. 149–95. (Academic Press: London.)

Stringer, A., and Wright, M.A. (1973). The effect of benomyl and some related compounds on Lumbricus terrestris and other earthworms. Pestic. Sci. 5, 165–70.

Svensson, B.H., Bostrom, U., and Klemedtson, L. (1986). Potential for higher rates of denitrification in earthworm casts than in the surrounding soil. Biol. Fertil. Soils 2, 147–9.

Swift, M.J. (1985). 'Tropical Soil Biology and Fertility (TSBF). Planning for Research'. Biol. Int. Special Issue No. 9. (IUBS: Paris.)

Swift, M.J., Heal, O.W., and Anderson, J.M. (1979). 'Decomposition in Terrestrial Ecosystems'. (Blackwell Scient. Publ.: Oxford.)

Syers, J.K., Sharpley, A.N., and Keeny, D.R. (1979). Cycling of nitrogen by surface-casting earthworms in a pasture ecosystem. Soil Biol. Biochem. 12, 181–5.

Syers, J.K., and Springett, J.A. (1984). Earthworms and soil fertility. Plant Soil 76, 93–104.

Tiedje, J.M., Sextone, A.J., Parkin, T.B., Revsbech, N.P., and Shelton, D.R. (1984). Anaerobic processes in soil. Plant Soil 76, 197–212.

Ulber, B. (1983). Einfluss von Onychiurus fimatus Gisin (Collembola, Onychiuridae) und Folsomia fimetaria (L) (Collembola, Isotomidae) auf Pythium ultimum Trow., einen Erreger des Wurzelbrandes der Zuckerrube. In 'New Trends in Soil Biology'. (Eds P. Lebrun, H.M. Andre, A. de Medts, C. Gregoire-Wibo and G. Wauthy). pp. 261–8. (Dieu-Brichart: Louvain-la-Neuve.)

Visser, S. (1985). Role of soil invertebrates in determining the composition of soil microbial communities. In 'Ecological Interactions in Soil'. (Eds A.H. Fitter, D. Atkinson, D.J. Read and M.J. Usher). pp. 297–317. (Blackwell Scientific Publications: Oxford.)

Wallace, H.R. (1962). The movement of nematodes in relation to some physical properties of soil. In 'Progress in Soil Zoology'. (Ed. P.W. Murphy) pp. 328–33. (Butterworths: London.)

Warnock, A.J., Fitter, A.H., and Usher, M.B. (1982). The influence of a springtail Folsomia candida (Insecta, Collembola) on the mycorrhizal association of leek Allium porrum and the vesicular-arbuscular mycorrhizal endophyte Glomus fasciculatum. New Phytol. 90, 285–92.

Whipps, J.M., and Lynch, J.M. (1986). The influence of the rhizosphere on crop productivity. Adv. Microb. Ecol. 9, 187–244.

Wielemaker, W.G. (1984). 'Soil Formation by Termites'. (Agricultural University: Wageningen.)

Wood, T.G. (1987). Termites and the soil environment. Biol. Fert. Soils (in press).

Wood, T.G., and Sands, W.A. (1977). The role of termites in ecosystems. In 'Production Ecology of Ants and Termites'. (Ed. M.V. Brian). pp. 245–92. (Cambridge University Press: Cambridge.)

Yeates, G.W. (1981). Soil nematode populations depressed in the presence of earthworms. Pedobiologia 22, 191–5.

Zlotin, R.I., and Khodashova, K.S. (1980). 'The Role of Animals in Biological Cycling of Forest-steppe Ecosystems'. (Dowden, Hutchinson and Ross: Stroudsberg, Penn.)

ROLE OF SOIL MICROFLORA IN NITROGEN TURNOVER

J.N. Ladd and R.C. Foster

ABSTRACT

Turnover of nitrogen in soils is closely linked to heterotrophic metabolism, energy generation, and its utilization for biosyntheses. Soil microflora play a dominant role in these processes. Studies of the distribution of immobilized ^{15}N in particle size fractions from dispersed soils, and of native organic N in soils with long histories of cultivation, indicate that during mineralization there is a preferential loss of N from clay-size components especially from particles of diameter <0.2 μm. The proportions of residual organic ^{15}N and ^{14}N accumulating in silt-size fractions and to a lesser extent coarse clay-size fractions increase with time. The microbial biomass is also concentrated in these fractions.

The use of isotope-labelled substrates has demonstrated close direct relationships between clay content of soils and the proportions of decomposing substrate C residual in microbial biomass. Such results are indicative of the important influence of adsorption reactions and perhaps of soil structure on the turnover of C and N through the microbial biomass pool. Electron microscopy studies have demonstrated the spatial heterogeneity of microorganisms and substrates in aggregated soils, and their visual evidence supports the view that physical factors, including pore space, influence the survival of the soil biota and their accessibility to substrates.

INTRODUCTION

The processes of mineralization and immobilization of N together constitute the internal heterotrophic subcycle of N in soils (Jansson 1958: Jansson and Persson 1982). Mineralization denotes the overall process by which organic N is transformed to NH_4^+-N, as a result of the concerted catabolic activities of the soil fauna and microflora. The opposing and intrinsically-linked process of immobilization denotes the conversion of inorganic N (NH_4^+, NO_3^-, NO_2^-) to organic forms due to synthetic reactions associated mainly with microbial growth and metabolism.

The key to our understanding of the relationships between both processes is the recognition of the close link between C and N metabolism in soils, and the turnover of C and N through the decomposer biomass pool (Fig. 1). Energy and growth substrates generated by heterotrophic metabolism are utilized to increase the biomass and hence N demand of decomposer populations. Cell metabolites and lytic products after cell death provide substrates for further turnover of organic and inorganic N via the biomass N pool.

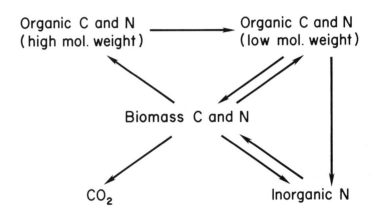

Fig. 1: Carbon and nitrogen turnover in soil.

Predictions of net mineralization or net immobilization of N during the early metabolism of substrates added to soils and based solely on the total C and N contents of the amendments, can be approximately correct. Indeed the general applicability of the C:N ratio "rule of thumb" tends to support the view that most, if not all, of the N of primary substrates (e.g. plant materials) entering soils is readily available for reactions within mineralization–immobilization turnover processes, and that the more stable, organic nitrogenous residues which accumulate in soils are essentially microbial in origin (Bartholomew 1965). Nevertheless, in different ecosystems, substrates will range widely in the proportions of their C readily metabolized by the soil biota. The competitive activities, chemical composition and survival of the soil biota will also vary, and thus influence efficiencies of C use, N demand, and the availability and nature of substrates resulting from biomass C and N turnover. Consequently some departures from the simple guide to net mineralization or immobilization based on C:N ratios of substrates are to be expected.

Twenty years ago McLaren (McLaren and Petersen 1967) remarked on several characteristics of soils which directly relate to the role of the soil microflora in N turnover. Firstly, soil microorganisms are highly diverse and many are opportunistic, with capacities for rapid growth and increases in specific activities to exploit temporary availability of substrates. Reaction rates will be directly influenced by temperature, moisture, salinity, etc., and by the amounts and amenability to biological attack of the diverse range of organic substrates which enter or are formed in soils. Secondly, the soil environment is heterogeneous with structural elements ranging from the gross particle to molecular dimensions. Within the soil matrix, organisms and their enzymes function in a range of changing physicochemical environments with point-to-point variation in substrate concentrations and at solid-liquid interfaces where there are gradients in pH and redox potential.

Considerations of the physical accessibility of substrates to enzymes and/or microorganisms, and of pore space in relation to the growth and survival of the soil biota, have in recent years enhanced interest in soil structure and its effect on biological activities, including C and N turnover. Electron microscope studies have demonstrated the spatial and morphological heterogeneity of microorganisms and substrates as they occur in soil in situ (Plate 1), and have contributed significantly to their characterization.

ORGANIC SUBSTRATES

Nature of the substrates

Plant tissues, exudates and leachates, and animal tissues and excreta comprise at any one time a small but biologically-labile fraction of soil organic C and N. The remainder is present in the cells of decomposer populations (overall of an intermediate biological lability) and in relatively stable organic residues of mainly microbial origin. Data for the amounts of readily-decomposable organic materials added annually to soils probably underestimate root-derived C and N. For example, Martin (1987) has reported that rhizosphere CO_2, i.e. CO_2 evolved from root respiration plus microbial degradation of root material during wheat growth to the flowering stage, was at least double that evolved due to decomposition of accumulated soil organic matter.

Up to 50% of the total organic C and N in soils can be chemically defined, but only as products of various degradative procedures (Tables 1 and 2). Thus analysis of products permits assumptions regarding the classes of compounds from which they were derived but does not distinguish their sources. Amino acids bound in proteins or peptides are consistently a major source of chemically defined organic N.

Location of organic substrates in soils

Dispersed soils

Depending upon soil type, the organic N content of particle size

Table 1: Soil organic carbon.

Compound			% of total organic C
1. Chemically defined			25–35
Carbohydrates		5–25	
Amino sugars	2–6		
Uronic acids	1–5		
Hexose sugars	4–12		
Pentose sugars	<5		
Proteins, peptides, amino acids		12–18	
Heterocyclic compounds		<1	
Lipids		2–6	
2. Humic materials			65–75

Table 2: Soil organic nitrogen.

Compound		% of total organic N
1. Chemically defined		35–55
Amino acids	30–45	
Amino sugars	5–10	
2. Acid–soluble, but unknown		10–20
3. Acid–insoluble		20–35

fractions either increases regularly with decreasing particle size, or maximizes in particles of c. 1–30 μm diameter (Chichester 1969; Ladd et al. 1977b; Turchenek and Oades 1979; Anderson et al. 1981; Christensen and Sørensen 1986). C/N ratios of fractions decrease with decreasing particle size.

Short–term (up to 5 months) and moderate term (5 years) studies of the distribution of immobilized ^{15}N in particle size fractions consistently demonstrate that clay size particles (<2 μm) are of higher ^{15}N atom % enrichments than are silt– and sand–size particles (Ladd et al. 1977b,c; Christensen and Sørensen 1986). The clay–size particles contain higher proportions of immobilized ^{15}N than they do of native organic [14]N (Table 3), whereas the reverse is true of the coarser size fractions. During periods of rapid net mineralization of N, the organic ^{15}N of the fine clay–size fractions (<0.2 μm) consistently decline whereas either gains or losses are recorded for the organic ^{15}N of silt– and coarse clay–size fractions, depending upon soil and carbon amendment (Ladd et al. 1977c).

Table 3: Percentage distribution of immobilized ^{15}N and native organic ^{14}N in particle size fractions from dispersed soils.

Soil	Isotope	Distribution (%) of org. ^{15}N and ^{14}N				Ref.
		<0.2 μm	0.2–2 μm	2–50 μm	>50 μm	
1. (13% clay)[1]	^{15}N	15.0	56.4	24.4	1.0	Ladd et al
	^{14}N	12.2	27.7	48.9	5.7	1977c
2. (47% clay)[1]	^{15}N	30.5	48.0	18.7	1.0	
	^{14}N	21.4	44.6	30.1	2.4	
		<2 μm	2–20 μm	20–2000 μm		
1. (12% clay)[2]	^{15}N	91.0	8.4	0.6		Christensen
	^{14}N	69.0	21.0	1–2		& Sørensen
						1986
2. (46% clay)[2]	^{15}N	88.8	11.3	n.d.		
	^{14}N	74	16.0	1–2		

1. Soils incubated with glucose and $^{15}NO_3^-$ for 160 days; 2. Soils incubated with hemicellulose and $^{15}NH_4^+$ for 5 years

The percentage distribution of ^{15}N and ^{14}N in particle size fractions suggests that in the long term, the silt fraction will accumulate an increasing proportion of the more stable nitrogenous residues. This view is supported by Tiessen and Stewart (1983) who demonstrated a preferential loss of organic matter from fine clay (<0.2 μm) material of a soil under long-term (60 years) cultivation. Thereafter the decline in organic matter of this fraction was small, reaching a new equilibrium level after 90 years of cultivation at approximately one half of its original C and N content. The proportions of soil organic matter residual in coarse clay- and fine silt-size fractions increased with time of cultivation.

Aggregated soils

The precise locations of organic compounds within natural undisturbed soils are largely unknown. Organic materials in aggregated soils are recognized at the ultrastructural level by their morphology, cytology and histochemistry. They may range in size from relatively unaltered histons of newly-incorporated plant and animal tissues (hundreds of microns in diameters) down to particles and fibrils of macromolecular size (1 nm in diameter). Fragments of tissues and cells can be recognized and identified (Plate 1). For example, plant cell wall fragments of submicron size can be distinguished from fragments of insect cuticle. Nevertheless, ultrastructural studies of organic matter in soils are largely limited to microbial cells, cell wall fragments, and insoluble, often polymeric materials such as microbial and root polysaccharides, polyphenolics and humic substances.

Detection of organic materials in soils by electron optical methods requires either that the material be coated with a heavy metal (SEM) or impregnated with heavy metal salts (TEM) (Foster and Martin 1981). The most commonly used "stains", Os and Mn, do not react with carbohydrates but mainly react with polyphenolics, unsaturated lipids and with amino acids (and hence proteins) which contain -SH or aromatic groups. Nevertheless, histochemical tests do exist for different classes of carbohydrates and for specific monosaccharides. Foster (1981a) used a Ru/Os test for microbial- and root-derived exocellular polysaccharides many of which are known to contain amino sugars. Further, neutral carbohydrates have been depicted not only in cells and cell wall fragments but as coatings on clay particles and as submicron-size deposits in microvoids within clay fabrics (Foster 1981b).

The most numerous cellular remnants in soils consists of cell wall fragments in various stages of physical comminution and biochemical degradation. Exocellular polysaccharides are usually the most easily broken down of the materials associated with cells and tissues; those from fungi may be more resistant to decay than those produced by bacteria and roots (Foster 1981a). Next to be degraded are the carbohydrate-rich layers of cell walls, unmasking phenolic groups. Remnants appear often as polylamellate structures composed of the more resistant lignified layers (Foster et al. 1983). How far the removal of nitrogenous compounds is concurrent with carbohydrate removal is unknown but probably accompanies it, although some preservation of cytoplasmic and cell wall N may result from tanning reactions.

Except for enzymes, there are no morphological or ultrahistochemical tests to locate nitrogenous compounds in soils, and none are available for those enzymes directly involved in N turnover processes. Enzymes such as phosphatases have been located in soil microorganisms, microbial exocellular polysacharide gels, root surface gels and plant tissue fragments (Foster and Martin 1981; Foster 1985). Since all techniques use heavy metals to confer electron density and hence detectability on the products of enzyme action, false positive reactions also occur due to non-specific adsorption of the heavy metals to humified organic matter. This has prevented conclusions on the occurrence and distribution of colloid-bound enzymes in the soil matrix.

It has been argued that enzymes acting exocellularly in soil microenvironments against substrates of high molecular weight must themselves be vulnerable to losses of activity, either due to proteolysis, or to adsorption to soil colloid surfaces, or to other reactions with soil humic substances. Continued activity against high molecular weight substrates would then depend on the availability of energy sources to allow enzyme synthesis, de novo (Ladd 1978). However, colloid-bound enzymes may be stabilized and protected from biological attack, and may retain some activity towards substrates of relatively low molecular weight. Evidence with added protein and peptide substrates supports this notion (Ladd and Paul 1973). Nevertheless, it is yet to be established that colloid-bound enzymes have an important role in N turnover processes, i.e. when acting on natural substrates in the concentrations and locations encountered in soil aggregates.

THE SOIL MICROFLORA

General description of the microorganisms

Up to 5% of soil organic N and 3% of soil organic C are present in cells and tissues of the soil biota. Of the soil microflora, bacteria (0.5–2 μm diameter) are the most numerous, whereas fungi (3–50 μm diameter) may account for most of the soil biomass (Anderson and Domsch 1978), especially in acidic soils. Jenkinson et al. (1976) reported that if spherical organisms in soils are divided logarithmically into equal groups according to organism volume, then each group contains the same total biovolume.

Reviews (Gray 1976; Lynch 1979, 1983; Foster 1987), summarise many features of the ecology of microorganisms in soil. The heterotrophic microflora of soils are very diverse; all known types of microorganisms occur, although the bacteria and fungi receive most attention. Variations occur in the regional and seasonal distribution of microorganisms in soil, in species diversity in the soil profile, and in the patterns of heterotrophic succession on decomposing plant residues.

Soil microflora are the prime decomposers of organic substrates, contributing more than 90% of the net energy flux in soil, and are the most important mediators of metabolic turnover of N. Organisms have been characterised as autochthonous, i.e. able to maintain low relatively-constant activities towards the more resistant components of soil organic matter; and zymogenous, i.e. able to respond rapidly to spasmodic influxes into soil of readily metabolizable organic substrates, but with the depletion of such energy sources, returning to dormant states of lowered metabolic activity.

Location of microorganisms in soil

Dispersed soils

Soil aggregates provide a range of microhabitats for the microflora. Hattori (1969) proposed that bacteria in aggregates could be categorised into two groups; (1) those on the surface and in the wider pores, viz. in the outer part of aggregates, and (2) those in narrower pores, viz. in the inner part of aggregates. Cells in the first category were counted and identified from aqueous suspensions of soils after repeated (10-20 times) shaking of the aggregates in sterile water and settling for 1 minute. Cells in the second category were counted and identified from aqueous suspensions after ultrasonically dispersing the repeatedly washed aggregates.

Bacteria in the outer parts of aggregates are more responsive to the supply of soluble substrates and to air-drying, are affected sooner by fumigation with ethylene dibromide, and are more sensitive to bactericides than are cells located internally in the aggregates (Hattori 1973).

Identification of microbial isolates indicate that fungi exist preferentially in the outer parts of aggregates, bacteria and actinomycetes preferentially in the inner parts. Bacteria in the outer parts are mainly spore formers, those present internally are mainly as vegetative forms.

Ladd et al. (1977a-c) used a more vigorous technique to disperse soils containing immobilized ^{15}N, after incubation of the soils with $^{15}NO_3^-$ and either glucose or wheat straw to promote microbial growth. Physical fractionation of soils demonstrated that microbial biomass ^{15}N and ^{14}N were associated with a light fraction (S.G. <1.59) in straw-amended soils, and a silt-size fraction (2-50 μm) in both glucose- and straw-amended soils. The coarse clay-size fraction (0.2-2.0 μm) of the glucose-amended soils also contained significant amounts of biomass ^{15}N.

In a later experiment soil was incubated with ^{14}C-, ^{15}N-labelled plant material, then dispersed and physically fractionated (Amato and Ladd 1980). The particle size fractions were assayed directly for biomass ^{14}C using the fumigation-incubation technique (Jenkinson and Powlson 1976). Dispersion and fractionation of the soil decreased biomass ^{14}C by 50%; most of that recovered was present in the coarse clay-size fraction, the least in the fine clay-size fraction.

Ahmed and Oades (1984) used ultrasound to disperse a sandy loam soil, with only 10% loss in ATP content, and a clay soil, with 70% loss in ATP. The recovered biomass (Table 4), based on ATP content, was concentrated in the finer soil fractions (mainly 0.5-5 μm), and especially in the silt-size fraction (2-5 μm).

Ahmed and Oades (1984) also fractionated soils which had been inoculated individually with strains of fungi and bacteria and incubated with ^{14}C-glucose. The distribution of residual organic ^{14}C in size fractions was independent of inoculum type but was affected by soil dispersion technique. Using a mild technique to partly disperse the soil it was found that a substantial proportion of residual ^{14}C remained in macroaggregates (>250 μm) due, they concluded, to the multiplication of added cells in pores associated with aggregate surfaces. More energetic dispersion methods caused a

Table 4: Distribution of ATP in fractions of different particle size obtained after ultrasonic dispersion of soils (adapted from Ahmed and Oades 1984).

Particle size	Sandy–loam soil	Clay soil
μ*m*	*% of total recovered ATP*	
>53	1.9	3.6
20–53	2.2	5.0
15–20	1.6	2.2
10–15	3.4	2.6
5–10	5.3	10.7
2–5	64.1	20.5
1–2	1.7	18.6
0.5–1	7.6	23.4
0.1–0.5	7.8	11.2
<0.1	4.5	1.9

redistribution of labelled microbial cells and metabolites in silt- and clay-sized fractions, until eventually cells were disrupted and cellular contents and debris were spread over particle surfaces.

The results obtained with dispersed soils are in general accord despite differences in soil properties, size range of fractions, and in the techniques applied for soil dispersion. Evidence indicates that microbial biomass which survives the fractionation procedure is concentrated in particles mainly in the 0.5–5 μm size range. However, studies of this type do not reveal the distribution of C and N immobilized in microbial cells in soil <u>in situ</u>. This is best achieved by microscopy studies of soil aggregates.

Aggregated soils

Electron microscopy studies of the soil microflora <u>in situ</u> require that the cytoplasmic structures of the organisms are stabilized by pre-fixing with organic aldehydes or by rapid freezing, and that their organic components are rendered electron-dense by using heavy metal reagents. In bulk soil, microorganisms are not uniformly distributed but congregate near suitable food sources, such as faecal material or remnants of plant and animal tissues, and in pores which usually are sufficiently large to contain them comfortably (Plate 1). Kilbertus (1980) indicated that the diameters of pores and of the bacteria or colonies contained within, were consistently of a ratio approximately 3:1.

Soil fungi are generally associated with decomposing plant materials and with root surfaces and, in bulk soil, are found in the larger pores between aggregates. Scanning electron microscopic (SEM) studies enhance the view that fungi dominate the soil microflora because during preparation of soils for

SEM, soils tend to fracture along surfaces of weakness between aggregates where fungi are located. Fungi growing in voids larger than the characteristic hyphal diameters assume normal dimensions and cylindrical shape, but in smaller voids fungal shapes become irregular and tend to conform to the space available (Foster 1987). Protozoa, e.g. amoebae (Plate 2a), and ciliates accumulate in pores within rhizospheres and along mycelial strands. Amoebae, although several microns in diameter, are able to produce pseudopodia less than a micron in diameter and thus reach into fine micropores where bacteria are concealed.

Bacteria and actinomycetes are the most numerous of the soil microflora but constitute a minority of the microbial biomass. Different types of bacteria can be distinguished by their morphology (shape, size, presence of flagellae), cell wall fine structure, cytoplasmic ultrastructure and the nature of exocellular polysaccharides (Foster 1987). In bulk soil, bacteria are found in and on faecal materials and cell wall remnants of animal and plant tissues (particularly those containing carbohydrates), and with dead (but not live) fungal hyphae, and in soil pores.

Bacteria tend to occupy pores within aggregates; their frequent ocurrence in these locations is revealed by transmission electron microscopy (TEM) of soil sections (Kilbertus 1980; Foster 1987). The pores may have single or multiple entrances and be connected by narrow convoluted channels (Plate 1); the bacteria may occur as single cells or, more often, in small colonies of one cell type (especially Bacillus and Arthrobacter sp.). Sometimes bacteria appear to be completely enclosed in a clay fabric, although they do not fill the pore voids (Kilbertus 1980). Individual bacteria or whole colonies may also become coated with close-fitting clay platelets which may assume different orientations to the bacterial surface (Kilbertus and Reisinger 1975; Foster et al. 1983).

Whereas fungi dominate the microbial biomass of bulk soil, bacteria dominate the rhizosphere biomass (van Vuurde and Schippers 1980). Rhizosphere bacteria predominantly are Gram negative Pseudomonads, and are of larger size than bacteria in the bulk soil, containing higher contents of storage carbohydrates (Foster 1985). Bacterial colonies in the rhizosphere are separated from each other by their own extracellular polysaccharides or by layers of compacted clay minerals (Foster et al. 1983).

SOIL MICROFLORA AND CARBON AND NITROGEN TURNOVER

Susceptibility of organic matter to degradation

Rates of C and N turnover in soils are determined by the activities, organization, and population dynamics of diverse biological communities. The susceptibility of organic matter to decomposition will be influenced by its solubility and by its chemical properties, for example its degree of polymerization and structural complexity, and by the extent of its induration

122

by polyphenolics or metals which act as enzyme inhibitors. Organic matter may also be physically protected from microbial decomposition. Separation of insoluble organic matter and decomposer organism by a few microns is sufficient to prevent induction of the appropriate exocellular enzymes required for substrate solubilization and subsequent absorption and metabolism by cells.

Mucilage may accumulate between root surface and compacted clay, the latter forming a zone of minimal voids through which bacteria are unable to penetrate, thus temporarily delaying mucilage decomposition (Foster et al. 1983). In other instances, mucilage may squeeze into voids of submicron size too small for even bacteria to enter (Foster 1981a,1985). It is commonly observed that organic particles such as cell wall fragments become coated by continuous cutans of dense clay which may be several microns thick (Plate 2b) (Foster et al. 1983). It seems likely that such organic material will resist microbial attack until the protective cutans are mechanically broken, perhaps after ingestion by soil macrofauna.

Microbial biomass C and N

An important feature of several simulation models of C and N turnover in soils is the passage of C and N through the microbial biomass pool, the turnover resulting from the growth, death and decay of decomposer organisms. Various techniques permit measurements of the amounts of C and N present exclusively in the soil biota (Jenkinson and Ladd 1981; Shen et al. 1984; Brookes et al. 1985, Amato and Ladd 1987). Such methods have been used in studies where biomass C and N respectively are treated as single, undifferentiated compartments of the soil organic pool, without discrimination between C and N of the different classes of soil microflora, or exclusion of C and N of micro- and meso-fauna.

A widely-used method for estimated biomass C and N in soils is a bioassay based on the flush of CO_2 and inorganic N derived from the decomposition of cellular material following cell death after soil fumigation (Jenkinson and Powlson 1976; Shen et al 1984). This technique, like others, gives approximate estimates only of biomass C and N in soils, since values are to some extent influenced both by cell type and soil properties. However, it has the advantage that it allows measurements of changes in the net amounts of biomass ^{14}C and ^{15}N in soils accompanying the decomposition of specific, isotope-labelled substrates.

Figures 2 and 3 respectively show that the amounts of residual organic C and N, and biomass C and N, derived from decomposing plant residues are influenced by both substrate and soil properties. Differences were established early and maintained over 3 years.

Maximal concentrations of biomass ^{14}C were generally present 0.5-1.0 years after incorporation of labelled plant materials in topsoils, and at times significantly later than the phase of rapid decomposition, which occurred soon (≤ 0.1 year) after commencement of the experiment. The same trends were obtained for both C and N, with one exception, viz. the N of wheat straw (C:N ratio, 48:1) remained essentially fully immobilized during the phase of rapid decomposition. By contrast, N of legume material (C:N ratio, 11:1) mineralized relatively rapidly from the outset.

Influence of substrate

Figure 2 compares the decomposition in a sandy loam soil of ^{14}C and ^{15}N-labelled wheat straw and legume tops material. Organic residues accounted for higher proportions of added wheat straw ^{14}C and ^{15}N than they did of legume material ^{14}C and ^{15}N. However, biomass ^{14}C from decomposing wheat straw accounted for less of input ^{14}C than did biomass ^{14}C derived from legume tops.

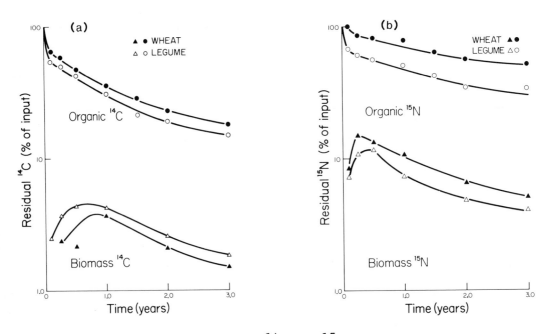

Fig. 2: Residual organic and biomass ^{14}C and ^{15}N in a sandy loam, accompanying the decomposition of (a) ^{14}C-labelled and (b) ^{15}N-labelled wheat straw and legume tops under field conditions (J.N. Ladd et al., unpublished data).

The differences in the proportions of input ^{14}C present in organic residues and biomass during the initial rapid decomposition phase would be explicable if wheat straw contained smaller proportions of its carbon in readily-decomposible forms than did legume material. However, two observations suggest that the differences, even during this early stage, were due in part to differences in turnover of C through decomposer populations.

Firstly, the differences were established early, and remained throughout the experiment. Secondly, wheat straw ^{15}N, unlike legume material ^{15}N, was completely immobilized during the period of rapid decomposition.

To speculate, the type and chemical nature of the decomposer microflora may have been influenced by the type of plant amendments, giving rise to differences in C utilization efficiencies, N demands, viability of populations, and decomposability of cell debris and products. Since N derived from wheat straw remained immobilized during the early stage of decomposition, it suggests that at that time N may have been limiting cell synthesis. Whilst this itself may not have substantially affected rates of C mineralization, it may well have led to increased proportions of the energy derived from heterotrophic metabolism in wheat straw-amended soils being lost as heat. By contrast, in legume-amended soils where N was not limiting synthesis reactions, there was probably potential for greater efficiency of C utilization. Whilst a higher efficiency of C use for cell synthesis, and perhaps a greater viability of decomposer populations in legume-amended soils may each have been factors promoting the observed greater biomass ^{14}C of these soils, they also would have led to a greater retention of ^{14}C in organic residues, which clearly was not the case. Thus the influences of these factors, if operating, need to be offset by the third process of the turnover cycle, viz. decomposition of microbial residues. A greater rate of decay of legume-derived, microbial residues (due perhaps to a lower C:N ratio of cell debris in these soils) would have led to the observed higher biomass ^{14}C and the lower total residual organic ^{14}C than observed in wheat straw-amended soils.

Influence of soil

The proportions of plant-derived ^{14}C and ^{15}N present in organic residues, and in the cells of decomposer organisms, were greater in a clay soil than in a sandy-loam soil (Fig. 3). The effects were apparent after 0.1y when the soils were first sampled, and remained throughout suggesting that the differences were due to soil conditions affecting the turnover of C and N through the soil biomass. The differences between the two soils could be related to a higher efficiency of substrate C (and hence N) use for cell synthesis, and/or a greater viability of decomposer cells in the clay soil compared with in the sandy-loam. The results are not solely explicable in terms of a greater stability of cell products in the clay soil since this would have resulted in lower proportions of substrate C and N being present as biomass C and N, not greater as observed.

The data in Fig. 3 exemplify the general observation that soils of high clay content retain greater proportions of applied substrate C and N immobilized in microbial biomass than do lighter textured soils. Sørensen's (1983) data are an exception. In most cases comparisons have been of a few soils only (e.g. Ladd et al. 1977a,1985; Sørensen 1983; Merckx et al. 1985; van Veen et al. 1985).

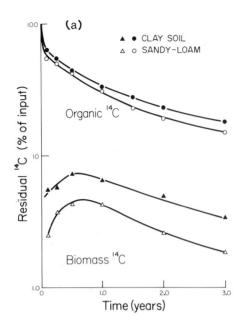
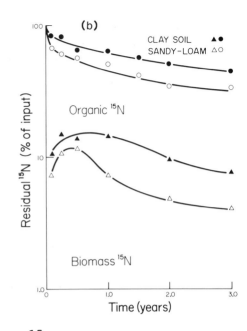

Fig. 3: Residual organic and biomass ^{14}C and ^{15}N in a clay soil and a sandy loam, accompanying the decomposition of (a) ^{14}C-labelled and (b) ^{15}N-labelled legume tops under field conditions (J.N. Ladd et al., unpublished data).

Latterly we have placed the relationship between soil texture and the proportions of substrate C and N residual in microbial biomass on a firmer basis by comparing results from 25 soils, ranging in clay content from 2 to 73%. Each soil was incubated with added ^{14}C-labelled glucose and plant material for about one year. Table 5 shows that residual biomass ^{14}C from the metabolism of each substrate was directly correlated with soil clay content (and cation exchange capacity) and soil pH, but not with soil organic matter content. Total biomass C was directly correlated with clay content, cation exchange capacity, and organic matter C and N content of the soils.

Clay particles are known to protect bacteria from predators and the effects of soil drying (Roper and Marshall 1974,1979; Marshall 1975). Elliott et al. (1984) have reiterated the important dual role of clay content in determining soil structure and, in poorly-structured soils, in influencing pore size distribution. Soils of finer texture generally have more smaller pores than do coarser textured soils. Elliott et al. (1980) demonstrated that energy flow in a soil microcosm, which included bacteria, amoebae and nematodes, was highly affected by soil texture. They postulated that in the finer textured soil there were fewer habitable pores for the larger predator, the nematodes. Consequently, in this soil, nematode growth became more dependent upon the presence and performance of the amoebae which were able to enter soil pores inaccessible to nematodes, to feed on the bacteria within, and to re-emerge as food for the nematodes.

Table 5: Correlations between biomass ^{14}C and some properties of 25 soils each incubated with ^{14}C-labelled leguminous plant material or glucose. (M. Amato and J.N. Ladd, unpublished data).

Substrate	Soil property (x)[1]	a[1]	b[1]	r	Significance
[14]C-labelled *Medicago littoralis* leaves	Clay content	0.25	0.10	0.91	<0.001
	Cation exchange capacity	0.67	0.009	0.94	<0.001
	Organic C content	1.99	0.51	0.13	N.S.
	pH	−7.06	1.34	0.65	<0.001
[14]C-labelled glucose	Clay content	0.13	0.18	0.89	<0.001
	Cation exchange capapcity	0.83	0.016	0.94	<0.001
	Organic C content	2.10	1.71	0.25	N.S.
	pH	−14.70	2.65	0.71	<0.001

1. $y = a + bx$ where y is biomass [14]C (% of input), and x is clay content (% by weight), cation exchange capacity ($m.mol^+/kg$), organic C content (% by weight), or pH.

Electron microscopy reveals that rhizospere bacteria at the surface of roots were lysed following fumigation with chloroform (Plate 3a), whereas cells surrounded by gels and polysaccharide, either their own or produced by the root, survived (Plate 3b) (Martin and Foster 1985). Similarly, microorganisms located deep within pores are not lysed after fumigation with chloroform vapour (Plate 3c) (Martin and Foster 1985). That some soil organisms survive fumigation under the conventional assay conditions is well-known. Paradoxically the activities of some survivors may be critical for the mineralization of C of neighbouring cells located nearer the entrances of pores (or in microcolonies coated with clay platelets) and which have been killed by fumigation. Such C may be inaccessible to attack by cells of an inoculum.

Data of the type shown in Figures 2 and 3 are useful in testing simulation models of C and N turnover in soils. Although C and N pools have been conceptualized and defined, measurements of the sizes of pools as discrete entities and of the rate constants for transfer between pools are not easily achieved. Improved methodology is needed to establish better parameter values, which in many cases are simply assumed. Biomass C and N clearly represent pools of biological significance in models describing C and N behaviour in soils. Short-term differences in the rates of C and N turnover in soils of contrasting texture can be ascribed to differences in the preservation of the decomposer populations and in the proportions of microbial decay products remaining in the near vicinity of surviving cells (van Veen et al. 1985,1987).

REFERENCES

Ahmed, M., and Oades, J.M. (1984). Distribution of organic matter and adenosine triphosphate after fractionation of soils by physical procedures. Soil Biol. Biochem. 16, 465–70.

Amato, M., and Ladd, J.N. (1980). Studies of nitrogen immobilization and mineralization in calcareous soils. V. Formation and distribution of isotope–labelled biomass during decomposition of ^{14}C– and ^{15}N–labelled plant material. Soil Biol. Biochem. 12, 405–11.

Amato, M., and Ladd, J.N. (1987). An assay for microbial biomass based on ninhydrin–reactive nitrogen in extracts of fumigated soils. Soil Biol. Biochem. in press.

Anderson, D.W., Saggar, S., Bettany, J.R., and Stewart, J.W.B. (1981). Particle size fractions and their use in studies of soil organic matter: I. The nature and distribution of forms of carbon, nitrogen and sulfur. Soil Sci. Soc. Am J. 45, 767–72.

Anderson, J.P.E., and Domsch, K.H. (1978). Mineralization of bacteria and fungi in chloroform–fumigated soils. Soil Biol. Biochem. 10, 207–13.

Bartholomew, W.V. (1965). Mineralization and immobilization of nitrogen in the decomposition of plant and animal residues. In 'Soil Nitrogen'. (Eds W.V. Bartholomew and F.E. Clark). pp. 285–306. (Am. Soc. Agron.: Madison, Wisc.)

Brookes, P.C., Landman, A., Pruden, G., and Jenkinson, D.S. (1985). Chloroform fumigation and the release of soil nitrogen: a rapid direct extraction method to measure microbial biomass nitrogen in soil. Soil Biol. Biochem. 17, 837–42.

Chichester, F.W. (1969). Nitrogen in soil organo–mineral sedimentation fractions. Soil Sci. 107, 356–63.

Christensen, B.T., and Sørensen, L.H. (1986). Nitrogen in particle size fractions of soils incubated for five years with ^{15}N–ammonium and ^{14}C–hemicellulose. J. Soil Sci. 37, 241–7.

Elliott, E.T., Anderson, R.V., Coleman, D.C., and Cole, C.V. (1980). Habitable pore space and microbial trophic interactions. Oikos 35, 327–35.

Elliott, E.T., Coleman, D.C., Ingham, R.E., and Trofymow, J.A. (1984). Carbon and energy flow through microflora and microfauna in the soil subsystem of terrestrial ecosystems. In 'Current Perspectives in Microbial Ecology'. (Eds M.J. King and C.A. Reddy). pp. 424–33. (Amer. Soc. Microbiol: Washington, D.C.)

Foster, R.C. (1981a). The ultrastructure and histochemistry of the rhizosphere. New Phytol. 88, 263–73.

Foster, R.C. (1981b). Polysaccharides in soil fabrics. Science 214, 665–7.

Foster, R.C. (1985). In situ localization of organic matter in soils. Quaestiones Entomologicae 21, 609–33.

Foster, R.C. (1987). Microenvironments of soil microorganisms. Biol. Fert. Soils (in press).

Foster, R.C. and Martin, J.K. (1981). In situ analysis of soil components of biological origin. In 'Soil Biochemistry'. (Eds E.A. Paul and J.N. Ladd). Vol. 5, pp. 75–110. (Marcel Dekker Inc.: New York.)

Foster, R.C., Rovira, A.D., and Cock, T.W. (1983). Ultrastructure of the root–soil interface. (Amer. Phytopath. Soc.: St Paul, Minnesota.)

Gray, T.R.G. (1976). Survival of vegetative microbes in soil. In 'The Survival of Vegetative Microbes'. (Cambridge University Press: Cambridge.), 26th Sympos. Soc. Gen. Microbiol., pp. 327–64.

Hattori, T. (1969). Fractionation of microbial cells in soil aggregates. Soil Biol. 11, 30–1.

Hattori, T. (1973). 'Microbial Life in the Soil'. (Marcel Dekker Inc.: New York.)

Jansson, S.L. (1958). Tracer studies on nitrogen transformations in soil. Ann. Roy. Agric. Coll., Sweden 24, 101–361.

Jansson, S.L., and Persson, J. (1982). Mineralization and immobilization of soil nitrogen. In 'Nitrogen in Agricultural Soils'. (Ed. F.J. Stevenson), pp. 229–52. (Am. Soc. Agron.: Madison, Wisc.)

Jenkinson, D.S., and Ladd, J.N. (1981). Microbial biomass in soil: measurement and turnover. In 'Soil Biochemistry'. (Eds Paul, E.A. and Ladd, J.N.). Vol. 5, pp. 415–71. (Marcel Dekker, Inc.: New York.)

Jenkinson, D.S., and Powlson, D.S. (1976). The effects of biocidal treatments on metabolism in soil. V. A method for measuring soil biomass. Soil Biol. Biochem. 8, 209–13.

Jenkinson, D.S., Powlson, D.S., and Wedderburn, R.W.M. (1976). The effects of biocidal treatments on metabolism in soil. III. The relationship between soil biovolume, measured by optical microscopy, and the flush of decomposition caused by soil fumigation. Soil Biol. Biochem. 8, 189–202.

Kilbertus, G. (1980). Etude des microhabitats contenus dans les agregats du sol, leur relation avec la biomasse bacteriene et la taille des procaryotes presents. Rev. Ecol Biol Sol. 17, 543–57.

Kilbertus, G., and Reisinger, O. (1975). Degradation du material vegetal activite in vitro et in situ de quelques microorganismes. Rev. Ecol. Biol. Sol. 12, 363–74.

Ladd, J.N. (1978). Origin and range of enzymes in soil. In 'Soil Enzymes'. (Ed. R.G. Burns), pp. 51–96. (Academic Press: London.)

Ladd, J.N., Amato, M., and Oades, J.M. (1985). Decomposition of plant materials in Australian soils. III. Residual organic and microbial biomass C and N from isotope–labelled legume material and soil organic matter, decomposing under field conditions. Aust. J. Soil. Res. 23, 603–11.

Ladd, J.N., Amato, M., and Parsons, J.W. (1977a). Studies of nitrogen immobilization and mineralization in calcareous soils. III. Concentration and distribution of nitrogen derived from the soil biomass. In 'Soil Organic Matter Studies'. Vol. I, pp. 301–10 (IAEA: Vienna.)

Ladd, J.N., Parsons, J.W., and Amato, M. (1977b). Studies of nitrogen immobilization and mineralization in calcareous soils. I. Distribution of immobilized nitrogen amongst soil fractions of different particle size and density. Soil Biol. Biochem. 9, 309–18.

Ladd, J.N., Parsons, J.W., and Amato, M. (1977c). Studies of nitrogen immobilization and mineralization in calcareous soils. II. Mineralization of immobilized nitrogen from soil fractions of different particle size and density. Soil Biol. Biochem. 9, 319–25.

Ladd, J.N., and Paul, E.A. (1973). changes in enzymic activity and distribution of acid–soluble amino acid–nitrogen in soil during nitrogen immobilization and mineralization. Soil Biol. Biochem. 5, 825–40.

Lynch, J.M. (1979). The terrestrial environment. In 'Microbial Ecology: A Conceptual Approach'. (Eds J.M. Lynch and N.J. Poole), pp. 67–91. (Blackwell Scientific Publications: Oxford.)

Lynch, J.M. (1983). 'Soil Biotechnology'. (Blackwell Scientific Publications: Oxford.)

Marshall, K.C. (1975). Clay mineralogy in relation to survival of soil bacteria. Ann. Rev. Phytopath. 13, 357–73.

Martin, J.K. (1987). Carbon flow through the rhizosphere of cereal crops – A review. Intercol Bulletin (Special edition), in press.

Martin, J.K., and Foster, R.G. (1985). A model system for studying the biochemistry and biology of the root–soil interface. Soil Biol. Biochem. 17, 261–9.

McLaren, A.D., and Peterson, G.H. (1967). Introduction to the biochemistry of terrestrial soils. In 'Soil Biochemistry'. (Eds A.D. McLaren and G.H. Peterson), Vol. 1, pp. 1–15. (Marcel Dekker Inc.: New York.)

Merckx, R., Hartog, A.d., and van Veen, J.A. (1985). Turnover of root–derived material and related microbial biomass formation in soils of different texture. Soil Biol. Biochem. 17, 565–9.

Roper, M.M., and Marshall, K.C. (1974). Modification of the interaction between Escherichia coli and bacteriophage in saline sediments. Microb. Ecol. 1, 1–13.

Roper, M.M., and Marshall, K.C. (1979). The survival of coliform bacteria in saline sediments. Australian Water Resources Council Tech. Paper No. 43.

Shen, S.M., Pruden, G., and Jenkinson, D.S. (1984). Mineralization and immobilization of nitrogen in fumigated soil and the measurement of microbial biomass nitrogen. Soil Biol. Biochem. 16, 437–44.

Sørensen, L.H. (1983). Size and persistence of the microbial biomass formed during the humification of glucose, hemicellulose, cellulose and straw in soils containing different amounts of clay. Plant Soil 75, 121–30.

Tiessen, H., and Stewart, J.W.B. (1983). Particle–size fractions and their use in studies of soil organic matter: II. Cultivation effects on organic matter composition in size fractions. Soil Sci. Soc. Am. J. 47, 509–14.

Turchenek, L.W., and Oades, J.M. (1979). Fractionation of organo–mineral complexes by sedimentation and density techniques. Geoderma 21, 311–43.

Veen, J.A. van, Ladd, J.N., and Amato, M. (1985). Turnover of carbon and nitrogen through the microbial biomass in a sandy loam and a clay soil incubated with [^{14}C(U)] glucose and [^{15}N] $(NH_4)_2SO_4$ under different moisture regimes. Soil Biol. Biochem. 17, 747–56.

Veen, J.A. van, Ladd, J.N., Martin, J.K., and Amato, M. (1987). Turnover of carbon, nitrogen and phosphorus through the microbial biomass in soils incubated with ^{14}C–, ^{15}N–, and ^{32}P–labelled bacterial cells. Soil Biol. Biochem. 19, 559–66.

Vuurde, J.W.L. van, and Schippers, B. (1980). Bacterial colonization of seminal wheat roots. Soil Biol. Biochem. 12, 559–65.

Legends for plates

Plate 1: Transmission electron micrograph (TEM) of an ultrathin section of a clay soil fixed in glutaraldehyde and osmium, showing irregular voids containing bacteria (B), cell wall remnants (CW), amorphous organic material (AO) and faecal material (FM).

Plate 2: (a) TEM showing an amoebae (A) producing fine pseudopodia (P) <1 μm in diameter which penetrate fine pores between cell remnants in a Lepidosperma rhizosphere. Bacteria (B) are scattered amongst cellular remnants. (b) Organic (cellular) material (OM) physically protected from soil microorganisms by inclusion in a dense mineral aggregate.

Plate 3: TEM of rhizosphere bacteria after chloroform treatment (as in the Jenkinson biomass method). (a) Bacteria (B) in the intercellular space not containing mucigel have all lysed: only cell wall ghosts remain. (b) Bacteria deep within mucigel survive although those at the surface show signs of autolysis (↟). (c) Bacterial cells deep within a pore (B) remain apparently alive whereas cells near the mouth of the pore are lysed (↟).

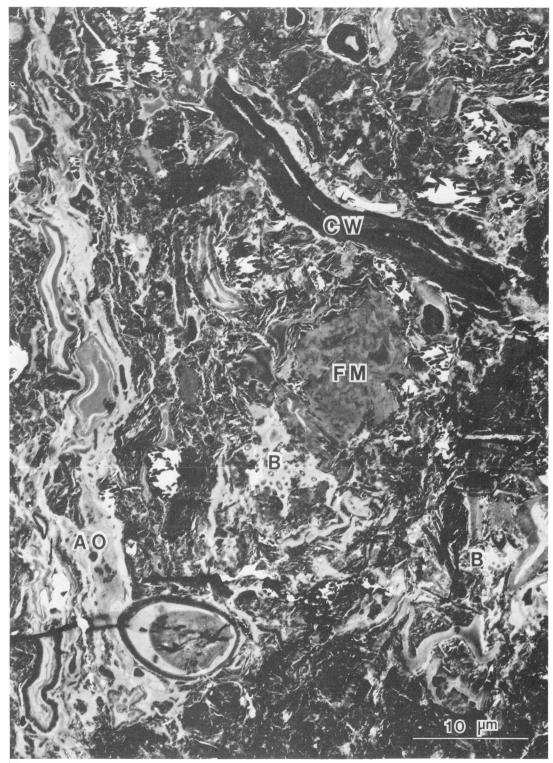

Plate 1

131

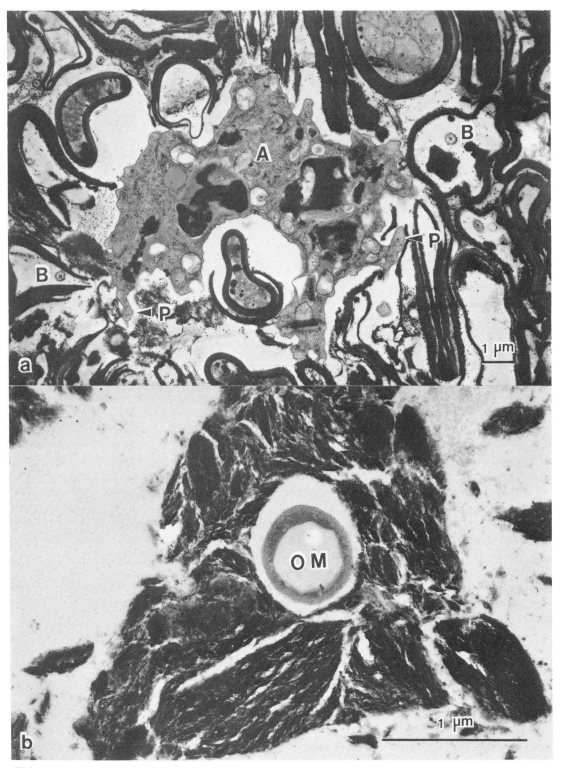

Plate 2

132

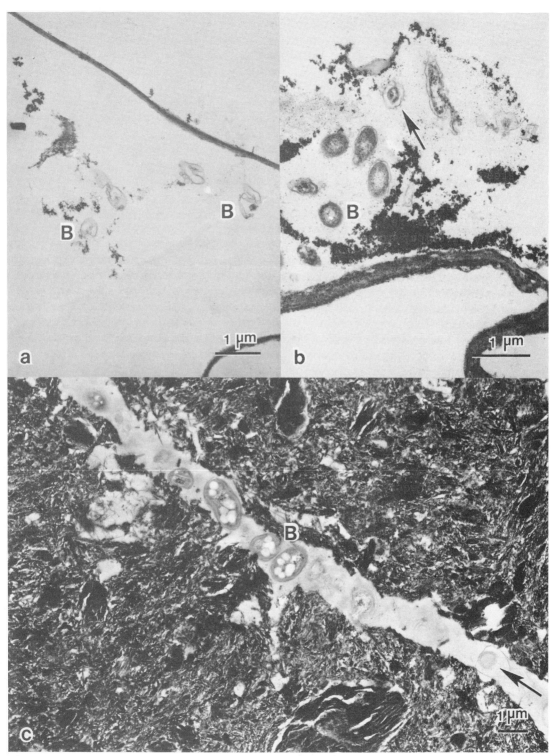

Plate 3

DECOMPOSITION OF SOIL ORGANIC NITROGEN

J.O. Skjemstad, I. Vallis and R.J.K. Myers

ABSTRACT

Recent research on the nature and susceptibility to decomposition of soil organic nitrogen has paid less attention to the chemical composition of soil organic matter and more to the association of the organic fractions with soil inorganic components and to soil structure or architecture. The inherent molecular recalcitrance of humic substances is aided by the formation of complexes with Ca, Fe, Al, polycations of Fe and Al and hydrated clay surfaces. Density and particle size separation are discussed in relation to chemical and biological components of soil organic matter and their role in N cycling. Spatial heterogeneity of soil may isolate soil microorganisms from decomposable substrate.

Notwithstanding the complex nature of soil N cycling, research has continued into the use of simple models of N mineralization with one or two pools of potentially mineralizable N each with characteristic decomposition rate constants estimated by laboratory incubation procedures to calculate N mineralization in the field. Results thus far have been variable and more research is needed before the method can become widely applicable.

INTRODUCTION

Organic matter is added to soils through the input of plant litter which acts as an energy and nutrient source for soil fauna and microflora. The actions of these decomposer organisms result in losses of organic matter through mineralization, the materials remaining being highly modified resistant parts of plant tissues and biosynthesised complex organic molecules (humic materials) varying greatly in chemical nature and molecular weight.

A proportion of the humic materials combines with the soil mineral components to form organo–mineral complexes of varying complexity and stability. These complexes are largely responsible for the increased short term stability of some organic fractions (Sorensen 1975). It has been demonstrated for example that the organic matter in clay soils is more resistant to mineralization than organic matter in light textured soils (Dalal and Mayer 1986a). There is also evidence that a high clay content retards the rate of loss of organic matter in the initial rapid phase of decomposition of added substrate (Ladd et al. 1985) and that the presence of clay minerals facilitates the formation of humic materials (Filip et al. 1972).

The equilibrium condition of soil in a natural stable ecosystem is subject to shifts due to seasonal and other short-term variations in microbial biomass and abiotic factors such as litter composition. Soil temperature, moisture and aeration can cause some shift in the equilibrium although overall levels of organic matter are retained. When cultivated, the equilibrium is quickly shifted resulting in a rapid decline in organic matter levels (Dalal and Mayer 1986a).

Although this decline can generally be defined by a simple equation, the overall rate of decline is the sum of a number of components declining at different rates. The study of soil organic matter and organic N dynamics is therefore a study of the processes which effect the turnover rates of a number of organic fractions or pools which are stabilized to varying degrees against mineralization. The three most significant long term moderators of turnover rates appear to be molecular recalcitrance, direct association of substrate with inorganic ions and/or clay surfaces and physical separation of substrate from the soil fauna and microflora.

Several simulation models express a number of organic pools in terms of chemical or physical stabilization. These pools are largely conceptual however it would be useful to identify more precisely the mechanisms responsible for increased stability and to isolate by chemical or physical means fractions dominated by these mechanisms. This paper discusses chemical and physical fractionation of soil organic matter, the changes in these fractions during decomposition of soil organic matter, and the use of laboratory incubation tests to predict N mineralization in the field.

CHEMICAL SEPARATIONS

A number of chemical agents have been used to extract organic fractions from soil. These include organic solvents, complexing agents, mineral acids and bases. The latter agents such as the sodium salts of tetraborate, pyrophosphate and hydroxide generally extract the most organic matter and have been extensively used. Humic acid, the acid insoluble fraction of alkali extracts, is considered to possess inherent molecular recalcitrance (Hayes and Swift 1978) brought about by a highly cross-linked, aromatic structure high in free radicles. In the soil, this recalcitrance is further aided by the formation of complexes with Ca, Fe, Al, polycations of Fe and Al and hydrated clay surfaces through the carboxyl groups which make up a significant proportion of these humic substances. Humic complexes have also been shown to protect enzymes (Burns et al. 1972) and regulate plant growth (Cacco and Dell'Agnola 1984).

The use of chemical extractants has proved most successful in the study of organic matter in spodic horizons but has been less successful in other soil types. This lack of success is mainly due to the relatively low extractability of organic matter in soils with higher clay contents and the risk of producing artifacts particularly in the highly alkaline environments produced by highly basic reagents.

The fraction remaining after extraction is termed humin. This fraction which has a closer C/N ratio than the soluble fractions comprises largely unaltered litter, the microbial biomass and cell fragments, humic molecules strongly associated with clay and in some soils, charcoal. Although charcoal is not regarded as organic, most methods of analysis for organic carbon will include a part of all of the charcoal fraction. The humin fraction has been shown to be the most stable both under woodland (Campbell et al. 1967) and under exploitive cropping (Skjemstad and Dalal 1987) although the chemical and physical heterogeneity of these fractions ensures that the properties of humin vary considerably from soil to soil.

PHYSICAL SEPARATIONS

Soils may also be fractionated according to physical properties such as the degree of association of the organic fractions with soil inorganic components at the fundamental particle or at a more complex aggregate level.

Density considerations

The density of soil physical components varies according to their chemistry and degree of interaction between organic and inorganic materials. Structural litter which is not associated with significant amounts of inorganic material can be separated at relatively low densities. As organic matter becomes more extensively modified by the humification process, the degree of association with inorganic components increases resulting in a range of organo-mineral complexes whose overall density varies according to the density of the mineral fraction and the carbon content of the resulting complex.

Densimetric separations are made using mixtures of organic solvents of differing densities such as ethanol/bromoform mixtures or concentrated aqueous solutions of highly soluble salts with a high density such as NaI and $ZnBr_2$. Separated factions with a range of densities are reported in the literature but the ranges that appear to be most useful are < 1.6, 1.6–2.0 and > 2.0 Mg/m^3. A number of pretreatments are also reported including fine grinding (Dalal and Mayer 1986c), boiling (Monnier et al. 1962) and ultrasonic vibration (Greenland and Ford 1964). The latter pretreatments ensure a higher yield of light fraction but may also include structural litter trapped within smaller aggregates. The light fraction (generally < 1.6 Mg/m^3) contains material with a high carbon content and wide C/N ratio. Microscopy (Spycher et al. 1983) and ^{13}C NMR spectroscopy (Skjemstad et al. 1986) have demonstrated that the organic matter in this fraction is structural litter which has not undergone significant humification. The intermediate fraction (1.6–2.0 Mg/m^3) consists of cell debris and humified material which are highly extractable with NaOH giving highly aromatic humic acids while the densest fraction (> 2.0 Mg/m^3) contains organic compounds which are more alkyl in nature, have a closer C/N ratio and are in more intimate contact with clay (Skjemstad and Dalal 1987). The humic acids extracted from the two densest fractions are respectively similar to the HA-A and HA-B fractions reported by Anderson and Paul (1984). Changes in the light fraction have been shown to be an early indicator of the consequences of cultural practice (Ford and Greenland 1968; Richter et al. 1975). Fractions with densities as high as 2.4 Mg/m^3 are reported in the literature. At higher densities however, clay mineralogy also influences the fractionation (Spycher and Young 1977) and although heavier fractions may be useful in separating organic materials within a soil, comparisons between soil types may be of limited value.

Particle size considerations

Particle size separations can be made on fundamental particles (clay, silt and sand) by sedimentation after dispersion or on water-stable aggregates. One difficulty with such separations is that the separation procedure may create artifacts. A considerable proportion of the organic matter separated with the coarser fractions may not be associated with the bulk of that fraction. Rather, a range of organic structures in varying degrees of association with the mineral fraction may occur in all size fractions.

Turchenek and Oades (1979) for example, demonstrated that in a number of soils, a considerable proportion of the organic matter in fractions > 0.5 μm could be separated at densities < 2.0 Mg/m^3 and therefore cannot be considered as being intimately associated with the bulk of that fraction. It is also clear that although the bulk of the microbial biomass occurs in fine silt/coarse clay fractions, a significant amount of ATP occurs in the finest clay fraction. McGill et al. (1975) suggest that the organic material in this fraction originated from cytoplasmic constituents liberated by the dispersion technique, although the presence of intact microorganisms in this fraction could not be excluded. A significant proportion of labelled C and N added to soils also appears in this fraction after a relatively short incubation time and is rapidly depleted on further incubation (Amato and Ladd 1980).

In some clay soils, only a small proportion of the carbon from the clay fraction has a density of < 2.0 Mg/m^3 and it is also less extractable (Skjemstad and Dalal 1987) suggesting that these humins are in much closer association with clay surfaces. Skjemstad and Dalal (1987), using ^{13}C NMR and

IR studies, also demonstrated that the clay–bound fraction from some soil types has a high alkyl and amide content but is low in carboxyl groups. These organic polymers therefore have a low charge density even at relatively high pH values. Unlike classical humic materials which have high carboxyl contents and react with clay surfaces and one another through bridging and exchange reactions (Tate and Theng 1980), these molecules have the potential to approach negatively charged surfaces more closely and interact directly through hydrogen bonding and Van der Waal's forces. Although these bonds are weak, they are accumulative and can result in relatively strong associations. This may in part explain the common observation of decreasing C/N ratio and increasing amounts of hydrolyzable N in soil fractions with decreasing particle size (Anderson et al. 1981).

Intercalation of organic acids such as fulvic or humic acids into layer silicate minerals is only possible at low pH where the negative charge density on the molecules are substantially reduced (Theng et al. 1986). The molecules considered above because of their alkyl–amide structure might be expected to behave in a similar manner to alkyl amines (Senkayi et al. 1985), alcohols (MacEwan and Wilson 1980) or pyrrolidones (Francis 1973) which can be readily intercalated into certain layer silicates at higher pH values. This expectation however has proved difficult to demonstrate and it is clear that a better knowledge of the structure of these organic molecules and their associated clay minerals is required.

In the soil, fundamental particles combine to form aggregates of varying size. Tisdall and Oades (1982) have suggested that these aggregates are important factors in retarding organic matter decomposition. They consider two classes of aggregate which are of most significance, microaggregates (those < 250 μm) and macroaggregates (those > 250 μm). Microaggregates are considered to be strongly cemented by organic matter, both humified and polysaccharide, polycations and amorphous inorganic binding agents. These are persistent and are not disrupted by normal cultural practices. That is, they are a characteristic of the soil rather than the management system. Macroaggregates consist of microaggregates held together by temporary or transient binding agents such as roots and fungal hyphal. Macroaggregates are therefore higher in organic carbon and have a larger labile N pool (Elliott 1986).

Soil architecture, through the spatial relationship of aggregates one to another, predetermines the size, distribution and geometry of soil pores. These factors in turn produce soil heterogeneity and may isolate soil organisms from potential substrate (J.N. Ladd and R.C. Foster, in these proceedings). McGill and Myers (1987) have suggested that large (> 10 μm) soil pores may normally be too dry to permit microbial activity while soil pores < 0.48 μm are small enough to exclude microorganisms. Substrate in the large and small pores would therefore be unavailable during drier periods. Reorganization of the soil fabric is therefore necessary to re–establish organism/substrate contact after local supplies have been depleted.

CHANGES DURING DECOMPOSITION

The partitioning of organic matter in a soil depends on environmental factors such as temperature, rainfall, disturbance and plant litter input as well as soil factors such as microbial biomass size and activity, clay content and mineralogy, aeration and soil pH. These factors in turn determine the distribution and turn-over rates of organic fractions within various chemically and physically stabilized pools.

Comminution of structural litter by soil fauna improves accessibility of the material to microorganisms. Further to this, earthworms appear to increase stabilized, clay-bound organic matter as well as increasing soil porosity and water-stable aggregates (J.M. Anderson, in these proceedings). The effect of cultivation through the input of mechanical energy is to reverse this process and reduce many favourable soil physical attributes. Reduction in the number of macroaggregates may provide a short-term increase in soil fertility through the release of nutrients from the labile organic matter which binds microaggregates into macroaggregates. Reorganization also improves substrate/organism contact and often infiltration and aeration. Prolonged exploitive cultivation will, however, result in a rapid decline in the number of macroaggregates along with a rapid decline in labile organic matter. The higher the level and frequency of tillage the more rapid is this decline.

The fractions that show the most rapid decline in organic matter are the light fraction (Dalal and Mayer 1986c) and the sand fraction (Dalal and Mayer 1986b). These organic fractions may often be one and the same and represent a labile pool of largely unaltered structural organic matter existing outside microaggregates and with a wide C/N ratio. Readily decomposable, microbially-derived polysaccharides also contribute to this labile pool (Tisdall and Oades 1982). Humified organic matter outside microaggregates declines more slowly owing to the recalcitrance of the humic structures. This material is represented by organic fractions of intermediate density and usually has a narrower C/N ratio. Skjemstad and Dalal (1987) suggest that molecular recalcitrance is largely responsible for the contrasting behaviour demonstrated by Dalal and Mayer (1986d) of two clay soils upon cultivation. In the case of a grey clay, lack of recalcitrance resulted in rapid loss of the lightest fractions with a resulting narrowing of the soil C/N ratio. In the contrasting case of a black earth, higher molecular recalcitrance in a substantial part of the humus resulted in less rapid decline in the lightest fractions but a widening of the soil C/N ratio as the less recalcitrant N-rich materials within these fractions appeared to be selectively mineralized.

The organic matter stabilized within microaggregates is the most resistant to mineralization (Elliott 1986). The overall stability of these fractions is dependent upon the wet strength of the smaller aggregates which in turn depends on clay mineralogy and the entropy of the internal organic and inorganic bonding mechanisms. The smaller aggregates (2–20 μm) contain much of the most stable organic matter (Tisdall and Oades 1982). This organic matter although strongly humified also contains a considerable proportion of the microbial biomass (Oades 1984) and is stabilized by both chemical and physical mechanisms. Although aggregates < 2 μm also appear to be stable (Tisdall and Oades 1982), a large proportion of the organic matter associated with fine clay may be rapidly depleted on cultivation (Tiessen and Stewart 1983). It is clear that the relationship between organic matter and fine clays requires further attention.

MEASUREMENT OF DECOMPOSABLE ORGANIC NITROGEN BY INCUBATION PROCEDURES

The preceding sections emphasize that soil organic N exists in a variety of materials, chemical forms and locations that differ in their susceptibility to decomposition and release of mineral N. It is not surprising, then, that many and varied chemical extraction and laboratory incubation procedures have been devised to attempt to measure the N mineralization capacity of soils. The literature prior to 1980 has been reviewed recently by Keeney (1982) and the present discussion emphasizes developments since then in the use of laboratory incubation tests to predict N mineralization in the field.

A practical disadvantage of most chemical and biological tests of soil N mineralization is that they provide information on relative rates of N mineralization, not information that can readily be used to estimate quantities of N mineralized under specified field conditions. A significant step towards overcoming this disadvantage was made some 15 years ago when Stanford and Smith (1972) proposed that for the purpose of assessing the long-term N mineralization capacity in cultivated soils a model that postulated a single source of potentially mineralizable N (N_0) that was mineralized according to first order kinetics would be adequate. They found that the size of N_0 varied widely between soils and their history (as expected), whereas the kinetic rate constant (k) was characteristic of a wide range of soils.

Stanford and Smith recognised that their model was an approximation in that net mineralization would continue, though slowly, after the total N mineralized exceeded the mineralization potential. However, they considered that this slow release of N from relatively inert sources would be insignificant compared with that from their potentially mineralizable N. Stanford and co-workers established relationships between mineralization rate and soil temperature and water content, and developed methods for estimating N_0 from short-term incubations and chemical extractions of soil (for a brief review, see Stanford 1982).

Later work has shown that the Arrhenius relationship between mineralization rate and temperature reported by Stanford et al. (1973) for U.S.A. soils might be a general one (Campbell et al. 1981). In the case of soil water content, however, Myers et al. (1982) found that relative mineralization rate was better related to relative available water content than to relative total water content. Further, although a direct linear relationship was appropriate for most soils, nine out of 32 Canadian soils and one of five Australian soils showed a curvilinear response.

The technique of estimating N_O and k has been refined by including soluble organic N removed by leaching the incubated soil (Smith et al. 1980; Beauchamp et al. 1986) and by the use of better mathematical analysis of the data (Smith et al. 1980; Talpaz et al. 1981; Reynolds and Beauchamp 1984). Prediction of N_O from soil chemical and taxonomic data has been attempted with moderate success (Jones et al. 1982; Jones 1984).

The validity of Stanford's original single exponential model has been questioned because it is at variance with the knowledge that several sources of organic N undergo mineralization simultaneously at different rates. Two-pool models, with one pool mineralizing rapidly and the other pool slowly, have fitted the data better in many cases. In these models either both pools are assumed to obey first-order kinetics (Molina et al. 1980; Cabrera and Kissel 1986; Deans et al. 1986), or the 'rapid' pool is assumed to follow zero-order kinetics (Jones 1984; Beauchamp et al. 1986). The rapidly decomposable pool is thought to be largely composed of soil microbial biomass killed by soil handling and pretreatment, especially air-drying (Richter et al. 1982; Beauchamp et al. 1986; Cabrera and Kissel 1986). In some cases a pool for easily decomposable plant residues may also be included (Richter et al. 1982). It seems that problems associated with air-dried soils can be avoided by using field moist soils (Beauchamp et al. 1986; Cabrera and Kissel 1986), two-pool models (Molina et al. 1980; Jones 1984; Beauchamp et al. 1986; Cabrera and Kissel 1986; Deans et al. 1986), or by discarding data for N mineralized during the first two weeks incubation (Stanford et al. 1974).

Further, recent research has yielded considerable variation in estimates of the constant k in the single exponential model (Juma et al 1984; Beauchamp et al. 1986; Deans et al. 1986).

Not all studies of N mineralization give a curvilinear relationship between the amount of N mineralized and time. Incubation of fresh, undried soil may give a linear relationship, i.e., the process may display zero-order kinetics (Tabatabai and Al-Khafaji 1980; Addiscott 1983). Where a curvilinear relationship exists, a hyperbolic equation may fit the data equally as well as a first-order equation, but give very different estimates of N_O and its half-life (Juma et al. 1984).

Limited testing of the ability to predict N mineralization in the field from N_O and k (with k modified according to soil moisture and temperature) has given good correlations in some cases (Stanford 1982) but variable results in others (Herlihy 1979). Considerable over-estimates have been obtained in some cases, and it has been proposed that better estimates might be obtained if N_O and k were estimated for undisturbed soil (Cabrera and Kissel 1986). The usefulness of the technique in the future will probably depend on careful standardisation of methods, on reliable rapid methods of estimating N_O (Keeney

1982; Stanford 1982) and on how well it can be adapted to different cropping systems (Smith 1981). It remains to be seen how well the method will compare with multi-compartment mechanistic models of N cycling for prediction of net N mineralization.

REFERENCES

Addiscott, T.M. (1983). Kinetics and temperature relationships of mineralization and nitrification in Rothamsted soils with differing histories. J. Soil Sci. 34, 343–53.

Amato, M., and Ladd, J.N. (1980). Studies of nitrogen immobilization and mineralization in calcareous soils – V. Formation and distribution of isotope-labelled biomass during decomposition of ^{14}C- and ^{15}N-labelled plant material. Soil Biol. Biochem. 12, 405–11.

Anderson, D.W., and Paul, E.A. (1984). Organo-mineral complexes and their study by radiocarbon dating. Soil Sci. Soc. Am. J. 48, 298–301.

Anderson, D.W., Saggar, S., Bettany, J.R., and Stewart, J.W. (1981). Particle size fractions and their use in studies of soil organic matter: I. The nature and distribution of forms of carbon, nitrogen and sulfur. Soil Sci. Soc. Am. J. 45, 767–72.

Beauchamp, E.G., Reynolds, W.D., Brasche-Villeneuve, D., and Kirby, K. (1986). Nitrogen mineralization kinetics with different soil pretreatments and cropping histories. Soil Sci. Soc. Am. J. 50, 1478–83.

Burns, R.G., Pukite, A.H., and McLaren, A.D. (1972). Concerning the location and persistence of soil urease. Soil Sci. Soc. Am. Proc. 36, 308–10.

Cabrera, M.L., and Kissel, D.E. (1986). Potentially mineralizable N in disturbed and undisturbed soil samples. Agronomy Abstracts p. 195.

Cacco, G., and Dell'Agnola, G. (1984). Plant growth regulator activity of soluble humic complexes. Can. J. Soil. Sci. 64, 225–8.

Campbell, C.A., Myers, R.J.K., and Weier, K.L. (1981). Potentially mineralizable nitrogen, decomposition rates and their relationship to temperature for five Queensland soils. Aust. J. Soil Res. 19, 323–32.

Campbell, C.A., Paul, E.A., Rennie, D.A., and McCallum, K.J. (1967). Applicability of the carbon-dating method of analysis of soil humus studies. Soil Sci. 104, 217–24.

Dalal, R.C., and Mayer, R.J. (1986a). Long term trends in fertility of soils under continous cultivation and cereal cropping in southern Queensland. 2. Total organic carbon and its rate of loss from the soil profile. Aust. J. Soil Res. 24, 281–92.

Dalal, R.C., and Mayer, R.J. (1986b). Long term trends in fertility of soils under continous cultivation and cereal cropping in southern Queensland. 3. Distribution and kinetics of soil organic carbon in particle-size fractions. Aust. J. Soil Res. 24, 293–300.

Dalal, R.C., and Mayer, R.J. (1986c). Long term trends in fertility of soils under continous cultivation and cereal cropping in southern Queensland. 4. Loss of organic carbon from different density fractions. Aust. J. Soil Res. 24, 301–9.

Dalal, R.C., and Mayer, R.J. (1986d). Long term trends in fertility of soils under continous cultivation and cereal cropping in southern Queensland. 5. Rate of loss of total nitrogen from the soil profile and changes in carbon nitrogen ratios. Aust. J. Soil Res. 24, 493–504.

Deans, J.R., Molina, J.A.E., and Clapp, C.E. (1986). Models for predicting potentially mineralizable nitrogen and decomposition rate constants. Soil Sci. Soc. Am. J. 50, 323–6.

Elliott, E.T. (1986). Aggregate structure and carbon, nitrogen, and phosphorus in native and cultivated soils. Soil Sci. Soc. Am. J. 50, 627–33.

Filip, Z., Haider, K., and Martin, J.P. (1972). Influence of clay minerals on the formation of humic substances by Epicoccum nigrum and Stachyleotrys chartarum. Soil Biol. Biochem. 4, 147–54.

Ford, G.W., and Greenland, D.J. (1968). The dynamics of partly humified organic matter in some arable soils. Trans. 9th Inter. Congr. Soil Sci. 2, 403–10.

Francis, C.W. (1973). Adsorption of polyvinylpyrolidone on reference clay minerals. Soil Sci. 115, 40–54.

Greenland, D.J., and Ford, G.W. (1964). Separation of partially humified organic materials from soil by ultrasonic dispersion. Trans. Intern. Congr. Soil Sci. 3, 137–48.

Hayes, M.H.B., and Swift, R.S. (1978). The chemistry of soil organic colloids. In 'The Chemistry of Soil Constituents'. (Eds D.J. Greenland and M.H.B. Hayes). pp. 179–320. (John Wiley: New York.)

Herlihy, M. (1979). Nitrogen mineralization in soils of varying texture, moisture and organic matter. I. Potential and experimental values in fallow soils. Plant Soil 53, 255–67.

Jones, C.A. (1984). Estimation of an active fraction of soil nitrogen. <u>Commun. Soil Sci. Plant</u> <u>Anal</u>. <u>15</u>, 23–32.

Jones, C.A., Ratliff, L.F., and Dyke, P.T. (1982). Estimation of potentially mineralizable soil nitrogen from chemical and taxonomic criteria. <u>Commun. Soil Sci. Plant Anal</u>. <u>13</u>, 75–86.

Juma, N.G., Paul, E.A., and Mary, B. (1984). Kinetic analysis of net nitrogen mineralization in soil. <u>Soil Sci. Soc. Am. J.</u> <u>48</u>, 753–7.

Keeney, D.R. (1982). Nitrogen — Availability indices. <u>In</u> 'Methods of Soil Analysis. Part 2. Chemical and Microbiological Properties'. 2nd edn. (Eds A.L. Page). <u>Agronomy</u> <u>9</u>, 711–33. (Am. Soc. Agron. — Soil Soc. Am.: Madison, U.S.A.)

Ladd, J.N., Amato, M., and Oades, J.M. (1985). Decomposition of plant material in Australian soils. III. Residual organic and microbial biomass C and N from isotope–labelled legume material and soil organic matter, decomposing under field conditions. <u>Aust. J. Soil Res</u>. <u>23</u>, 603–11.

MacEwan, D.M.C., and Wilson, M.J. (1980). Interlayer and intercalation complices of clay minerals. <u>In</u> 'Crystal Structures of Clay Minerals and Their X–ray Identification'. (Eds G.W. Bundley and G. Brown). pp. 197–248. (Mineralogical Society: London.)

McGill, W.B., and Myers, R.J.K. (1987). Controls on dynamics of soil and fertilizer nitrogen. <u>In</u> 'Soil Fertility and Organic Matter as Critical Components of Production Systems'. (Eds R.F. Follett, J.W.B. Stewart and C.V. Cole). pp. 73–99. (Soil Sci. Soc. Am. and Am. Soc. Agron.: Madison, U.S.A.)

McGill, W.B., Shields, J.A., and Paul, E.A. (1975). Relation between carbon and nitrogen turnover in soil organic fractions of microbial origin. <u>Soil Biol. Biochem</u>. <u>7</u>, 57–63.

Molina, J.A.E., Clapp, C.E., and Larson, W.E. (1980). Potentially mineralizable nitrogen in soil : The simple exponential model does not apply for the first 12 weeks of incubation. <u>Soil Sci. Soc. Am. J.</u> <u>44</u>, 442–3.

Monnier, G., Tusc, L., and Jeanson–Luusinang, L. (1962). Une methode de fractionement densimetrique par centrifugation des matieses organiques du sol. <u>Ann. Agron</u>. <u>13</u>, 55–63.

Myers, R.J.K., Campbell, C.A., and Weier, K.L. (1982). Quantitative relationship between net nitrogen mineralization and moisture content of soils. <u>Can. J. Soil Sci</u>. <u>62</u>, 111–24.

Oades, J.M. (1984). Soil organic matter and structural stability: mechanisms and implications for management. <u>Plant Soil</u>. <u>76</u>, 319–37.

Reynolds, W.D., and Beauchamp, E.G. (1984). Comments of 'Potential errors in the first–order model for estimating soil nitrogen mineralization potentials.' <u>Soil Sci. Soc. Am. J.</u> <u>48</u>, 698.

Richter, J., Nuske, A., Habenicht, W., and Bauer, J. (1982). Optimized N–mineralization parameters of loess soils from incubation experiments. <u>Plant Soil</u> 68, 379–88.

Richter, M., Mizuno, I., Arangue, S., and Uriarte, S. (1975). Densimetric fractionation of soil organo–mineral complexes. <u>J. Soil Sci</u>. <u>26</u>, 112–23.

Senkayi, A.L., Dixon, J.B., Hossner, L.R., and Kippenburger, L.A. (1985). Layer charge evaluation of expandable soil clays by an alkylammonium method. <u>Soil Sci. Soc. Am. J.</u> <u>49</u>, 1054–60.

Skjemstad, J.O., and Dalal, R.C. (1987). Spectroscopic and chemical differences in organic matter of two Vertisols subjected to long periods of cultivation. <u>Aust. J. Soil Res</u>. <u>25</u>, 323–35.

Skjemstad, J.O., Dalal, R.C., and Barron, P.F. (1986). Spectroscopic investigations of cultivation effects on organic matter of Vertisols. <u>Soil Sci. Soc. Am. J.</u> <u>50</u>, 354–9.

Smith, J.L., Schnabel, R.R., McNeal, B.L., and Campbell, G.S. (1980). Potential errors in the first–order model for estimating soil nitrogen mineralization potentials. <u>Soil Sci. Soc. Am. J.</u> <u>44</u>, 996–1000.

Smith, S.J. (1981). Field nitrogen mineralization. Agronomy Abstracts, p. 190.

Sorensen, L.H. (1975). The influence of clay on the rate of decay of amino acid metabolites synthesized in soils during decomposition of cellulose. <u>Soil Biol. Biochem</u>. 7, 171–7.

Spycher, G., Sollins, P., and Rose, S. (1983). Carbon and nitrogen in the light fraction of a forest soil. Vertical distribution and seasonal patterns. <u>Soil Sci</u>. <u>135</u>, 79–87.

Spycher, G., and Young, J.L. (1977). Density fractionation of water–dispersible soil organic–mineral particles. <u>Commun. Soil Sci. Pl. Anal</u>. <u>8</u>, 37–48.

Stanford, G. (1982). Assessment of soil nitrogen availability. <u>In</u> "Nitrogen in Agricultural Soils". (Ed. F.J. Stevenson). <u>Agronomy</u>. <u>22</u>, 651–88.

Stanford, G., Carter, J.N., and Smith, S.J. (1974). Estimates of potentially mineralizable soil nitrogen based on short–term incubations. <u>Soil Sci. Soc. Am. Proc</u>. 38, 99–102.

Stanford, G., Frere, M.H., and Schwaninger, D.H. (1973). Temperature coefficient of soil nitrogen mineralization. <u>Soil Sci</u>. <u>115</u>, 321–3.

Stanford, G., and Smith, S.J. (1972). Nitrogen mineralization potentials of soils. <u>Soil Sci. Soc. Am. Proc</u>. <u>36</u>, 465–72.

Tabatabai, M.A., and Al–Khafaji, A.A. (1980). Comparison of nitrogen and sulfur mineralization in soils. <u>Soil Sci. Soc. Am. J.</u> <u>44</u>, 1000–6.

Talpaz, H., Fine, P., and Bar-Yosef, B. (1981). On the estimation of N-mineralization parameters from incubation experiments. Soil Sci. Soc. Am. J. 45, 993-6.

Tate, K.R., and Theng, B.K.G. (1980). Organic matter and its interactions with inorganic soil constituents. In 'Soils with Variable Charge'. (Ed. B.K.G. Theng), pp. 225-49. (N.Z. Soc. Soil Sci.: Palmerston North.)

Theng, B.K.G., Churchman, G.J., and Newman, R.H. (1986). The occurrence of interlayer clay-organic complexes in two New Zealand soils. Soil Sci. 142, 262-6.

Tiessen, H., and Stewart, J.W.B. (1983). Particle-size fractions and their use in studies of soil organic matter. II. Cultivation effects on organic matter composition in size fractions. Soil Sci. Soc. Am. J. 47, 509-14.

Tisdall, J.M., and Oades, J.M. (1982). Organic matter and water-stable aggregates in soils. J. Soil Sci. 33, 141-63.

Turchenek, L.W., and Oades, J.M. (1979). Fractionation of organo-mineral complexes by sedimentation and density techniques. Geoderma 21, 311-43.

MECHANISTIC AND PRACTICAL MODELLING OF NITROGEN MINERALIZATION-IMMOBILIZATION IN SOILS

J.J. Neeteson and J.A. Van Veen

ABSTRACT

Two types of soil nitrogen models can be distinguished: mechanistic and practical models. The purpose of mechanistic models is to gain a better understanding of the processes involved and their interactions. Practical models are developed for predictive and management purposes.

Examples of the treatment of the nitrogen mineralization-immobilization process in both mechanistic and practical models are given. The two types of models appear to develop in opposite directions. Whereas mechanistic models are becoming more refined, practical models are increasingly simplified.

INTRODUCTION

Frissel and Van Veen (1982) classified soil N models according to their purpose. They distinguished scientific models, predictive models and models for management purposes. Scientific models are constructed by a bottom-up process: all processes and mechanisms that might be important are described in detail using all available knowledge. The mathematical expressions describing these details are mounted in a large framework, the simulation model. The purpose of scientific models is to gain a better understanding of the processes involved and their interactions. Predictive and management models are constructed by a top-down process: the models are developed by analyzing results of earlier years, and not by a conceptual consideration of the processes. The purpose of predictive models could be to predict, for instance, optimum N fertilizer application rate. Models for management purposes are based on predictive models, but they include various options. They can be used, for instance, to indicate the implications of various measures with respect to the use of slurry. This classification of models could give the erroneous impression that scientific models are unable to predict and that predictive and management models do not deal with science. On the contrary, models often include fundamentally described processes for predictive and management purposes.

In this paper we propose to discriminate between mechanistic and practical models. Mechanistic models are developed to improve scientific understanding, whereas practical models are developed for practical purposes. Examples will be given of the treatment of N mineralization—immobilization in both types of models.

MECHANISTIC MODELS

Early models of mineralization and immobilization of N described net processes, i.e. changes in the amount of mineral N (Kirkham and Bartholomew 1955). These models often included a so-called turning point of net mineralization or net immobilization. This turning point was set by the C/N ratio of the decomposing organic material. When the C/N ratio was smaller than 20–25, net mineralization occurred. Nitrogen was assumed to be immobilized at higher C/N ratios (Schaffer et al. 1969; Parnas 1975; Bosatta 1981).

In later models the impact of the quality of organic matter on the mineralization and immobilization of N was more explicitly described. The model of Beek and Frissel (1973) distinguished specific organic compounds such as proteins and lignins as a substrate for the mineralization process. The separation of organic products into specific chemical compounds was meant to describe the heterogeneity of organic materials as substrate for the microorganisms, which were supposed to be the key organisms responsible for the fate of N in the soil. To describe in more detail the availability of organic materials as a substrate for microorganisms, Van Veen and Frissel (1981) developed a new version of the Beek and Frissel model. This model consisted of submodels for nitrification, denitrification, ammonia volatilization, ammonium fixation on clay minerals, leaching, and mineralization and immobilization including an explicit description of the growth of the microbial biomass. Figure 1 shows the detailed outline of the mineralization—immobilization submodel. Mineralization and immobilization of N were considered to be controlled by the growth and activity of the total microbiological biomass in soil. Growth and activity of the biomass were assumed to be limited by the availability of C and N as a substrate. Both crop residues and soil organic matter were divided into several compounds (Fig. 1). The crop residues were split into three pools. Of these, crop residue carbon was divided into sugars and other easily decomposable carbohydrates, and into slowly decomposable material, mostly consisting of hemicellulose. The third crop residue pool consisted of all easily decomposable materials which contain N, e.g. proteins. Soil organic matter was divided into four pools: microbial biomass, decomposable active materials, resistant active materials and old organic matter. The active materials contain lignins and microbial products such as amino acids.

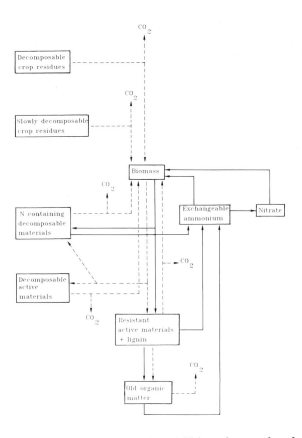

Fig. 1: Layout of the mineralization–immobilization submodel of Van Veen and Frissel (1981) for nitrogen (———) and carbon (– – –) in soils.

It became clear that the availability of organic materials as a substrate for microorganisms is not only determined by their biochemical constitution, but also by their localization in the soil. Soil type and changes in the soil structure, for instance due to soil tillage, were shown to have a dramatic impact on the availability of organic matter as a substrate and thereby on the decomposability and mineralization of organic materials. Van Veen and Paul (1981) presented a model to simulate soil organic matter turnover in grassland soils after cultivation. An outline of the model is shown in Figure 2. In the model soil organic matter was divided into three major fractions: microbial biomass, and decomposable and recalcitrant soil organic matter (Fig. 2). An important feature of the model was that both decomposable and recalcitrant organic matter consisted of a physically and a not-physically protected part. It was assumed that protection occurs by means of adsorption on clay minerals and by entrapment in soil aggregates. Obviously, protection leads to a decrease in the decomposition rate of the fraction involved. Cultivation diminished the protective action of soil through the breakdown of soil structure.

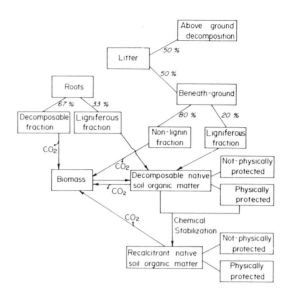

Fig. 2: Layout of the mineralization–immobilization model from Van Veen and Paul (1981). [After Van Veen and Paul 1981].

Jenkinson and Ladd (1981) reviewed data which pointed to a dependency of the biomass ^{14}C and ^{15}N formation during the decomposition of $^{14}C–^{15}N$–labelled crop residues on the texture and structure of a soil. This was further studied in incubation experiments with different $^{14}C–$, $^{15}N–$, $^{32}P–$labelled materials in two soils, a clayey and sandy soil at controlled temperature and moisture conditions (Van Veen et al. 1985, 1987). Among other analyses, at regular intervals the decline in residual organic ^{14}C and the net immobilization or mineralization of N were measured. With respect to the turnover of C, the soils differed significantly. From day 10 until the end of the experiment net decomposition rates in the sandy soil were about twice as high as in the clay soil (Fig. 3). The effect of soil type on the net rates of immobilization or mineralization of N was consistent with its effect on C metabolism. After an initial period (about 10 days) of immobilization, net mineralization of N took place in both soils, but sooner in the sandy loam soil, and at higher rates despite the much higher organic–N content of the clay soil (Fig. 4). To describe these observations Van Veen's model was adapted (Van Veen et al. 1985). An important characteristic of the adapted version is that microbial biomass consisted of a protected and an unprotected fraction. When the protection capacity of the soil was larger than the total biomass, the entire microbial population was considered to be protected. The death rate of protected organisms was taken as 0.5% per day. Biomass formed in excess of a soil's protection capacity is assumed to decay at a much higher rate, about 70% per day. The authors succeeded in describing the observed effect of soil type on the turnover of C and N (Figs. 3 and 4). They concluded that soils might have characteristic preserving capacities of

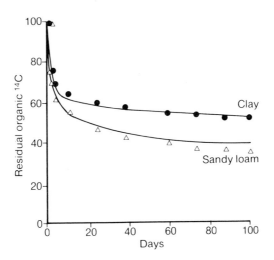

Fig. 3: Residual organic ^{14}C (% of input), calculated from $^{14}CO_2$ evolution, in clay or sandy loam soil after incubation with $[^{14}C]$ glucose for varying lengths of time under continuously moist conditions. [After Van Veen et al. 1985]. Measured ($\triangle$, ●), simulated (——).

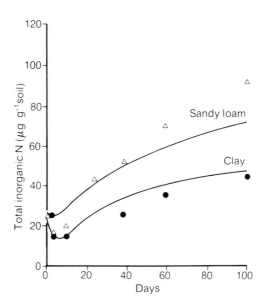

Fig. 4: Total inorganic nitrogen concentrations in clay or sandy loam soil after incubation with $(^{15}NH_4)_2SO_4$ for varying lengths of time under continously moist conditions. [After Van Veen et al. 1985]. Measured ($\triangle$, ●), simulated (——).

organic matter and organisms depending on the soil's texture and structure. The preserving capacities of clayey soils are often larger than those of sandy soils, which results in a lower rate of N mineralization.

PRACTICAL MODELS

The development of practical models has mainly been independent of that of the mechanistic models. The aim of practical soil N models is usually to predict nitrate leaching with respect to the contamination of groundwater, and to predict the supply of soil N to crops in order to assess the optimum N fertilizer application rate.

In most West-European countries N fertilizer recommendations for arable crops are based on the amount of mineral N which is already present in the soil profile. The higher the amount of soil mineral N, the lower the recommended rate of N fertilizer. This method of recommendation implies that each year at about the same time, which is early spring, most farmers want to know the amount of mineral N in their soils. It is obvious that this creates organizational difficulties for the laboratories, the results not being available before the proper time of fertilizer application. Such difficulties could be overcome if soil samples were taken and analyzed at any time in autumn or winter, and the amount of soil mineral N in spring then calculated on the basis of this early measurement. In the Netherlands, Zandt and De Willigen (1981) developed a model to calculate soil mineral N in spring based on measurement in the preceding autumn. Only two soil N processes were included in the model: leaching and mineralization-immobilization. Denitrification, ammonia volatilization and fixation of ammonium by clay minerals were ignored. Moreover, it was assumed that all mineral N was in the form of nitrate, because in Dutch soils the rate of nitrification generally is much higher than the rate of mineralization. Leaching was calculated on the basis of daily values of the precipitation surplus according to Burns's model (Burns 1974). With respect to the process of mineralization-immobilization only two pools were distinguished: old organic matter and crop residues. Based on results of Kortleven (1963) it was assumed that each year 2% of the old orgganic matter was decomposed. Zandt and De Willigen (1981) also assumed a release of 2% of the N in the old organic matter. Kortleven (1963) also found that during the first year a fixed fraction of the crop residues was decomposed. Kolenbrander (1974) related the size of this fraction to the type of crop residues (see also Van Dijk 1982). Zandt and De Willigen (1981) used this concept in their model to describe the decomposition of crop residues and assumed that in the second year the remainder of the crop residues became old organic matter. The model correctly predicted the course of soil mineral N from November to June on a loamy soil with an organic-matter content of 2.2% (Fig. 5). Because only two processes were included in the model, and one of them — leaching of nitrate — was checked by using chloride as a tracer, it is reasonable to assume that the model correctly predicted the other process, i.e. mineralization-immobilization of N, for this data set.

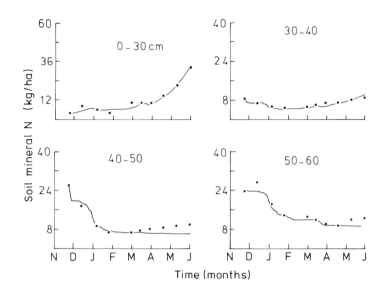

Fig. 5: Simulated (———) and measured (•) time course of soil mineral nitrogen changes. [After Zandt and De Willigen 1981].

Greenwood et al. (1985a,b) attempted to determine the extent to which data obtained in 11 N fertilizer experiments with potato could be described in terms of simple but widely applicable equations. They aimed to provide a better basis for prediction of response of potato to N fertilizer. The experiments were conducted on three widely different sites (Table 1).

Table 1: Soil characteristics (0–40 cm) and experimental years of the research described by Greenwood et al. (1985a,b).

Experiment and soil type	Fraction < 16 μm	Organic matter	Experimental years
	%	%	
PO (sand)	0	4.9	1969 – 1972
PZ (sandy loam)	12	1.5	1969 – 1972
IB (clay)	54	3.3	1970 – 1972

Detailed measurements of soil mineral N, and the weights and N contents of foliage and tubers were carried out. Two soil N processes were included in the model to describe the fate of soil mineral N: leaching and apparent N mineralization. Leaching was described according to Burns's model (Burns 1974). Without loss of N from the soil/plant system due to leaching, or due to other processes, the apparent amount of N mineralized ΔM_N over a given period was given by

$$\Delta M_N = \Delta N_S + \Delta U_P + \Delta U_R \qquad (1)$$

where ΔN_S is the change with time in the amount of soil mineral N, ΔU_P is the increment in the amount of N in the potato foliage plus tubers, and ΔU_R is the increment in the amount of N in the fibrous roots. ΔN_S and ΔU_P were measured in the experiments, but ΔU_R was not. Based on the maximum recoveries of N fertilizer in the experiments it was estimated that ΔU_R equalled 0.25 ΔU_P, which on substitution in Equation (1) gives

$$\Delta M_N = \Delta N_S + 1.25 \ \Delta U_P \qquad (2)$$

For each experiment M_N was calculated on numerous occasions during the growing period. M_N closely fitted a straight line when plotted against time for plots from which N fertilizer had been withheld (e.g. Fig. 6). The slopes of the

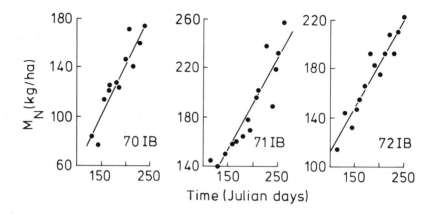

Fig. 6: Relationship between apparent nitrogen mineralization (M_N) and time on plots receiving no nitrogen fertilizer. [After Greenwood et al. 1985a].

lines are a measure of the rate of apparent mineralization. Values of the slopes in 9 of the 11 experiments are given in Table 2. The values of two experiments are not given because most likely substantial amounts of N had leached in these experiments. The apparent mineralization rate was remarkably similar in 8 of the 9 experiments, and averaged 0.78 kg N/ha/day (Table 2). The experiments covered soils with textures ranging from sands to clays with organic-matter contents that differed by a factor of more than three (Table 1). These unexpected findings most probably can be explained by the cropping history in the experiments. Previous cropping had been similar for more than 10 years. For all experiments the crop rotation was potatoes–wheat(or rye)–oats–potatoes. The relative constancy in the apparent mineralization rate thus suggests that nearly all organic matter was old and was decomposing slowly, and that contents of the readily decomposable fraction were similar on all sites. This is in agreement with Janssen's concept of "old" and "young" soil organic matter (Janssen 1984).

Table 2: Apparent mineralization rates on plots from which nitrogen fertilizer had been withheld (Greenwood et al. 1985a,b).

Experiment	Year	Apparent mineralization rate
		kg N/ha/day
PO (sand)	1969	0.69
	1970	1.03
	1971	0.71
PZ (sandy loam)	1969	0.75
	1970	0.84
	1971	1.59
IB (clay)	1970	0.80
	1971	0.71
	1972	0.72
Average (PZ 1971 excluded)		0.78

Zandt et al. (1986) used data from long-term experiments on a sandy soil, a sandy loam and a clay to compare the above-mentioned methods of Zandt and De Willigen (1981) and Greenwood et al. (1985a,b) for the description of N mineralization. In 9 spring and 7 autumn periods soil mineral N was measured at approximately 10-day intervals. In these periods no crop was grown and no fertilizer N was applied. The first measurement in each experiment was used to initialize the models. A total of 377 comparisons of measured and simulated values of soil mineral N could be made with each model. A simulated value was supposed to be correct when the deviation from the measured value was less than 20 kg N/ha in the 0-60 cm layer. The results of the calculations (Table 3) showed that the method of Greenwood et al. was superior

Table 3: Percentages of successful simulations (Zandt et al. 1986). Total number of simulation calculations given in brackets.

Method of calculation of N mineralization	Soil type		
	Sandy (128)	Sandy loam (126)	Clay (123)
Zandt and De Willigen (1981)	48	75	50
Greenwood et al. (1985a,b)	84	79	85

to the method of Zandt and De Willigen. The method of Zandt and De Willigen failed to give good predictions on the sandy soil and the clay soil. Due to the relatively high organic matter contents of these soils (Table 1) the method of Zandt and De Willigen (1981), which relates the rate of N mineralization to the organic matter content, overestimated N mineralization.

CONCLUSIONS

The present trend in developing mechanistic models is opposite to that in practical models. Whereas mechanistic models are becoming more refined, practical models are increasingly simplified. Since various research groups all over the world increasingly direct their attention to the role of the soil fauna in N cycling, it is to be expected that mechanistic models will become even more refined. On the other hand, it is quite possible that the assumption of a fixed rate of apparent mineralization in practical models is too simple to be generally applicable. Maybe more than only one organic-N pool should be distinguished, as was done in the mineralization-immobilization submodel of Van Keulen and Seligman's model to simulate the growth of spring wheat (Van Keulen and Seligman 1987). However, after comparing various models with the same data set De Willigen and Neeteson (1985) concluded that for practical purposes the simplest models were the best in describing the N cycle in the soil.

Now the question arises how the opposite trend in development of mechanistic and practical models should be judged. In our opinion, it is a logical, realistic situation which reflects the differences in goals for which the models are developed. To improve scientific understanding, mechanistic models should be as refined as possible using as much information as is available. The relevant processes should therefore be described fundamentally and in great detail. Practical models, however, need to be as simple as possible. Ultimately, these models will be used by farmers or by their technical advisors. The practical models should therefore require a limited number of, easily-obtained, input parameters.

This paper might give the impression that mechanistic and practical models are always developed seperately. Obviously this is not so, practical models can also be obtained by simplifying mechanistic models in such a way that they give reasonably good predictions and that the input parameters can be easily obtained.

REFERENCES

Beek, J., and Frissel, M.J. (1973). 'Simulation of Nitrogen Behaviour in Soils'. (PUDOC: Wageningen.)

Bosatta, E. (1981). Plant-soil system of an old Scots pine forest in Central Sweden. In 'Simulation of Nitrogen Behaviour of Soil-Plant Systems'. (Eds M.J. Frissel and J.A. Van Veen) pp. 155–70. (PUDOC: Wageningen.)

Burns, I.G. (1974). A model for predicting the redistribution of salts applied to fallow soils after excess rainfall or evaporation. J. Soil Sci. 25, 165–78.

De Willigen, P., and Neeteson, J.J. (1985). Comparison of six simulation models for the nitrogen cycle in the soil. Fert. Res. 8, 157–71.

Frissel, M.J., and Van Veen, J.A. (1982). A review of models for investigating the behaviour of nitrogen in soil. Phil Trans. R. Soc. Lond. 296, 341–9.

Greenwood, D.J., Neeteson, J.J., and Draycott, A. (1985a). Response of potatoes to N fertilizer: Quantitative relations for components of growth. Plant Soil 85, 163–83.

Greenwood, D.J., Neeteson, J.J., and Draycott, A. (1985b). Response of potatoes to N fertilizer: Dynamic model. Plant Soil 85, 185–203.

Janssen, B.H. (1984). A simple method for calculating decomposition and accumulation of "young" soil organic matter. Plant Soil 76, 297–304.

Jenkinson, D.S., and Ladd, J.N. (1981). Microbial biomass in soil: measurement and turnover. In "Soil Biochemistry". (Eds E.A. Paul and J.N. Ladd) Vol 5., pp. 415–71. (Marcel Dekker: New York.)

Kirkham, D., and Bartholomew, W.V. (1955). Equations for following nutrient transformations in soil, utilizing tracer data: II. Soil Sci. Soc. Amer. Proc. 19, 189–92.

Kolenbrander, G.J. (1974). Efficiency of organic manure in increasing soil organic matter content. Trans. 10th Int. Congr. Soil Sci. 2, pp. 129–36.

Kortleven, J. (1963). Kwantitatieve aspecten van humusopbouw en humusafbraak. Versl. Landbouwk. Onderz. No. 69.1.

Parnas, H. (1975). Model for decomposition of organic material by microorganisms. Soil Biol. Biochem. 7, 161–9.

Schaffer, M.J., Dutt, G.R., and Moore, W.J. (1969). Predicting changes in nitrogenous compounds in soil–water systems. Water Pollution Control Research Series 13030 ELY, pp. 15–28.

Van Dijk, H. (1982). Survey of Dutch soil organic matter research with regard to humification and degradation rates in arable land. In 'Soil Degradation'. (Eds D. Boels, D.B. Davies and A.E. Johnston). pp. 133–44. (Balkema: Rotterdam.)

Van Keulen, H., and Seligman, N.G. (1987). 'Simulation of Water Use, Nitrogen Nutrition and Growth of a Spring Wheat Crop'. Simulation Monographs (PUDOC: Wageningen.)

Van Veen, J.A., and Frissel, M.J. (1981). Simulation model of the behaviour of N in soil. In 'Simulation of Nitrogen Behaviour of Soil–Plant Systems'. (Eds M.J. Frissel and J.A. Van Veen) pp. 126–44. (PUDOC: Wageningen.)

Van Veen, J.A., Ladd, J.N., and Amato, M. (1985). Turnover of carbon and nitrogen through the microbial biomass in a sandy loam and a clay soil incubated with $[^{14}C(U)]$glucose and $[^{15}N(NH_4)_2SO_4$ under different moisture regimes. Soil Biol. Biochem. 17, 747–56.

Van Veen, J.A., Ladd, J.N., Martin, J.K., and Amato, M. (1987). Turnover of carbon, nitrogen and phosphorus through the microbial biomass in soils incubated with $^{14}C-$, $^{15}N-$, ^{32}P–labelled bacterial cells. Soil Biol. Biochem. 19, 559–66.

Van Veen, J.A., and Paul, E.A. (1981). Organic carbon dynamics in grassland soils. 1. Background information and computer simulation. Can. J. Soil. Sci. 61, 185–201.

Zandt, P.A., and De Willigen, P. (1981). Simulatie van de stikstofverdeling in de grond in winter en voorjaar. Inst. Bodemvruchtbaarheid, Rapp. 4–81.

Zandt, P.A., De Willigen, P., and Neeteson, J.J. (1986). Simulatie van de stikstofhuishouding van de grond in voor– en najaar. Inst. Bodemvruchtbaarheid, Rapp. 9–86.

IMPORTANCE OF AMMONIA VOLATILIZATION

AS A LOSS PROCESS

J.R. Freney and A.S. Black

ABSTRACT

The development of new methods to measure ammonia volatilization in the undisturbed field environment has enabled the assessment of the importance of NH_3 loss and the factors which control loss in different agricultural systems. Simplified micrometeorological techniques have been developed which require less labour and skill; one method does not require a source of power, pumps, flowmeters or anemometers and is therefore ideal for use in remote locations.

Grazed pastures appear to be one of the main sources of atmospheric NH_3. Much of the NH_3 comes from the excreta of animals, but some is derived from fertilizer applications; losses ranged from 9 to 86% of the applied fertilizer nitrogen.

Ammonia losses following fertilizer application to upland crops and flooded rice varied from 0-50% of the applied nitrogen. The extent of loss varied with fertilizer form, method and time of application, stage of crop growth, solar radiation, windspeed, floodwater pH and other factors.

Loss of fertilizer nitrogen by NH_3 volatilization can be reduced by suitable fertilizer management practices including coatings, split applications, deep placement, and additions of acids, soluble salts, urease inhibitors and algicides.

INTRODUCTION

Little understanding of the processes involved in the transport of nitrogenous gases from terrestrial and aquatic systems to the atmosphere had been achieved until recently because of the lack of suitable techniques to measure rates of emission. In the case of NH_3 exchange this problem has now been overcome with the development of micrometeorological methods (Denmead 1983). These methods allow the determination of rates of NH_3 emission or absorption in the undisturbed field environment, and have enabled workers to assess the importance of NH_3 loss, and the factors which control loss, in different agricultural systems.

Consequently, NH_3 volatilization has received considerable attention in recent years; it has been discussed at a number of symposia and numerous reviews have been written on the topic (e.g. Nelson 1982; Beauchamp 1983; Freney et al. 1983; Harper et al. 1983b; Fenn and Hossner 1985; Fillery and Vlek 1986). We have decided, therefore, not to review the whole field of work on NH_3 volatilization, but to be selective, and concentrate on the more recent work.

FACTORS AFFECTING AMMONIA LOSS

Ammonia is constantly being formed in soils from the biological degradation of organic compounds and is being added in increasing amounts as fertilizer. As a result of biological, chemical and physical processes NH_3 is exhaled into the atmosphere and the amount emitted is influenced by the ecosystem, soil and water characteristics, prevailing weather conditions, cropping procedures and fertilizer techniques. The reactions which determine NH_3 volatilization from soil may be represented (Simpson 1981; Vlek et al. 1981) as:

$$\begin{array}{cc} \text{Source 1} & \text{Source 2} \\ \downarrow & \downarrow \end{array}$$

$$\text{Adsorbed } NH_4^+ \rightleftharpoons NH_4^+ \text{ in solution} \rightleftharpoons NH_3 \text{ gas in solution} \rightleftharpoons NH_3 \text{ gas in soil} \rightleftharpoons NH_3 \text{ gas in atmosphere} \qquad [1]$$

where Source 1 may be an NH_4^+ based fertilizer (e.g. ammonium sulfate) and Source 2 could be anhydrous NH_3. The rate of NH_3 volatilization may be controlled by the rate of removal and dispersion of NH_3 into the atmosphere, by changing the concentration of NH_4^+ or NH_3 in solution, or by displacing any of the equilibria in some other way.

The driving force for NH_3 volatilization from a moist soil or a solution is normally considered to be the difference in NH_3 partial pressure between that in equilibrium with the liquid phase and that in the ambient atmosphere (Koelliker and Miner 1973). The equilibrium NH_3 vapour pressure is controlled by the temperature and NH_3 concentration in solution, which is affected by the NH_4^+ concentration, temperature and pH. Further discussion on these basic reactions is presented in Moeller and Vlek (1982), Beauchamp (1983) and Freney et al. (1983).

Other processes and factors which affect NH_3 volatilization are the buffer capacity and cation exchange capacity of the soil, presence of calcium carbonate, water content of soil, leaching, rate of nitrification, soil texture, urease activity, presence of plants and plant residues, fertilizer form, and environmental conditions such as radiation, atmospheric NH_3 concentration, temperature and wind speed. The importance of these factors is reviewed by Faurie and Bardin (1979a,b), Terman (1979), Freney et al. (1981a, 1983), Nelson (1982) and Fenn and Hossner (1985), and additional information is given in the recent publications of Nelson et al. (1980), Craig and Wollum (1982), Denmead et al. (1982a), Knop (1982), Rodgers (1983), Torello and Wehner (1983), Vlek and Carter (1983), Ezzeldin Ibrahim et al. (1984), Ferguson et al. (1984), Kumar and Wagenet (1984), Stumpe et al. (1984), Vallis and Gardner (1984), Black et al. (1985b, 1987a,b), Bouwmeester et al. (1985), Freney et al. (1985b), O'Toole et al. (1985), Tomar et al. (1985), Bacon et al. (1986), Cai et al. (1986), Ellington (1986), Ferguson and Kissel (1986), Hoult and McGarity (1986), Keller and Mengel (1986), McInnes et al. (1986a,b), de Preez and Burger (1986) and Reynolds and Wolf (1987).

Some of these factors have been incorporated into models to describe the volatilization of NH_3 from urine patches, and from soil following the application of fertilizer (Parton et al. 1981; Sherlock and Goh 1985a,b; Rachhpal – Singh and Nye 1986a,b,c).

DETERMINATION OF AMMONIA EMISSION

During the past ten years significant advances in techniques for the measurement of gaseous emission of NH_3 from field soils have been made (Denmead 1983; Harper 1988). Despite these advances many workers continue to use techniques which have a poor theoretical base and which provide results difficult to interpret.

The methods used for measuring NH_3 emission in the field can be grouped into four categories:

(i)   Those which employ open acid traps or acid-saturated filter papers (Luebs et al. 1973; Mahendrappa and Ogden 1973; Sudhakara and Prasad 1986).

(ii) Difference methods, which include isotopic techniques (Marshall and DeBell 1980; Haunold and Blochberger 1982; Sherwood 1983; Rodgers and Pruden 1984; Black et al. 1985a; San 1986).

(iii) Those which employ enclosures to measure changes in gas concentration at the soil or water surface (Sherlock and Goh 1984; Black et al. 1985a; Bacon et al. 1986).

(iv) Micrometeorological methods which measure the vertical flux density of NH_3 in the free air above the surface (Beauchamp 1983; Denmead 1983; Black et al. 1985a; McInnes et al. 1985).

Acid-trap methods

If open acid traps or acid-saturated filter papers are used to sorb NH_3 from the atmosphere strong concentration gradients are developed which increase the diffusion rate and amount of NH_3 collected. It is impossible with this technique to determine the rates of emission.

Difference methods

Difference methods which rely on the recovery of N as urea, NH_4^+ and NO_3^- (e.g. Black et al. 1985a) or labelled N (e.g. Marshall and DeBell 1980) tend to be unreliable and inaccurate. For these techniques to work immobilization of N must be negligible (Black et al. 1985a), and the extent of other loss processes such as denitrification, leaching and runoff must be quantified. In some ecosystems, e.g. in a flooded rice field, NH_3 volatilization and denitrification proceed concurrently, immediately after fertilizer addition (Galbally et al. 1987). So, even if it is known that leaching and run-off losses do not occur, it cannot be assumed that denitrification is insignificant at any stage and that the difference method can be safely employed. As with all difference methods, errors in the determination of the other components of the balance result in an inaccurate estimation of NH_3 loss (Harper 1988).

Enclosure methods

Enclosure methods are the most popular because of their relative simplicity, their suitability for small experimental areas, and the low sensitivity required for measuring gas concentrations. However, there are problems with their use and a great deal of uncertainty exists in relating enclosure measurements to a field system (Denmead 1983; Harper 1988).

Static enclosures permit measurement of gas flux for short periods only. As soon as NH_3 diffuses into the enclosure, the NH_3 concentration at the soil or water surface increases, the concentration gradient decreases and the flux decreases (Ryden and Rolston 1983). This deficiency can be minimized by the use of continuous flow enclosures in which the air space enclosed is swept by a flow of air drawn from the atmosphere outside the enclosure. The flux density of the gas is calculated from the difference between the concentration of NH_3 in the air entering and leaving the chamber (Ryden and Rolston 1983).

The use of enclosures has also been criticised because (i) they may shield surfaces from pressure fluctuations which may influence gaseous emission from soils, (ii) continuous flow enclosures may induce mass flow of NH_3 from the soil, or alter soil drying rates which indirectly affect NH_3 emission (Black et al. 1985a), (iii) they may induce elevated temperatures within the cover, (iv) NH_3 may be retained on their walls and on air pipes or be dissolved by free water anywhere in the system, (v) they eliminate the natural environment, and NH_3 volatilization is strongly influenced by rain, dew formation and wind speed (Denmead 1983). In addition, the area within the enclosure may not be representative of the field as a whole because of spatial variability in soil properties.

Attempts have been made to overcome these problems, such as covering the treated area for short periods only (Kissel et al. 1977) or by using "wind tunnel" enclosures so that air flow through the enclosure could be matched to wind speed in the field (Vallis et al. 1982; Lockyer 1984; Ryden and Lockyer 1985). These modified enclosures were found to have little effect on sward conditions, air and soil temperature, and soil water content. Problems may be encountered with condensation on the inner surfaces, and counter-gradient fluxes (Harper 1988).

Because conditions inside the enclosure cannot be matched to those outside, and because of the marked effect of the environment on NH_3 loss, there is a great deal of uncertainty in relating enclosure measurements to the field situation (Denmead 1983; McInnes et al. 1986b; Harper 1988). Nevertheless, despite all these objections Black et al. (1985a) found that the cumulative NH_3 loss determined by an enclosure method was essentially the same as that determined by a micrometeorological method. The hourly flux rates, however, were quite different.

Micrometeorological methods

Micrometeorological methods are to be preferred in principle as they do not disturb the environmental or soil processes which influence gas exchange, they allow continuous measurements (which permit the investigation of environmental effects) and they provide a measurement of flux integrated over a large area (Denmead 1983). The latter overcomes the spatial variability problem.

While there are three general kinds of micrometeorological techniques available, eddy correlation, gradient diffusion and mass balance (Denmead 1983), only gradient diffusion and mass balance methods have been used to study emission of NH_3 from soil and water surfaces.

Gradient diffusion

Gradient diffusion methods are based on the knowledge that gases diffuse along their mean concentration gradients. In this case the flux density (F) can be expressed in the terms:

$$F = - K \frac{\partial c}{\partial z} \qquad [2]$$

where c is concentration, z is height above the surface and K is a transport coefficient for the gas. The theory underlying this method and details of the measurements required are given in Denmead (1983). This method has been used successfully to determine the exchange of NH_3 between soils, waters, plants and the atmosphere (e.g. Denmead et al. 1978; Dabney 1982; Hutchinson et al. 1982; Harper et al. 1983a,b; Dabney and Bouldin 1985). The main difficulty with this approach is the requirement for a relatively large uniform area so that equilibrium profiles are developed.

Mass balance

This approach is probably the most convenient and simplest of the micrometeorological techniques. It has a sound theoretical basis, has no special requirement for equilibrium profiles and is most successful when the experimental area is small. It is important, however, that the surface flux from the experimental area be much greater than that from the surrounding area.

The horizontal flux of NH_3 across a vertical plane of unit width on the downwind edge of a treated area is equated with the surface flux from a strip of similar width upwind. The horizontal flux at any height is the product of horizontal wind speed and NH_3 concentration. The total flux is obtained by integrating that product over the atmospheric layer modified by NH_3 emission (Denmead 1983). The flux density (F) can be expressed as:

$$F = \frac{1}{x} \int_{0}^{Z} \overline{uc} \, dz \qquad [3]$$

where x is the distance travelled by the wind over the treated area, Z is the height of the modified layer in the atmosphere, u is wind speed and c is NH_3 gas concentration in excess of the background. To evaluate the integral the profiles of u and c must be precisely defined, which means that u and c must be determined at a number of heights extending through the modified layer.

This technique has been used to determine NH_3 emission from upland and flooded fields following additions of fertilizer, urine, manure and sewage sludge (Denmead et al. 1977, 1982a,b; Beauchamp et al. 1978, 1982; Wilson et al. 1982, 1983); Fillery et al. 1984, 1986a; Leuning et al. 1984; Ryden and McNeill 1984; Simpson et al. 1984, 1985; Black et al. 1985a; Freney et al. 1985a, b; McInnes et al. 1985, 1986a, b; Cai et al. 1986; Fillery and De Datta 1986; Harper et al. 1987; Simpson and Freney 1987).

The mass balance method requires no special form for the wind profile nor any corrections for thermal stratification. If, however, the wind does have a log- or power law-profile, theory exists which makes it possible to predict the surface flux from only one measurement of gas concentration and windspeed at a particular height (called ZINST) above the soil or water surface (Wilson et al. 1982). Denmead (1983) and McInnes et al. (1985) discuss this abbreviated approach and the results of Denmead (1983), Wilson et al. (1983), Freney et al. (1985a, 1987) and McInnes et al. (1985) indicate that the fluxes determined agree well with those measured by the full profile mass balance method. The abbreviated method requires less equipment, time and labour than the full profile method. It can be simplified even further by using a sampling device which provides a measure of the horizontal transport of NH_3, i.e. it provides an estimate of uc in eqn [3] (Leuning et al. 1985). By using a circular plot and mounting the intake of the sampler at ZINST the flux of NH_3 over a set period can be determined from one NH_3 analysis. The technique has no requirement for power, pumps, flowmeters or anemometers (Freney et al. 1987).

Bulk aerodynamic approach

Another simplified micrometeorological method has been proposed for assessing NH_3 losses from flooded rice fields (Freney et al. 1985a, 1987). The driving force for NH_3 volatilization from a solution is considered to be the difference between the NH_3 gas concentration in equilibrium with the liquid phase and that in the ambient atmosphere. Increasing wind speed increases the rate of volatilization from flooded systems by promoting more rapid transport of NH_3 away from the water surface. The combined effect of these influences on volatilization rate have been incorporated into a bulk aerodynamic formula of the type:

$$F = ku_z (c_O - c_z) \qquad [4]$$

where k is a constant, u_z is the mean wind speed at a reference height, c_O is the mean NH_3 concentration in equilibrium with that in the liquid phase (calculated from ammoniacal N, pH and temperature, Leuning et al. 1984) and c_z is the mean NH_3 concentration in the atmosphere at the reference height. To use this technique, knowledge is required of ammoniacal N concentration, pH and temperature of the floodwater, and wind speed at a fixed height in the atmosphere. The bulk aerodynamic approach is not as accurate as the abbreviated mass balance method (Freney et al. 1985a), but it is very useful for survey purposes or for comparing NH_3 emission rates from a large number of fertilizer treatments within a relatively small area (De Datta et al. 1986).

Few comparisons of the different techniques for assessing NH_3 volatilization have been made and little information on their accuracy and precision is available (Marshall and DeBell 1980; Black et al. 1985a; Harper 1988).

EXTENT OF AMMONIA LOSS FROM DIFFERENT ECOSYSTEMS

Numerous laboratory and greenhouse measurements have demonstrated the potential for NH_3 loss from soils, plants and water surfaces, but there have been few measurements of NH_3 loss in the field with non-disturbing, reliable techniques. We confine our discussion to the loss measurements made in the field.

Grasslands

Ammonia can be lost from grasslands in four different ways, (i) by direct emission from growing plants (Lemon and van Houtte 1980; Farquhar et al. 1983), (ii) from decomposing litter (Denmead et al. 1976), (iii) from fertilizer and the excreta of grazing animals (Woodmansee et al. 1978, 1981; Vallis et al. 1982; Harper et al. 1983b; Simpson and Steele 1983; Schimel et al. 1985), and (iv) by biomass burning (Crutzen et al. 1979).

Very few direct measurements of NH_3 loss from growing plants have been made (Farquhar et al. 1979, 1980; Lemon and van Houtte 1980; Dabney and Bouldin 1985). As Farquhar et al. (1983) explain, this loss occurs when the partial pressure of NH_3 in the substomatal cavities exceeds that in the atmosphere. Schimel et al. (1985) calculated the potential loss rate of NH_3 from senescing vegetation on a shortgrass steppe and concluded that it was an order of magnitude greater than the total rate of loss by all other processes.

Grazed pastures are one of the main sources of atmospheric NH_3. Much of the NH_3 comes from the excreta of animals (Galbally et al. 1980; Vallis et al. 1982, 1985; Ball and Keeney 1983; Harper et al. 1983b; Sherwood 1983) but some is derived from N fertilizer applications (Catchpoole et al. 1983a; Harper et al. 1983a, b; Black et al. 1984, 1985b; Sherlock and Goh 1984; O'Toole et al. 1985). Nitrogen fertilizers are used in pasture systems to either sustain growth in pure grass swards or to overcome seasonal N limitations in grass/ legume swards. In New Zealand, a common practice is to apply N fertilizer about 1 week after grazing; this can result in large losses of NH_3 if the applied urea falls on a urine patch (Black et al. 1984). Losses by NH_3 volatilization from urine–affected and unaffected soils were 27% and 7% of the applied N, respectively. The effect will vary with stocking rate and period of grazing as these control the proportion of the soil surface which is affected by urine. Ammonia losses from pastures may vary from 9 to 86% of the applied N (Catchpoole et al. 1983 a, b; Black et al. 1985b; Ryden and Lockyer 1985).

In some countries, farmers burn vegetation at the end of a dry season so that, after rain, palatable green growth will be available for stock feed. During burning, 80–90% of the plant N may be lost as NH_3, N_2O and NO (Crutzen et al. 1979).

Crops

Losses of N by NH_3 volatilization after fertilizer application to upland crops range from 0–50% of the N applied (Rodgers and Pruden 1984; Freney et al. 1985b; Bacon et al. 1986; Ellington 1986; Keller and Mengel 1986). As would be expected, the loss varies with fertilizer form, method of application and stage of growth. For example, when NH_3 was applied in irrigation water to a corn crop, average volatilization losses were 7% of the

N present in the irrigation water per hour for a short crop (0.9 m) compared with 1% per hour from the tall crop (2.1 m) (Denmead et al. 1982a). Loss rates were much lower when urea was applied in the irrigation water in place of NH_3 (Freney et al. 1985b).

Bacon et al. (1986) used flow chambers to investigate the effects of initial soil moisture, rainfall, stubble retention and fertilizer management on the loss of NH_3 after fertilization of irrigated wheat. Ammonia loss from urea broadcast 0–3 days after irrigation was c.15% of the added N, but was only 4% when urea was applied 6 days after irrigation.

Ammonia losses from flooded rice, after fertilization with urea or ammonium sulfate, can also vary widely (0–47% of the applied N) depending on the method and time of application, solar radiation, windspeed, floodwater pH, and other factors (Freney et al. 1981b; Fillery et al. 1984; Simpson et al. 1984; Cai et al. 1986; Fillery and De Datta 1986; Simpson and Freney 1987). The high losses (47% of applied N, Fillery et al. 1984) occurred when urea was broadcast into the floodwater after transplanting, under conditions of high solar radiation and high wind speeds. However, when solar radiation was reduced (e.g. by cloudiness) the growth of photosynthetic organisms was inhibited and floodwater pH remained low; this resulted in low NH_3 losses (c. 9% of applied N, Cai et al. 1986).

No NH_3 loss was detected when urea (80–100 kg N/ ha) was broadcast onto the surface of either an acid soil or an alkaline soil immediately prior to permanent flooding of dry-seeded rice crops (Hoult and Bacon 1984; Humphreys et al. 1987). Similarly, when ammonium sulfate was drilled into an acid, cultivated, pasture soil prior to flooding, negligible loss by ammonia volatilization was detected (by a micrometeorological technique) after flooding (Denmead et al. 1979).

Forests

Ammonia loss from undisturbed forest floors should be low because of the low pH in that environment. However, if the system is disturbed by fertilization, clear cutting or other management practices then NH_3 may be lost. There have been some studies on NH_3 loss from fertilizer applications to forest soils (Freney et al. 1983) but to the authors' knowledge none have been made with micrometeorological techniques. Results using other methods suggest that losses following urea application vary from 4 to 48% of the N applied (Marshall and DeBell 1980; Ryabukha 1980; Voronkova 1981; Craig and Wollum 1982; Prusinkiewicz and Jósefkowicz-Kotlarz (1982); Freney et al. 1983; Foster and Beauchamp 1986). The real importance of NH_3 volatilization in this ecosystem will not be known until more reliable techniques are used for estimating NH_3 loss.

MANAGEMENT PRACTICES TO REDUCE AMMONIA LOSS

Even though it is recognised that grazing animals constitute the main source of atmospheric NH_3 (e.g. Healy et al. 1970; Galbally et al. 1980) no attempts have been made to reduce NH_3 emissions from this source. Most of the research effort has been devoted to reducing gaseous emissions from applied fertilizers so that the economic return of farmers can be improved. Many of the techniques designed to reduce loss of N after fertilizer application have been reviewed recently (Hauck 1983; Fenn and Hossner 1985; Brandon and Wells 1986; De Datta 1986; Youngdahl et al. 1986) so we will briefly discuss some of the recent findings with respect to the main fertilizer used, i.e. urea.

The important factors which affect NH_3 volatilization, such as moisture, urease activity, pH, soil surface conditions, have now been elucidated and researchers have attempted to manipulate these factors to control NH_3 loss.

Urea hydrolysis (Vlek and Carter 1983; Ezzeldin Ibrahim et al. 1984) and NH_3 loss (Stumpe et al. 1984; Black et al. 1987b) vary only slightly when urea is applied to soil at moisture contents within the plant available range. Thus, for rainfed crops, manipulation of soil moisture within the plant available range does not appear to be a feasible strategy to reduce NH_3 loss. However, for irrigated crops there is scope for reducing NH_3 loss (Fenn and Hossner 1985); e.g. addition of water after urea application reduces NH_3 volatilization and the extent of reduction depends on the degree of hydrolysis which has occurred (Black et al. 1987b).

The presence of plants and plant residues on the soil surface can influence NH_3 volatilization by acting as a source of urease, by influencing the conditions of the underlying soil and by acting as medium through which NH_3 must pass before being lost to the atmosphere (Nelson et al. 1980; Frankenberger and Tabatabai 1982; Torello and Wehner 1983; Vallis and Gardener 1984; Tomar et al. 1985; Bacon et al. 1986; Hoult and McGarity 1986). The apparent high urease levels and associated NH_3 losses from residues would appear to have implications where no-tillage systems of crop establishment are used. There is evidence that NH_3 volatilization is greater from no-tilled than ploughed soil (Mengel et al. 1982; Eckert et al. 1986). Thus some disturbance of the soil surface in no-till situations to facilitate location of urea away from urease sources would assist the retention of NH_3.

As pH significantly influences the NH_3 volatilization process, control of pH is an important management option. When lime is applied to ameliorate acid soil conditions it should be incorporated into the soil or added well before any N fertilizer application to prevent increased NH_3 loss (Matocha 1976; Adams 1986). Likewise when crop residues are burnt on the soil surface, the ash, which is alkaline, should be incorporated into the soil prior to fertilizer application (Raison and McGarity 1978; Bacon et al. 1986). In flooded rice fields NH_3 losses are closely linked to diurnal pH fluctuations of the water associated with the variation in microbial photosynthetic activity (Mikkelsen et al. 1978; Fillery et al. 1986b). The fluctuations can be controlled and NH_3 loss reduced by the addition of algicides to the floodwater (Bowmer and Muirhead 1987). Reducing the solution pH by addition of acids to irrigation water would seem to be a useful practice where cheap acids are available as industrial by-products and water-run NH_3 is used (Miyamoto et al. 1975; Yahia and Stroehlein 1980).

Modification of urea fertilizer, by addition of soluble salts of calcium and potassium, results in the formation of calcium carbonate, reducing pH elevation and NH_3 loss (Fenn and Hossner 1985). Incorporation of potassium salts reduces loss by initially displacing calcium ions from exchange sites enabling calcium carbonate formation. Rappaport and Axley (1984) reported that incorporation of potassium chloride into an acid soil reduced NH_3 loss and increased forage yield of sudan grass and maize. The ratio of potassium chloride to urea required for maximum effect ranged between 0.6 and 1.0.

Use of sulfur-coated urea rather than conventional urea reduced NH_3 volatilization. This was presumably because the urea dissolved more slowly and the subsequent rise in pH was less (Mason 1985). However, the benefits of coating in terms of reduced volatilization are not always translated into improved crop yields (Craswell and Vlek 1979; Tejeda et al. 1980; Torello et al. 1983; Cao et al. 1984; De Datta 1986). Impregnation of urea into a sulfur matrix did not reduce volatilization in a field trial in New Zealand (Black et al. 1985b).

It has been suggested that, in dryland situations, the environment around the urea granule and thus the magnitude of NH_3 loss may be manipulated by altering the granule size. However, Buresh et al. (1984) reported that granule size had no effect on NH_3 loss from broadcast urea, and Black et al. (1987a) found no significant difference in the proportion of N lost from materials ranging from powder to 5.6 mm diameter granules when applied to short pasture. However, losses were increased when 8 mm diameter granules were applied.

If the hydrolysis of urea to NH_3, by soil urease, could be retarded, then the applied urea would have a greater opportunity to penetrate the soil where it could be retained or absorbed directly by plants. Gaseous NH_3 loss should then be decreased. Thus a simple way of increasing fertilizer efficiency might be to add a urease inhibitor with the urea. Considerable effort has produced a range of chemicals which inhibit or reduce urease activity (see Sahrawat 1980; Mulvaney and Bremner 1981). The inhibitors include phenols and quinones, antimetabolites, hydroxamates, substituted ureas and phenyl ureas, pesticides, heterocyclic mercaptans and metallic compounds.

To be successful, inhibitors must fulfil several criteria, viz. (i) they must cost less than the N lost by volatilization (Gautney et al. 1984), (ii) they must be capable of being incorporated into urea and be stable during manufacture and storage (Gautney et al. 1984, 1986), (iii) they must inhibit urease activity (Mulvaney and Bremner 1981; Martens and Bremner 1984; Bremner and Chai 1986), (iv) they must reduce volatilization under field conditions (Byrnes et al. 1983; Buresh et al. 1984; Simpson et al. 1985; Fillery and De Datta 1986; Sherlock et al. 1987), (v) they must result in increased crop or pasture yields (Broadbent et al. 1985; Joo and Christians 1986; Schlegal et al. 1986) and, (vi) inhibitors must be environmentally safe.

CONCLUSIONS

Ammonia losses following fertilizer application to crops and pastures can be large, but they vary with fertilizer form, method and time of application, stage of crop growth, solar radiation and other factors. Consequently prospects for reductions in NH_3 emissions are good. Already some reduction in NH_3 loss has been achieved by management practices, including deep placement of fertilizer and the use of urease inhibitors and algicides. Further reductions are possible when an understanding of the factors affecting urease inhibition and algicide use in the field has been achieved.

REFERENCES

Adams, S.N. (1986). The interaction between liming and forms of nitrogen fertilizer on established grassland. J. Agric. Sci. 106, 509–13.

Bacon, P.E., Hoult, E.H., and McGarity, J.W. (1986). Ammonia volatilization from fertilizers applied to irrigated wheat soils. Fert. Res. 10, 27–42.

Ball, P.R., and Keeney, D.R. (1983). Nitrogen losses from urine–affected areas of a New Zealand pasture under contrasting seasonal conditions. Proc. 14th Int. Grassl. Cong., Lexington 1981, pp. 342–4.

Beauchamp, E.G. (1983). Nitrogen loss from sewage sludges and manures applied to agricultural lands. In 'Gaseous Loss of Nitrogen from Plant–Soil Systems'. (Eds J.R. Freney and J.R. Simpson.) pp. 181–94. (Martinus Nijhoff/Dr W. Junk Publishers: The Hague, Netherlands.)

Beauchamp, E.G., Kidd, G.E., and Thurtell, G. (1978). Ammonia volatilization from sewage sludge applied in the field. J. Environ. Qual. 7, 141–6.

Beauchamp, E.G., Kidd, G.E., and Thurtell, G. (1982). Ammonia volatilization from liquid dairy cattle manure in the field. Can. J. Soil Sci. 62, 11–9.

Black, A.S., Sherlock, R.R., Cameron, K.C., Smith, N.P., and Goh, K.M. (1985a). Comparison of three field methods for measuring ammonia volatilization from urea granules broadcast on to pasture. J. Soil Sci. 36, 271–80.

Black, A.S., Sherlock, R.R., and Smith, N.P. (1987a). Effect of urea granule size on ammonia volatilization from surface applied urea. Fert. Res. 11, 87–96.

Black, A.S., Sherlock, R.R., and Smith, N.P. (1987b). Effect of timing of simulated rainfall on ammonia volatilization from urea, surface applied to soil of varying initial moisture content. J. Soil Sci. (in press).

Black, A.S., Sherlock, R.R., Smith, N.P., Cameron, K.C., and Goh, K.M. (1984). Effect of previous urine application on ammonia volatilization from 3 nitrogen fertilisers. N.Z. J. Agric. Res. 27, 413–6.

Black, A.S., Sherlock, R.R., Smith, N.P., Cameron, K.C., and Goh, K.M. (1985b). Effect of form of nitrogen, season, and urea application rate on ammonia volatilisation from pastures. N.Z. J. Agric. Res. 28, 469–74.

Bouwmeester, R.J.B., Vlek, P.L.G., and Stumpe, J.M. (1985). Effect of environmental factors on ammonia volatilization from urea–fertilized soil. Soil Sci. Soc. Am. J. 49, 376–81.

Bowmer, K.H., and Muirhead, W.A. (1987). Inhibition of algal photosynthesis to control pH and reduce ammonia volatilization from rice floodwater. Fert. Res. 13, 13–30.

Brandon, D.M., and Wells, B.R. (1986). Improving nitrogen fertilization in mechanized rice culture. Fert. Res. 9, 161–70.

Bremner, J.M., and Chai, H.S., (1986). Evaluation of N–butyl phosphorothioic triamide for retardation of urea hydrolysis in soils. Commun. Soil Sci. Plant Anal. 17, 337–51.

Broadbent, F.E., Nakashima, T., and Chang, C.Y. (1985). Performance of some urease inhibitors in field trials with corn. Soil Sci. Soc. Am. J. 49, 348–51.

Buresh, R.J., Vlek, P.L.G., and Stumpe, J.M. (1984). Labeled nitrogen fertilizer research with urea in the semi arid tropics. I. Greenhouse studies. Plant Soil 80, 3–19.

Byrnes, B.H., Savant, N.K., and Craswell, E.T. (1983). Effect of a urease inhibitor phenyl phosphorodiamidate on the efficiency of urea applied to rice. Soil Sci. Soc. Am. J. 47, 270–4.

Cai, G.X., Zhu, Z.L., Trevitt, A.C.F., Freney, J.R., and Simpson, J.R. (1986). Nitrogen loss from ammonium bicarbonate and urea fertilizers applied to flooded rice. Fert. Res. 10, 203–15.

Cao, Z.H., De Datta, S.K. and Fillery, I.R.P. (1984). Effect of placement methods on floodwater properties and recovery of applied nitrogen (^{15}N-labeled urea) in wetland rice. Soil Sci. Soc. Am. J. 48, 196–203.

Catchpoole, V.R., Harper, L.A., and Myers, R.J.K. (1983a). Annual losses of ammonia from a grazed pasture fertilized with urea. Proc. 14th Int. Grassl. Cong., Lexington 1981, pp. 344–7.

Catchpoole, V.R., Oxenham, D.J., and Harper, L.A. (1983b). Transformation and recovery of urea applied to a grass pasture in south–eastern Queensland. Aust. J. Exp. Agric. Anim. Husb. 23, 80–6.

Craig, J.R., and Wollum, A.G. (1982). Ammonia volatilization and soil nitrogen changes after urea and ammonium nitrate fertilization of Pinus taeda L. Soil Sci. Soc. Am. J. 46, 409–14.

Craswell, E.T., and Vlek, P.L.G. (1979). Fate of fertilizer nitrogen applied to wetland rice. In 'Nitrogen and Rice'. pp. 175–192. (International Rice Research Institute: Los Baños, Philippines.)

Crutzen, P.J., Heidt, L.E., Krasnec, J.P., Pollock, W.H., and Seiler, W. (1979). Biomass burning as a source of atmospheric gases CO, H_2, N_2O, NO, CH_3Cl and COS. Nature (Lond.) 282, 253–6.

Dabney, S. (1982). Ammonia fluxes over an alfalfa (Medicago sativa) field. Ph.D. dissertation. Cornell University (Diss. Abstr 42/12B).

Dabney, S.M., and Bouldin, D.R. (1985). Fluxes of ammonia over an alfalfa field. Agron. J. 77, 572–8.

De Datta, S.K. (1986). Improving nitrogen fertilizer efficiency in lowland rice in tropical Asia. Fert. Res. 9, 171–86.

De Datta, S.K., Trevitt, A.C.F., Obcemea, W.N., Freney, J.R., and Simpson, J.R. (1986). Comparison of total N loss and ammonia volatilization in lowland rice using simple techniques. Agron. Abs., p. 197.

Denmead, O.T. (1983). Micrometeorological methods for measuring gaseous losses of nitrogen in the field. In 'Gaseous Loss of Nitrogen from Plant–Soil Systems'. (Eds J.R. Freney and J.R. Simpson.) pp. 133–57. (Martinus Nijhoff/Dr W. Junk Publishers: The Hague, Netherlands.)

Denmead, O.T., Freney, J.R., and Simpson, J.R. (1976). A closed ammonia cycle within a plant canopy. Soil Biol. Biochem. 8, 161–4.

Denmead, O.T., Freney, J.R., and Simpson, J.R. (1979). Nitrous oxide emission during denitrification in a flooded field. Soil Sci. Soc. Am. J. 43, 716–8.

Denmead, O.T., Freney, J.R., and Simpson, J.R. (1982a). Dynamics of ammonia volatilization during furrow irrigation of maize. Soil Sci. Soc. Am. J. 46, 149–55.

Denmead, O.T., Freney, J.R., and Simpson, J.R. (1982b). Atmospheric dispersion of ammonia during application of anhydrous ammonia fertilizer. J. Environ. Qual. 11, 568–72.

Denmead, O.T., Nulsen, R., and Thurtell, G.W. (1978). Ammonia exchange over a corn crop. Soil Sci. Soc. Am. J. 42, 840–2.

Denmead, O.T., Simpson, J.R., and Freney, J.R. (1977). A direct field measurement of ammonia emission after injection of anhydrous ammonia. Soil Sci. Soc. Am. J. 41, 1001–4.

Eckert, D.J., Dick, W.A., and Johnson, J.W. (1986). Responses of no–tillage corn grown in corn and soybean residues to several nitrogen fertilizer sources. Agron. J. 78, 231–5.

Ellington, A. (1986). Ammonia volatilization losses from fertilizers applied to acid soil in the field. Fert. Res. 8, 283–96.

Ezzeldin Ibrahim, M., Awadalla, E.A., Badr El–Din, M.M., and Kassim, A.S. (1984). Effect of rate of urea application and soil moisture on the behaviour of urea in soil. Z. Pflanzen Bodenk. 147, 177–86.

Farquhar, G.D., Firth, P.M., Wetselaar, R., and Weir, B. (1980). On the gaseous exchange of ammonia between leaves and the environment: determination of the ammonia compensation point. Plant Physiol. 66, 710–4.

Farquhar, G.D., Wetselaar, R., and Firth, P.M. (1979). Ammonia volatilization from senescing leaves of maize. Science (Wash. D.C.) 203, 1257–8.

Farquhar, G.D., Wetselaar, R., and Weir, B. (1983). Gaseous nitrogen losses from plants. In 'Gaseous Loss of Nitrogen from Plant–Soil Systems'. (Eds J.R. Freney and J.R. Simpson.) pp. 159–80. (Martinus Nijhoff/Dr W. Junk Publishers: The Hague, Netherlands.)

Faurie, G., and Bardin, R. (1979a). La volatilization de l'ammoniac. I. Influence de la nature du sol et des composes azotes. Ann. Agron. 30, 363–85.

Faurie, G., and Bardin, R. (1979b). La volatilization de l'ammoniac. II. Influence des facteurs climatiques et du couvert vegetal. Ann. Agron. 30, 401–14.

Fenn, L.B., and Hossner, L.R. (1985). Ammonia volatilization from ammonium or ammonium — forming nitrogen fertilizers. In 'Advances in Soil Science'. (Ed. B.A. Stewart) Vol. 1, pp 123–69. (Springer–Verlag: New York.)

Ferguson, R.B., and Kissel, D.E. (1986). Effect of soil drying on ammonia volatilization from surface–applied urea. Soil Sci. Soc. Am. J. 50, 485–90.

Ferguson, R.B., Kissel, D.E., Koelliker, J.K., and Basel, W. (1984). Ammonia volatilization from surface–applied urea: Effect of hydrogen ion buffering capacity. Soil Sci. Soc. Am. J. 48, 578–82.

Fillery, I.R.P., and De Datta, S.K. (1986). Ammonia volatilization from nitrogen sources applied to rice fields. I. Methodology, ammonia fluxes, and nitrogen –15 loss. Soil Sci. Soc. Am. J. 50, 80–6.

Fillery, I.R.P., De Datta, S.K., and Craswell, E.T. (1986a). Effect of phenylphosphorodiamidate on the fate of urea applied to wetland rice fields. Fert. Res. 9, 251–63.

Fillery, I.R.P., Roger, P.A., and De Datta, S.K. (1986b). Ammonia volatilization from nitrogen sources applied to rice fields. II. Floodwater properties and submerged photosynthetic biomass. Soil Sci. Soc. Am. J. 50, 86–91.

Fillery, I.R.P., Simpson, J.R., and De Datta, S.K. (1984). Influence of field environment and fertilizer management on ammonia loss from flooded rice. Soil Sci. Soc. Am. J. 48, 914–20.

Fillery, I.R.P., and Vlek, P.L.G. (1986). Reappraisal of the significance of ammonia volatilization as an N loss mechanism in flooded rice fields. Fert. Res. 9, 79–98.

Foster, N.W., and Beauchamp, E.G. (1986). Nitrogen release from urea and sulfur–coated urea in jack pine forest humus. Soil Sci. Soc. Am. J. 50, 226–9.

Frankenberger, W.T., and Tabatabai, M.A. (1982). Amidase and urease activities in plants. Plant Soil 64, 153–66.

Freney, J.R., Denmead, O.T., Watanabe, I., and Craswell, E.T. (1981b). Ammonia and nitrous oxide losses following applications of ammonium sulfate to flooded rice. Aust. J. Agric. Res. 32, 37–45.

Freney, J.R., Leuning, R., Simpson, J.R., Denmead, O.T., and Muirhead, W.A. (1985a). Estimating ammonia volatilization from flooded rice fields by simplified techniques. Soil Sci. Soc. Am. J. 49, 1049–54.

Freney, J.R., Simpson, J.R., and Denmead, O.T. (1981a). Ammonia volatilization. In 'Terrestrial Nitrogen Cycles'. (Eds F.E. Clark and T. Rosswall.) Ecol. Bull. (Stockholm) 33, 291–302.

Freney, J.R., Simpson, J.R., and Denmead, O.T. (1983). Volatilization of ammonia. In 'Gaseous Loss of Nitrogen from Plant–Soil Systems'. (Eds J.R. Freney and J.R. Simpson.) pp. 1–32. (Martinus Nijhoff/Dr W. Junk Publishers: The Hague, Netherlands.)

Freney, J.R., Simpson, J.R., Denmead, O.T., Muirhead, W.A., and Leuning, R. (1985b). Transformations and transfers of nitrogen after irrigating a cracking clay soil with a urea solution. Aust. J. Agric. Res. 36, 685–94.

Freney, J.R., Trevitt, A.C.F., Zhu, Z.L., Cai, G.X., and Simpson, J.R. (1987). Methods for estimating volatilization of ammonia from flooded rice fields. Acta Pedol. Sin. 24, 142–51.

Galbally, I.E., Freney, J.R., Denmead, O.T., and Roy, C.R. (1980). Processes controlling the nitrogen cycle in the atmosphere over Australia. In 'Biogeochemistry of Ancient and Modern Environments'. (Eds P.A. Trudinger, M.A. Walter and B.J. Ralph.) pp. 319–25. (Aust. Acad. Sci.: Canberra, Australia.)

Galbally, I.E., Freney, J.R., Muirhead, W.A., Simpson, J.R., Trevitt, A.C.F., and Chalk, P.M. (1987). Emission of nitrogen oxides (NO_x) from a flooded soil fertilized with urea: relation to other nitrogen loss processes. J. Atmos. Chem. 5, 343–66.

Gautney, J., Barnard, A.R., Penney, D.B., and Kim, Y.K. (1986). Solid state decomposition kinetics of phenyl phosphorodiamidate. Soil Sci. Soc. Am. J. 50, 792–7.

Gautney, J., Kim, Y.K., and Gagen, P.M. (1984). Feasibility of cogranulating the nitrogen loss inhibitors dicyandiamide, thiourea, phenyl phosphorodiamidate, and potassium ethyl xanthate with urea. Ind. Eng. Chem. Prod. Res. Dev. 23, 483–9.

Harper, L.A. (1988). Comparison of methods to measure ammonia volatilization in the field. In 'Ammonia Volatilization from Urea Fertilizers'. (Eds D.E. Kissel and B.R. Bock), in press. (Tennessee Valley Authority Publ.: Muscle Shoals, Alabama.)

Harper, L.A., Catchpoole, V.R., Davis, R., and Weier, K.L. (1983a). Ammonia volatilization: soil, plant, and microclimate effects on diurnal and seasonal fluctuations. Agron. J. 75, 212–8.

Harper, L.A., Catchpoole, V.R., and Vallis, I. (1983b). Ammonia loss from fertilizer applied to tropical pastures. In 'Gaseous Loss of Nitrogen from Plant–Soil Systems'. (Eds J.R. Freney and J.R. Simpson.) pp. 195–214. (Martinus Nijhoff/Dr W. Junk Publishers: The Hague, Netherlands.)

Harper, L.A., Sharpe, R.R., Langdale, G.W., and Giddens, J.E. (1987). Nitrogen cycling in a wheat crop: Soil, plant and aerial nitrogen transport. Agron. J. 79, (in press).

Hauck, R.D. (1983). Agronomic and technological approaches to minimising gaseous nitrogen losses from croplands. In 'Gaseous Loss of Nitrogen from Plant–Soil Systems'. (Eds J.R. Freney and J.R. Simpson.) pp. 285–312. (Martinus Nijhoff/Dr W. Junk Publishers: The Hague, Netherlands.)

Haunold, E., and Blochberger, K. (1982). Gaseous nitrogen losses after ammonium application. Bodenkultur 33, 95–105.

Healy, T.V., McKay, H.A.C., Pilbeam, A., and Scargill, D. (1970). Ammonia and ammonium sulfate in the troposphere over the United Kingdom. J. Geophys. Res. 75, 2317–21.

Hoult, E.H., and Bacon, P.E. (1984). N volatilization losses from an Australian rice bay. In 'National Soils Conference Proceedings'. p. 386. (Aust. Soc. Soil Sci. Inc.: Brisbane, Australia.)

Hoult, E.H., and McGarity, J.W. (1986). The measurement and distribution of urease activity in a pasture system. Plant Soil 93, 359–66.

Humphreys, E., Freney, J.R., Muirhead, W.A., Denmead, O.T., Simpson, J.R., Leuning, R., Trevitt, A.C.F., Obcemea, W.N., Wetselaar, R., and Cai, G.X. (1987). Loss of ammonia after application of urea at different times to dry–seeded, irrigated rice. Fert. Res. (in press).

Hutchinson, G.L., Mosier, A.R., and Andre, C.E. (1982). Ammonia and amine emissions from a large cattle feedlot. J. Environ. Qual. 11, 288–93.

Joo, Y.K., and Christians, N.E. (1986). The response of Kentucky bluegrass turf to phenylphosphorodiamidate (PPD) and magnesium applied in combination with urea. J. Fert. Issues. 3, 30–3.

Keller, G.D., and Mengel, D.B. (1986). Ammonia volatilization from nitrogen fertilizers surface applied to no–till corn. Soil Sci. Soc. Am. J. 50, 1060–3.

Kissel, D.E., Brewer, H.L., and Arkin, G.F. (1977). Design and test of a field sampler for ammonia volatilization. Soil Sci. Soc. Am. J. 41, 1133–8.

Knop, K. (1982). Gaseous nitrogen losses in the form of ammonia from nitrogen fertilizers. Rostlinná Výroba. 28, 935–46.

Koelliker, J.K., and Miner, J.R. (1973). Desorption of ammonia from anaerobic lagoons. Trans. Am. Soc. Agric. Eng. 16, 148–51.

Kumar, V., and Wagenet, R.J. (1984). Urease activity and kinetics of urea transformations in soils. Soil Sci. 137, 263–9.

Lemon, E., and Van Houtte, R. (1980). Ammonia exchange at the land surface. Agron J. 72, 876–83.

Leuning, R., Denmead, O.T., Simpson, J.R., and Freney, J.R. (1984). Processes of ammonia loss from shallow floodwater. Atmos. Environ. 18, 1583–92.

Leuning, R., Freney, J.R., Denmead, O.T., and Simpson, J.R. (1985). A sampler for measuring atmospheric ammonia flux. Atmos. Environ. 19, 1117–24.

Lockyer, D.R. (1984). A system for the measurement in the field of losses of ammonium through volatilisation. J. Sci. Food Agric. 35, 837–48.

Luebs, R.E., Davis, K.R., and Laag, R.E. (1973). Enrichment of the atmosphere with nitrogen compounds volatilized from a large dairy area. J. Environ. Qual. 2, 137–41.

McInnes, K.J., Ferguson, R.B., Kissel, D.E., and Kanemasu, E.T. (1986a). Ammonia loss from applications of urea – ammonium nitrate solution to straw residue. Soil Sci. Soc. Am. J. 50, 969–74.

McInnes, K.J., Ferguson, R.B., Kissel, D.E., and Kanemasu, E.T. (1986b). Field measurements of ammonia loss from surface applications of urea solution to bare soil. Agron. J. 78, 192–6.

McInnes, K.J., Kissel, D.E., and Kanemasu, E.T. (1985). Estimating ammonia flux : A comparison between the integrated horizontal flux method and theoretical solutions of the diffusion profile. Agron. J. 77, 884–9.

Mahendrappa, M.K., and Ogden, E.D. (1973). Patterns of ammonia volatilization from a forest soil. Plant Soil. 38, 257–65.

Marshall, V.G., and De Bell, D.S. (1980). Comparison of four methods of measuring volatilization losses of nitrogen following urea fertilization of forest soils. Can. J. Soil Sci. 60, 549–63.

Martens, D.A., and Bremner, J.M. (1984). Urea hydrolysis in soils : Factors influencing the effectiveness of phenylphosphorodiamidate as a retardant. Soil Biol. Biochem. 16, 515–9.

Mason, M.G. (1985). Sulfur–coated urea as a source of nitrogen for cereals in Western Australia. Aust. J. Exp. Agric. 25, 913–21.

Matocha, J.E. (1976). Ammonia volatilization and nitrogen utilization from sulfur–coated ureas and conventional nitrogen fertilizers. Soil Sci. Soc. Am. J. 40, 597–601.

Mengel, D.B., Nelson, D.W., and Huber, D.M. (1982). Placement of nitrogen fertilizers for no-till and conventional till corn. Agron. J. 74, 515-8.

Mikkelsen, D.S., De Datta, S.K., and Obcemea, W.N. (1978). Ammonia volatilization losses from flooded rice soils. Soil Sci. Soc. Am. J. 42, 725-30.

Miyamoto, S., Ryan, J., and Stroehlein, J.L. (1975). Sulfuric acid for the treatment of ammoniated irrigation water. I Reducing ammonia volatilization. Soil Sci. Soc. Am. J. 39, 544-8.

Moeller, M.B., and Vlek, P.L.G. (1982). The chemical dynamics of ammonia volatilization from aqueous solution. Atmos. Environ. 16, 709-17.

Mulvaney, R.L., and Bremner, J.M. (1981). Control of urea transformations in soil. In 'Soil Biochemistry'. (Eds E.A. Paul and J.N. Ladd.) pp. 153-96. (Marcel Dekker : New York)

Nelson, D.W. (1982). Gaseous losses of nitrogen other than through denitrification. In 'Nitrogen in Agricultural Soils'. (Ed F.J. Stevenson.) pp.327-63. (American Society of Agronomy: Madison, Wisconsin.)

Nelson, K.E., Turgeon, A.J., and Street, J.R. (1980). Thatch influence on mobility and transformation of nitrogen carriers applied to turf. Agron. J. 72, 487-92.

O'Toole, P., McGarry, S.J., and Morgan, M.A. (1985). Ammonia volatilization from urea-treated pasture and tillage soils: effects of soil properties. J. Soil Sci. 36, 613-20.

Parton, W.J., Gould, W.D., Adamson, F.J., Torbit, S., and Woodmansee, R.G. (1981). NH_3 volatilization model. In 'Simulation of Nitrogen Behaviour of Soil-Plant Systems'. (Eds M.J. Frissel and J.A. van Veen.) pp. 233-44. (PUDOC: Wageningen, Netherlands.)

Preez, C.C. De, and Burger, R. du T. (1986). A proposed mechanism for the volatilization of ammonia from fertilized neutral to alkaline soils. South Afr. J. Plant Soil 3, 31-4.

Prusinkiewicz, Z., and Jozefkowicz-Kotlarz, J. (1982). Dynamics of ammonia volatilization from the urea applied in fertilization of poor forest soils and the possibility of reducing the nitrogen losses by simultaneous application of potassium chloride. Roczniki Gleboznawcze. 33, 19-35.

Rachhpal-Singh, and Nye, P.H. (1986a). A model of ammonia volatilization from applied urea. I. Development of the model. J. Soil Sci. 37, 9-20.

Rachhpal-Singh, and Nye, P.H. (1986b). A model of ammonia volatilization from applied urea. II. Experimental testing. J. Soil Sci. 37, 21-9.

Rachhpal-Singh, and Nye, P.H. (1986c). A model of ammonia volatilization from applied urea. III. Sensitivity analysis, mechanisms, and applications. J. Soil Sci. 37, 31-40.

Raison, R.J., and McGarity, J.W. (1978). Effect of plant ash on nitrogen fertilizer transformations and ammonia volatilization. Soil Sci. Soc. Am. J. 42, 140-3.

Rappaport, B.D., and Axley, J.H. (1984). Potassium chloride for improved urea fertilizer efficiency. Soil Sci. Soc. Am. J. 48, 399-401.

Reynolds, C.M., and Wolf, D.C. (1987). Effect of soil moisture and air relative humidity on ammonia volatilization from surface applied urea. Soil Sci. 143, 144-51.

Rodgers, G.A. (1983). Effect of dicyandiamide on ammonia volatilization from urea in soil. Fert. Res. 4, 361-7.

Rodgers, G.A., and Pruden, G. (1984). Field estimation of ammonia volatilization from 15N-labelled urea fertilizer. J. Sci. Food Agric. 35, 1290-3.

Ryabukla, E.V. (1980). Gaseous losses of ammonia when Scots pine plantations are fertilized. Agrokhimiya No. 5, 17-22.

Ryden, J.C., and Lockyer, D.R. (1985). Evaluation of a system of wind tunnels for field studies of ammonia loss from grassland through volatilization. J. Sci. Food Agric. 36, 781-8.

Ryden, J.C. and McNeill, J.E. (1984). Application of the micrometeorological mass balance method to the determination of ammonia loss from a grazed sward. J. Sci. Food Agric. 35, 1297-1310.

Ryden, J.C., and Rolston, D.E. (1983). The measurement of denitrification. In 'Gaseous Loss of Nitrogen from Plant-Soil Systems'. (Eds J.R. Freney and J.R. Simpson.) pp. 91-132. (Martinus Nijhoff/Dr W. Junk Publishers: The Hague, Netherlands.)

Sahrawat, K.L. (1980). Control of urea hydrolysis and nitrification in soil by chemicals — prospects and problems. Plant Soil 57, 335-52.

San, C.H. (1986). The simple open method of measuring urea volatilization losses. Plant Soil 92, 73-9.

Schimel, D.S., Parton, W.J., Ademsen, F.J., Woodmansee, R.G., Senft, R.L., and Stillwell, M.A. (1985). The role of cattle in the volatile loss of nitrogen from a shortgrass steppe. Biogeochemistry 2, 39-52.

Schlegel, A.J., Nelson, D.W., and Sommers, L.E. (1986). Field evaluation of urease inhibitors for corn production. Soil Sci. Soc. Am. J. 78, 1007-12.

Sherlock, R.R., Black, A.S., Spackman, W.A., and Smith, N.P. (1987). Relative effectiveness of N-Butylphosphorothioic triamide (NBPT) and phenylphosphorodiamidate (PPD) in reducing ammonia volatilization from broadcast urea. Intern. Symp. 'Advances in Nitrogen Cycling in Agricultural Ecosystems'. Brisbane, Australia, May 1987, Poster abstr., pp. 69-70.

171

Sherlock, R.R., and Goh, K.M. (1984). Dynamics of ammonia volatilization from simulated urine patches and aqueous urea applied to pasture. I. Field experiments. Fert. Res. 5, 181–95.

Sherlock, R.R., and Goh, K.M. (1985a). Dynamics of ammonia volatilization from simulated urine patches and aqueous urea applied to pasture. II. Theoretical derivation of a simplified model. Fert. Res. 6, 3–22.

Sherlock, R.R., and Goh, K.M. (1985b). Dynamics of ammonia volatilization from simulated urine patches and aqueous urea applied to pasture. III. Field verification of a simplified model. Fert. Res. 6, 23–36.

Sherwood, M. (1983). Fate of nitrogen applied to grassland in animal wastes. Proc. 14th Int. Grassl. Cong., Lexington 1981, pp. 347–50.

Simpson, J.R. (1981). A modelling approach to nitrogen cycling in agro-ecosystems. In 'Nitrogen Cycling in South-East Asian Wet Monsoonal Ecosystems'. (Eds R. Wetselaar, J.R. Simpson and T. Rosswall.) pp. 459–66. (Aust. Acad. Sci.: Canberra, Australia.)

Simpson, J.R., and Freney, J.R. (1987). Interacting processes in gaseous nitrogen loss from urea applied to flooded rice fields. In 'Urea Technology and Utilisation'. (Malaysian Soc. Soil Sci.: Kuala Lumpur, Malaysia.) (in press).

Simpson, J.R., Freney, J.R., Muirhead, W.A., and Leuning, R. (1985). Effects of phenylphosphorodiamidate and dicyandiamide on nitrogen loss from flooded rice. Soil Sci. Soc. Am. J. 49, 1426–31.

Simpson, J.R., Freney, J.R., Wetselaar, R., Muirhead, W.A., Leuning, R., and Denmead, O.T. (1984). Transformations and losses of urea nitrogen after application to flooded rice. Aust. J. Agric. Res. 35, 189–200.

Simpson, J.R., and Steele, K.W. (1983). Gaseous nitrogen exchanges in grazed pastures. In 'Gaseous Loss of Nitrogen from Plant-Soil Systems'. (Eds J.R. Freney and J.R. Simpson.) pp. 215–36 (Martinus Nijhoff/Dr W. Junk Publishers: The Hague, Netherlands.)

Stumpe, J.M., Vlek, P.L.G., and Lindsay, W.L. (1984). Ammonia volatilization from urea and urea phosphates in calcareous soils. Soil Sci. Soc. Am. J. 48, 921–6.

Sudhakara, K., and Prasad, R. (1986). Ammonia volatilization losses from prilled urea, urea supergranules (USG) and coated USG in rice fields. Plant Soil 94, 293–5.

Tejeda, H.R., Hong, C.W., and Vlek, P.L.G. (1980). Comparison of modified urea fertilizers and estimation of their availability coefficient using quadratic models. Soil Sci. Soc. Am. J. 44, 1256–62.

Terman, G.L. (1979). Volatilization losses of nitrogen as ammonia from surface-applied fertilizers, organic amendments, and crop residues. Adv. Agron. 31, 189–223.

Tomar, J.S., Kirby, P.C., and MacKenzie, A.F. (1985). Field evaluation of the effects of a urease inhibitor and crop residues on urea hydrolysis, ammonia volatilization and yield of corn. Can. J. Soil Sci. 65, 777–87.

Torello, W.A., and Wehner, D.J. (1983). Urease activity in a Kentucky Bluegrass turf. Agron. J. 75, 654–6.

Torello, W.A., Wehner, D.J., and Turgeon, A.J. (1983). Ammonia volatilization from fertilized turfgrass stands. Agron. J. 75, 454–6.

Vallis, I., and Gardener, C.J. (1984). Short-term nitrogen balance in urine-treated areas of pasture on a yellow earth in the subhumid tropics of Queensland. Aust. J. Exp. Agric. Anim. Husb. 24, 522–8.

Vallis, I., Harper, L.A., Catchpoole, V.R., and Weier, K.L. (1982). Volatilization of ammonia from urine patches in a subtropical pasture. Aust. J. Agric. Res. 33, 97–107.

Vallis, I., Peake, D.C.I., Jones, R.K., and McCown, R.L. (1985). Fate of urea-nitrogen from cattle urine in a pasture-crop sequence in a seasonally dry tropical environment. Aust. J. Agric. Res. 36, 809–17.

Vlek, P.L.G., and Carter, M.F. (1983). The effect of soil environment and fertilizer modifications on the rate of urea hydrolysis. Soil Sci. 136, 56–63.

Vlek, P.L.G., Fillery, I.R.P., and Burford, J.R. (1981). Accession, transformation, and loss of nitrogen in soils of the arid region. Plant Soil 58, 133–75.

Voronkova, A.B. (1981). Nitrogen losses in ammonia form from surface-applied fertilizers in spruce forests. Soviet Soil Sci. 13, 58–64.

Wilson, J.D., Catchpoole, V.R., Denmead, O.T., and Thurtell, G.W. (1983). Verification of a simple micrometeorological method for estimating the rate of gaseous mass transfer from the ground to the atmosphere. Agric. Meteorol. 29, 183–9.

Wilson, J.D., Thurtell, G.W., Kidd, G.E., and Beauchamp, E.G. (1982). Estimation of the rate of gaseous mass transfer from a surface source plot to the atmosphere. Atmos. Environ. 16, 1861–7.

Woodmansee, R.G., Dodd, J.L., Bowman, R.A., Clark, F.E., and Dickinson, C.E. (1978). Nitrogen budget of a shortgrass prairie ecosystem. Oecologia (Berl.) 34, 363–76.

Woodmansee, R.G., Vallis, I., and Moth, J.J. (1981). Grassland nitrogen. _In_ 'Terrestrial Nitrogen Cycles'. (Eds F.E. Clark and T. Rosswall) Ecol. Bull. (Stockholm) _33_, 443-62.

Yahia, T.A., and Stroehlein, J.L. (1980). Effect of sulfuric acid on ammonia volatilization under furrow irrigation. _Libyan J. Agric._ _9_, 173-82.

Youngdahl, L.J., Lupin, M.S., and Craswell, E.T. (1986). New developments in nitrogen fertilizers for rice. _Fert. Res._ 9, 149-60.

DENITRIFICATION AT DIFFERENT TEMPORAL AND GEO-GRAPHICAL SCALES : PROXIMAL AND DISTAL CONTROLS

P.M. Groffman, J.M. Tiedje, G.P. Robertson and S. Christensen

ABSTRACT

Study of denitrification in soil is hindered by the fact that each of the primary, or proximal, factors controlling denitrification is affected by a wide range of physical and biological factors (distal factors). As the spatial and temporal scale of investigation increases, one must focus on distal rather than proximal factors to gain an understanding of denitrification at the particular scale of study. Further, understanding proximal and distal controls of denitrification at small scales of investigation is essential for understanding denitrification at higher scales. Mechanistic relationships between the proximal factors controlling denitrification and a wide range of distal factors (including soil water, organic matter distribution and decomposition, soil texture, plant community composition and organismal competition) are important as foci of study at organismal, microsite, field, landscape, regional and global scales of investigation.

INTRODUCTION

Through denitrification, nitrogenous oxides, principally nitrate and nitrite, are reduced to dinitrogen gases (Tiedje 1987) and this activity constitutes a potential source of loss of N from the ecosystem. Most denitrification in soil is carried out by bacterial respiratory processes where nitrogen oxide reduction is coupled to electron transport phosphorylation (Payne 1981). Research over the last 100 years has addressed the dynamics of denitrification at scales of resolution ranging from the gene to the globe. This paper aims to discern linkages between the factors controlling denitrification at different scales of investigation and to show how this understanding can aid the design and interpretation of denitrification research.

Denitrification in soil is primarily controlled by three main factors: oxygen, nitrate and carbon. Any of these primary factors can exert fundamental, limiting control of denitrification at the cellular level. This study of denitrification in soil is hindered because the primary, or proximal, factors controlling it are affected by many physical and biological factors (distal factors). As the scale of investigation increases above the cellular level, it usually becomes necessary to focus on distal rather than proximal factors as controllers of denitrification. Understanding proximal and distal controls at different scales of investigation is important in the design, execution and interpretation of denitrification experiments, and can be crucial to evaluating the significance of denitrification at the scale of study.

In this paper, we examine the relationships between distal and proximal factors controlling denitrification. Our thesis is that as the temporal and spatial scale of investigation increases, distal factors become increasingly more significant as foci of study. Even though the activity of denitrifying organisms is always directly controlled by proximal factors, at different scales of investigation, different factors control the dynamics of the proximal factors. The investigator must focus on the proper factors if the study is to be successful in characterizing patterns of activity in time and space, and in quantifying the significance of denitrification at the scale required.

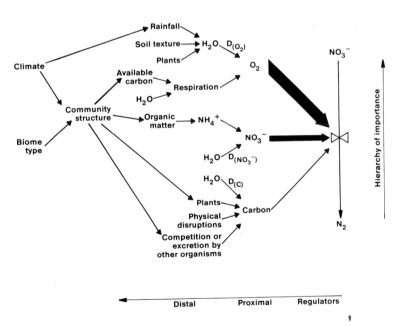

Fig. 1: Relationships between proximal and distal controlling factors of denitrification (Adapted from Tiedje 1987).

175

Figure 1 presents a simplified illustration of the mechanistic relationships between proximal and distal factors controlling denitrification. A more exhaustive set of linkages, including indirect and interactive effects, would be too complex to diagram. Figure 2 shows how the controlling factors change as the scale of investigation increases. Below, we discuss the mechanistic relationships presented in figure 1 and give examples of how the scale of investigation affects the significance of the different controlling factors.

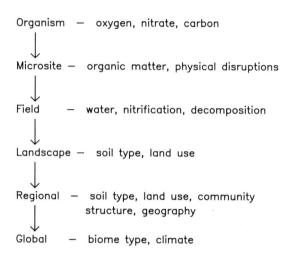

Fig. 2: Factors controlling denitrification at different scales of investigation.

RELATIONSHIPS BETWEEN PROXIMAL AND DISTAL FACTORS

Oxygen

The level of O_2 available to microorganisms in any habitat is a product of supply and demand. Oxygen flow in soil is hindered primarily by slow diffusion through water in porous media. Oxygen flow through air is c. $10^{-3}/cm^2/s$ compared to flow through particle free water of c. $10^{-5}/cm^2/s$ (Revsbech et al. 1980). The next distal factors controlling O_2 in soil are the factors that control soil water. Rainfall is an obvious controller of soil moisture, as are plants, which are a major sink for water. Soil texture is also important because soil particle size affects the adhesion of water molecules to soil particles and the flow of water through the soil matrix (Papendick and Campbell 1980).

The main sink for O_2 in soil is respiration by plant roots and by aerobic microorganisms. Consumption rates as high as 70 mM $O_2/cm^2/h$ occur in soils (Singh and Gupta 1977) and can often exceed O_2 supply rates. The distal factors affecting soil respiration are complex. Available C is the substrate for microbial respiration and is controlled by the decomposition of plant materials, exudation and/or sloughing of material from plant roots, and decomposition of microbial cells (Helal and Sauerbeck 1986). Soil water affects respiration by controlling the activities of decomposer organisms (Griffin 1980) and plant roots (Smucker 1984). Sharp changes in soil water (drying and rewetting events) and other physical disturbances increase C availability by physical rearrangement of soil particles (Adu and Oades 1978) and by killing microbial cells (van Veen et al. 1985; Kieft et al. 1987).

Nitrate

Nitrate is similar to O_2 in that the level available to denitrifiers in soil is a function of supply, demand and diffusion. Nitrate is produced by nitrification or added in fertilizer; nitrate sinks include plant uptake, microbial uptake, dissimilatory reduction to ammonium, groundwater, and denitrification. Due to its diverse fates, distal control of NO_3 can be quite complex. In N-fertilized soils, NO_3 is not likely to be limiting to denitrification since denitrifiers generally have a high affinity for NO_3 (Myrold and Tiedje 1985). In unfertilized soils, the supply of ammonium from mineralization is often a dominant control of nitrification and therefore of NO_3 availability (Robertson 1982). Nitrogen mineralization in unfertilized soils is controlled by interactive effects of soil type, climate and plant community composition (Pastor et al. 1984; Pastor and Post 1986).

Soil water affects NO_3 supply by controlling the activities of mineralizing and nitrifying organisms (Fisher and Gosz 1986). Water also facilitates diffusion and leaching of NO_3 through the soil matrix. In many water saturated environments, NO_3 levels are extremely low due to inhibition of nitrification by anaerobiosis.

Carbon

As heterotrophs, the electron accepting denitrifying process is controlled by the supply of electron donor, usually organic C. The factors controlling C availability to microorganisms are discussed above in relation to soil respiration and O_2 status. The effect of C on denitrifiers is more commonly expressed through indirect effects on O_2 status than on direct effects of substrate availability. By stimulating heterotrophic activity carbon availability reduces O_2 availability and thus favors denitrifiers over strictly aerobic organisms that cannot assimilate C under anaerobic conditions. Because competition for C is low when O_2 is limiting, C is probably the least common factor directly regulating denitrification in soils.

PROXIMAL AND DISTAL CONTROL OF DENITRIFICATION AT DIFFERENT SCALES O INVESTIGATION

Organism

When studying denitrification at the organismal level we need to focus only on the three proximal controllers. Regulation at this level is best studied using pure culture techniques under highly controlled laboratory conditions. It is now possible to extend these mechanistic studies to the molecular level, a likely area of progress in the near future. It is necessary to quantitatively understand the effects of proximal factors on denitrification at the organismal level to be able to confidently use distal factors as predictors of activity at higher spatial or temporal scales; recent data from our laboratory illustrate this point.

Table 1 shows rates of NO_3 consumption by a denitrifying organism (Pseudomonas fluorescens) and a barley plant (Hordeum vulgare) at different O_2 levels. Activity of both organisms was measured in a temperature controlled continuous culture vessel equipped with a polarographic O_2 sensing electrode. These cultivation conditions allowed for vigorous mixing and rigid control of solution O_2 concentrations. Only very small amounts of O_2 (0.05%) completely inhibited denitrification by P. fluorescens.

Table 1: Nitrate reduction by a plant (Hordeum vulgare) and a denitrifier (Pseudomonas fluorescens) at different oxygen concentrations (data from Christensen and Tiedje 1987).

Oxygen level	Denitrification	Plant uptake
%, v/v	$\mu g\ NO_3^--N/mg\ biomass/h$	
20	0	1.85
10	0	1.15
5	0	0.52
1	0	0
0.05	0	0
0.01	0.15	0
0	3.5	0

In some previous studies, denitrification activity could be measured at considerably higher (0.35 to 2.0%) O_2 concentrations (Nelson and Knowles 1978; Dunn et al. 1979). Preliminary work in our study showed that unless the culture was vigorously mixed (greater than 1000 rpm) anaerobic microsites developed in the culture vessel even though the O_2 electrode indicated significant concentrations of O_2 in the medium. Development of anaerobic sites in culture vessels may account for the high O_2 tolerance of denitrifiers reported in other studies. Our data also show that plants and denitrifiers do not directly compete for NO_3 at the organismal level. There was a wide range of intermediate O_2 concentrations at which neither the plant nor the denitrifier could take up NO_3.

These results at this level have important implications for study of denitrification at higher scales of investigation. The low O_2 tolerance of the denitrifier shows that for denitrification to occur in soil there must be virtually a complete absence of O_2. Also, at the organismal level plants and denitrifiers did not directly compete for NO_3, whereas at higher spatial and temporal scales of investigation, their competition for NO_3 can be a major factor controlling denitrification. Organismal level studies show that a range of habitats must be present in soils where plants and denitrifiers are both active. These studies thus serve as guides for investigation of the spatial heterogeneity of the soil environment as it affects denitrification at the microsite and field scale.

Microsites

The concept of microsites within a spatially heterogeneous environment has important application to study of denitrification in soil. The persistence and activity of denitrifying enzymes in apparently well aerated soil has been attributed to microsites of anaerobiosis within the soil matrix. Research and modelling efforts have addressed the presence of microsites of denitrification activity by focusing on the occurrence of anaerobic zones in soil aggregates (Smith 1980; Sexstone et al. 1985b; Lefelaar 1986). These studies have been only moderately successful at explaining the occurrence of microsite 'hotspots' of denitrification. This lack of success is due partly to the difficulty of studying soil aggregates in situ, but perhaps more fundamentally, it is because analysis of O_2 dynamics within the soil matrix does not account for the C and NO_3 factors that also control denitrification.

In addition to studying aggregates, another approach to understanding the occurrence of hotspots of denitrification in soil is to focus on the distribution and decomposition of organic C. Parkin (1987) showed that nearly 100% of the denitrification in a soil core could be attributed to a decomposing plant leaf that represented 0.08% of the mass of the core. Research in our laboratory (S. Christensen and T.M. Tiedje, unpublished data) confirmed that denitrification activity could be traced to microsites of decaying plant material and also showed that hotspots could be created by injecting concentrated portions of a non-diffusible C source (such as dead microbial cells) into soil. These results suggest that, at the microsite scale, it is useful to focus on the distribution and decomposition of organic C to understand the distribution of activity centers of denitrification in soil.

It is not surprising that organic C is a strong distal controller of denitrification at the microsite scale. Fresh plant residues or other readily decomposable compounds can provide the C necessary for a strong respiratory sink for O_2, can be a source of NO_3 through mineralization of organic N compounds present in the residue and, finally, provide reductant for the denitrifier.

The fact that organic C can provide all the necessary proximal factors for denitrification makes it a more important distal controller of denitrification activity at the microsite level than soil aggregate structure. While aggregate structure can produce anaerobic volumes of soil, C and NO_3 may not be available to denitrifiers in the aggregate centers. A promising area of research is in understanding the relationships between aggregate formation and organic C dynamics. Macro-aggregate formation and stability are tied to the dynamics of readily decomposable C compounds (Chaney and Swift 1986; Elliott 1986). Biological activity (including denitrification) associated with aggregates should thus be studied in the context of C availability and aggregate formation and turnover.

Field scale

Most investigations of denitrification have been at the field or plot scale, comparing different agricultural treatments, soil types or forest communities. Several reviews or summaries of large data sets have been published (Colbourn and Dowdell 1984; Rolston et al. 1984; Aulakh and Rennie 1986) and we can now draw some conclusions about the key factors controlling denitrification at the field scale. In agricultural soils, soil water appears to be the dominant factor controlling denitrification, since soil NO_3 levels are usually high in these soils due to fertilization. In non-agricultural soils where competition for available N is high, soil NO_3 production is frequently the key limiting factor. The role of soil C in controlling denitrification at the field scale appears to be important in some cases, but this factor has been less well studied than either NO_3 or water.

Rolston et al. (1984) modelled an extensive denitrification data set using three main factors: soil water, NO_3 concentration and available C. Soil water had the strongest effects on denitrification and NO_3 concentration the least. The dynamics of available C was the most difficult factor to model, since there is no simple direct measurement of this parameter. Other studies have also found soil water to be the dominant factor controlling denitrification in field soils (Mosier et al. 1983, 1986; Burton and Beauchamp 1985; Aulakh and Rennie 1986). Rainfall, as the next distal controller of soil moisture has also been shown to be a strong controller of denitrification at the field scale (Sexstone et al. 1985a; Duxbury and McConnaughey 1986).

The role of available C in controlling denitrification at the field scale is poorly characterized. Several models have been developed to predict the flux of available C in soil (Juma and Paul 1981; van Veen and Frissel 1981; Molina et al. 1983) but these have generally been considered to be too complicated to apply to denitrification studies (Rolston et al. 1984). The relatively dramatic effects of soil water and NO_3 on denitrification activity have also inhibited study of available C control of denitrification. However, Sexstone et al. (1985a) suggested that lack of available C can limit denitrification in field soils at certain times. They observed a lack of denitrification response to rainfall during autumn in soils that had previously shown sharp responses to rainfall. Since soil NO_3 levels were always high, they concluded that the non-responsive soils were depleted of available C.

Recent research in our laboratories shows how seasonal patterns of C availability can affect denitrification at the field scale. In a temperate forest soil in Michigan (Table 2), denitrification rates showed a decreasing response to additions of water and NO_3 through the summer (from July to August), indicating that available C was becoming depleted. In November, following C input to soil by litterfall, water plus NO_3-amended rates were about three orders of magnitude higher than they were in August. At least some tropical forest soils also show evidence of C limitation of denitrification (Table 2). High temperature, moisture, and NO_3 availability in tropical soils make conditions favorable for denitrification, which can be stimulated by C inputs as discussed previously. Patterns of C availability thus may be an important controller of denitrification at the field scale in both temperate and tropical forests. Studying available C control of denitrification by measuring the response of denitrification to different amendments appears to be useful for characterizing these patterns. Further understanding of distal control of available C levels in soil is dependent on increasing our understanding of decomposition and microbial turnover processes.

Table 2: Denitrification response to amendments in temperate and tropical forest soils.

| | Denitrification | | | |
Ecosystem	June	July	August	November
		$ng/cm^2/h$		
Temperate upland forest[1]				
Unamended	5.2	1.5	0.4	5.9
Plus water	2.9	2.7	1.0	1.6
Plus water and nitrate	113.7	23.6	4.2	1068
Tropical lowland rainforest[2]				
Plus water	110.7			
Plus water and nitrate	91.6			
Plus water and glucose	1064			

1. Groffman and Tiedje (1987a);   2. Robertson and Tiedje (1987)

Nitrate can also be a key factor controlling denitrification in soils, especially in unfertilized, non-agricultural soils with actively growing vegetation, where vigorous competition for mineral N between plants and microorganisms can severely restrict NO_3 availability. Distal control of denitrification in these soils is often centered on nitrification. In the temperate forest soil in Table 2, moisture additions did not increase denitrification rates at any time, suggesting that NO_3 was limiting activity. Nitrification potential was the strongest predictor of denitrification in a range of forest soils in Michigan and North Carolina (Table 3). Because nitrification has been shown to be linked to patterns of forest community composition and landscape position (Robertson and Vitousek 1981; Pastor et al. 1984; Zak et al. 1986), understanding how nitrification exerts distal control over denitrification at the field scale allows us to understand increasingly distal controls of denitrification at landscape and regional levels.

Table 3: Predictors of denitrification in forest soils.

Site	Coefficient of determination[1]	Probability level
12 Michigan sites[2]		
Nitrification potential	0.356	0.041
CO_2 production	0.488	0.049
Percent moisture	0.531	0.094
8 clearcut North Carolina treatment plots[3]		
Nitrification potential	0.88	0.001
Nitrate pool size	0.55	0.05
Old-field (successional) site (238 samples)[4]		
Nitrate pool size	0.286	0.001
CO_2 production	0.433	0.001

1. Coefficient of determination (r^2) values are additive going down the column for the Michigan and the old-field sites; 2. Robertson and Tiedje (1984); 3. Robertson et al. (1987); 4. Robertson et al. (1988)

One poorly characterized area of denitrification investigation in agricultural soils is activity during the non-crop season. Estimates of annual N loss to denitrification are generally considered to be non-significant in an agronomic context (Colbourn and Dowdell 1984; Mosier et al. 1986), but the vast majority of agricultural field studies have only measured activity between planting and harvest of a summer crop. This crop season-based sampling often eliminates the periods of highest soil water from the investigation. Available C is also likely to be high in the 'off' season, due to residue input from harvest, plowing, or disruption by freeze/thaw cycles.

Soil NO_3 levels are likely to be high during this time due to a lack of plant uptake. Several studies have shown that most of the annual N loss to denitrification occurs during the non–crop season (Groffman 1985; Aulakh and Rennie 1986; Mahli and Nyborg 1986). Mahli and Nyborg (1986) from a number of experiments in Canada found an average of almost 50 kg N/ha were denitrified during winter. Clearly, studies of denitrification in agricultural field soils need to consider non–crop season activity.

Landscape

There are few data on denitrification at scales of investigation larger than the field scale. Landscape and regional scale data are needed for calculation of large scale N budgets and for atmospheric chemistry questions affected by soil–atmosphere N gas flux (Robertson 1986). We have begun to experimentally address landscape scale dynamics of denitrification by focusing on distal, landscape scale factors that influence denitrification. Such studies form the basis for regional and global scale studies of denitrification.

We have studied landscape scale dynamics of denitrification using an approach based on soil classification. Soil texture and drainage are landscape scale parameters that have strong effects on denitrification, largely through their influence on soil water. Drainage class is a general indicator of soil wetness and soil texture has strong affects on the ability of soil to hold and transmit water. Texture also affects denitrification because fine textured soils have smaller pores that more easily become anaerobic than the large pores present in coarse textured soils (Papendick and Campbell 1980).

Soils are organized in coherent patterns across the landscape that are useful as guides for landscape scale experimental designs. Soil texture is controlled by parent material, which is related to broad scale geological and physiographic features. Soils of similar texture are often found in adjacent topographic positions and thus differ in soil drainage. The coherent patterns of soil type in the landscape, and the fact that soils are well mapped for large regions of the world, make soil texture and drainage potentially very useful parameters for study of denitrification at the landscape scale.

Table 4 presents estimates of annual N loss to denitrification for nine forest soils of different texture and drainage in Michigan. Soils of similar texture were located in adjacent slope positions in catenas. Catenas of different texture were associated with different glacial landforms – moraine, till plain, outwash plain. Annual N loss to denitrification was determined by repeated sampling using soil cores over the course of a year.

Soil texture and drainage appear to be very strong predictors of denitrification activity at the landscape scale. We quantified soil drainage by calculating a Drainage Index parameter developed by soil geographers (Hole

Table 4: Annual N loss to denitrification, soil drainage (Drainage Index) and soil texture (percent sand) for forested soils in Michigan.

Soil type	Denitrification N loss[1]	Drainage Index[2]	Soil texture[2]
	kg N/ha/y		*% sand*
Loam			
Well drained	10	40	26
Somewhat poorly drained	11	50	42
Poorly drained	24	70	18
Clay loam			
Well drained	18	44	34
Somewhat poorly drained	17	64	35
Poorly drained	40	74	15
Sandy loam			
Well drained	0.6	40	76
Somewhat poorly drained	0.8	60	74
Poorly drained	0.5	61	77

1. Groffman and Tiedje (1987a); 2. Groffman and Tiedje (1987b)

1978; Schaetzl 1986). Drainage Index is largely based on soil drainage class but is modified by soil great group, sub-group and textural factors. The index provides a measure of general soil wetness and is a continuous numerical variable suitable for regression analysis. Percent sand (from mechanical analysis) was a useful variable for regression analysis of soil texture. In a multiple regression model, we were able to explain 86% of the variability in annual N loss to denitrification with soil texture (% sand) and drainage (Drainage Index).

The results in Table 4 suggest that large scale studies of denitrification may be productive. We were able to explain a much higher percentage of the variability in annual N loss to denitrification at the landscape scale than we could at the field scale (within each of our 9 soils), for example, over 80% of the variability in annual N loss to denitrification at the landscape scale was explained by distal factors such as soil texture and drainage, whereas we were seldom able to explain more than 50% of the variability in daily or hourly rates of denitrification at the field scale using more proximal factors such as soil water, NO_3 or CO_2 production. Increasing the scale of investigation in both time (annual loss rather than hourly or daily rates) and space (landscape rather than field scale) may be useful for overcoming the variability problems frequently encountered in denitrification research (Folonoruso and Rolston 1984; Robertson and Tiedje 1984; Burton and Beauchamp 1985).

Regional

To address regional dynamics of denitrification, we need to extrapolate data from landscape studies, and to investigate land use and plant community variables as distal controllers of denitrification. Few data address the effects of land use on denitrification, for example comparing adjacent forest and agricultural areas. There is however, much data on denitrification in different agricultural management systems that can be assembled for regional scale studies, or that can at least serve as a guide for the type of data needed. While there is relatively little data on denitrification in most natural communities (Bowden 1986), we have accumulated data on denitrification in successional communities in both tropical and temperate forests (Table 5) that can serve as a guide for the collection of data for regional scale studies.

Table 5: Denitrification rates in a temperate and a tropical forest sere. Values are means (+ standard error).

Successional stage	Denitrification rate	
	Tropical[1]	Temperate[2]
	$ng/cm^2/h$	
Early	24.3(2.2)	1100(370)
Mid	4.8(1.2)	15(8)
Late	15.0(3.5)	36(28)

1. Robertson and Tiedje (1987); 2. Robertson and Tiedje (1984).

Studying denitrification in different plant communities within a successional sere appears to be a useful approach to characterizing regional scale dynamics of denitrification in forest communities. Table 5 shows that in both a temperate and a tropical sere, denitrification is relatively high in early and late successional communities and is low in mid-successional communities. This pattern of denitrification is consistent with overall successional patterns of nutrient cycling and loss as outlined by Vitousek and Reiners (1975) and Bormann and Likens (1979). Nitrogen losses and soil NO_3 levels in mid-successional, aggrading forests are often very low due to intense competition for N between and among plants and microorganisms. Early and late successional communities may have higher N losses than mid-successional communities due to less intense levels of competition and a more patchy distribution of vegetation, although patterns in early succession may vary with type of disturbance (Gorham et al. 1979). For regional scale studies, data on successional patterns of denitrification can be combined with data on the effects of soil type on denitrification to produce estimates of denitrification in forest components of a regional area.

Extrapolating data from the landscape to a regional scale can be done using geographic information systems (GIS). These systems contain information about soils, land use, hydrology and other factors for areas ranging from tens to thousands of square kilometers. Simple algebra can be used to incorporate biological functions into the GIS database to produce large scale estimates of biological processes like denitrification. We are currently working with a simple GIS for the State of Michigan that contains information about land use and soil type. We are using this GIS to extrapolate our landscape scale data to produce estimates of denitrification from forest soils for a large area of Michigan. To do this, all soils in the region are assigned to one of nine groups represented by the nine soils in our original study. Thus, each soil mapping unit is assigned one of nine estimates of annual N loss to denitrification. The GIS can calculate the areal extent of each soil type under forest cover and can thus calculate regional estimates of annual N loss to denitrification. While this estimate requires broad extrapolation of our landscape data, it is an improvement on previous regional scale studies of N cycling processes. Our nine groups contain two orders of magnitude of variability in annual N loss estimates and provide considerably higher resolution than any previous studies that have assigned single values to much larger study areas. Future work in this area will incorporate successional information to our models for forest soils and will address the effects of other land uses on denitrification within our study region.

To test the validity of our broad scale extrapolations, we are developing methods to verify landscape or regional scale patterns of denitrification. Figure 3 shows that the annual mean (nine sample dates) of denitrification enzyme activity (DEA) was strongly related to annual N loss to denitrification among the nine soils in our landscape scale study. While it is quite laborious to measure annual N loss to denitrification, DEA can be quantified relatively easily since is varies very little in time and space relative to actual denitrification rate. Within any one soil in our landscape study, DEA varied by a factor of 2 or 3 over the course of the year while actual denitrification rate varied by 2 or 3 orders of magnitude. DEA can thus be measured only a few times, at a large number of sites, to verify extrapolation of landscape patterns of denitrification to a regional scale. We measured DEA (on one occasion only) in 18 soils taken over a wide area of Michigan similar to those in our landscape scale study. We found that the strong patterns of denitrification with soil texture and drainage that we observed at the landscape scale held up at a regional scale, and thus extrapolating results from our landscape scale studies to a regional level was justified.

Our success at characterizing landscape and regional scale dynamics of denitrification is encouraging for developing methods for characterizing large scale dynamics of microbial processes in general. Figure 4 shows that DEA was reasonably closely related to microbial biomass C in the group of 18 soils that we sampled across a wide area of Michigan, and in turn microbial biomass C was strongly related to soil texture (Fig. 5, see also Merckyx et al. 1985; van Veen et al. 1985). Microbial biomass C and N content are also related to

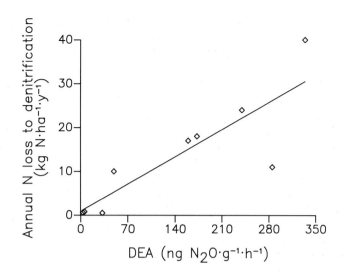

Fig. 3: Annual N loss to denitrification versus denitrification enzyme activity (DEA, phase I assay, Smith and Tiedje 1979) for nine forest soils of different texture and drainage classes in Michigan, USA. (Adapted from Groffman and Tiedje 1987b).

other landscape scale parameters such as land use, succession and soil drainage. Easily measured, low variance parameters like DEA and microbial biomass C and N content, that are strongly related to biological fluxes such as denitrification and soil respiration may be very useful for studying

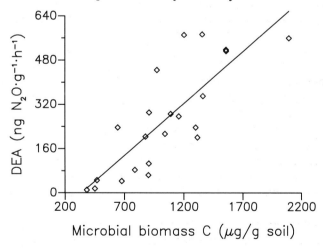

Fig. 4: Denitrification enzyme activity (DEA) versus microbial biomass carbon content (chloroform fumigation–incubation method, Jenkinson and Powlson 1976) in 18 forest soils of different texture and drainage classes in Michigan, USA. (Adapted from Groffman et al. 1987).

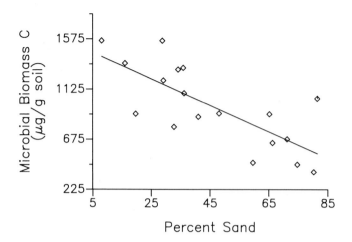

Fig. 5: Microbial biomass carbon content versus percent sand in 18 forest soils of different texture and drainage classes in Michigan, USA. (Adapted from Groffman et al. 1987).

regional scale dynamics of biogeochemical processes in general.

A potentially useful tool for landscape and regional scale studies of microbial processes is remote sensing. Figure 6 shows a conceptual model of soil water–plant productivity–N cycling and N loss relationships and how remote sensing can be used to study these relationships. Soil water can be remotely–sensed by microwave radiometry (Schmugge 1983), and plant production (Goward et al. 1985) and N content (Spanner et al. 1985) can be remotely–sensed using space satellites. We are currently investigating the use of remotely–sensed soil water as a predictor of denitrification at the landscape scale. Remotely–sensed data may be especially useful for large scale studies when combined with GIS data on soil type and hydrology.

Global

Global estimates of N cycle processes, including denitrification, have been compiled (Soderlund and Svensson 1976; Rosswall 1981; Banin et al. 1984). The approaches used assembled data from different ecosystem or biome types and matched these data with estimates of the areal extent of each biome to produce global estimates. Since the available sets of denitrification data are scattered in time and space, highly variable in methodology, and often very limited in spatial and/or temporal scope, the current global estimates of denitrification are highly uncertain. Global scale estimates of biogeochemical processes like denitrification are important for understanding the effects of land use changes and other anthropogenic effects on atmospheric chemistry and climate.

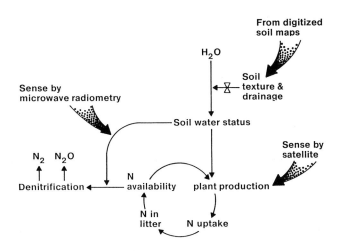

Fig. 6: Model of denitrification control and how remote sensing could improve large scale estimates of the process.

There are several obvious routes to improve our understanding of the global dynamics of denitrification. Fairly good data sets of relative comparisons of biome productivity have been assembled (Webb et al. 1983) and there are strong links between biomass productivity and climate and microbial processes (Heal and Ineson 1984). Our landscape scale studies have shown that there are strong relationships between microbial processes in general and denitrification. If enough landscape or regional scale denitrification data sets become available, then relatively accurate global N budgets could be produced. An organized effort towards producing these data sets, perhaps within the aegis of existing global scientific programs could be worthwhile.

ACKNOWLEDGEMENTS

The preparation of this paper and much of our work reported herein was supported by the U.S. National Science Foundation, principally grant BSR–83–18387. We would like to thank Stephen Simkins, Chuck Rice and Els Aerssens for contributing to discussions relating to this paper.

REFERENCES

Adu, J.K., and Oades, J.M. (1978). Physical factors influencing decomposition of organic materials in soil aggregates. Soil Biol. Biochem. 10, 109–15.

Aulakh, M.S., and Rennie, D.A. (1986). Nitrogen transformations with special reference to gaseous N losses from zero–tilled soils of Saskatchewan, Canada. Soil Till. Res. 7, 157–71.

Banin, A., Lawless, J.G., and Whitten, R.C. (1984). Global N_2O cycles–terrestrial emissions, atmospheric accumulation and biospheric effects. Adv. Space Res. 4, 207–16.

Bormann, F.H. and Likens, G.E. (1979). 'Pattern and Process in a Forested Ecosystem.' (Springer–Verlag: New York.)

Bowden, W.B. (1986). Gaseous nitrogen emissions from undisturbed terrestrial ecosystems: An assessment of their impacts on local and global nitrogen budgets. Biogeochemistry 2, 249–79.

Burton, D.L., and Beauchamp, E.G. (1985). Denitrification rate relationships with soil parameters in the field. Commun. Soil Sci. Plant Anal. 16, 539–49.

Chaney, K., and Swift, R.S. (1986). Studies on aggregate stability. I. Re–formation of soil aggregates. J. Soil Sci. 37, 329–35.

Christensen, S., and Tiedje, J.M. (1987). Oxygen sensitivity of denitrifying pseudomonads and barley roots for nitrate utilization. Appl. Environ. Microbiol. (in press).

Colbourn, P., and Dowdell, R.J. (1984). Denitrification in field soils. Plant Soil 76, 213–26.

Dunn, G.M., Herbert, R.A., and Brown, C.M. (1979). Influence of oxygen tension on nitrate reduction by a Klebsiella sp. growing in chemostat culture. J. Gen. Microbiol. 112, 379–83.

Duxbury, J.M., and McConnaughey, P.K. (1986). Effect of fertilizer source on denitrification and nitrous oxide emissions in a maize–field. Soil Sci. Soc. Am. J. 50, 644–8.

Elliott, E.T. (1986). Aggregate structure and carbon, nitrogen, and phosphorus in native and cultivated soils. Soil Sci. Soc. Am. J. 50, 627–33.

Fisher, F.M., and Gosz, J.R. (1986). Effects of trenching on soil processes and properties in a New Mexico mixed–conifer forest. Biol. Fertil. Soils 2, 35–42.

Folonoruso, O.A., and Rolston, D.E. (1984). Spatial variability of field–measured denitrification gas fluxes. Soil. Sci. Soc. Am. J. 48, 1213–9.

Gorham, E., Vitousek, P.M., and Reiners, W.A. (1979). The regulation of chemical budgets over the course of terrestrial ecosystem succession. Ann. Rev. Ecol. Syst. 10, 53–88.

Goward, S.N., Tucker, C.J., and Dye, D.G. (1985). North American vegetation patterns observed with the NOAA–7 advanced very high resolution radiometer. Vegetatio 64, 3–14.

Griffin, D.M. (1980). Water potential as a selective factor in the microbial ecology of soils, In 'Water Potential Relations in Soil Microbiology'. (Soil Science Society of America: Madison, WI). SSSA Spec. Publ. No. 9, pp. 141–51.

Groffman, P.M. (1985). Nitrification and denitrification in conventional and no–tillage soils. Soil Sci. Soc. Am. J. 49, 329–34.

Groffman, P.M., Tiedje, J.M., Mokma, D.L., and Simkins, S. (1987). Regional scale estimates of denitrification in north temperate forest soils. Biogeochemistry (in press).

Groffman, P.M., and Tiedje, J.M. (1987a). Denitrification in north temperate forest soils. I. Spatial and temporal patterns. Soil Biol. Biochem. (in press).

Groffman, P.M., and Tiedje, J.M. (1987b). Denitrification in north temperate forest soils. II. Relationships between denitrification and microbial biomass parameters. Soil Biol. Biochem. (in press).

Heal, O.W., and Ineson, P. (1984). Carbon and energy flow in terrestrial ecosystems: Relevance to microflora. In 'Current Perspectives in Microbial Ecology'. (Eds M.J. Klug and C.A. Reddy), pp. 394–404. (American Society of Microbiology: Washington.)

Helal, H.M., and Sauerbeck, D. (1986). Effect of plant roots on carbon metabolism of soil microbial biomass. Z. Pflanzenernaehr. Bodenk. 149, 181–8.

Hole, F.D. (1978). An approach to landscape analysis with emphasis on soils. Geoderma 21, 1–23.

Jenkinson, D.S., and Powlson, D.S. (1976). The effects of biocidal treatments on metabolism in soil. V. A method for measuring soil biomass. Soil Biol. Biochem. 8, 209–13.

Juma, N.G., and Paul, E.A. (1981). Use of tracers and computer techniques to assess mineralization and immobilization of soil nitrogen. In 'Simulation of Nitrogen Behaviour of Soil–Plant Systems'. (Eds M.J. Frissel and J.A. van Veen) pp. 145–54. (PUDOC: Wageningen, The Netherlands.)

Kieft, T.L., Soroker, E., and Firestone, M.K. (1987). Microbial biomass response to a rapid increase in water potential when dry soil is wetted. Soil Biol. Biochem. 19, 119–26.

Lefelaar, P.A. (1986). Dynamics of partial anaerobiosis, denitrification and water in a soil aggregate. Soil Sci. 142, 352–8.

Malhi, S.S., and Nyborg, M. (1986). Increase in mineral N in soils during winter and loss of mineral N during early spring in north–central Alberta. Can. J. Soil Sci. 66, 397–409.

Merckx, R., Hartog, A. den, and Veen, J.A. van (1985). Turnover of root–derived material and related microbial biomass formation in soils of different texture. Soil Biol. Biochem. 17, 565–9.

Molina, J.A.E., Clapp, C.E., Shaffer, M.J., Chicester, F.W., and Larson, W.E. (1983). NCSOIL, a model of nitrogen and carbon transformations in soil: Description, calibration, and behavior. Soil Sci. Soc. Am. J. 47, 85–91.

Mosier, A.R., Guenzi, W.D., and Schweizer, E.E. (1986). Soil losses of dinitrogen and nitrous oxide from irrigated crops in northeastern Colorado. Soil Sci. Soc. Am. J. 50, 344–8.

Mosier, A.R., Parton, W.J., and Hutchinson, G.L. (1983). Modelling nitrous oxide production from cropped and native soils. In 'Environmental Biogeochemistry'. (Ed. R. Hallberg), pp. 229–42. (Ecological Bulletins: Stockholm.)

Myrold, D.D., and Tiedje, J.M. (1985). Diffusional constraints on denitrification in soil. Soil Sci. Soc. Am. J. 49, 651–7.

Nelson, L.M., and Knowles, R. (1978). Effect of oxygen and nitrate on nitrogen fixation and denitrification by _Azospirillum brasiliense_ grown in continuous culture. Can. J. Microbiol. 24, 1395–403.

Papendick, R.I., and Campbell, G.S. (1980). Theory and measurement of water potential. In 'Water Potential Relations in Soil Microbiology'. (Soil Science Society of America: Madison, Wisconsin.), SSSA Spec. Publ. No. 9, pp. 1–22.

Parkin, T.B. (1987). Soil microsites as a source of denitrification variability. Soil Sci. Soc. Am. J. (in press).

Pastor, J., Aber, J.D., and McClaugherty, C.A. (1984). Aboveground production and N and P cycling along a nitrogen mineralization gradient on Blackhawk Island, Wisconsin. Ecology 65, 256–68.

Pastor, J., and Post, W.M. (1986). Influence of climate, soil moisture, and succession on forest carbon and nitrogen cycles. Biogeochemistry 2, 3–27.

Payne, W.J. (1981). 'Denitrification'. (John Wiley and Sons: New York.)

Revsbech, N.P., Sorensen, J., Blackburn, T.H., and Lomholt, J.P. (1980). Distribution of oxygen in marine sediments measured wtih microelectrodes. Limnol. Oceanogr. 25, 403–11.

Robertson, G.P. (1982). Nitrification in forested ecosystems. Phil. Trans. R. Soc. Lond. B 296, 445–57.

Robertson, G.P. (1986). Nitrogen: Regional contributions to the global cycle. Environment 28, 16–21.

Robertson, G.P., Huston, M.A., Evans, F.C., and Tiedje, J.M. (1988). Spatial complexity in a successional plant community: Patterns of nitrogen availability. Ecology (in press).

Robertson, G.P., and Tiedje, J.M. (1984). Denitrification and nitrous oxide production in old growth and successional Michigan forests. Soil Sci. Soc. Am. J. 48, 383–9.

Robertson, G.P., and Tiedje, J.M. (1987). Denitrification in a lowland tropical rain forest. Nature (in press).

Robertson, G.P., and Vitousek, P.M. (1981). Nitrification in primary and secondary succession. Ecology 62, 376–86.

Robertson, G.P., Vitousek, P.M., Matson, P.A., and Tiedje, J.M. (1987). Denitrification in a clearcut Loblolly pine (_Pinus taeda_ L.) plantation in the southeastern US. Plant Soil 97, 119–29.

Rolston, D.E., Rao, P.S.C., Davidson, J.M., and Jessup, R.E. (1984). Simulation of denitrification losses of nitrate fertilizer applied to uncropped, cropped and manure–amended field plots. Soil Sci. 137, 270–9.

Rosswall, T. (1981). The biogeochemical nitrogen cycle. In 'Some Perspectives of the Major Biogeochemical Cycles'. (Ed. G.E. Likens), pp. 25–50. (J. Wiley and Sons: New York.)

Schaetzl, R.J. (1986). Soilscape analysis of contrasting glacial terrains in Wisconsin. Ann. Assoc. Am. Geog. 76, 414–25.

Schmugge, T. (1983). Remote sensing of soil moisture with microwave radiometers. Trans. ASAE. 26, 748–53.

Sexstone, A.J., Parkin, T.B., and Tiedje, J.M. (1985a). Temporal response of soil denitrification rates to rainfall and irrigation. Soil Sci. Soc. Am. J. 49, 99–103.

Sexstone, A.J., Revsbech, N.P., Parkin, T.P., and Tiedje, J.M. (1985b). Direct measurement of oxygen profiles and denitrification rates in soil aggregates. Soil Sci. Soc. Am. J. 49, 645–51.

Singh, J.S., and Gupta, S.R. (1977). Plant decomposition and soil respiration in terrestrial ecosystems. Bot. Rev. 43, 449–528.

Smith, K.A. (1980). A model of the extent of anaerobic zones in aggregated soils, and its potential application to estimates of denitrification. J. Soil Sci. 31, 263–77.

Smith, M.S., and Tiedje, J.M. (1979). Phases of denitrification following oxygen depletion in soils. Soil Biol. Biochem. 11, 262–7.

Smucker, A.J.M. (1984). Carbon utilization and losses by plant root systems. In 'Roots, Nutrient and Water Influx, and Plant Growth'. (Eds S. Barber and D.R. Bouldin), pp. 27–46. (Amer. Soc. Agron: Madison, WI.)

Soderlund, R., and Svensson, B.H. (1976). The global nitrogen cycle. In 'Nitrogen, Phosphorus, and Sulfur–Global Cycles'. (Eds B.H. Svensson and R. Soderlund), pp. 23–73. (Ecological Bulletins: Stockholm.)

Spanner, M.A., Petersen, D.L., Acevedo, W., and Matson, P. (1985). High–resolution spectrometry of leaf and canopy chemistry for biogeochemical cycling. In 'Proceedings of the Airborne Imaging Spectrometer Data Analysis Workshop'. (Eds G. Vane and A.F.H. Goetz.), pp. 92–9. (Jet Propulsion Laboratory Publication 85–41: Pasadena, CA.)

Tiedje, J.M. (1987). Ecology of denitrification and dissimilatory nitrate reduction to ammonium. In 'Environmental Microbiology of Anaerobes'. (Ed. A.J.B. Zehnder) in press. (John Wiley & Sons: New York.)

Veen, J.A. van, and Frissel, M.J. (1981). Simulation model of the behavior of N in soil. *In* 'Simulation of Nitrogen Behavior of Soil-Plant Systems'. (Eds M.J. Frissel and J.A. van Veen), pp. 126-44. (PUDOC: Wageningen, The Netherlands.)

Veen, J.A. van, Ladd, J.N., and Amato, M. (1985). Turnover of carbon and nitrogen through the microbial biomass in a sandy loam and a clay soil incubated with [^{14}C(U) glucose and [^{15}N](NH$_4$)$_2$SO$_4$ under different moisture regimes. *Soil Biol. Biochem.* **17**, 747-56.

Vitousek, P.M., and Reiners, W.A. (1975). Ecosystem succession and nutrient retention: A hypothesis. *Bioscience* **25**, 376-81.

Webb, W.L., Lauenroth, W.K., Szarek, S.R., and Kinerson, R.S. (1983). Primary production and abiotic controls in forests, grasslands, and desert ecosystems in the United States. *Ecology* **64**, 134-51.

Zak, D.R., Pregitzer, K.S., and Host, G.E. (1986). Landscape variation in nitrogen mineralization and nitrification. *Can. J. For. Res.* **16**, 1258-63.

LEACHING

R.E. White

ABSTRACT

The rate of nitrate leaching is governed by the volume rate of water flow across the bottom of the root zone and the concentration of nitrate ions in that water. Water fluxes are highly variable because of the episodic nature of inputs, the heterogeneity of flow pathways in the soil and the non-uniform extraction of water by plants. Soil solution concentrations are similarly spatially and temporally variable because of N transformations and losses, fertilizer inputs, return of excreta and plant uptake. Because of this complexity and variability, deterministic models, such as the convection-dispersion equation (CDE), which use single value parameters, are of limited worth for describing field-scale transport of nitrate. One approach to solving this problem is to retain the mechanistic framework of the CDE, but to introduce stochasticity into the key parameters − pore water velocity v and dispersion coefficient D. Alternatively, a priori assumptions of transport mechanisms can be abandoned and the fate of nitrate within a defined volume of soil treated stochastically. Sample realizations of the probability density function of nitrate travel times between chosen entry and exit surfaces are observed and yield a space and time-averaged transport volume. This, together with a measure of the mean soil solution nitrate concentration, forms the basis of a predictive model of nitrate leaching. Both approaches have scope for dealing with the spatial and temporal variability of nitrate transport in the field, and assessment of their relative merits awaits further testing using quality field data.

INTRODUCTION

Leaching involves the transport of N in water-soluble forms out of a defined soil volume, usually the root zone, into a subsoil region that may or may not contain a watertable. The watertable is the upper boundary of perched or regional groundwater, but in either case, movement of solutes can occur within the groundwater so that subsequently they may appear in surface waters. Flow of water and solutes through soil to groundwater in response to rainfall is predominantly one-dimensional, although on sloping land two and three-dimensional flow can occur, including interflow through the vadose zone above the watertable. In agricultural land, water and solutes moving through the root zone can be intercepted by drainage systems (e.g. mole and tile drains) which usually discharge directly into surface waters.

Although more than 90% of soil N is in insoluble organic forms, much of the remainder (mineral N) is present as NO_3^- in solution. Ammonium ions produced through ammonification of organic N compounds, or by the hydrolysis of urea (in urine), or added in fertilizers, are generally rapidly oxidized by chemoautotrophic bacteria to nitrate. In soils of pH < 5 and high organic matter content, heterotrophic nitrifying organisms may achieve the same result using organic N compounds (Focht and Verstraete 1977; Adams 1986). Hence, N is leached from soil almost entirely as NO_3^-, and even when large amounts of slurry, animal manure or sewage sludge are added to soil, the bulk of the N that is leached is nitrate (Anon. 1983).

This review concentrates on nitrate leaching through soil to surface water and groundwater. The topic is emotive, not only because making good the loss of N is expensive, but also because nitrate in potable waters at concentrations > 10 (Environmental Protection Agency 1976) or 11.3 mg N/L (European Community 1980) is considered unacceptable for health reasons. Several other reviews of N leaching have appeared in recent years (Wild and Cameron 1980; Nielsen et al. 1982; Cameron and Scotter 1986; Keeney 1986) so the treatment will be selective. First, I shall address the factors of soil, climate, vegetation system and management which appear to influence the rate of nitrate leaching. An attempt will be made to extract generalizations from appropriate lysimeter and field leaching measurements. Next, in the light of these generalizations about factors influencing nitrate leaching, I shall examine some of the approaches used in modelling nitrate leaching under field conditions, and the problems encountered. Finally, the more promising avenues for future progress in the prediction of nitrate leaching will be identified.

MEASUREMENT OF THE RATE OF NITRATE LEACHING

A priori, the two major determinants of the rate of NO_3^- leaching are the quantity of water passing across a surface of interest (the bottom of the root zone or the watertable) per unit time and the concentration of NO_3^- in that water. Measurement of these two variables presents conceptual and technical problems arising from the difficulty of access for sampling (especially for measurements deep in the subsoil), and the spatial and temporal variability of both water flux and NO_3^- concentration: so much so that Keeney (1986) pessimistically concluded that there were few options available for estimating losses in solution from the root zone. The options are:

(i) Lysimeters, where the soil volume is precisely defined and inputs and outputs of water and NO_3^- can be easily measured (although N transformations in the soil can present problems). Wild and Cameron (1980) have discussed the advantages and disadvantages of lysimeter studies.

(ii) Sampling the soil solution using porous cups installed at various depths, combined with measurements of water flux from a soil water balance, or calculated from Darcy's Law using measured values of hydraulic conductivity and head gradient. The latter are usually unsatisfactory when transient, unsaturated flow occurs through the soil. A variation on this option is to sample the soil solution using small cores which are then extracted for the measurement of NO_3^- (Mohammed et al. 1984; Field et al. 1985).

(iii) Monitoring flow rate and NO_3^- concentrations in tile drainage. Intermittent sampling during drainflow is unsatisfactory (Kolenbrander 1969; Cooke and Williams 1970) because the concentration can vary unpredictably with flow rate (Wild and Cameron 1980; White 1985a). Consequently, samples should be taken at a frequency proportional to the flow rate (Hood 1976; Haigh and White 1986) so that the measured nitrate concentration is automatically weighted with respect to flow volume. Nevertheless, Macduff and White (1984) found that when drainage from a small agricultural catchment was sampled bi-weekly over several months, the quantity of N leached calculated from unweighted concentrations was only 14% less than the 32.2 kg N/ha calculated from weighted concentrations. A further complication is that the drains may not intercept all the percolating water and therefore do not properly integrate the variability in leaching that occurs over a whole field (Thomas and Barfield 1974).

(iv) Monitoring NO_3^- concentrations and stream flow rates in catchments: this is a more problematic method because of uncertainties about contributions from surface runoff and channel precipitation (usually small), the impermeability of the catchment substratum, and the delay in reaching the measuring point for contributions from different parts of the catchment, which might have different N inputs. However, where such qualifications are immaterial, or can be accounted for, catchment monitoring provides useful data for NO_3^- leaching on a large scale (Wild and Cameron 1980; White et al. 1983).

(v) Groundwater sampling, combined with a water balance for the vadose zone: in this case, output from a one-dimensional system is superimposed on three-dimensional transport in part or all of the groundwater system. Deconvolution of such data is extremely complex, but good progress is being made using a probabilistic modelling approach (Dagan 1984; Sposito and Jury 1987). However, because the emphasis of this review is on NO_3^- transport through soil, the problems of groundwater sampling will not be discussed.

FACTORS INFLUENCING THE RATE OF NITRATE LEACHING

The rate of NO_3^- leaching is influenced primarily by climatic factors acting directly to determine the net downward flux of water, or indirectly through their effect on the soil NO_3^- content, and hence on the NO_3^- concentration of the leaching water. Seasonal variability in rainfall and evaporation, temperature effects, the type of land use, soil properties – particularly structure and texture, and the use of fertilizers interact in complex ways to determine the relative influence of climate and the size of the soil NO_3^- pool on leaching.

Climate

Rainfall and evaporation

Studies in Northern Hemisphere regions where winters are cold and wet show that NO_3^- leaching becomes negligible by early summer, when plant growth accelerates and evapotranspiration generally exceeds rainfall (Anon. 1983). Exceptions occur when heavy spring rain falls on land that has recently been fertilized: e.g., Macduff and White (1984) found that 18.7 out of an annual total of 48.7 kg N/ha was leached from mixed arable and pasture land in one week in May when 48 mm of rain fell. The arable area, which had comprised roughly one quarter of the total, had been fertilized with 85 kg N/ha six days before. During winter, leaching of NO_3^- is correlated with excess winter rainfall (van der Paauw 1962); in Britain, this is the amount of rain falling from the time the soil regains "field capacity" to the end of February (MAFF 1979). NO_3^- concentrations in the drainage from structured clay soils are commonly highest in the first flows in autumn and early winter and subsequently decline (Harris et al. 1984; Haigh and White 1986). This is often most marked if the previous summer has been unusually hot and dry (Williams 1976; Garwood and Tyson 1977; Foster and Walling 1978). However, monitoring of stream flows (Tomlinson 1971; White et al. 1983) indicates that NO_3^- concentrations may increase as flow rates increase during the winter, and subsequently decline as flow rates decrease in the spring and summer. A possible explanation is that low summer flows reflect the influence of slow drainage from subsoil and groundwater reserves that have low NO_3^- concentrations. When rain falls in the autumn, several factors can operate: (i) soil NO_3^- not taken up by the crop becomes vulnerable to leaching; (ii) a flush of mineralization as the dry soil rewets (Birch 1958) augments the pool of residual NO_3^-, particularly in the A horizon where ammonifying and nitrifying organisms are most plentiful; and (iii) depending on the soil's structure, some of the rain infiltrating the soil, particularly during storms, may leach NO_3^- rapidly from the A horizon by pathways between aggregates which bypass much of the soil in the B horizon (Smettem et al. 1983; White et al. 1983). In this case, significant quantities of N can be leached before a soil reaches its winter mean water content (Table 1). However, when there is no enrichment of NO_3^- in the A horizon, either because fertilizer has not been recently applied or conditions do not favour the mineralization of organic N,

Table 1: Quantities of NO_3-N leached from a catchment of arable land and grassland in southern England in relation to fertilizer applied and soil moisture status (White et al. 1983).

Period	Moisture status	Fertilizer applied	Leached nitrate	
			Amount	Percent of total
		kg N/ha	*kg N/ha*	*%*
1.9.78– 1.2.79	deficit	5	5	28
2.2.79–31.5.79	WMWC[1]	94	13	72
1.6.79–31.8.79	deficit	45	0	0
Totals		144	18	100
1.9.79–13.12.79	deficit	4	7	22
14.12.79–3.4.80	WMWC	89	24	75
4.4.80–31.8.80	deficit	42	1	3
Totals		135	32	100

1. Winter mean water content

NO_3^- concentrations in the drainage are diluted when bypassing flow of rainwater occurs in the soil (Coles and Trudgill 1985; White and Bramley 1986).

Soil texture and structure

The field capacity or volumetric water content, θ, above which water moves freely under the influence of gravity, is generally greater in clay soils than sandy soils. For a rate of water application q (mm/h), therefore, water will percolate at a higher mean velocity v (= q/θ) through a sandy soil than a clay, and given the same period of leaching, NO_3^- should be displaced to a greater depth in the sandy soil. Reporting on lysimeter and field experiments in Holland, Kolenbrander (1969) noted that the amount of nitrate leached decreased from c. 60 kg N/ha/y for soils with < 10% clay to c. 10 kg N/ha/y for soils with > 35% clay. Cooke (1976), summarizing data collected for the OECD, reported NO_3^- leaching to decrease from 59 to 22 to 6 kg N/ha/y for sandy soils, light clays and clays respectively. On the other hand, drainage measurements made on lysimeters of undisturbed soil under grass and fertilized with 400 kg N/ha as $Ca(NO_3)_2$ indicated losses of 44, 39 and 18 kg N/ha from a clay, silt loam and sandy soil, respectively, in the first year (Dowdell et al. 1980).

It is oversimplistic, therefore, to assume that NO_3^- always leaches more rapidly from sandy than clay soils, particularly because the effect of texture is much modified by structure and the microscale distribution of NO_3^- in the soil (White 1985b). Field soils exhibit varying degrees of preferential flow of water down cracks and channels between aggregates (Anderson and Bouma 1977; Scotter 1978) which affects the rate of NO_3^- leaching. For example, Cunningham and Cooke (1958), Wild (1972) and Thomas and Phillips (1979) have suggested that when NO_3^- is held within soil aggregates it will be protected from leaching when bypass flow occurs: others (Addiscott and Cox 1976; Barraclough et al. 1983) have observed that if N fertilizer has been recently applied, or if soil-generated NO_3^- is held on the outside of aggregates, bypassing flow causes it to leach faster than it would by uniform displacement. Shaffer et al. (1979) applied water with and without NO_3^- to a soil surface and sampled the drainage with suction cups at several depths, or directly from large channels (macropores) at a depth of 1.2 m. The macropore drainage responded quickly to a change in concentration of the applied water whereas the concentration of the matrix flow (sampled by the suction cups) changed only slowly with time. Similarly, Shuford et al. (1977) found that NO_3^- was leached from all depths down to 1 m in a well-drained silt loam following single applications of 89 and 178 mm of water. The recovery of added NO_3^- down to 1.35 m was less than expected if uniform displacement had occurred, suggesting that some may have moved by preferential flow below this depth.

Thus, in most soils and especially structured clay soils, NO_3^- from fertilizer and that generated in the large pore system can be rapidly leached during rain or irrigation. Nitrate within aggregates is relatively protected during individual events, but between events it diffuses to the aggregate exteriors whence it can be leached. Analysis of leaching data for undisturbed soils in large cores, lysimeters or drained field plots indicates that the part of the soil volume apparently effective in NO_3^- transport during individual rainfall events, the fractional transport volume θ_{st}, can range from c. 0.05 m^3/m^3 to a value comparable to the field capacity (Smettem et al. 1983; Scotter et al. 1984; White et al. 1984,1986; White 1987).

Irrigation

To prevent the accumulation of soluble salts in an irrigated soil, water surplus to the crop's requirements is supplied to provide a leaching fraction, LF (defined as the ratio of the depth of drainage water to the depth of water applied). Ideally, LF should be < 0.1, but in practice, because of uncertainty about the soil moisture deficit to be eliminated and the non-uniform distribution of applied water, LF can be as high as 0.3 to 0.4. High LFs associated with inefficient irrigation practices can lead to excessive leaching of NO_3^- (Letey et al. 1977). Sprinkler or spray irrigation gives a more uniform distribution of applied water than flood or furrow irrigation and hence better control of the leaching fraction. Trickle or drip irrigation, which aims to match the water supplied to the crop's transpirational demand, is probably the most efficient method of applying water and hence of controlling leaching losses. Soluble N can be dissolved in the irrigation water and applied directly to the root zone through the drip emitters (Clothier et al. 1986).

However, when irrigation is primarily supplementary to rainfall rather than essential for plant survival, the improved growth and uptake of N may offset the greater percolation of water through the root zone so that leaching losses are not increased. For example, Webster and Dowdell (1984) studied NO_3^- leaching from grassed lysimeters containing clay loam and silt loam soils receiving 400 kg N/ha/y and 730 and 670 mm of rain, respectively. Supplementing the summer rainfall by 20% annually over 4 years did not increase the leaching loss of N above the average values of 41 and 15 kg/ha/y in the clay loam and silt loam respectively.

The pool of nitrate available for leaching

Temperature

Temperature affects the rate of ammonification of soil organic N and the rate of oxidation of NH_4^+ to NO_3^-. Depending on the soil and its environment, the optimum temperature for nitrification lies between 20 and 40°C, with activity generally insignificant below 5°C (Schmidt 1982). However, Macduff and White (1985) found that nitrification was nearly as rapid in the top 10 cm of an Evesham clay soil (Aquic Eutrochrept) at 4°C as at 10°C. Temperature also affects plant growth rates and hence the capacity of plants to absorb water and NO_3^- from the soil.

Fertilizer use and soil type

The amount and type of N fertilizer used can have a major influence on the amount of NO_3^- available for leaching. Many experiments on arable crops and grassland show that the apparent recovery of fertilizer N in the plant in the first year after application is in the range 30 to 70%, with an average of 50% (Anon. 1983). Labelling the fertilizer with ^{15}N enables a more precise estimate of N recovery to be made, in the first year at least, and often gives values lower than these apparent recoveries, which underestimate the contribution of non-fertilizer sources to N uptake by the crop (Dowdell et al. 1980). As expected, the proportion of fertilizer N that is leached is inversely related to the recovery of N by the plants; e.g., fertilizer N recovered in grass grown on clay and silt loam soils in lysimeters over 4 years was 52 and 63%, and that leached was 7% and 4% respectively (Webster and Dowdell 1984). Even without fertilizer application, NO_3^- leaching losses can be large if natural NO_3^- levels, probably of geological origin, are high, as in the soils on the west side of the San Joaquin Valley in California (Letey et al. 1977). Furthermore, past applications of fertilizer N which have become incorporated into the organic N fraction may, on mineralization, provide the main source of nitrate for leaching (Mackenzie and Viets 1974). An example is the Evesham clay soil which had a total N content of 0.94% in the 2-10 cm layer, a C/N ratio of 8 and a net mineralization rate under simulated field conditions of c. 350 kg N/ha/y (Macduff and White 1985). Up to 43% of a current year's application of fertilizer N could be leached in this soil under grazing (Haigh and White 1986); but more importantly, there was great variation in the year to year amounts of N leached in this, and other soils under similar management (Field et al. 1985), because of changing weather conditions which affect N transformations in the soil, plant uptake of N and the proportion of rainfall going to drainage. This poses severe problems for the development of predictive models of nitrate leaching.

Soil horizons high in sesquioxides and low in organic matter can sometimes exhibit positive adsorption of NO_3^- (Black and Waring 1976) which retards NO_3^- leaching. But the majority of soils have a net negative charge on their colloid surfaces so that the non-specifically adsorbed NO_3^- anion is repelled from the surfaces. Negative adsorption and the exclusion of NO_3^- from very fine pores in soils of this kind may cause a NO_3^- pulse to move slightly faster than the accompanying water (Cameron and Wild 1982a), but the effect is usually insignificant compared to the bypassing effect created by the soil's structure (White et al. 1984; Dyson and White 1987).

Land use and management

Leaching of NO_3^- is likely to be greatest under bare fallow. A 10-year comparison of NO_3^- profiles under pasture, fallow-wheat and continuous fallow in western Canada (Rennie et al. 1976, cited by Legg and Meisenger 1982) revealed leaching losses of 0 (pasture), 500 (fallow-wheat) and 1082 (continuous fallow) kg N per ha to 3.6 m depth. Vegetation reduces the propensity for NO_3^- to be leached, but annual crops are generally less effective than perennials such as trees and grass. Cooke and Williams (1970) and Cooke (1976) reported that NO_3^- leaching from arable land was greater than from grassland. Using OECD figures scaled to a fertilizer input of 100 kg N/ha/y and 250 mm of drainage, Cooke (1976) quoted average leaching losses of 4 kg N/ha/y from grassland and 23 kg N/ha/y from cultivated soils. Nitrate leaching losses of 52 - 93 kg N/ha/y have been estimated for arable land over Chalk aquifers in southern England, but losses from grassland fertilized at rates up to 400 kg N/ha/y were thought not to exceed 20 kg N/ha/y (Anon. 1983). Burns and Greenwood (1982) suggested that 42% of the fertilizer N applied to arable soils in England and Wales each year was lost by leaching. For the 5.18 M ha of crops fertilized at an average rate of 135 kg N/ha in 1980-81, this represents a loss of 294 kt N/y.

Leaching losses from forest soils, on the other hand, usually do not exceed inputs from the atmosphere and by biological fixation (Johnson et al. 1976; White et al. 1983). But when the trees are felled and regrowth is suppressed, leaching losses increase markedly; e.g., over a 10 year period, N loss by leaching from an undisturbed hardwood forest on acid soils in New Hampshire was 43 kg/ha compared to c. 499 kg/ha from an area that was clear-felled and regrowth suppressed (Likens et al. 1978).

The view, based mainly on data from cut swards, that NO_3^- leaching from grassland is negligible at fertilizer rates up 250 kg N/ha/y (Dowdell et al. 1980) now needs to be modified in the light of other evidence derived from moderately to intensively managed, grazed grassland (Table 2). The main

Table 2: Leaching losses (kg N/ha/y) from grazed pasture land

Soil type	Grazing regime	Sward and N applied	Nitrogen leached
Loam (Ryden et al. 1984)	9.3 steers/ha (rotational)	ryegrass + 420 kg N/ha (NH_4NO_3)	162[1]
Sandy loam (Field et al. 1985)	periodic mobs of sheep	ryegrass, no N	42[2]
		white clover + ryegrass, no N	73[2]
		ryegrass + 110 kg N/ha (LAN)	100[2]
Friable clay (Steele et al. 1984)	8 steers/ha (rotational)	white clover + ryegrass, no N	88
		white clover + ryegrass, 172 kg N/ha (urea)	193
Clay (Haigh & White 1986)	periodic sheep and cattle up to 2.7 stock units/ha	ryegrass + 154 kg N/ha (NH_4NO_3)	33[2]

1. This loss was approx. 5.5 times that of a cut sward fertilized at the same rate; 2. Averaged over 2 seasons.

effect of grazing seems to be the concentration of soluble N in small volumes of soil under urine patches. The N applied in urine (70 to 90% urea) ranges from 30 to 60 g/m^2, equivalent to 300 to 600 kg N/ha (Ball and Ryden 1984). Urea in soil rapidly hydrolyzes to ammonium carbonate, which decomposes to release NH_4^+ ions. The low C/N ratio in the urine-affected soil retards microbial immobilization of N (Carran et al. 1982) and the local concentration of mineral N greatly exceeds plant uptake capacity. NH_4^+ can be lost by volatilization as NH_3, but nitrification also produces very high concentrations of NO_3^- (Ball and Ryden 1984) which is vulnerable to leaching. In the U.K., virtually all the faeces and urine from sheep and about half that from cattle are deposited directly on to grass (Anon. 1983). The quantity of

N returned (1020 kt in 1978) was comparable to that applied as fertilizer at that time (1155 kt). The Royal Society Report (Anon. 1983) emphasizes that there is a lack of quantitative data on the partitioning of excretal N between the atmosphere, drainage water and soil organic matter in U.K. agriculture, and this is true of other countries as well.

Ploughing of old grassland also leads to accelerated leaching of NO_3^- because of the increased rate of mineralization of organic N (Cooke 1976; Foster et al. 1982). For example, Ryden et al. (1984) estimated that 275 to 310 kg/ha of N in excess of arable crop requirements were mineralized in one year in a deep friable soil on Chalk, when it was ploughed out of grass. To argue, as does the Royal Society Report, that it is "not possible to distinguish with certainty the relative importance of fertilizing and ploughing in enhancing nitrate leaching" is to beg the question: the important point is that whenever organic N builds up under grass, ploughing the grass renders much more N vulnerable to leaching through a change in the rate of transformation of N to mineral forms in the soil.

QUANTITATIVE PREDICTION OF THE RATE OF NITRATE LEACHING

The difficulties associated with the prediction of NO_3^- leaching in the field arise from several causes:

(i) Inputs of water and mineral N are spatially and temporally variable. This is particularly true of a fertilized soil under natural rainfall growing grass that is grazed.

(ii) Microbial transformations and plant uptake of N take place during and between leaching events so that the pool of nitrate available for leaching changes with time.

(iii) The movement of water through the soil is generally unsteady because of episodic inputs and changes in water content induced by evaporation and transpiration; it is also spatially variable because of the heterogeneous nature of the pore space. Notwithstanding these difficulties, the majority of the attempts to quantify nitrate leaching in the past 25 years have used deterministic models based on both simple and complex mechanisms (Addiscott and Wagenet 1985).

Deterministic models

The pioneering work of Taylor (1953), who studied the spread of a solute "slug" under the combined action of diffusion and velocity variations in a fluid flowing slowly through a capillary tube, laid the foundations for the convection–dispersion model of solute transport in porous media (Brigham et al. 1961). The partial differential equations for convective transport of the solute and its diffusion in response to a concentration gradient are combined, and, subject to the appropriate initial and boundary conditions, are solved to determine the movement and distribution of NO_3^- in the soil (Gardner 1965). The combined equation which has been developed and applied to many cases of solute transport in soil (Biggar and Nielsen 1967, 1980; Nielsen et al. 1982, 1986) is called the convection–dispersion equation, CDE. For one–dimensional transport of a non–adsorbed solute solute through a defined volume of soil, the CDE has the form:

$$\Theta \frac{\partial C}{\partial t} = D\Theta \frac{\partial^2 C}{\partial z^2} - q \frac{\partial C}{\partial z} \tag{1}$$

where C is the concentration of NO_3^- in the soil solution, t is time, z is the vertical distance travelled, D is the dispersion coefficient, and q and Θ have been defined. Nitrate adsorption need not be considered for most soils, but the effect of NO_3^- formation, or loss through denitrification, can be included by adding a term R (dimensions $M/L^3/T$), which may be positive or negative, to the right–hand–side of (1). However, the dispersion coefficient obtained by solving (1) can have a wide range of values, depending on the experimental conditions (Nielsen et al. 1982), and emerges as an adjustable parameter based on optimizing the fit of the CDE to a set of effluent NO_3^- concentrations measured in a leaching experiment (Wild and Cameron 1980; Dyson and White 1987). A further refinement is the "two–region" CDE, introduced by Coats and Smith (1964) and popularized in soil science by van Genuchten and Wierenga (1976), in which the fluid volume is divided into a "mobile" and "immobile" phase, with mass transfer between the phases at a rate proportional to the concentration difference between them. The introduction of additional adjustable parameters increases the power of the CDE to simulate leaching data for a range of experimental conditions. However, the case of unsteady flow of solutions through heterogeneous soil profiles is far removed from Taylor's (1953) original experimental system, and the validity of applying (1) to such systems has been questioned (Sposito et al. 1979) on the grounds that no one has yet proved the existence of the functions relating concentration and velocity to time and distance, which are implicit in the use of the CDE as a model of macroscopic solute transport in soil.

Other deterministic approaches to modelling NO_3^- leaching that embody different mechanisms have included: (i) the piston displacement model, involving uniform displacement of the resident soil solution by the influent solution without mixing or dispersion (Wild and Babiker 1976), or with dispersion (Rose et al. 1982); (ii) layer models, in which the influent water mixes with all the resident soil solution in a defined layer (Burns 1974), or with only part of this water (Addiscott 1977, 1981), and piston flow occurs from one layer to the next below. Allowance is made for evaporation (Burns, Addiscott loc. cit.) and for fast flow down preferential pathways (Addiscott 1984). Where a distinction between a mobile volume, which mixes with the influent water and participates in solute transport, and an immobile volume, which interacts with the mobile phase only by diffusion, is made, the line is drawn at some arbitrary soil moisture suction, e.g. 200 kPa (Addiscott 1977), or 8 kPa (Nkedi–Kizza et al. 1982). Such models have been used with mixed success to predict NO_3^- leaching in the field (Addiscott 1977; Burns 1980; Cameron and Wild 1982b; Barraclough et al. 1983), but none really addresses the underlying problems of spatial and temporal variability in the NO_3^- supply and soil hydraulic properties governing water movement.

Stochastic models

Nielsen et al. (1982) concluded that in view of the generally disappointing outcome in predicting rates of NO_3^- leaching in the field, "deterministic models must give way to stochastic models, or at least to stochastic parameters included in the deterministic models". On the premise that the variability in field–scale solute transport can be most efficiently described as "random" (Sposito et al. 1986), the observations of one leaching experiment are recognized as a single sample realization of a stochastic transport process that comprises an ensemble of all possible realizations of that process. Thus, instead of the assumption implicit in the CDE that the pore water velocity v and dispersion coefficient D are single value parameters for a given set of conditions, v and D may be considered as random functions having skewed frequency distributions (Amoozegar–Fard et al. 1982). In this example, simulations of solute breakthrough data for soil columns showed that the variability in D was much less important than that in v, indicating that velocity variations rather than diffusion effects made the overriding contribution to variations in solute flux. This was consistent with the approach of Dagan and Bresler (1979) and Bresler and Dagan (1979) for vadose zone transport, and Gelhar and Axness (1983) for three–dimensional transport in aquifers, who used the known spatial variability in hydraulic conductivity to generate a random function for pore water velocity v in a stochastic formulation of the CDE. Only Wagenet and Rao (1983) appear to have attempted to model the distribution of NO_3^- in a field soil using a stochastic CDE, with limited success.

As an alternative to invoking randomness in the parameters of an existing deterministic model, such as the CDE, Jury (1982) and Jury et al. (1986) have argued that, when the mechanisms governing solute transport may be unknown, the mass rate of solute output from a defined volume of soil at an observation time t can be treated as a stochastic function of the input at an earlier time t'. In respect of NO_3^- in particular, there is much uncertainty about the way in which transport takes place – where the ions originate, what pathways they follow, what physicochemical effects and biological transformations they experience, and at what points they exit the soil volume. One may postulate, therefore, the existence of a fluid volume within the soil that is effective in transporting NO_3^- during the time of observation, but which is highly irregular in shape, variable in space and time, and difficult to measure and describe mathematically – this is the transport volume, V_{st}. Whatever the mechanisms and transformations by which transport is realized, the passage of the solute molecules through the soil can be described by a conditional probability density function (pdf) of solute lifetimes in the transport volume

– the transfer function (Jury et al. 1986). The key variables of the lifetime pdf are (i) the time t' at which a solute molecule first appears in the transport volume, and (ii) the interval $t - t'$ ($= \tau$) during which it exists in the transport volume. The lifetime pdf, $g(\tau|t')$ then contains the information necessary to develop a predictive model of solute transport, given some simplifying assumptions. For example, if $Q_{in}(t')$ is the mass rate of solute input divided by the total mass input, and $Q_{out}(t)$ is the mass rate of output similarly normalized, then the equation:

$$Q_{out}(t) = {}_0\!\int^t g(\tau|t') Q_{in}t' \, dt' \qquad (2)$$

relates the input between time 0 and t' to the output observed at time t. When the form of the conditional lifetime pdf is assumed to be independent of the input time t', $g(\tau|t')$ becomes simply the pdf of solute lifetimes $g(t - t')$. Usually, solute output can only be monitored at one exit surface, e.g. a drain network, so that losses of solute by other means from the transport volume are ignored and $g(t - t')$ then becomes a pdf of solute travel times. If the output is monitored over a period that is long compared to the input time t', $g(t - t')$ reduces to $g(t)$ which is equal to the normalized mass rate of solute output at the exit surface, $Q_{ex}(t)$. White (1987) has applied this analysis to the mass rates of NO_3^- loss from 1 ha of mole and tile–drained field over several seasons. $g(t)$ for both NO_3^- and Cl^- conformed to lognormal distributions which could be characterized by their moments (μ and σ^2) and the fractional transport volume Θ_{st} for each rainfall event calculated from the equation (White et al. 1986):

$$\Theta_{st} = q_0 \, \bar{t} \, / \, L \qquad (3)$$

In (3), q_0 is the mean rainfall rate (mm/h), L is the mean mole drain depth (mm) and $\bar{t}$(h) can be calculated either as the mean travel time ($= \exp(\mu + \sigma^2/2)$) or as the median travel time ($= \exp(\mu)$). Θ_{st} values calculated from NO_3^- and Cl^- travel time pdfs were very similar for individual events.

For the 11 events monitored, under different conditions of rainfall and antecedent soil water content, the limited variation in Θ_{st} (0.016 – 0.102 m^3/m^3) enabled a space and time–averaged value of Θ_{st} ($= 0.052$) to be obtained. V_{st} was then calculated and used in the equation (White 1987):

$$N = 0.001 \, C_0 \, V_{st} \, [1 - \exp(-10d_0t/V_{st})] \qquad (4)$$

In (4), N is the quantity of N leached in kg/ha, C_0 is the mean concentration of NO_3^- in the soil solution at time 0, d_0 is the mean drainage rate (mm/h) during a monitoring period lasting for t hours. Good agreement was obtained between values calculated from (4) and measured amounts of N leached, suggesting that the predictive power of the transfer function approach is considerable, provided a space and time–averaged value of Θ_{st} can be determined for a particular soil.

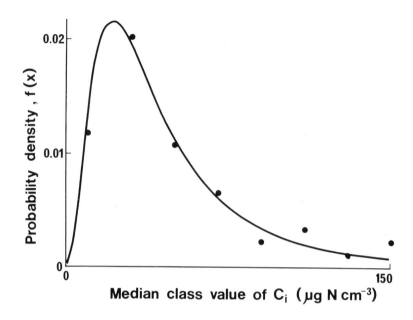

Fig. 1: Frequency distribution for a sample (n = 54) of soil solution nitrate concentrations from a grazed pasture. Fitted line according to the equation $f(x) = (\sqrt{2}\pi\sigma x)^{-1} \exp[-(\ln x - \mu)^2/2\sigma^2]$ where x = soil solution concentration, C_i (White and Bramley 1986).

A major problem in applying (4), or any model of field-scale NO_3^- leaching, lies in obtaining a reliable estimate of C_o. Grazing and fertilizing, in particular, create extreme spatial variability of mineral N in the surface soil (Thompson and Coup 1940; Ball and Ryden 1984; Macduff and White 1984). Most of this variability occurs over short distances (< 0.4 m) and requires that large numbers of samples be collected, each of which is as large as practicable, in order that means with low standard errors can be obtained (White et al. 1987). An example is given in Figure 1 of a frequency distribution for a sample population of 54 soil NO_3^- values from a 1 ha grazed field, showing a large spread of values that conforms closely to a lognormal function (White and Bramley 1986). The amount of NO_3^- leached in this field, when measured over 12 h under standard conditions on 12 individual large cores, or microlysimeters (Richter and Jury 1986), was very variable (1 – 100 mg N/core or 0.24 – 24 kg N/ha). N loss could only be modelled satisfactorily when the cores were treated as an assembly, having an average transport volume defined from field measurements (White 1987), and an initial soil NO3-concentration equal to the estimated arithmetic mean of the population of values from which the sample in Fig. 1 was drawn (White and Bramley 1986).

CONCLUSIONS

Leaching is a process by which N, predominantly as NO_3^-, is lost from the soil-plant-animal system. The extent of the loss ranges from insignificant in many natural forest and ungrazed grassland systems, to being the major component of the overall loss of N from intensively managed, grazed grasslands and N-fertilized, cropped soils.

The estimation of leaching losses is difficult because both the supply of water for leaching and the reservoir of NO_3^- in the soil are influenced by such a variety of factors - physical, biological and anthropogenic. Extensive and frequent sampling of water in the soil is required to overcome the spatial and temporal variability in NO_3^- leaching, and many of the studies conducted so far are inadequate in this respect.

Attempts to model NO_3^- leaching have involved mainly deterministic models, based on assumed mechanisms of solute transport such as piston displacement, or convection and diffusion, either combined or separated between "mobile" and "immobile" water phases. For a restricted range of conditions, e.g. minimum biological transformations and spatial variability, this kind of model can simulate NO_3^- leaching satisfactorily. Such conditions rarely pertain in the field, however, and the utility of deterministic models with single value parameters is then limited. The versatility of deterministic models can be improved by introducing stochasticity into the key parameters, but the robustness of the parameter distribution functions under varying conditions remains to be determined.

The alternative of treating the transport and transformations of NO_3^- in a defined volume of soil as a completely stochastic process is conceptually simple, and encompasses with economy the complexity of the various mechanisms in a single transfer function - the probability density function of NO_3^- lifetimes, and its special case for entry and exit through monitored surfaces, the pdf of NO_3^- travel times, $g(t-t' | t')$. Given the frustrating complexity of the transport problem in the field (Nielsen et al. 1986), the transfer function approach, which is not committed to mechanisms postulated a priori, is intuitively attractive as the basis of a predictive model for NO_3^- leaching. However, there is a shortage of quality data for field situations over extended periods that can be used to assess the relative merits of different ways of applying probability concepts to field-scale solute transport (Sposito et al. 1986). Nor should particular deterministic models be abandoned in situations, such as heavily fertilized vegetable crops on uniform, light textured soils (Burns 1974, 1980), where they have proved successful. Indeed, no reasonable theoretical approach to modelling NO_3^- leaching should be rejected until testing against reliable field results has been completed.

REFERENCES

Adams, J.A. (1986). Identification of heterotrophic nitrification in strongly acid larch humus. Soil Biol. Biochem. 18, 339–41.

Addiscott, T.M. (1977). A simple computer model for leaching in structured soils. J. Soil Sci. 28, 554–63.

Addiscott, T.M. (1981). Leaching of nitrate in structured soils. In 'Simulation of the Nitrogen Behaviour of Soil–plant Systems'. (Eds M.J. Frissel and J.A. van Veen) pp. 245–53. (Pudoc: Wageningen.)

Addiscott, T.M. (1984). Modelling the interaction between solute leaching and intra–ped diffusion in clay soils. In 'Water and Solute Movement in Heavy Clay Soils'. (Eds. J. Bouma and P.A.C. Raats) pp. 279–97. (ILRI Publication 37: Wageningen.)

Addiscott, T.M., and Cox, D. (1976). Winter leaching of nitrate from autumn–applied calcium nitrate, ammonium sulphate, urea and sulphur–coated urea in bare soil. J. Agric. Sci., Camb. 87, 381–9.

Addiscott, T.M., and Wagenet, R.J. (1985). Concepts of solute leaching in soils: a review of modelling approaches. J. Soil Sci. 36, 411–24.

Amoozegar–Fard, A., Nielsen, D.R., and Warrick, A.W. (1982). Soil solute concentration distributions for spatially varying pore water velocities and apparent diffusion coefficients. Soil Sci. Soc. Am. J. 46, 3–9.

Anderson, J.L. and Bouma, J. (1977). Water movement through pedal soils. II. Unsaturated flow. Soil Sci. Soc. Am. J. 41, 419–23.

Anon. (1983). 'The Nitrogen Cycle of the United Kingdom'. A Study Group Report. (The Royal Society: London.)

Ball, P.R., and Ryden, J.C. (1984). Nitrogen relationships in intensively managed temperate grasslands. Pl. Soil 76: 23–33.

Barraclough, D., Hyden, M.J. and Davies, G.P. (1983). Fate of fertilizer nitrogen applied to grassland I. Field leaching results. J. Soil Sci. 34, 483–97.

Biggar, J.W., and Nielsen, D.R. (1967). Miscible displacement and leaching phenomenon In 'Irrigation of Agricultural Lands' (Eds R.M. Hagen et al.) Agron. Monog. No. 11, pp. 254–274. (Am. Soc. Agron.: Madison.)

Biggar, J.W., and Nielsen, D.R. (1980). Mechanisms of chemical movement in soils. In 'Agrochemicals in Soils' (Eds A. Banin and U. Kafkafi) pp. 213–227. (Pergamon: Oxford.)

Birch, H.F. (1958). The effect of soil drying on humus decomposition and nitrogen availability. Pl. Soil 10, 9–31.

Black, A.S., and Waring, S.A. (1976). Nitrate leaching and adsorption in a Krasnozem from Redland Bay, Qld. III. Effect of nitrate concentration on adsorption and movement in soil columns. Aust. J. Soil Res. 14, 189–95.

Bresler, E., and Dagan, G. (1979). Solute dispersion in unsaturated heterogeneous soil at field scale. II. Applications. Soil Sc. Soc. Am. J., 43, 467–72.

Brigham, W.E., Reed, P.W., and Dew, J.N. (1961). Experiments on mixing during miscible displacement in porous media. Soc. Pet. Eng. J. 1, 1–8.

Burns, I.G. (1974). A model for predicting the redistribution of salts applied to fallow soils after excess rainfall or evaporation. J. Soil Sci. 25, 165–78.

Burns, I.G. (1980). A simple model for predicting the effects of leaching of fertilizer nitrate during the growing season on the nitrogen fertilizer need of crops. J. Soil Sci. 31, 175–85.

Burns, I.G., and Greenwood, D.J. (1982). Estimation of the year to year variations in nitrate leaching in different soils and regions of England and Wales U.K. Agric. Environ. 7, 35–46.

Cameron, K.C., and Scotter, D.R. (1986). Nitrate leaching losses from temperate agricultural systems In 'Nitrogen Cycling in Agricultural Systems of Temperate Australia' (Aust. Soc. Soil Sci. Symp.: Wagga Wagga.)

Cameron, K.C., and Wild, A. (1982a). Comparative rates of leaching of chloride, nitrate and tritiated water under field conditions. J. Soil Sci. 33, 649–57.

Cameron, K.C., and Wild, A. (1982b). Prediction of solute leaching under field conditions: an appraisal of three methods. J. Soil Sci. 33, 659–69.

Carran, R.A., Ball, P.R., Theobold, P.W. and Collins, M.E.G. (1982). Soil nitrogen balances in urine–affected areas of pasture under two moisture regimes in Southland. N.Z. J. Exp. Agric. 10, 377–81.

Clothier, B.E., Sauer, T.J., and Green, S.R. (1986). Nitrogen movement in soil during fertigation with urea. Proc. 13th Int. Congr. Soil Sci., Symp. I/II, Hamburg.

Coats, K.H., and Smith, B.D. (1964). Dead–end pore volume and dispersion in porous media. Soc. Pet. Eng. J. 4, 73–84.

Coles, N., and Trudgill, S. (1985). The movement of nitrate fertilizer from the soil surface to drainage waters by preferential flow in weakly structured soils, Slapton, S. Devon. Agric., Ecosyst. and Environ. 13, 241–59.

Cooke, G.W. (1976). A review of the effects of agriculture on the chemical composition and quality of surface and underground waters In 'Agriculture and Water Quality' MAFF Tech. Bull. 32, pp. 5–57. (HMSO: London.)

Cooke, G.W., and Williams, R.J.B. (1970). Losses of nitrogen and phosphorus from agricultural land. Water Treatment and Examination. 19, 253–74.

Cunningham, R.K., and Cooke, G.W. (1958). Soil nitrogen. II Changes in levels of inorganic nitrogen in a clay loam soil caused by fertiliser application, by leaching and uptake by grass. J. Sci. Food Agric. 9, 317–24.

Dagan, G. (1984). Solute transport in heterogeneous porous formations. J. Fluid Mech. 145, 151–77.

Dagan, G., and Bresler, R. (1979). Solute dispersion in unsaturated heterogeneous soil at field scale. I. Theory. Soil Sci. Soc. Am. J. 43, 461–7.

Dowdell, R.J., Morrison, J., and Hood, A.E.M. (1980). The fate of fertilizer nitrogen applied to grassland: uptake by plants, immobilization into soil organic matter and losses by leaching and denitrification. In 'The Role of Nitrogen in Intensive Grassland Production'. Proc. Int. Symp. Europ. Grassland Fed., pp. 129–136. (Pudoc: Wageningen.)

Dyson, J.S., and White, R.E. (1987). A comparison of the convection–dispersion equation and transfer function model for predicting chloride leaching through an undisturbed structured clay soil. J. Soil Sci. 38, 157–72.

Environmental Protection Agency (1976). 'National Interim Primary Drinking Water Regulations'. USEPA Report (EPA 570/9–76–003) (U.S. Government Printing Office: Washington, D.C.)

European Community (1980). Council directive on the quality of water for human consumption. Official Journal No. 80/778, EEC L229, 11.

Field, T.R.O., Ball, P.R., and Theobold, P.W. (1985). Leaching of nitrate from sheep–grazed pastures. Proc. N.Z. Grassld Assoc. 46, 209–14.

Focht, D.D., and Verstraete, W. (1977). Biochemical ecology of nitrification and denitrification. Adv. Microb. Ecol. 1, 135–214.

Foster, I.D.L., and Walling, D.E. (1978). The effects of the 1976 drought and autumn rainfall on stream solute levels. Earth Surf. Process. 3, 393–406.

Foster, S.S.D., Cripps, A.C., and Smith–Carington, A. (1982). Nitrate leaching to groundwater in Britain from permeable agricultural soils. Phil. Trans. Roy. Soc. London, Series B, Biol. Sci. 447–89.

Gardner, W.R. (1965). Movement of nitrogen in soil. In 'Soil Nitrogen'. (Eds. W.V. Bartholomew and F.E. Clark). Agron. Monogr. No. 10, pp. 550–572. (Am. Soc. Agron.: Madison.)

Garwood, E.A., and Tyson, K.C. (1977). High loss of nitrogen in drainage from soil under grass following a prolonged period of low rainfall. J. Agric. Sci., Camb. 89, 767–8.

Gelhar, L.W., and Axness, C.L. (1983). Three–dimensional stochastic analysis of macrodispersion in aquifers. Water Resour. Res. 19, 161–80.

Haigh, Rosalyn, A., and White, R.E. (1986). Nitrate leaching from a small underdrained grassland clay catchment. Soil Use Manage. 2, 65–70.

Harris, G.L., Goss, M.J., Dowdell, R.J., Howse, K.R., and Morgan, A. (1984). A study of mole drainage with simplified cultivation for autumn–sown crops on a clay soil. 2. Soil water regimes, water balances and nutrient loss in drainage water 1978–80. J. Agric. Sci., Camb. 102, 561–81.

Hood, A.E.M. (1976). Nitrogen, grassland and water quality in the United Kingdom. Outlook on Agric. 8, 320–7.

Johnson, A.H., Bouldin, D.R., Goyette, E.A., and Hedges, A.M. (1976). Nitrate dynamics in Fall Creek, New York. J. Environ. Qual. 5, 386–91.

Jury, W.A. (1982). Simulation of solute transport using a transfer function model. Water Resour. Res., 18, 363–8.

Jury, W.A., Sposito, G., and White, R.E. (1986). A transfer function model of solute transport through soil. 1. Fundamental concepts. Water Resour. Res. 22, 243–7.

Keeney, D. (1986). Sources of nitrate to groundwater. CRC Critical Reviews in Environ. Control 16, 257–304.

Kolenbrander, G.J. (1969). Nitrate content and nitrogen loss in drainwater. Neth. J. Agric. Sci. 17, 246–55.

Legg, J.O., and Meisinger, J.J. (1982). Soil nitrogen budgets. In 'Nitrogen in Agricultural Soils'. (Ed. F.J. Stevenson). Agron. Monogr. No. 22, pp. 503–566. (Am. Soc. Agron.: Madison.)

Letey, J., Blair, J.W., Devitt, D., Lund, L.J., and Nash, P. (1977). Nitrate—nitrogen in effluent from agricultural tile drains in California. Hilgardia 45, 289—319.

Likens, G.E., Bormann, F.H., Pierce, R.S., and Reiners, W.A. (1978). Recovery of a deforested ecosystem. Replacing biomass and nutrients lost in harvesting northern hardwoods may take 60 to 80 years. Science 199, 492—6.

Macduff, J.H., and White, R.E. (1984). Components of the nitrogen cycle measured for cropped and grassland soil—plant systems. Pl. Soil 76, 35—47.

Macduff, J.H., and White, R.E. (1985). Net mineralization and nitrification rates in a clay soil measured and predicted in permanent grassland from soil temperature and moisture measurements. Pl. Soil 86, 151—72.

Mackenzie, A.J., and Viets, F.G. (1974). Nutrients and other chemicals in agricultural drainage waters. In 'Drainage for Agriculture' (Ed. J. van Schilfgaarde) Agron. Monogr. No. 17, pp. 489—508, (Am. Soc. Agron.: Madison.)

MAFF (1979). 'Fertilizer Recommendations'. 2nd Ed. Bull. 209A, Ministry of Agriculture, Fisheries and Food (HMSO: London.)

Mohammed, I.H., Scotter, D.R., and Gregg, P.E.H. (1984). The short—term fate of urea applied to barley in a humid climate. I. Experiments. Aust. J. Soil Res. 22, 173—80.

Nielsen, D.R., Biggar, J.W., and Wierenga, P.J. (1982). Nitrogen transport processes in soil. In 'Nitrogen in Agricultural Soils' (Ed. F.J. Stevenson). Agron. Monogr. No. 22, pp. 423—448. (Am. Soc. Agron.: Madison.)

Nielsen, D.R., van Genuchten, M.Th., and Biggar, J.W. (1986). Water flow and solute transport processes in the unsaturated zone. Water Resour. Res. 22, 89S—108S.

Nkedi—Kizza, P., Rao, P.S.C., Jessup, R.E., and Davidson, J.M. (1982). Ion exchange and diffusive mass transfer during miscible displacement through an aggregated Oxisol. Soil Sci. Soc. Am. J. 46, 471—6.

Rennie, D.A., Racz, G.J., and McBeath, D.K. (1976). 'Nitrogen Losses'. Proc. Western Canada Nitrogen Symp., pp. 325—353. (Alberta Agric.: Edmonton.)

Richter, G., and Jury, W.A. (1986). A micro—lysimeter field study of solute transport through a structured sandy loam soil. Soil Sci. Soc. Am. J. 50, 863—8.

Rose, C.W., Chichester, F.W., Williams, J.R., and Ritchie, J.T. (1982). A contribution to simplified models of field solute transport. J. Environ. Qual. 11, 146—50.

Ryden, J.C., Ball, P.R., and Garwood, E.A. (1984). Nitrate leaching from grassland. Nature, London 311, 50—2.

Schmidt, E.L. (1982). Nitrification in soil. In 'Nitrogen in Agricultural Soils'. (Ed. F.J. Stevenson), Agronomy. Monogr. No. 22, pp. 253—288. (Am. Soc. Agron.: Madison.)

Scotter, D.R. (1978). Preferential solute movement through larger soil voids. I. Some computations using simple theory. Aust. J. Soil Res. 16, 257—67.

Scotter, D.R., Mohammed, I.H., and Gregg, P.E.H. (1984). The short—term fate of urea applied to barley in a humid climate. II. A simple computer model. Aust. J. Soil Res. 22, 181—90.

Shaffer, K.A., Fritton, D.D., and Baker, D.E. (1979). Drainage water sampling in a wet, dual—pore soil system. J. Environ. Qual. 8, 241—6.

Shuford, J.W., Fritton, D.R., and Baker, D.E. (1977). Nitrate—nitrogen and chloride movement through undisturbed field soil. J. Environ. Qual. 6, 255—9.

Smettem, K.R.J., Trudgill, S.T., and Pickles, A.M. (1983). Nitrate loss in soil drainage waters in relation to by—passing flow and discharge on an arable site. J. Soil Sci. 34, 499—509.

Sposito, G., Gupta, V.K., and Bhattacharya, R.N. (1979). Foundational theories of solute transport in porous media: a critical review. Adv. Water Resour. 2, 59—68.

Sposito, G., and Jury, W.A. (1987). Stochastic modelling of solute movement in the vadose and groundwater zones. U.S.—Japan seminar 'New Approaches in Hydrology', Hawaii.

Sposito, G., Jury, W.A., and Gupta, V.K. (1986). Fundamental problems in the stochastic convection—dispersion model of solute transport in aquifers and field soils. Water Resour. Res., 22, 77—88.

Steele, K.W., Judd, M.J., and Shannon, P.W. (1984). Leaching of nitrate and other nutrients from a grazed pasture. N.Z. J. Agric. Res. 27, 5—11.

Taylor, G.L. (1953). Dispersion of soluble matter in solvent flowing slowly through a tube. Proc. Roy. Soc. 219, 186—203.

Thomas, G.W., and Barfield, B.J. (1974). The unreliability of tile effluent for monitoring subsurface nitrate—nitrogen losses from soils. J. Environ. Qual. 3, 183—5.

Thomas, G.W. and Phillips, R.E. (1979). Consequences of water movement in macropores. J. Environ. Qual. 8, 149—52.

Thompson, F.B., and Coup, M.R. (1940). Studies on nitrate and ammonia in soils under permanent pasture. N.Z. J. Sci. Technol. 22A, 72—8.

Tomlinson, T.E. (1971). Nutrient losses from agricultural land. Outlook on Agric. 6, 272—8.

210

van der Paauw, F. (1962). Effect of winter rainfall on the amount of nitrogen available to crops. Pl. Soil 16, 361–80.

van Genuchten, M.Th., and Wierenga, P.J. (1976). Mass transfer in sorbing porous media I. Analytical solutions. Soil Sci. Soc. Am. Proc. 40, 473–80.

Wagenet, R.J., and Rao, B.K. (1983). Description of nitrogen movement in the presence of spatially–variable soil hydraulic properties. Agric. Water Manage. 6, 227–42.

Webster, C.P., and Dowdell, R.J. (1984). Effect of drought and irrigation on the fate of nitrogen applied to cut permanent grass swards in lysimeters: leaching losses. J. Sci. Food Agric, 35, 1105–11.

White, R.E. (1985a). A model for nitrate leaching in undisturbed structured clay soil during unsteady flow. J. Hydrol. 79, 37–51.

White, R.E. (1985b). The influence of macropores on the transport of dissolved and suspended matter through soil. Adv. Soil Sci. 3, 95–120.

White, R.E. 1987. A transfer function model for the prediction of nitrate leaching under field conditions. J. Hydrol. 92, 207–22.

White, R.E., and Bramley, R.G.V. (1986). Spatial variability of nitrate leaching in soil. In 'Surface Soil Management', NZSSS–ASSS Joint Conference pp. 98–103. (Forest Res. Inst.: Rotorua.).

White, R.E., Dyson, J.S., Haigh, R.A., Jury, W.A., and Sposito, G. (1986). A transfer function model of solute transport through soil. 2. Illustrative applications. Water Resour. Res. 22, 248–54.

White, R.E., Haigh, R.A., and Macduff, J.H. (1987). Frequency distributions and spatially dependent variability of ammonium and nitrate concentrations in soil under grazed and ungrazed grassland. Fertil. Res. 11 193–208.

White, R.E., Thomas, G.W., and Smith, M.S. (1984). Modelling water flow through undisturbed soil cores using a transfer function model derived from [3]HOH and Cl transport. J. Soil Sci. 35, 159–68.

White, R.E., Wellings, S.R., and Bell, J.P. (1983). Seasonal variations in nitrate leaching in structured clay soils under mixed land use. Agric. Water Manage. 7, 391–410.

Wild, A. (1972). Nitrate leaching under bare fallow at a site in northern Nigeria. J. Soil Sci. 23, 315–24.

Wild, A., and Babiker, I.A. (1976). The asymmetric leaching pattern of nitrate and chloride in a loamy sand under field conditions. J. Soil Sci. 27, 460–6.

Wild, A., and Cameron, K.C. (1980). Soil nitrogen and nitrate leaching. In 'Soils and Agriculture'. Critical Reports on Applied Chemistry, Vol. 2, pp. 35–70. (Blackwell Sci. Publ.: Oxford.).

Williams, R.J.B. (1976). The chemical composition of rain, land drainage and borehole water from Rothamsted, Broom's Barn, Saxmundham and Woburn Experimental Stations In 'Agriculture and Water Quality'. MAFF Tech. Bull. 32, pp. 174–200. (HMSO: London.).

EROSION AND RUNOFF OF NITROGEN

C.W. Rose and R.C. Dalal

ABSTRACT

This review considers the processes of water and (briefly) wind erosion, and their consequences for loss of nitrogen. Loss of soluble forms of N in runoff is considered, but more emphasis is given to the loss of N in eroded sediment. Generally the latter is quantitatively the more important of these two means of N loss. Losses are separately considered for total N, mineralizable N, for N in the microbial biomass, and mineral N. A theoretical framework, based on recent developments in understanding the size or settling velocity characteristics of sediment eroded by different mechanisms, is presented for the loss of total N in sediment. This framework promises to provide a better understanding of the enrichment ratio of plant nutrients.

Research on the effect of soil erosion on N transforming processes is reviewed, indicating the need for further work in this area. Using a new method of plotting experimental results, the effect of soil erosion on crop yield is illustrated for data from eastern Australia.

INTRODUCTION

Water and wind erosion reduce productivity of a soil in a number of ways. The first is by reducing plant-available soil water capacity; secondly, by increasing the losses of plant-nutrient elements; thirdly, by degrading soil structure; fourthly, by exposure of sub-soil which is a poorer environment for crop growth; and fifthly, by increasing the non-uniformity in topography and soil characteristics (Williams <u>et al</u>. 1981).

Erosion losses of total N from cultivated crop systems vary widely from 1 to 100 kg N/ha/year, but they generally exceed those from pasture or rangeland. Most of this loss of N (usually > 90%) occurs as forms of organic matter in eroded sediments and organic suspensions. Typically only 1–2% of N

212

in soil is in a form which is susceptible to loss in solution by leaching, in runoff, by denitrification or volatilization. Thus erosion is a most important cause of long-term fertility decline of soils.

White (1986) reviewed studies in which more than 90% of nutrients removed in erosion were in sediments rather than in the runoff water. Since soil organic matter and its associated N pool typically increases rapidly towards the soil surface, it is particularly prone to loss by erosion. Furthermore, organic N and total N have a high enrichment ratio, so that the loss of N is exaggerated relative to the loss of soil (Sharpley 1985; White 1986). Hence the loss of N with eroded sediment receives significant emphasis in this paper.

The implications of wind erosion for N loss are covered in far less detail than those of water erosion. This is largely due to the paucity of research in this area and does not imply that the effect of wind erosion on soil N is of limited significance.

RUNOFF AND SOIL EROSION

Runoff is involved in loss of both soluble and insoluble forms of N. The latter can be transported with the eroded sediment.

Relation between runoff and soil loss

The sediment flux (q_s) for unit width is related to the flux of water (q) across the same unit width ($m^3/m/s$) and the sediment concentration (c, kg/m^3) by:

$$q_s = qc \ (kg/m/s) \qquad (1)$$

It will now be assumed that the erosion processes at work are those characterised by loss of soil in sheets or in rills, prior to gully formation, and that no mass movement under gravity occurs. The possible effect of seepage forces will also be ignored (Onstad and Moldenhauer 1975).

With these assumptions the processes which add sediment to overland flow are rainfall detachment and the entrainment process resulting from the mutual shear stresses between the land surface and overland flow. The resultant sediment concentration at any position and time depends not only on the rate of processes adding sediment to the overland flow, but also on the rate of deposition. Deposition is a continually occurring process which withdraws (however temporarily) sediment from the ponded or flowing water above the soil surface. Thus soil particles or aggregates are commonly removed by one or other of the processes mentioned, then deposited, to be subsequently removed again. Thus entrainment subsequent to initial removal by overland flow might be described as re-entrainment. Because the deposition process is highly size-selective (due to vast differences in the settling velocity of sediment), the process of re-entrainment operates on material with a different size distribution from that initially entrained.

Thus the operative mechanisms at work in erosion processes have consequences for the size distribution of sediment either left behind as a deposit or transported further on. Since the distribution of any element such as nitrogen, as well as its various forms, is also a function of sediment size, then the significance of understanding the size-sorting effects associated with erosion and deposition can be appreciated.

The sediment concentratio c (L,t) at any distance (L) down a surface, and at time t can be written as:

$$c (L,t) = A + B \qquad\qquad (2)$$

where A = net contribution to sediment concentration of rainfall detachment over deposition, and B = net contribution of entrainment and re-entrainment processes over deposition.

The theory of Rose et al. (1983a) and Rose (1985) provides a description of the dependence of sediment concentration on measurable factors for different erosion mechanisms. In the field it is common for sediment at early times during a runoff event to be dominantly removed by rainfall detachment, and later, in a substantial erosion event, for entrainment/re-entrainment to be the dominant erosion mechanism (Rose et al. 1983b). It is useful to describe the size-distribution in any sediment in terms of a number of size classes distinguished by a distribution of individual settling velocities denoted v_i.

For term A in Eqn (2) the distribution of sediment concentration in size class i (c_i) resulting from rainfall of rate P is given (Rose et al. 1983a) by:

$$c = \sum_{i=1}^{I} c_i = \frac{a_1 C_e P}{QI} \sum_{i=1}^{I} \frac{1}{1 + v_i/Q} \qquad\qquad (3)$$

where a_1 = detachability of soil by rainfall where the power of P is taken to be unity, C_e = exposure fraction of the soil surface to rainfall detachment, Q = runoff rate per unit area of plane, and I = arbitrarily set maximum number of size classes i into which the settling velocity distribution is divided for computational purposes.

For small-sized sediment, term v_i/Q in Eqn (3) will be small compared to unity. Thus virtually all fine material detached will be found present in the suspension contributing to c_i. In contrast, if v_i/Q is large (commonly of order 10^2 for larger water-stable aggregates) then the concentration in the suspension of such aggregates will be very low indeed compared to their relative representation in the eroding material. Term B in Eqn (4) will now be considered.

Without attempting to distinguish between earlier entrainment and subsequent re-entrainment of deposited material, the component B (Eqn 4) of concentration can be expressed (Rose 1985) as:

$$c = 2700 \ SC_r \ \eta \ (1 - \Omega_o/\Omega) \quad (kg \ m^{-3}) \quad\quad\quad (4)$$

where 2700 is a number assuming SI units which makes assumptions regarding particle density amongst other factors. Also S = slope of the land surface, the sine of the angle of inclination, C_r = fraction of the soil surface exposed to overland flow, η = efficiency of the net entrainment process, Ω = the stream power of overland flow (the rate of working of the shear stresses), and Ω_o = threshold value of Ω for the entrainment/re-entrainment process to commence.

It is an implicit assumption in Eqn (4) that the size distribution of sediment being eroded is unaltered by the entrainment/re-entrainment processes. This assumption has experimental support.

Soil erosion terminology

Especially for soils high in silt, rill frequency is so high that the only erosion process between rills is rainfall detachment, and the processes of entrainment and re-entrainment are restricted to rills. Thus rainfall detachment is an "interrill process" and entrainment/re-entrainment a "rill process" (e.g. Foster 1981). However if rill frequency is low, or rills uncommon, this identity between visually-obvious erosion features and a particular erosion process no longer applies, since entrainment/re-entrainment is not restricted to rills. However, this identity is commonly assumed in the erosion literature of the USA, so this terminology will be employed in referring to such literature.

Sediment characteristics

Onstad and Moldenhauer (1975) compared the characteristics of sediment from rill and interrill flow. As shown by Eqn (3), rainfall detachment (or interrill erosion) is accompanied by substantial particle selectivity. This can be enhanced if aggregate breakdown occurs during rainfall detachment. In rills there is minimal particle selectivity and the eroded soil is removed en masse and Eqn (4) applies. Therefore, selectivity in the erosion process was assumed by Foster and Meyer (1975) to occur either in interrill regions or during deposition of the eroded sediment.

It follows from Eqn (3) and (4), and it is found experimentally, that sediment sizes from interrill areas are generally enriched in smaller particles and aggregates than those eroding from rills. Hereafter the term particle will be used to include aggregates.

215

"Enrichment ratio" is the ratio of the concentration of a constituent of eroded sediment to its concentration in the bulk soil. Thus the enrichment ratio is greater than unity for sediment from rainfall detachment or following net sediment deposition because of the predominance of finer fractions (clay and silt). Soils high in sand have large enrichment ratios (compared to clay soils) because their sediment is weakly aggregated, and the clay in their sediment is in small particles, not readily deposited. However, in high-clay soils, clay in the sediment is evenly distributed among most particle classes so that even when large particles are deposited, much clay is also deposited, resulting in only slight clay enrichment (Foster <u>et al</u>. 1985). This would also be expected to hold for N in forms closely associated with clay (Fig. 1). Fig. 1 shows the form of curve fitted to experimental data obtained on soils with a wide range of clay content.

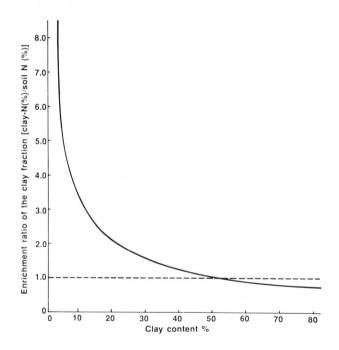

Fig. 1: Enrichment ratios of the clay fraction N as function of clay contents of soils. Clay fractions selectively lost in eroded sediments from sandy soils may be enriched many times depending on the clay content of the original soil (R.C. Dalal, <u>unpublished data</u>).

Since the specific surface areas of fine particles are very large, the sediment eroded by rainfall detachment is enriched in adsorbed N (such as NH_4^+-N) as well as clay-associated N (such as organic N). Therefore, the losses of N relative to sediment yield differ considerably, depending upon the dominant soil erosion process. Furthermore, as discussed later, losses of N also occur when unassociated organic matter is lost in erosion and runoff because of its low density and low settling velocity.

Selective removal of fine organic matter and clay particles also occurs in wind erosion (Daniel and Langham 1936; Marsh 1981). However, in general less is known about the enrichment ratios of wind-eroded particles. Once a soil partcile is initially dislodged, it moves downwind by suspension, saltation, or surface creep, depending upon its size, shape and density (Lyles 1977). The bulk of total transport (50–80%) is by saltation (0.2–1mm size particles), but suspension accounts for most of the loss of fine material, and consequently this process has a high enrichment ratio for plant nutrients, contributing strongly to the loss of N from the soil. Quantitative estimates of N losses by these processes do not appear to be available.

NITROGEN CONCENTRATION AND ENRICHMENT RATIO OF ERODED SEDIMENT

This section considers the effect of erosion on loss of the various forms of N. Whilst the major losses in terms of total amount are in forms sorbed to or intimately associated with the soil, losses in overland flow of soluble mineral forms of N are also briefly considered.

Total nitrogen

Total N in eroded sediments depends on many factors including the concentration in the source, the mix and extent of erosion processes and their effect on selective sorting of particle sizes. Generally, the N concentration in eroded sediment is higher than that of the parent soil, and the enrichment ratio (ER) declines as sediment yield increases (Fig. 2).

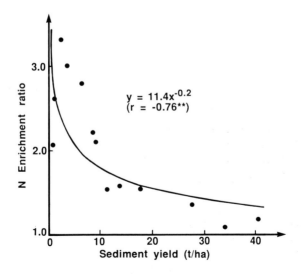

Fig. 2: Sediment yield and its N enrichment ratio in a sandy loam soil. (R.C. Dalal, R.J. Loch and L.A. Warrell, unpublished data).

Relationships of the type in Fig. 2 are the basis of current models of sorbed chemical runoff (e.g. CREAMS, Knisel 1980). Enrichment ratio (ER) is fitted by:

$$ER = A \ (SED)^{-B}$$

or

$$\ln (ER) = \ln A - B \ln (SED) \tag{5}$$

where SED is accumulated soil loss. For both N and P, Menzel (1980) suggests A = 7.4, and B = 0.2 for a range of soils, with SED in kg/ha. This value of B agrees with that in Fig. 2, but A differs.

Thus the total N yield in sediment (SED–N, kg N/ha) is calculated from:

$$SED\text{–}N = SOIL\text{–}N \ x \ SED \ x \ ER \tag{6}$$

where SOIL–N is the N concentration in the soil (kg N/kg soil) and SED is the sediment yield (kg/ha).

Since total N concentration varies between soils by as much as twenty times (2×10^{-4} to 4×10^{-3} kg N/kg soil), it is necessary, to measure total N for individual sites. ER may vary from c.1 to 10, depending upon sediment yield (Fig. 2), delivery ratio and soil type (Fig. 3). Thus, total N loss by eroded sediments may vary from < 1 kg N/ha/yr to > 100 kg N/ha/yr (Frere et al. 1980).

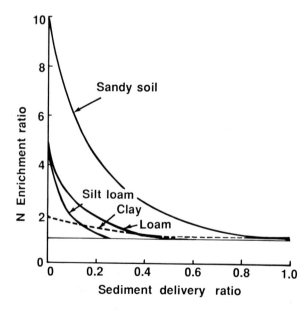

Fig. 3: Relationship between sediment delivery ratio and enrichment ratio (calculated from Foster et al. 1985).

The soil erosion factors which affect sediment yield and sediment characteristics also affect the total N in the sediment. Unfortunately, all of these effects on sediment N concentration have not been determined.

Foster et al. (1985) indicate that the effect of slope length and gradient on the N concentration and ER of sediments may be related to their effect on sediment delivery ratio. As slope increases, sediment delivery ratio increases which may thereby decrease N enrichment ratio (Fig. 3). The dominant erosion processes differ between rill and interrill areas, the effect of which is that N enrichment ratios for sediment from interrill regions is some 2 to 3 times greater than for sediment from rills (Fig. 4).

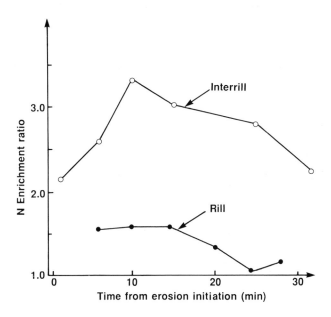

Fig. 4: N enrichment ratios of the eroded sediment from Emu Creek sandy loam (total N, 0.072%), having a slope of 6%. Plot-sizes for interrill and rill erosion were 3 x 4 m and 22.5 x 4 m (Dalal and Loch 1984).

The concentration of total N varies with the size of aggregate or particle. Table 1 gives an example indicating a bimodal distribution. In light-textured soils N concentration is generally highest in the clay fraction.

There are similar features in particles eroded by wind. Table 2 shows that the total N concentration of wind eroded particles <0.053 mm may be twice that of the whole soil or that of the same size fraction of the uneroded soil. Obviously, the smaller the particle, the longer it remains in suspension resulting in greater particle selectivity and higher ER.

219

Table 1: Total N concentration in sediment of different aggregate size classes.

Aggregate size classes	Total N[1]	Total N[2]
mm	*%*	*%*
2–1	0.242	0.202
1–0.5	0.240	0.191
0.5–0.21	0.233	0.203
0.21–0.05	0.328	0.213
0.05–0.035	0.235	0.135
0.035–0.020	0.094	0.090
0.020–0.010	0.224	0.172
0.010–0.002	0.397	0.352
<0.002	0.512	0.536

1. Alberts et al. (1981). Antecedent soil moisture condition: dry. Obtained from 1.8 m length with 27% surface cover using a rainfall simulator on a Typic Arguidolls, having 5% slope; 2. Alberts et al. (1983). Antecedent soil moisture condition: dry. Obtained from 3.0 x 10.7 m plots with 6% slope on a Typic Hapludolls.

Table 2: Total N and mineral N contents[1] of various size fractions of the whole soil and wind–eroded particles (adapted from Hagen and Lyles 1985).

Size fraction	Total N	NO_3^-–N	NH_4^+–N
		mg/kg soil	
Whole soil			
<0.053 mm	1185	16.0	9.2
0.053–0.25 mm	1335	17.4	9.0
>0.25 mm	1250	15.4	8.0
Eroded particles			
<0.053 mm	2385	39.8	32.8
0.10–0.15 mm	1275	23.5	5.0
0.29–0.42 mm	1050	18.1	4.6
0.59–0.84 mm	1450	22.7	9.2

1. Total N calculated from OM content assuming OM/Total N ratio of 20. Enrichment ratios of the eroded particles (<0.053 mm) were 2.0, 2.5 and 3.6 for total N, NO_3^-–N and NH_4^+–N, respectively.

Tillage affects aggregate–size distribution and sediment yield, and thus influences the ER for N of eroded sediments. Zero–till may reduce sediment yield compared to chisel plough and conventional till, but its ER for N of eroded sediments is higher than that from the other two tillage practices (Table 3).

Table 3: Effect of tillage practices on soil loss and N enrichment ratio of eroded sediment[1] (adapted from Young et al. 1985).

Tillage practices	Soil loss	N enrichment ratio
	tonne/ha	
Conventional	13.9	1.61
Chisel	17.4	1.43
Zero-till	7.3	2.03

1. Rainfall simulator run on a Typic Haplaudalfs having 6% slope, plot size 4.1 x 10.7 m.

Differences in total N loss among tillage practices may also be due to different surface stubble cover left after tillage (Fig. 5). Conventional cultivation (moldboard plough–disc–harrow) leaves a minimum of stubble cover on the soil surface and thus represents the most erodible tillage situation. Increasing stubble cover reduces sediment concentration, and hence total N loss by reducing the particle detachment effects of rainfall and runoff, and by reducing runoff transport capacity (Onstad and Moldenhauer 1975). Thus losses in total N are an inverse function of percentage of the surface area covered by stubble (Fig. 5). Rose and Freebairn (1985) found sediment yield declines by almost two–thirds with only 10% of the surface area under stubble, although the decrease in total N loss may be much less, depending on soil type and sediment delivery ratio (Fig. 3).

Besides stubble, vegetative cover by crops also reduces sediment yield and hence total N loss. Its effectiveness depends upon the tillage required, the type of vegetation, and the distribution of erosive rainfall relative to crop cover conditions (Onstad and Moldenhauer 1975). This is illustrated for two Darling Downs soils (Table 4).

Total sediment and total N loss decreased as the stubble cover increased from bare fallow to zero tillage. Summer crops provided limited surface contact cover against runoff from the high intensity summer storms. Note that compared to Table 4 values, estimated soil N loss would be slightly higher if the relationship of Fig. 2 had been used, but much lower if the Menzel (1980) relationship (ER = 7.39 SED$^{-0.2}$) had been used.

Estimates of soil N losses under 'conventional' cultivation in eastern Australia increased rapidly from south to north, associated with the increase in the incidence of higher rainfall intensity (Fig. 6). This indicates the special need for erosion control in the tropics and subtropics if soil

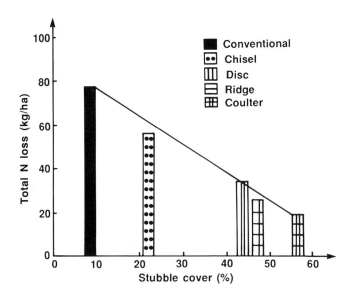

Fig. 5: Effect of maize stubble cover and tillage practices on N loss in sediment eroded from Ida silt loam, Iowa (adapted from Barisas et al. 1978).

Table 4: Annual soil loss and estimated total N loss from two vertisols under different tillage and stubble management on the Darling Downs, Queensland[1].

Tillage and stubble management	Soil loss		Total N loss	
	Irving clay	Acland clay	Irving clay	Acland clay
	t/ha/yr		*kg/ha/yr*	
Bare fallow	48	31	77	77
Stubble incorporated	14	7	28	21
Stubble mulched	3	3	8	10
Zero tillage	1	1	2	3
Summer crop	29	16	52	45

1. Soil loss data are means for 5 years during 1977–82 for Acland (Typic Chromusterts) and 6 years during 1976–82 for Irving (Typic Pellusterts). Slope 5–7% (from Freebairn and Wockner 1983). Maximum sediment delivery: 100 t/ha. Enrichment ratio calculated from sediment delivery ratio for clay soils (Fig. 3) and total N loss using Eqn (6). Total N concentrations for Irving and Acland clay 1.31×10^{-3} kg N/kg soil and 1.86×10^{-3} kg N/kg soil. Crop: wheat (winter), sunflower (summer). (D.M. Freebairn, unpublished data).

productivity is to be sustained. Even to maintain short–term soil productivity, 'conventional' clean cultivation must be replaced by 'conservation' cultivation.

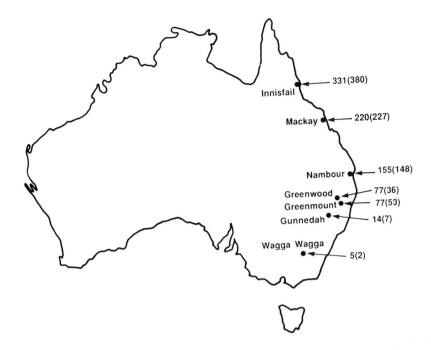

Fig. 6: Estimated annual soil total–N losses (kg/ha/yr) under 'conventional' cultivation in eastern Australia, using the relationships in Eqn (5); from estimated soil N content of 10^{-3} kg N/kg soil and soil loss values (t/ha/yr) from Freebairn 1982 (shown in brackets). Soil N losses for Greenwood and Greenmount are taken from Table 4.

Avnimelich and McHenry (1984) investigated the enrichment of total N in soil to its deposited state in water reservoirs from large catchments. There was a distinct tendency for the ER to be highest for catchment soils possessing the lowest N, and it tended towards unity for catchment soils higher in N, in general agreement with Fig. 1.

The N (and P) loss in eroded sediment is not only detrimental to soil productivity but can be a burdensome cost at the point of delivery off–site. Water with excess sediment requires treatment to provide potable water. More advanced stages of eutrophication, where both nitrates and phosphates are involved, can also lead to economic costs associated with the control of explosive growth of algae and other life forms in lakes and streams (Boughton 1983).

A theoretical framework for interpreting loss of total N in sediment

A theoretical framework is outlined to aid understanding and interpreting experimental results and the empirical relationship (Eqn 5) discussed in the previous sub–section.

It will be assumed that total N (or more exactly organic N) is closely associated or bound together with soil material of different size or fall velocity fractions. Organic N may be in finely divided colloidal form or humus; it may also be in forms essentially sorbed to mineral soil particles, or as larger, less-decomposed organic matter fragments. It is possible to separate these forms, and any soil with which they may be closely associated, into a size distribution of water-stable aggregates or particles using a top-entry or settling tube (e.g. Hairsine and McTainsh 1986), or by bottom-withdrawal tube (Anon. 1943; Lovell and Rose 1986). The separation is then in terms of fall velocity, v_i, the characterisitic seen earlier as directly relevant to erosion and deposition processes.

With soil material fractionated in this way, it is assumed that most inorganic N will have been dissolved from the particles into the water through which the soil has settled. If so, total N on soil fractions would be dominantly organic N. Under B in Fig. 7 is illustrated the distribution of mass of element (organic N) per mass of soil as a function of fall velocity (v_i).

Whilst the clay in sediment is the site of adsorption of most nutrients and chemicals, this fraction can be transported in either primary or aggregated form. Measurement of the distribution illustrated at B in Fig. 7 should be carried out on eroded sediment, but may vary little with the mix of erosion mechanisms. The distribution of total N concentration with aggregate size can vary substantially with different soils, and unlike the distribution illustrated at B in Fig. 7 may increase with v_i, as found for a silt loam by Alberts and Moldenhauer (1981).

Earlier it was discussed that for any soil a particular mass distribution of sediment with v_i results from any particular mix of soil erosion mechanisms. Under A in Fig. 7 two possible distributions are illustrated.

The product of the two fractions at A and B in Fig. 7 leads to the distribution at C:

$$\frac{\text{kg soil in } \Delta v_i}{\text{kg soil}} \quad \times \quad \frac{\text{kg organic N in } \Delta v_i}{\text{kg soil in } \Delta v_i} \quad = \quad \frac{\text{kg organic N in } \Delta v_i}{\text{kg soil}}$$

$$A \qquad \times \qquad B \qquad = \qquad C$$

Summing the area under the distribution at C (Fig. 7) gives the value of kg element (viz. organic N)/kg soil in runoff for the particular erosion mechanism yielding the distribution at A. The alternative distributions (a) and (b) at A in Fig. 7 will result in different values of ER. The value of ER for any particular mix of erosion mechanisms is the area under distribution C for this mix divided by the area under this distribution for the original uneroded soil.

224

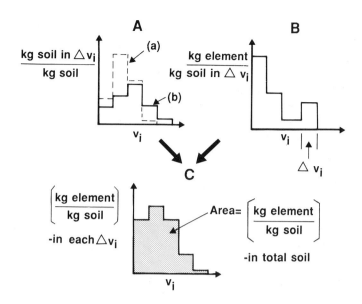

Fig. 7: Schematic diagram of the conceptual basis of a model of loss of sorbed chemicals in runoff. Velocity v_i is particle settling velocity (logarithmic scale). In A, (a) might refer to rainfall detachment and (b) to entrainment/re-entrainment mechanisms. The form of B is experimentally determined for any particular soil and element. Distribution C is the product of distributions A and B.

Under rainfall the mix of erosion mechanisms at work changes with time. For example, early in the erosion event rainfall detachment may be significant, and as illustrated by Eqn (3), when this mechanism predominates the size distribution of sediment initially is much finer than the distribution of the soil as a whole. As is illustrated at B in Fig. 7, relative element concentrations are typically much higher for finer fractions of soil. Thus, especially when this erosion mechanism is dominant, an ER significantly > 1 would be expected. This prediction has wide experimental support (e.g. Knisel 1980; Dalal and Loch 1984).

As an erosion event proceeds, rainfall detachment is commonly increasingly overshadowed by the processes of entrainment and re-entrainment, for which the size distribution of sediment in overland flow is very similar to that of the original soil, as implied by Eqn (4). The data of Alberts et al. (1980) illustrates this change in the size distribution of soil particles. In their experiments erosion was dominated by entrainment/re-entrainment in rills, and by rainfall detachment for inter-rill flow. These authors found that a much greater proportion of larger aggregates was transported in rills compared with the finer sediment typical of inter-rill flow.

Even though as erosion proceeds sediment is dominantly removed by entrainment and re-entrainment, fine sediment associated with rainfall detachment will still be a component, however small, of the total sediment load. This can be because of the presence of inter-rill areas with typically quite shallow flow. However, even if rills are absent, rainfall detachment can still play a role because of the roughness of natural land surfaces, perhaps exaggerated by cultivation, which normally ensures that some of the soil surface is not submerged by overland flow and is thus subject to rainfall impact.

As a result of these shifts in erosion processes with time it would be expected that ER will decrease towards unity as illustrated in Fig. 2. Different measured chemical components do not necessarily show the same ER, perhaps due to different characteristics at B in Fig. 7, or for other reasons such as different densities of particles or aggregates with which each chemical component is associated.

The model illustrated in Fig. 7 may provide a theoretical and process-based interpretation of such empirical relations as Eqn (5) which provide the basis for current models of chemical runoff associated with soil erosion. It is clear from experimental data (e.g. Sharpley 1985), and the above discussion that parameters A and B in Eqn (5) will be affected by management, soil type, chemical element or compound of interest, and by the change in mix of erosion processes. This approach thus has potential in evaluating the effect of eroson on soil fertility.

Mineralizable nitrogen

Total N loss by erosion generally reflects a potential loss in soil productivity. Mineral N and mineralizable N losses, however, reflect an immediate loss, which are better correlated with the shorter-term reductions in crop yield.

Potentially mineralizable N (PM-N), as a measure of mineralizable N (Stanford and Smith 1972) represents 10% of total N in Inceptisols and Spodosols, 15% in Aridosols and Vertisols, 18% in Alfisols and Entisols and 24% in Ultisols. The ER of PM-N is higher than that of total N in Table 5. Unfortunately, information is lacking to derive ER sediment yield relationships for PM-N in different soils.

Microbial biomass nitrogen

Soil microbial biomass is an agent of transformation of the added and native organic matter and a small, but labile, source as well as sink of N. Although it contains only 2-5% of the total N in soil, it may provide a biological index of soil productivity loss due to erosion. Since the microbial biomass size is predominantly in the coarse clay and fine silt size range ($1 - 5 \times 10^{-6}$m), the selective loss of fine particles in eroded

Table 5: Effect of crop rotation on losses of soil and potentially mineralizable N (PM–N) and enrichment ratios of PM–N and total N in the eroded sediment[1] (adapted from Young et al. 1985).

Rotation system[2]	Soil loss	Loss of PM–N[7]	Enrichment ratio PM–N	Enrichment ratio Total N
	t/ha	kg/ha		
Maize –1[3]	20.6	7.4	2.26	1.76
Maize –2[4]	18.0	6.7	2.37	1.55
Soybeans[5]	13.5	5.3	2.50	2.02
Fallow[6]	14.4	5.8	2.53	1.84

1. Rainfall simulator run on a Udic Haploborolls, 7% slope, plots 4.1 x 10.7m; 2. Rotation = maize–maize–soybean; 3. 1st year maize; 4. 2nd year maize; 5. 3rd year soybean; 6. Continuous fallow; 7. PM–N, potentially mineralizable N (Stanford et al. 1974).

sediments would be expected to result in a proportionately larger loss of microbial biomass that that in the whole soil, but no ER values are yet available.

Assessment of erosion losses of microbial biomass N have been made from comparison of microbial biomass N in the surface layer of cultivated and uncultivated rangeland soils at three slope positions (Table 6). Larger losses of microbial biomass N occurred from the summit position than the footslopes for the Haploborolls where, because of its coarse nature, the ER of N, would be expected to be high (Fig. 3).

Mineral N

Mineral N (NH_4^+ + NO_3^- + NO_2^-) is commonly less than 2% of total N in soil, and in eroded sediments rarely exceeds 10% of the total N loss, viz. — <1 to 10 kg mineral N/ha. These forms of N are soluble and thus can also be lost in runoff water, but the amount again is commonly small. However, because these forms of N are readily available for uptake by plants, they are important losses in the short term, even if small in amount compared to losses of organic N by soil erosion.

Substantial losses of N in runoff appear to be limited to situations where high rates of surface broadcast fertilizer have been applied and followed by substantial rainfalls on sloping land (e.g. Moe et al. 1967, 1968). Under mature oil palm in Malaysia, Maene et al. (1979) showed annual total N loss in runoff water to be nearly twice that lost through sediment transport. The loss in runoff was 11% of the N applied as fertilizer broadcast on the surface or between fronds of oil palm.

Table 6: Microbial biomass N in the top (0.01m) layer of (A) uncultivated and (B) cultivated rangeland soils developed on three slope positions (adapted from Schimel et al. 1985).

Soil type	Position	Slope	Microbial biomass N		
			A	B	(A–B)/A
		%	*mg/kg soil*		*%*
Coarse Haploborolls	Summit	1–7	190	73	62
	Backslope	4–6	179	75	58
	Footslope	2	201	118	41
Fine Haploborolls	Summit	1–7	206	112	46
	Backslope	5.5	195	149	24
	Footslope	4	231	129	44

Mineral N concentrations in eroded sediments vary widely with type of soil, fertilizer added, sorption characteristics (especially for NH_4^+-N), and erosivity factors. For example, eroded sediments from soils high in vermiculite and micaceous minerals, are expected to be high in sorbed NH_4^+-N, and a substantial loss of mineral (and fixed) NH_4^+-N can occur this way.

In tropical and sub-tropical regions land clearing also results in a substantial loss of mineral N in runoff. Kang and Lal (1981) found large runoff and nutrient losses for three months following various forms of land clearing in Nigeria. Also, Hunter and Lawrence (personnel communication) observed appreciable losses of soluble N in runoff from land clearing in Central Queensland following burning during the first year, with much smaller losses in later years under more stable soil conditions.

Study of the physico-chemical processes involved in the loss of inorganic N in overland flow appears to be limited, despite observations on its magnitude and the effects of management (e.g. Schuman et al. 1973; Burwell et al. 1975). The degree of soil incorporation and solubility of fertilizer N should have an important effect on this form of loss (Moe et al. 1968). However, loss to overland flow is limited by the dominantly downward bulk movement of water in infiltration. Possible lateral movement of water beneath the soil surface at slopes and for soil conditions which favour interflow would be expected to enhance this form of loss of inorganic N.

Losses of mineral N in wind-eroded particles may be substantial (Hagen and Lyles 1985). Because mineral N is concentrated mostly in the top layer, wind-blown fine particles with an ER for N of 2-3 may carry large amounts (Table 2).

EFFECT OF SOIL EROSION ON NITROGEN TRANSFORMING PROCESSES

Erosion changes water intake, storage and transmission through the soil profile and hence soil air, soil temperature, organic C and N substrates and mineral N; it can expose soil layers containing high amounts of $CaCO_3$, salts or acidity. There is a paucity of information on the interaction of these effects with N transformation processes, so the following discussion is mainly speculative. Attempts have been made, however, to simulate these effects in models such as CREAMS (Knisel 1980) and EPIC (Williams et al. 1985).

Mineralization – immobilization

Mineralization in a soil or eroded sediment is described by a first-order process, that is:

$$N = N_0 (1 - exp (-kt)) M$$

where N is the mineralized N, N_0 is the mineralization potential or potentially mineralizable N (PM–N), k is a temperature-adjusted mineralization rate constant, t is time and M is the soil moisture content expressed as a fraction of field capacity (Smith et al. 1980).

Erosion reduces potentially mineralizable N (N_0) in soil through the loss of sediments enriched in PM–N by 2–3 times (Table 5). The mineralization rate constant (k) remains largely unaffected, but moisture (M) may vary for a number of reasons including selective loss of fine particles from the soil. As a result, the net amount of N mineralized in an eroded soil may be considerably decreased. Also, sub-soil layers can be exposed which normally have reduced amounts of mineralizable N.

Immobilization, or rather net immobilization, occurs in the presence of readily assimilable organic substrates and mineral N. In immobilization-mineralization turnover studies of erosion-affected soil and off-site eroded sediments, an active ('labile') phase of organic matter should be considered. The net effect of immobilization-mineralization turnover depends upon soil texture, structure, chemical composition of substrates, climate, cropping system and crop residue management, among other factors, such as microbial biomass (Jansson and Persson 1982). Most of these factors are affected by and affect soil erosion. In addition, the growing crop utilizes mineral N in direct competition with micro-organisms and supplies organic substrates through root exudates and root debris and decaying plant matter. Its effect on immobilization-mineralization turnover in erosion-affected soils should be considered.

Denitrification

Denitrification is the microbial reduction of nitrate or nitrite to gaseous N either as molecular N or as an oxide(s) of N.

When runoff occurs surface soil water contents will be high, and hence denitrification by facultative anaerobic micro-organisms is possible. Quantitative measurement of denitrification rate is not simple as the dynamics of O_2 concentration need to be understood, as well as those of the denitrification process. The latter is not well understood though commonly represented as an enzymatic process of the Michaelis-Menten type (Tanji and Gupta 1978) involving nitrate reductase. Nitrite reductase, nitric oxide reductase and nitrous oxide reductase are also involved.

Denitrification activity is affected by readily available C sources (labile soil organic matter, plant residues, manure, root exudates, etc.), NO_3 or NO_2 concentration, O_2 concentration (hence soil moisture characteristics, air-filled pores, partial pressure of O_2 and redox potentials), temperature and soil pH (Firestone 1982). Most of these are affected by soil erosion to some extent, especially on slope-positions which change from steep to flat, resulting in deposition of sediment enriched in organic C and NO_2^--N or NO_3^--N and providing a microbial environment at least temporarily depleted in O_2.

Volatilization

Gaseous losses of fertilizer N can occur through NH_3 volatilization from soils whenever free NH_3 is present near the surface. This process can occur rapidly in calcareous soils, e.g. after addition of $(NH_4)_2SO_4$ fertilizer. Increase in soil pH and decrease in cation-exchange capacity (CEC) will also increase NH_3 volatilization.

In many soils of arid and semi-arid regions, $CaCO_3$ increases with depth. Water and wind erosion may remove the top soil layer, and application of ammonium-producing and ammonia-cal fertilizers to the exposed subsoil layers of high $CaCO_3$ content would result in substantial NH_3 volatilization (Fenn and Kissel 1973). Also, pH and salinity of subsoil layers of soils from these regions are generally higher than the top soil, and if exposed by erosion, increased NH_3 volatilization losses would result.

Since CEC is the most important factor governing the loss of NH_3 following application of urea and $(NH_4)_2SO_4$ to soil (Gasser 1964), reduction in CEC due to selective loss of organic matter and clay particles in soil erosion would result in increased NH_3 volatilization losses in eroded soils, especially those low in CEC in their original state.

EFFECT OF SOIL EROSION ON CROP YIELD

In many soils, especially those with strongly developed profiles and those originally supporting non-grassland vegetation, crop yields decrease with the amount of erosion. This is illustrated (Fig. 8) for soils in a semi-arid environment, where wheat yields declined linearly with increasing loss of top-soil plotted on a logarithmic scale. The rate of decline was more

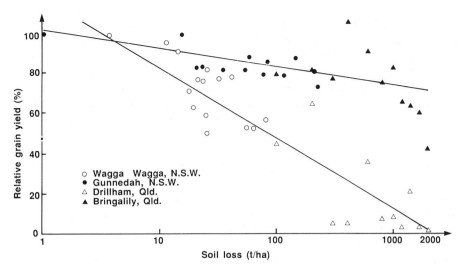

Fig. 8: Relationship between soil loss and relative wheat grain yield under dryland conditions in eastern Australia. Circles represent cumulative soil loss under natural rainfall over 26 years (Gunnedah) and 30 years (Wagga Wagga) followed by 2 and 4 wheat crops, respectively (from Aveyard 1983). Triangles represent soil removal by desurfacing, followed by a wheat crop (P.J. White, <u>unpublished data</u>). Soil profiles at Gunnedah and Bringalily are medium to deep whereas the others are shallow, with either texture-contrast (Wagga Wagga) or saline subsoil (Drillham).

moderate for soils with medium to deep profiles than for the shallower soils with a texture-contrast profile or a saline subsoil. Fig. 8 indicates that percentage yield decline is approximately linearly related to the logarithm of soil loss for any particular soil.

By the time annual soil loss reaches 10 t/ha (corresponding to about 1 mm depth of soil), yield reduction is about 20% for sites with a shallow topsoil. The results obtained by desurfacing seem, for both soil types, consistent with the general trend obtained with erosion under natural rainfall. Thus, if interpreted in this way, desurfacing experiments may not lead to such unrealistic results as commonly supposed.

Few attempts, however, have been made to identify the yield-limiting effects of soil erosion and to quantify losses in crop production. Although, in soils of strongly developed profiles, the loss of plant-available water capacity especially in the semi-arid regions may be the most vital factor limiting crop yield, the loss of plant nutrients (especially N) has also reduced crop yields (Eck et al. 1965). Aveyard (1983) reported a decline in wheat grain protein with the increasing amount of top soil removed. The proportion of total soil N (0–0.15 m depth) recovered by a crop has been found to decrease as the depth of soil removed increases (Eck 1968).

Fertilizer addition can partly compensate for loss of N in eroded soil although thereby increasing production costs, and possibly causing off-site environmental damage. Eck et al. (1965) were able to produce optimum grain sorghum yields on a Pullman silty clay loam after top-soil removal up to 0.3 m depth by application of N fertilizers (up to 320 kg N/ha), but where 0.4 m top-soil was removed, even heavy fertilization produced only 80% of the yields from similarly fertilized, undisturbed soil. Mbagwu et al. (1984) also found that fertilizer did not restore the yields of maize and cowpea on an Ultisol in Nigeria, where 0.05-0.20 m of soil were removed.

Lyles (1977) investigated wind erosion effects on wheat yields in the U.S.A. and found a yield reduction of c. 2 kg grain/ha/mm of soil loss for wheat-fallow rotations. The yield-limiting effects of wind erosion were not identified. Marsh (1981) attributed yield reductions from wind-eroded soils in Western Australia to the selective removal of N-enriched soil particles and organic matter.

CREAMS (Knisel 1980) and Nitrogen-Tillage-Residue Management (Shaffer 1985) models may be used to simulate the effects of management, supplemental N, and water on crop productivity from eroded soils. Williams et al. (1981) suggested a nutrient cycling model component that for N included leaching, runoff losses, crop residues, volatilization, denitrification, immobilization, mineralization, nitrification, crop uptake, rainfall contributions, and fertilizer. Erosion loss of N is another component, important both to on-site productivity and off-site environmental impacts.

CONCLUSIONS

In conclusion, many long-term studies on restoring productivity to eroded soils have shown that it is more practical and economical to use management practices that minimise soil erosion than to adopt the practices required to restore eroded soil (Dormaar et al. 1986).

Short-term yield reductions due to leaching or runoff losses of soluble N can be quite important. However, the loss of N with eroded sediment is generally much larger and this effect will have the greater long-term impact on yields.

Therefore the major challenge in reducing loss of N is to reduce soil loss. Retention of crop residues as a surface mulch has been, and will continue to be, one of the most effective protective measures against soil erosion. Interest will continue in the advantages and limitations of other soil conserving practices such as minimizing tillage, contour banks, etc.

Recent advances in understanding the relationship between the various soil erosion mechanisms and the size-or settling-velocity distribution of resulting sediment holds promise for better understanding of the enrichment ratio for nutrients which, like organic forms of N, are closely associated with or sorbed to eroded sediments.

A method of data presentation is illustrated which appears useful for bringing together in a comparable way the results of soil erosion on crop yield. This methodology should be more widely tested.

The literature supports the view that studies of the relationship between soil loss, nutrient decline, and crop yield are urgently required, especially in tropical and sub-tropical croplands, where huge losses of soil and N are possible (Lal 1984; and Fig. 6).

ACKNOWLEDGEMENTS

We are grateful to the following for permitting us to use their unpublished data: Mr D.M. Freebairn, Mrs H. Hunter, Mr P.A. Lawrence, Mr R.J. Loch, Mr L.A. Warrell and Dr P.J. White. Discussions with Mr P.B. Hairsine, Mr C.J. Lovell and Mr A.P.B. Proffitt on the settling velocity characteristics associated with various erosion mechanisms were most helpful.

REFERENCES

Alberts, E.E., and Moldenhauer, W.C. (1981). Nitrogen and phosphorus transported by eroded soil aggregates. Soil Sci. Soc. Am. J. 45, 391–6.

Alberts, E.E., Moldenhauer, W.C., and Foster, G.R. (1980). Soil aggregates and primary particles transported in rill and interrill flow. Soil Sci. Soc. Am. J. 44, 590–5.

Alberts, E.E., Neibling, W.H., and Moldenhauer, W.C. (1981). Transport of sediment nitrogen and phosphorus through cornstalk residue strips. Soil Sci. Soc. Am. J. 45, 1177–84.

Alberts, E.E., Wendt, R.C., and Piest, R.F. (1983). Physical and chemical properties of eroded soil aggregates. Trans. ASAE 26, 465–71.

Anon. (1943). A study of new methods for size analysis of suspended sediment samples. Report No. 7: Office of Indian Affairs, Bureau of Reclamation T.V.A., Corps of Eng. Geol. Surv., Dept. of Agr. and Iowa Institute of Hydraulic Res., University of Iowa.

Aveyard, J.M. (1983). Soil erosion: productivity research in New South Wales to 1982. Soil Cons. Serv. NSW Tech. Bull. No. 24.

Avnimelech, Y., and McHenry, J.R. (1984). Enrichment of transported sediments with organic carbon, nutrients, and clay. Soil Sci. Soc. Am. J. 48, 259–66.

Barisas, S.G., Baker, J.L., Johnson, H.P., and Laflen, J.M. (1978). Effect of tillage systems on runoff losses of nutrients: a rainfall simulation study. Trans. ASAE 21, 893–7.

Boughton, W.C. (1983). Non–point source pollution in Australia. In 'Workshop on Non–point Sources of Pollution in Australia, Proceedings'. (Aust. Gov. Publ. Service: Canberra).

Burwell, R.E., Timmons, D.R., and Holt, R.F. (1975). Nutrient transport in surface runoff as influenced by soil cover and seasonal periods. Soil Sci. Soc. Am. Proc. 39, 523–8.

Dalal, R.C., and Loch, R.J. (1984). The effect of erosion processes on nutrient enrichment of sediments eroded from a sandy loam soil. National Soils Conf., Brisbane, Australia, p. 244.

Daniel, H.A., and Langham, W.H. (1936). The effect of wind erosion and cultivation on the total nitrogen and organic matter content of soils in the Southern High Plains. J. Am. Soc. Agron. 28, 587–96.

Dormaar, J.F., Lindwall, C.W., and Kozub, G.C. (1986). Restoring productivity to an artificially eroded dark brown chernozemic soil under dryland conditions. Can. J. Soil Sci. 66, 273–85.

Eck, H.V. (1968). Effect of topsoil removal on nitrogen supplying ability of Pullman silty clay loam. Soil Sci. Soc. Am. Proc. 32, 686–91.

Eck, H.V., Hauser, V.L., and Ford, R.H. (1965). Fertilizer needs for restoring productivity on Pullman silty clay loam after various degrees of soil removal. Soil Sci.Soc.Am.Proc.29,209–13.

Fenn, L.B., and Kissel, D.E. (1973). Ammonia volatilization from surface applications of ammonium compounds on calcareous soils: 1. General theory. Soil Sci. Soc. Am. Proc. 37, 855–9.

Firestone, M.K. (1982). Biological denitrification. In 'Nitrogen in Agricultural Soils'. (Ed. F.J. Stevenson). Am. Soc. Agron., Agron. Monog. No. 22, pp. 289–326.

Foster, G.R. (1982). Modelling the erosion process. In 'Hydrologic Modelling of Small Watersheds'. (ed. C.T. Haan). Am. Soc. Agr. Eng. Monogr. No. 5, pp. 297–379.

Foster, G.R., and Meyer, L.D. (1975). Mathematical simulation of upland erosion using fundamental erosion mechanics. Proc. Sediment Yield Workshop, USDA Sedimentation Lab., Oxford, Miss. RS-S-40, pp. 190–207.

Foster, G.R., Young, R.A., and Neibling, W.H. (1985). Sediment composition for nonpoint source pollution analyses. Trans. ASAE 29, 133–9, 146.

Freebairn, D.M. (1982). Soil erosion in perspective. Div. Land Util. Tech. News (Queensl. Dep. Primary Ind.) Vol. 6 No. 1, pp. 12–5.

Freebairn, D.M., and Wockner, G.H. (1983). Soil erosion control research provides management answers. Queensl. Agric. J. 103, 227–34.

Frere, M.H., Ross, J.D., and Lane, L.J. (1980). The nutrient submodel. In 'CREAMS: A Field-scale Model for Chemicals, Runoff, and Erosion from Agricultural Management Systems'. (Ed. W.G. Knisel). USDA Conserv. Res. Rep. No. 26, pp.65–87.

Gasser, J.K.R. (1964). Some factors affecting the losses of ammonia from urea and ammonium sulfate applied to soils. J. Soil Sci. 15, 258–72.

Hagen, L.J., and Lyles, L. (1985). Amount and nutrient content of particles produced by soil aggregate abrasion. In 'Erosion and Soil Productivity'. pp. 117–29. (Am. Soc. Agric. Eng.: St Joseph, Michigan.)

Hairsine, P.B., and McTainsh, G.H. (1986). The Griffith Tube: A simple settling tube for the measurement of settling velocity of aggregates. AES Working Paper 3/86. Griffith University, Nathan, Brisbane, Australia.

Jansson, S.L., and Persson, J. (1982). Mineralization and immobilization of soil nitrogen. In 'Nitrogen in Agricultural Soils'. (Ed. F.J. Stevenson). Am. Soc. Agron., Agron. Monogr. No.22, pp. 229–52.

Kang, B.T. and Lal, R. (1981). Nutrient losses in water runoff from agricultural catchments. In 'Tropical Agricultural Hydrology'. (Eds. R. Lal and E.W. Russell). pp. 153–61 (John Wiley and Sons: London).

Knisel, W.G. (1980). 'CREAMS. A Field-scale Model for Chemicals, Runoff and Erosion from Agricultural Management Systems'. USDA Conserv. Res. Rep. No. 26.

Lal, R. (1984). Soil erosion from tropical arable lands and its control. Adv. Agron. 37, 183–248.

Lovell, C.J., and Rose, C.W. (1986). Measurement of the settling velocities of soil aggregates using a Modified Bottom Withdrawal Tube. AES Working Paper 4/86. Griffith University, Nathan, Brisbane, Australia.

Lyles, L. (1977). Wind erosion: Processes and effect on soil productivity. Trans. ASAE 20, 880–4.

Maene, L.M., Thong, K.C., Ong, T.S., and Mokhtaruddin, A.M. (1979). Surface wash under mature oil palm. Proc. Symp. 'Water in Agriculture in Malaysia', Kuala Lumpur, pp 203–16.

Marsh, B.a'B. (1981). Wind erosion in relation to tillage systems. In 'Proc. National Workshop on Tillage Systems for Crop Production.', Roseworthy, S. Aust., pp. B22–25.

Mbagwu, J.S.C., Lal, R., and Scott, T.W. (1984). Effects of desurfacing of alfisols and ultisols in southern Nigeria: 1. Crop performance. Soil Sci. Soc. Am. J. 48, 828–33.

Menzel, R.G. (1980). Enrichment ratios for water quality modelling. In 'CREAMS: A Field-scale Model for Chemicals, Runoff and Erosion from Agricultural Management Systems'. (Ed. W.G. Knisel) USDA Conserv. Res. Rep. No. 26, pp. 486–92.

Moe, P.G., Mannering, J.V., and Johnson, C.B. (1967). Loss of fertilizer nitrogen in surface runoff water. Soil Sci. 104, 389–94.

Moe, P.G., Mannering, J.V., and Johnson, C.B. (1968). A comparison of nitrogen losses from urea and ammonium nitrate in surface runoff water. Soil Sci. 105, 428–33.

Onstad, C.A., and Moldenhauer, W.C. (1975). Watershed soil detachment and tranportation factors. J. Environ. Qual. 4, 29–33.

Rose, C.W. (1985). Developments in soil erosion and deposition models. Adv. Soil Sci. 2, 1–63.

Rose, C.W. and Freebairn, D.M. (1985). A new mathematical model of soil erosion and deposition processes with applications to field data. In 'Soil Erosion and Conservation'. (Eds. S.A. El-Swaify, W.C. Moldenhauer and A. Lo), pp. 549–57. (Soil Cons. Soc. Am: Iowa).

Rose, C.W., Williams, J.R., Sander, G.C., and Barry, D.A. (1983a). A mathematical model of soil erosion and deposition processes. I. Theory for plane land element. Soil Sci. Soc. Am. J. 47, 991–5.

Rose, C.W., Williams, J.R., Sander G.C., and Barry D.A. (1983b). A mathematical model of soil erosion and deposition processes. II. Application to data from an arid-zone catchment. Soil Sci. Soc. Am. J. 47, 996–1000.

Schimel, D.S., Coleman, D.C., and Horton, K.A. (1985). Soil organic matter dynamics in paired rangeland and cropland toposequences in north Dakota. Geoderma 36, 201–14.

Schuman, G.E., Burwell, R.E., Piest, R.F. and Spomer, R.G. (1973). Nitrogen losses in surface runoff from agricultural watershed on Missouri Valley loess. J. Environ. Qual. 2, 299–302.

Shaffer, M.J. (1985). Simulation model for soil erosion–productivity relationships. J. Environ. Qual. 14, 144–50.

Sharpley, A.N. (1985). The selective erosion of plant nutrients in runoff. Soil Sci. Soc. Am. J. 49, 1527–34.

Smith, S.J., Kissel, D.E., and Williams, J.R. (1980). Nitrate production, uptake, and leaching. In 'CREAMS: A Field-scale Model for Chemicals, Runoff, and Erosion from Agricultural Management Systems'. (Ed. W.G. Knisel) USDA Conserv. Res. Rep. No. 26, pp. 493–508.

Stanford, G., Carter, J.N., and Smith, S.J. (1974). Estimates of potentially mineralizable soil nitrogen based on short-term incubations. Soil Sci. Soc. Am. Proc. 38, 99–102.

Stanford, G., and Smith, S.J. (1972). Nitrogen mineralization potentials of soils. Soil Sci. Soc. Am. Proc. 36, 465–72.

Tanji, K.K., and Gupta, S.K. (1978). Computer simulation modelling for nitrogen in irrigated croplands. In 'Nitrogen in the Environment' Vol. 1. (Eds. D.R. Nielson and J.G. MacDonald), pp. 79–130. (Academic Press Inc.: New York).

White, P.J. (1986). A review of soil erosion and agricultural productivity with particular reference to grain crop production in Queensland. J. Aust.Inst. Agric. Sci. 52, 12–22.

Williams, J.R., Allmaras, R.R., Renard, K.G., Lyles, L., Moldenhauer, W.C., Langdale, G.W., Meyer, L.D., Rawls, W.J., Darby, G., Daniels, R., and Magelby, R. (1981). Soil erosion effects on soil productivity: a research perspective. J. Soil Water Cons. 36, 82–90.

Williams, J.R., Putman, J.W., and Dyke, P.T. (1985). Assessing the effect of soil erosion on productivity with EPIC. In 'Erosion and Soil Productivity'. pp. 215–26. (Amer. Soc. Agric. Eng.: St. Joseph, Michigan.)

Young, R.A., Olness, A.E., Mutchler, C.K., and Moldenhauer, W.C. (1985). Chemical and physical enrichment of sediment from cropland. In 'Erosion and Soil Productivity'. pp. 107–16. (Amer. Soc. Agric. Eng.: St. Joseph, Michigan.)

PART III

THE NITROGEN CYCLE IN DIFFERENT SYSTEMS

NITROGEN CYCLING IN WETLAND RICE SOILS

I. Watanabe, S.K. De Datta and P.A Roger

ABSTRACT

Nitrogen cycling in wetland rice soils is unique : (a) wetland rice is more dependent on soil N than dryland crops, (b) N accumulates at the soil surface through the activities of the photosynthetic aquatic biomass, and (c) inorganic nitrogen is not stable at the soil surface and in the floodwater due to denitrification and ammonia volatilisation.

Factors stimulating soil organic N mineralisation are known. However, methods of predicting soil N release to rice need improvement. Soil microbial biomass is an important source of N to wetland rice in mineral N–deficient conditions. Photosynthetic biomass in the floodwater causes N accumulation at the soil surface by immobilisation and N_2-fixation, but stimulates ammonia loss by increasing floodwater pH. Fauna in soil and floodwater play a role in recycling of N. Aerodynamic measurement of ammonia volatilisation loss accompanied by ^{15}N balance in the field revealed that some loss is mediated by a mechanism other than ammonia volatilisation, probably denitrification. Point placement of granular urea, use of urease inhibitors, and proper timing of N fertiliser application reduce fertiliser N losses and increase N fertiliser efficiency.

INTRODUCTION

About 75% of the 143 M ha of rice land are lowlands (wetlands), where rice grows in flooded fields during part or all of the cropping period.

Flooding favours rice environments by : (1) bringing the soil pH near to neutral; (2) increasing availability of nutrients, especially P and Fe, (3) maintaining soil N ; (4) stimulating N_2-fixation; (5) depressing soil–borne diseases; (6) supplying nutrients from irrigation water; (7) suppressing weeds, especially those of C_4 type, and (8) preventing water percolation and soil erosion.

By 1950's most of the unique characteristics of nitrogen transformations in wetland rice soils were described by Mitsui (1954). They are:

(i) N fertility is higher in the flooded rice soils than in the upland soils.
(ii) The surface of the flooded soil and the floodwater accumulate N by recycling or by biological N_2-fixation.
(iii) Surface-applied ammonium-N is unstable due to nitrification-denitrification. NH_4^+ is stable in reduced soil.

The recent proceedings of the symposium "Nitrogen Economy of Flooded Rice Soil" (De Datta and Patrick 1986) contains a wealth of updated knowledge on N transformations in flooded rice soils and several reviews are available (Anon. 1979 ; Watanabe et al. 1981a ; Patrick 1982 ; Savant and De Datta 1982). Knowledge and technology of biological nitrogen fixation have also recently been reviewed (Watanabe and Roger 1984 ; Roger and Watanabe 1986 ; Watanabe 1986). Therefore, to avoid repetition, this review focuses on recent knowledge of the N cycle in wetland rice soils, and introduces some untranslated Japanese work.

RELEASE AND MAINTENANCE OF AVAILABLE SOIL NITROGEN

Factors affecting the release of soil N

N balance studies in long-term fertility trials have demonstrated that N supplied by the soil is an extremely important component of rice production (Watanable et al. 1981a). Despite this importance, very little research effort is expended on studying how to use soil N more effectively (Bouldin 1986). Early research, however, identified most factors favouring soil N mineralisation. It was shown that drying of soil stimulates ammonia formation and renders soil organic N more decomposable (Shioiri et al. 1941 ; Harada 1959). Other factors reported to stimulate soil N mineralisation were temperature elevation from $30^{\circ}C$ to $40^{\circ}C$, liming, solubilisation of soil colloids by NaF, Na_2HPO_4 and Na oxalate (Harada 1959), mechanical destruction of soil aggregates (Harada et al. 1964 ; Hayashi and Harada 1964), and root growth (Hayashi and Harada 1964).

Drying of fallow soil has long been practiced by Chinese and Japanese farmers. Soil desiccation during the dry season in the tropics may stimulate soil N mineralisation the following wet season. Ventura and Watanabe (1978, 1984) showed that air-drying the soil during the dry season temporarily depressed N uptake at early stage, probably due to the accumulation of toxic substances, but increased total N uptake of rice measured at harvest. Flooding of dried soil for 2-4 weeks before transplanting eliminated the growth retarding effect, and the increase in soil N supply due to previous drying was not affected. The stimulation of N mineralisation was more than denitrification loss of the nitrate accumulated during the dry period.

Puddling of the wet fields is a characteristic of land preparation in South and Southeast Asia. It incorporates weeds, prevents water seepage and may also stimulate soil organic matter breakdown. Soil N uptake by rice at early growth stages was faster in a transplanted, wet-puddled field than in a direct-seeded, uncultivated field (Ooyama 1975 ; Nonoyama and Nishi 1981). This was ascribed to the stimulation of soil N decomposition by mechanical destruction of soil aggregates, an effect observed under laboratory conditions by Hayashi and Harada (1964). However, Yoshino and Onikura (1980) found no difference in soil N mineralisation between soils collected before and after puddling and incubated under anaerobic conditions. Sharma and De Datta (1985) observed differences in chemical changes of puddled and non-puddled soils after flooding and the N uptake by rice, but no difference was found in the inorganic N content of soil between the two methods.

The effects of various farming practices on soil N mineralisation need to be re-examined.

Microbial biomass as a source of available N

Harada (1959) found that the soil N fractions that are made decomposable by air-drying, liming, temperature elevation and other factors, were included in labile fractions solubilised in sodium chloride. Further studies showed that microbial cell walls are major constituents of the soil organic matter that becomes decomposable by air-drying (Kai and Wada 1979). Microbial biomass is now regarded as a major channel through which organic nutrients are transferred to crop plants (Jenkinson and Ladd 1981).

The chloroform fumigation method has added significantly to the knowledge of the role of the microbiomass in cultivated soils (Jenkinson and Ladd 1981). Marumoto (1984) found that in oven-dried and rewetted rice soils, 66% of the N mineralised during 28 days of incubation came from the newly killed (chloroform fumigated) microbial biomass. In soils where ^{15}N-labelled ammonium N was newly immobilised due to the addition of glucose, ^{15}N abundance in the N mineralised from dried and rewetted soils was smaller than that from chloroform fumigated soils, indicating that air-drying made older immobilised N more mineralisable than chloroform fumigation (Inubushi and Watanabe 1987). The contributions of microbial biomass to the organic fractions rendered decomposable by methods other than drying are not known.

The ^{15}N abundance in the N absorbed by rice at late growth stages (when ammonium N in soil was almost depleted), or by rice in non-fertilised N plots was similar to that of flush N after chloroform fumigation (Inubushi and Watanabe 1986). This suggests that microbial biomass in flooded rice soil is the N source for rice when mineral N is deficient.

There are few estimates of microbial biomass in flooded soils. The limited data obtained from wetland soils before or after flooding (Marumoto 1984 ; Hasebe et al. 1985) show higher ratios of microbial biomass C to total soil C (4-8%) than reported for upland arable lands (Jenkinson and Ladd 1981). Total microbial biomass may be larger in flooded soil because of the aquatic microbial community, especially microalgae.

Replenishment of microbial biomass and mineralisable N

If an energy source is not added to soil and new biomass is not synthesized, microbial biomass must decline after releasing nutrients. Watanabe and Inubushi (1986) observed that microbial biomass measured by chloroform fumigation (flush N + mineralised N without fumigation) increased at the soil surface and decreased in the puddled layer during flooding. Microbial biomass N in planted plots declined slightly more than that in fallow plots. The difference was, however, smaller than the absorption of soil N by the rice plant. This suggests that the replenishment of microbial biomass is related to the activity of the rice root. Inubushi and Watanabe (1986) estimated the residence time of microbial biomass N (or available N) to be 33 days, which suggests that the turnover of microbial biomass is much faster in tropical-wetland soils than in temperate upland soils (Jenkinson and Ladd 1981).

The significance of the microbial biomass accummulation at the soil surface and the corresponding enrichment in mineralisable N is discussed later.

Predicting mineralisation of soil N

The heavy dependence of rice on N mineralised from soil organic matter emphasizes the need for methods to assess N supply. There are two aspects of the prediction of available (mineralisable) N : (1) the estimation of the total amount of soil N available to the rice crop, and (2) the rate of supply of the mineralisable N. The first is necessary to determine the amount of fertiliser N for a given rice yield and the second to determine the timing of its split applications.

The amount of ammonium-N released during incubation of air-dried or wet soil under flooded conditions at $30^{\circ}C$ for 2-4 weeks has been widely used as a biological index of N availability. Incubation of wet samples is preferable to the incubation of air-dried soil, because the N release pattern of moist-soil is closer to that observed in field. After air-dried soil is flooded, N mineralisation usually exhibits a very active phase followed by a slow phase. The mineralisation rate during the slow phase is not necessarily proportional to that of non-dried soil (Inubushi et al. 1985a, b). Two to four times more N is released by incubation of air-dried samples as compared with wet samples (Yoshino and Dei 1977 ; Inubushi et al. 1985a).

Total N content of the soil could be the simplest estimate of its mineralisable N. Sahrawat (1983a) reported a high correlation coefficient between mineralisable N and total N in Philippine soils (r = 0.94). However, much lower values were reported for soils from Japan (Shioiri 1948, r = 0.52) China (Zhu et al. 1984, r = 0.56) and South and Southeast Asia (Kawaguchi and Kyuma 1977, r = 0.58), indicating that total N is not an accurate index of mineralisable N. In addition to total N, various chemical methods have been proposed to replace incubation methods (Sahrawat 1983a,b;Keeney and Sahrawat 1986), but none is widely used.

Flooded incubation of wet soils collected before transplanting has been used to predict the N release pattern during rice growth (see review by Dei and Yamasaki 1979).

Yoshino and Dei (1977) proposed the following equation for calculating mineralised N as a function of incubation period and temperature:

$$Y = K[(t-15).D/a]^n \qquad (1)$$

where Y = N mineralised after D days of incubation ; K = Y at the end of the incubation ; t = temperature in $^{\circ}$C ; a = (t-15) x total number of days of incubation. The parameter n is related to the pattern of N release.

Yoshino and Onikura (1980) used 13 non-N fertiliser field plots to examine the validity of Yoshino and Dei's equation. N uptake of rice and ammonia present in the soil were summed to calculate soil N mineralisation in the field. Samples of the puddled layer collected before flooding and transplanting were incubated under flooded conditions at 30°C to estimate the parameters of N release. N release in the field was predicted by applying the measured soil temperature in the field to the equation. The time-sequence patterns of estimated soil N mineralisation fairly represented the N uptake pattern in the field. The amounts of N taken up by rice at the harvest, however, differed from the estimated values which were 72 to 148% of the N actually mineralised in the field up to harvest. Cai and Zhu (1983) also reported discrepancies between the predicted amount of mineralised N and N uptake by the plant.

Shiga and Ventura (1976) compared the N uptake pattern of rice grown in the field or in pots in the greenhouse with the N release pattern during the flooded incubation of wet soil samples at 30°C. The greenhouse results agreed well with the N release pattern of flooded incubation. During the early dry season, there were large discrepancies between N release estimated from Yoshino and Dei's equation (1977) and the actual N uptake by rice in the field, probably due to lower soil temperature during the dry season. Gao et al. (1984) applied Yoshino and Dei's method to estimate N release in the field and the response of early rice to N fertiliser. Soils were divided into 3 groups according to parameters K and n. Rice yield and response to N-fertiliser were correlated with the pattern of N mineralisation during flooded incubation.

Konno and Sugihara (1986) and Sugihara et al. (1986) developed another approach in which the parameters are: decomposable N, the kinetic constant of soil N mineralisation, and temperature change of the kinetic constant (expressed as apparent activation energy in Arrhenius's law) as shown in the equation:

$$N = N_0[1-e^{-k.t}] \qquad (2)$$

where N = N mineralised at time t ; N_0 = N mineralised at infinite time = decomposable N; k, kinetic constant = $A^{-Ea/R.T.}$; Ea = apparent activation energy; R = gas constant; T = temperature in ^oK.

Yamamoto et al (1986) applied this equation in two sandy loam gley wetland soils in Kyushu, Japan. The estimate of soil N mineralisation fit well to the sum of N uptake in rice and ammonium N in soil.

Both approaches excluded the contribution of subsoil and available N derived from photoautotrophic organisms on the soil surface. Yoshino and Onikura (1980) ascribed the discrepancy between actual values and predicted values to higher N fertility of subsoil. Although Ventura and Watanabe (1984) showed the importance of N contributed from subsoil to N uptake, the role of subsoil in supplying N to rice has not been recognised in most predictions of N supply. Still, prediction methods of soil N mineralisation need improvement and field verification.

Part of the non-exchangeable ammonium is released during rice growth and absorbed by rice (Keerthisinghe et al. 1984, 1985) but the quantity of N absorbed by rice from this fraction is not yet known.

Varietal difference in the ability to absorb soil N

Research on the genotypic differences in soil N utilisation by rice is limited. Collaborative research between IRRI and University of California (Broadbent et al. 1987) has identified rice varieties that consistently

Table 1: Rice varietal differences in nitrogen utilisation efficiency (adapted from Broadbent et al. 1987).

Rice variety	Growth duration	N utilisation efficiency Rank in 3 seasons[1]
	days	
IR13429-150-3	110	1, 1, 3
IR18349-135-2	110	2, 3, 5
IR8608-167-1	100	21, 23, 24
IR42(check)	130	9, 13, 17

1. Out of 24 rice varieties tested.

produce maximum grain yield with minimum fertiliser input (Table 1). Some are substantially better than IR42, an established variety with a reputation for good performance in poor soils. This study also suggests that varieties that mature in less than 100 days depend primarily on fertiliser N whereas late-maturing varieties depend primarily on soil N. The consistent performance of outstanding genotypes demonstrate that genetic improvement in N utilisation efficiency is practical.

Differences in soil N uptake among rice varieties were demonstrated by either the N uptake in non N-fertilised plots or the uptake of unlabelled N in plots where labelled fertiliser was applied. The ^{15}N technique suggests that rice genotypic differences in soil N uptake are not associated with varietal differences in ability to promote nitrogen fixation (Watanabe et al. 1987).

ROLE OF AQUATIC PHOTOSYNTHETIC COMMUNITIES IN N TRANSFORMATION

After flooding, particularly when N and P fertilisers are applied, there is an upsurge in growth of photoautotrophic organisms in the floodwater and at the soil surface. Shioiri and Mitsui (1935) suggested four roles for these organisms: (1) supplying organic matter to the soil surface; (2) immobilising N (reservoir effect); (3) supplying O_2; and (4) fixing N_2 (at that time, N_2-fixation was regarded as symbiosis between N_2-fixing bacteria and algae).

In 1943, Shioiri and Harada briefly reported the possible effect on ammonia loss of high floodwater pH resulting from algal activity, but they detected no ammonia loss. It is only recently that the role of the photosynthetic aquatic biomass in stimulating ammonia volatilisation by increasing pH has been demonstrated and extensively studied (Mikkelsen et al. 1978); Fillery and Vlek 1986).

The role of aquatic biomass in N transformations was not covered in most early reviews, except for N_2-fixation. Recent reviews (Roger and Watanabe 1984; Roger 1986; Roger et al. 1987) cover this subject more fully. Ecology of the floodwater was reviewed by Watanabe and Furusaka (1980) and Watanabe and Roger (1985). The transformations and transfers of N as affected by the photosynthetic aquatic biomass are shown schematically in Figure 1.

Aquatic biomass productivity

The photosynthetic aquatic biomass in flooded rice fields is composed of prokaryotic and eukaryotic algae, and vascular macrophytes (aquatic cormophytes) free-floating or growing on soil, plants, and organic debris. Roger (1986) showed that the photosynthetic aquatic biomass is usually a few hundred kg dry weight/ha and rarely exceeds 1 t dry weight/ha. At 2.5% N content, an average aquatic biomass of 200 kg dry matter/ha would correspond to only 5 kg biomass N/ha. Biomass N rarely exceeds 10-20 kg/ha.

Daily productivity at early growth is approximately 1 g C/m^2 (Yamagishi

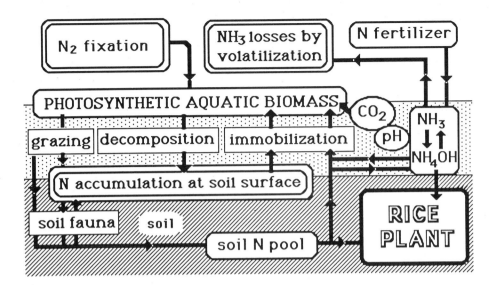

Fig. 1: Nitrogen dynamics in relation to the activity of the photosynthetic aquatic biomass in a rice field (after Roger 1986).

et al. 1980b; Vaquer 1984). Productivity decreases as the canopy density increases except when macrophytic algae emerge (Vaquer 1984). Recorded productivities are 50–60 g C/m^2 in 90 days (Saito and Watanabe 1978) and 70 g C/m^2 in 144 days (Yamagishi et al. 1980b). These values are similar to those encountered in eutrophic lakes.

Surface accumulation of N

Accumulation of total N at the soil surface was observed only when the surface was exposed to light. Reported values range from a few kg N/ha (App et al. 1984) to 35 kg N/ha per crop (Ono and Koga 1984).

Nitrogen may come from the atmosphere through biological fixation, from the floodwater through the trapping of N by aquatic biomass, and from the soil through absorption by plants or ingestion by invertebrates. A decrease of N accumulation from 35 to 26 kg N/ha occurred when the surface soil was isolated from the deeper soil by placing it in a Petri dish (Ono and Koga 1984). This means that the recycling of soil N to the surface is equivalent to 1/3 of the N originated from the photosynthetic aquatic biomass.

Watanabe and Inubushi (1986) and Inubushi and Watanabe (1986) determined the accumulation of microbial biomass N by chloroform fumigation (N mineralised without fumigation was not subtracted). Surface accumulation of microbial biomass N was demonstrated only in soils exposed to light. Microbial biomass N in the 0–1 cm soil layer accounted for 10–20% of that in the 0–15 cm layer. This percentage may be underestimated, because algal debris are brought down to a deeper layer by soil fauna (Grant and Seegers 1985b).

Chlorophyll–like substances also accumulated at the surface in parallel with microbial biomass N (Watanabe and Inubushi 1986). A correlation between chlorophyll–type compounds and mineralisable N was reported by Inubushi et al. (1982) and Wada et al. (1982), indicating that photosynthetic biomass contributes significant quantities of available N and has an important role to play in maintaining the fertility of wetland soils. The ^{15}N abundance in surface–accumulated microbial biomass N was similar to that in aquatic weeds, but higher than that in floodwater blue–green algae (Inubushi and Watanabe 1986), suggesting that recycled soil N was the major source of the surface enrichment in available N in this small plot experiment.

Nitrogen immobilisation and recycling

Nitrogen immobilisation by the photosynthetic biomass reduces N losses and N pollution of the environment. Shioiri and Mitsui (1935) measured an immobilisation rate of 10–26% of N applied in pots. Using a gas lysimeter, Vlek and Craswell (1979) estimated immobilisation at 18–30% for surface–applied urea and 0.4–6% for ammonium sulfate in 3 weeks. Immobilisation at 18–41% was recorded using ^{15}N–labelled urea (Vlek et al. 1980). When ^{15}N ammonium sulfate was mixed with the soil of the puddled layer, only 5% or less was immobilised (Inubushi and Watanabe 1986). Aquatic weeds greatly stimulated the removal of N from N–contaminated (NO_3^-, NH_4^+, and organic N) irrigation water (Ito and Masujima 1980).

Recycling of N in the floodwater community is mediated by grazing invertebrates (Grant et al. 1986; Roger 1986). Ingestion and excretion rates of Heterocypris luzonensis (Ostracoda) were determined in the laboratory and used to estimate N ingestion and excretion in a field population. The calculated ingestion rate of 190 g N/ha per day (of which 120 g was excreted as NH_3) is much smaller than the estimated assimilation rate (1 kg N/ha per day) by the photosynthetic aquatic biomass. Burrowing of soil fauna such as tubificids facilitates the downward movement of surface–accumulated organic debris (Grant and Seegers 1985 a,b). This may explain the low difference in ^{15}N recovery between surface–applied or incorporated blue–green algae observed in the field (Tirol et al. 1982) and low total N accumulation at the soil surface (App et al. 1984). The roles of soil fauna in N cycling in floodwater and puddled soil needs to be re–examined.

Availability of photosynthetic biomass N to rice

Experiments on N recovery from ^{15}N–labelled algae and aquatic macrophytes were summarised by Roger (1986).

Recovery of blue–green algal N varied from 13 to 50% depending on the nature of the algal material (fresh vs. dried), the method of application (surface—applied vs. incorporated), and the presence or absence of soil fauna. Highest recovery (50%) was obtained when fresh material was incorporated into a soil depleted of fauna (Wilson et al. 1980). Lowest recovery was obtained when dried material was spread on the surface of a soil rich in tubificids. Their activity made soil N available to rice through mineralisation and thereby reduced the recovery of algal N (Grant and Seegers 1985a). A residual effect of algal N was observed in the second rice crop in which 4 to 7% of algal N was recovered (Tirol et al. 1982; Grant and Seegers 1985a).

The ^{15}N recovery from Azolla and aquatic macrophytes ranged from 12 to 34% (Roger 1986), averaging 29% when incorporated. Just as rice absorbs more inorganic N when the feriliser is incorporated into the soil, it absorbs more N from the photosynthetic aquatic biomass when the biomass is incorporated into the soil, than when it decomposes on the soil surface.

Experiments with soil labelled with ^{15}N fertiliser also provide data on N available from the photosynthetic aquatic biomass. Ventura and Watanabe (1983) showed that when the soil surface was exposed to light, ^{15}N in the rice plant was more diluted by unlabelled N. Nitrogen gains in the soil—plant system were higher than when the soil surface was not exposed to light. The contribution of photodependent N_2—fixation on N uptake, estimated by ^{15}N dilution method, accounted for 20% of total N gains.

Effect of the photosynthetic aquatic biomass on N losses by ammonia volatilisation

Photosynthetic CO_2 depletion by aquatic communities increases floodwater pH and stimulates NH_3 volatilisation. The suppression of algal growth by Cu^{++} (Mikkelsen et al. 1978) and deep—placement of N—fertiliser (Cao et al. 1984) decreases diurnal variation of pH and N losses.

Fillery et al. (1986) estimated the photosynthetic biomass in fields where N losses were evaluated. One week after fertiliser application, a limited and uneven growth of algae (about 100 kg fresh weight/ha) was observed and pH at noon ranged from 7.8 (no visible algal growth) to 10.5 in the vicinity of algal colonies. Despite the low algal biomass, N losses were 30—40%. Although this high loss may be related to the natural alkalinity of the floodwater at the site, it indicates that large algal populations are not required to increase floodwater pH to levels that promote rapid N losses.

Various factors affect dissolved carbon dioxide (DIC) concentration, which determines the pH at a given temperature, as shown by the following equation (Yamagishi et al. 1980a):

$$d(DIC)/dt = -Pg + R + CER + F_2 \qquad (3)$$

where Pg = photosynthetic activity; R = respiration activity; CER = carbon exchange rate (positive when CO_2 is absorbed from atmosphere); and F_2 =

CO_2 transfer from soil to floodwater. All are expressed in g C/m^2.

CER decreases with temperature increase (Yamagishi et al. 1980a). This may be another parameter that increases ammonia volatilisation at higher temperature.

Azolla, floating on water, absorbs CO_2 from the atmosphere and inhibits photosynthesis of phytoplankton in the water by hindering light penetration. The pH of floodwater under Azolla is, therefore, stable and frequently below 8. These results indicate a potential for combined use of Azolla and chemical N to reduce losses by volatilization.

FATE OF FERTILISER NITROGEN

Many reviews are available on the fate of fertiliser N applied to wetland rice and the inefficiency of surface-applied N fertilisers (see De Datta et al. 1983; Vlek and Byrnes 1986). Figure 2 summarises recent ^{15}N balance studies on N losses and recoveries by the plant in field microplots. The variation of soil recovery was smaller than the variations of plant recovery and loss. Recoveries of N in plant higher than 60% were obtained only by deep placement of urea supergranules in flooded soils (Craswell et al. 1985; De Datta et al. 1983; Cao et al. 1984, Chen and Zhu 1982). When N fertilisers were applied at panicle initiation or maximum tiller number stage, N

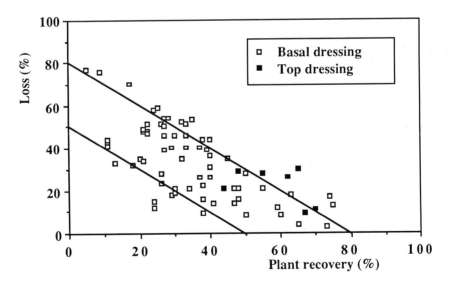

Fig. 2: Distribution of ^{15}N recovery to plant and loss. Basal dressing: [15]N applied before transplanting or just after seedling establishment. Top dressing: [15]N applied at panicle initiation. From Chen and Zhu (1982), Craswell et al. (1985), Mo and Qian (1983), Yamamuro (1986), Fillery et al. (1986), Katyal et al. (1985), Cao et al. (1984), and De Datta et al. (1983).

recoveries by the rice plant were larger and losses smaller than with early application (Fig. 2).

N losses from surface-applied fertiliser

Two pathways of gaseous losses of N from flooded soil are known; viz. ammonia volatilisation, which has been quantified by direct measurements, and denitrification, which has been primarily estimated by indirect methods.

The extent of NH_3 volatilisation loss following urea application to flooded rice and the problems of relating loss measurements in chambers to the field, prompted many researchers to measure this loss by micrometeorological techniques. In the Asian tropics, Freney et al. (1981) were the first to use these techniques to assess field NH_3 volatilisation loss following fertiliser application. However, they used ammonium sulfate, which is not now a common fertiliser for rice. Studies on NH_3 volatilisation were summarised by Mikkelsen and De Datta (1979) and Fillery and Vlek (1986).

A recent study in the Philippines, of N losses following application of urea to rice (De Datta et al. 1986) determined total N loss by ^{15}N balance techniques and NH_3 loss by measuring NH_3 concentration in the floodwater and windspeed in the atmosphere 0.8 m above the floodwater (Table 2). NH_3 loss was 31% when N fertiliser was applied onto floodwater 10 days after transplanting — a common practice for many Asian rice farmers. The NH_3 loss occurred during the first 8 days after fertiliser application.

Table 2: Relation between total N loss and NH_3 volatilisation. Mabitac, Laguna, Philippines, 1985 late dry season. (De Datta et al. 1986).

Application method	Fertiliser rate	Water depth	Total N loss	NH_3 loss	Estimated denitrification[1]
	kg/ha	cm	%	%	%
Researchers' split[2]	53	0	33	6	27
Researchers' split[2]	53	5	54	22	32
Farmers' split[3]	53	5	60	27	33
Researchers' split[2]	80	0	32	7	25
Researchers' split[2]	80	5	58	27	31
Farmers' split[3]	80	5	59	31	28
LSD (0.05)	–	–	13	8	–
Farmers' split in circle[3,4]	80	5	55	31	24

1. Total N loss minus NH_3 loss; 2. 2/3 basal incorporated without standing water + 1/3 at 5–7 days before panicle initiation; 3. 2/3 at 10 days after transplanting into standing water + 1/3 at booting stage; 4. Unreplicated treatment where NH_3 loss was directly measured in the circle.

The estimated total N loss after 30 days was 55% of applied N (Table 2). No ^{15}N was detected in the soil below 15 cm and no overflow of floodwater from microplots was allowed so losses by leaching and runoff were negligible. Therefore, the 24% difference in the loss (Table 2) was probably due to denitrification. This needs to be confirmed by direct measurement.

Because nitrification is much slower than denitrification, the rate of loss by denitrification is determined by the nitrification rate. Focht (1979) assumed that nitrification in the oxidized soil is controlled by the O_2 supply to the soil surface and calculated a maximum nitrifying activity of 500 g N/ha per day. In situ soil core measurement of the nitrification rate by ^{15}N dilution (Watanabe et al. 1981b) showed a maximum of 1200 g N/ha per day, which is much lower than the potential NH_3 loss. The measurement of N_2 or N_2O evolved from the ^{15}N labelled fertiliser probably would be the best method to estimate denitrification loss. Data show that denitrification loss should not be underestimated. Ammonia or ammonium in the floodwater and at the soil surface could be lost by denitrification if NH_3 volatilisation was blocked.

Cultural practices to improve fertiliser N efficiency in flooded soil.

In recent years, a deeper understanding of the mechanisms causing poor N utilisation has helped to develop cultural practices to improve N fertiliser use efficiency in lowland rice. De Datta et al. (1987) suggested that applying N onto the floodwater between transplanting and early tillering, a common practice of Southeast Asia farmers, is wasteful. Appropriate timing of

Table 3: Effects of N source and application method on grain yield of transplanted rice in farmers' fields. Nueva Ecija, Philippines, 1986 dry season. (S.K. De Datta, IRRI, unpublished data).

Treatment[1]	Fertiliser applied	Grain yield[2]
	kg N/ha	t/ha
Control	0	4.3
Farmers' split, PU[3]	58	5.6
Researchers' split, PU[4]	58	6.0
Point-placement, USG	58	6.1
Press wedge, USG	58	5.3
Plunger auger, PU	58	5.7
Farmers' split, PU	87	5.7
Researchers' split, PU	87	6.3
Standard error		0.1

1. PU = prilled urea, USG = urea supergranules. 2. Average of 3 farms. 3. One half topdressed at 15 days after transplanting + 1/2 topdressed at post panicle initiation. 4. Two thirds basal broadcast and incorporated without standing water + 1/3 topdressed 5-7 days before panicle initiation.

N application and proper water management minimizes N loss and maximizes N use efficiency in lowland rice. In 1986 dry season trials in three farmers' fields in the Philippines, researchers' timing at two N levels increased grain yield by 0.4 – 0.6 t/ha over farmers' timing (Table 3).

Various urea and modified urea products are now available for extensive testing on lowland rice. These include urea of different granule sizes and slow-, and controlled-release fertilisers. The potential of urease inhibitors to reduce NH_3 volatilisation loss was also extensively tested. Results suggest that the urease inhibitor phenylphosphorodiamidate (PPD) can reduce NH_3 loss a little but seldom shows significant increase in grain yield (De Datta et al. 1983; Fillery and De Datta 1986).

Deep placement of fertiliser either by hand or machine was also promising. However, tests of machine deep placement have given inconsistent results.

CONCLUSION

Whereas N supply to rice by the soil is an extremely important component of rice production, little effort is expended in the tropics on studying how to use soil N more efficiently. A few predictive models for soil N mineralisation have been developed but have not yet been tested in the tropics.

The photosynthetic aquatic biomass plays a major role in N cycling in wetland soils. It stimulates NH_3 losses by increasing floodwater pH, but also enriches the soil surface with N by N_2 fixation, and N immobilisation and recycling. Quantitative data on the two last processes are still meagre. Soil microbial biomass is an important source of N to wetland rice in mineral N-deficient conditions.

NH_3 volatilisation loss from wetland rice soils has been quantified. Total N losses were measured by ^{15}N balance technique. Denitrification loss was estimated by the differences between total N and NH_3 losses as affected by various N fertiliser management practices. Future challenge is to directly measure both NH_3 volatilisation and denitrification losses in relation to total nitrogen losses using ^{15}N balance technique.

Based on these basic researches, cultural practices have been developed which minimise NH_3 and apparent denitrification losses and maximise fertiliser N use efficiency in wetland rice. Cultural practices to increase availability of soil nitrogen in wetland rice fields have to be developed.

REFERENCES

Anonymous (1979). 'Nitrogen and Rice'. (Intern. Rice Res. Inst.: Manila, Philippines), p. 499.

App, A., Santiago, T., Daez, C., Menguito, C., Ventura, W., Tirol, A., Po, J., Watanabe, I., De Datta, S.K., and Roger, P.A. (1984). Estimation of the nitrogen balance for irrigated rice and the contribution of phototrophic nitrogen fixation. Field Crops Res. 9, 17–27.

Bouldin, D.R. (1986). The chemistry and biology of flooded soils in relation to the nitrogen economy in rice fields. In 'Nitrogen Economy of Flooded Rice Soils'. (Eds S.K. De Datta and W.H. Patrick, Jr.) pp. 1–15. (Martinus Nijhoff Publisher: Dordrecht.)

Broadbent, F.E., De Datta, S.K., and Laureles, E.V. (1987). Measurement of nitrogen utilization in rice genotypes. Agron. J. In press.

Cai, Gui–xin, and Zhu Zhao–liang (1983). Effect of rice growth on the mineralization of soil nitrogen. Acta Pedologia Sinica 20, 272–8.

Cao Zhi–hong, De Datta, S.K., and Fillery, I.R.P. (1984). Nitrogen–15 balance and residual effects of urea–N in wetland rice field as affected by deep placement techniques. Soil Sci. Soc. Amer. J. 48, 203–8.

Chen Rong–ye, and Zhu Zhao–liang (1982). Studies on fate of nitrogen fertilizer. 1. The fate of nitrogen fertilizer in paddy soils. Acta Pedologia Sinica 19, 122–30.

Craswell, E.T., De Datta, S.K., Weeraratne, C.S., and Vlek, P.L.G. (1985). Fate and efficiency of nitrogen fertilizer applied to wetland rice. The Philippines. Fert. Res. 6, 49–63.

De Datta, S.K., Fillery, I.R.P., and Craswell, E.T. (1983). Results from recent studies on nitrogen fertilizer efficiency in wetland rice. Outlook Agric. 12, 125–34.

De Datta, S.K., Obcemea, W.N., Chen, R.Y., Calabio, J.C., and Evangelista, R.C. (1987). Effect of water depth on nitrogen use efficiency and nitrogen–15 balance in lowland rice. Agron. J. 79, 210–6.

De Datta, S.K., and Patrick, W.H. (1986). 'Nitrogen Economy of Flooded Rice Soils'. (Martinus Nijhoff Publisher: Dordrecht.)

De Datta, S.K., Trevitt, A.C.F., Obcemea, W.N., Freney, J.R., and Simpson, J.R. (1986). Comparison of total N loss and ammonia volatilization in lowland rice using simple techniques. Annual Meeting American Society of Agronomy, New Orleans, USA, 30 Nov–5 Dec 1986, pp. 197.

Dei, Y., and Yamasaki, S. (1979). Effect of water and crop management on the nitrogen supplying capacity of paddy soils. In 'Nitrogen and Rice'. (Intern. Rice Res. Inst.: Manila, Philippines.), pp. 451–64.

Fillery, I.R.P., and De Datta, S.K. (1986). Ammonia volatilization from nitrogen sources applied to rice fields: I. Methodology, ammonia fluxes and nitrogen–15 loss. Soil Sci. Am. J. 50, 80–6.

Fillery, I.R.P., Roger, P.A., and De Datta, S.K. (1986). Ammonia volatilization from nitrogen sources applied to rice field: II. Floodwater properties and submerged photosynthetic biomass. Soil Sci. Soc. Am. J. 50, 86–91.

Fillery, I.R.P., and Vlek, P.L.G. (1986). Reappraisal of the significance of ammonia volatilization as an N loss mechanism in flooded rice fields. In 'Nitrogen Economy of Flooded Rice Soils'. (Eds S.K. De Datta and W.H. Patrick, Jr.) pp. 79–98. (Martinus Nijhoff Publisher: Dordrecht.)

Focht, D.D. (1979). Microbial kinetics of nitrogen losses in flooded soils. In 'Nitrogen and Rice'. (Intern. Rice Res. Inst.: Manila, Philippines.), pp. 119–34.

Freney, J.R., Denmead, O.T., Watanabe, I., and Craswell, E.T. (1981). Ammonia and nitrous oxide losses following applications of ammonium sulfate to flooded rice. Aust. J. Agric. Res. 32, 37–45.

Gao Jia–hua, Zhang Yun, Huan Dong–mai, Wu Jing–min, and Pan Zun–pu (1984). Nitrogen mineralization pattern and nitrogen efficiency in paddy soil. Acta Pedologica Sinica 21, 341–9.

Grant, I.F., Roger, P.A., and Watanabe, I. (1986). Ecosystem manipulation for increasing biological N_2 fixation by blue–green algae (Cynanobacteria) in lowland rice fields. Biol. Agric. and Hort. 3, 299–315.

Grant, I.F., and Seegers, R. (1985a). Tubificid role in soil mineralization and recovery of algal nitrogen by lowland rice. Soil Biol. Biochem. 17, 559–63.

Grant, I.F., and Seegers, R. (1985b). Movement of straw and algae facilitated by tubificids (Oligochaeta) in lowland rice soil. Soil Biol. Biochem. 17, 729–30.

Harada, T. (1959). The mineralization of native organic nitrogen in paddy soils and mechanism of its mineralization. Bull. Nat. Inst. Agric. Sci. B9, 123–200.

Harada, T., Hayashi, R., and Chikamoto, A. (1964). Effect of physical pretreatment of soils on the mineralization of native organic nitrogen in paddy soils. J. Sci. Soil Manure, Jpn 35, 21–4.

Hasebe, A., Kanazawa, S., and Takai, Y. (1985). Microbial biomass carbon measured by Jenkinson's fumigation method. Soil Sci. Plant Nutr. 31, 349–59.

Hayashi, R., and Harad, T. (1964). Effect of cropping on mineralization of native organic nitrogen in paddy soils. J. Sci. Soil Manure, Japan 35, 123–6.

Inubushi, K., Wada, H., and Takai, Y. (1982). Easily decomposable organic matter in paddy soils. II. Chlorophyll type compounds in Apg horizons. Jpn J. Soil Sci. Plant Nutr. 53, 277–82.

Inubushi, K., Wada, H., and Takai, Y. (1985a). Easily decomposable organic matter in paddy soils. V. Ammonification in the submerged soil as related with the mineralisable nitrogen. Jpn J. Soil Sci. Plant Nutr. 56, 404–8.

Inubushi, K., Wada, H., and Takai, Y. (1985b). Easily decomposable organic matter in paddy soils. VI. Kinetics of nitrogen mineralization in submerged soils. Soil Sci. Plant Nutr. 31, 563–72.

Inubushi, K., and Watanabe, I. (1986). Dynamics of available nitrogen in paddy soils. II. Mineralized N of chloroform-fumigated soil as a nutrient source for rice. Soil Sci. Plant Nutr. 32, 561–78.

Inubushi, K., and Watanabe, I. (1987). Microbial biomass nitrogen in anaerobic soil as affected by N immobilization and N_2-fixation. Soil Sci. Plant Nutr. 33, 213–24.

Ito, A., and Masujima, H. (1980). Studies on the purification of polluted water in paddy fields. J. Sci. Soil Manure, Japan 51, 478–85.

Ito, O., and Watanabe, I. (1985). Availability to rice plants of nitrogen fixed by Azolla. Soil Sci. Plant Nutr. 31, 91–104.

Jenkinson, D.S., and Ladd, J.N. (1981). Microbial biomass in soil: measurement and turnover. In 'Soil Biochemistry 5'. (Ed. E. Paul) pp. 415–71. (Marcel Dekker: New York.)

Kai, H., and Wada, K. (1979). Chemical and biological immobilization of nitrogen in paddy soil. In 'Nitrogen and Rice'. (Intern. Rice Res. Inst.: Manila, Philippines.), pp. 157–74.

Katyal, J.C., Singh, B., Vlek, P.L.G., and Craswell, E.T. (1985). Fate and efficiency of nitrogen fertilizers applied to wetland rice. II. Punjab, India. Fert. Res. 6, 279–90.

Kawaguchi, K., and Kyuma, K. (1977). 'Paddy Soils in Tropical Asia'. (Univ. Hawaii Press: Honolulu.).

Keeney, D.R., and Sahrawat, K.L. (1986). Nitrogen transformation in flooded rice soils. In 'Nitrogen Economy of Flooded Rice Soils'. (Eds S.K. De Datta and W.H. Patrick.) pp. 15–39. (Martinus Nijhoff Pub.: Dordrecht.)

Keerthisinghe, G., De Datta, S.K., and Mengel, K. (1985). Importance of exchangeable soil NH_4^+ in nitrogen nutrition of lowland rice. Soil Sci. 140, 194–201.

Keerthisinghe, G., Mengel, K., and De Datta, S.K. (1984). The release of nonexchangeable ammonium (^{15}N labelled) in wetland rice soils. Soil Sci. Soc. Amer. J. 48, 291–4.

Konno, K., and Sugihara, S. (1986). Temperature index for characterizing biological activity in soil.Its application to decomposition of soil organic matter. Bull. Nat. Inst. Agro-Evn. Sci., Japan. 1, 51–68.

Marumoto, T. (1984). Mineralization of C and N from microbial biomass in paddy soil. Plant Soil 76, 165–73.

Mikkelsen, D.S., and De Datta, S.K. (1979). Ammonia volatilization from wetland rice soils. In 'Nitrogen and Rice'. (Intern. Rice Res. Inst.: Manila, Philippines.), pp. 157–74.

Mikkelsen, D.S., De Datta, S.K., and Obcemea, W. (1978). Ammonia volatilization losses from flooded rice soils. Soil Sci. Soc. Amer. J. 42, 725–30.

Mitsui, S. (1954). 'Inorganic Nutrition, Fertilisation and Soil Amelioration for Lowland Rice'. (Yokendo: Tokyo.)

Mo Shu-xun, and Qian Ju-fang (1983). Studies on the transformation of nitrogen of milk vetch in red earth and its availability to rice plant. Acta Pedologia Sinica 20, 12–22.

Nonoyama, Y., and Nishi, H. (1981). Edaphological studies on non-tillage direct sowing cultivation of paddy rice. Bull. Chugoku Nat. Agric. Expt. Stn. E18, 1–55.

Ono, S., and Koga, S. (1984). Natural nitrogen accumulation in a paddy soil in relation to nitrogen fixation by blue-green algae. Jpn J. Soil Sci. Plant Nutr. 55, 465–70.

Ooyama, N. (1975). Nitrogen supplying patterns of paddy soils for rice plant in temperate area in Japan. J. Sci. Soil Manure, Japan 46, 297–302.

Patrick, W.H. (1982). Nitrogen transformations in submerged soils. In 'Nitrogen in Agricultural Soils'. Agronomy Monograph No. 22, pp 449–65 (ASA–CSSA–SSSA: U.S.A.)

Roger, P.A. (1986). Effect of algae and aquatic macrophytes on nitrogen dynamics in wetland rice fields. Congr. Int. Soil Sci. Soc., Hamburg, Germany, pp. 13–21.

Roger, P.A., Grant, I.F., Reddy, P.M., and Watanabe, I. (1987). The photosynthetic aquatic biomass in wetland rice fields and its effect on nitrogen dynamics. In 'Efficiency of Nitrogen Fertilizers for Rice'. (Intern. Rice Res. Inst.: Manila, Philippines.), pp. 43–68.

Roger, P.A., and Watanabe, I. (1984). Algae and aquatic weeds as source of organic matter and plant nutrients for wetland rice. In 'Organic Matter and Rice'. (Intern. Rice Res. Inst.: Manila, Philippines.), pp. 147–68.

Roger, P.A., and Watanabe, I. (1986). Technologies for utilizing biological nitrogen fixation in wetland rice: potentialities, current usage, and limiting factors. Fert. Res. 9, 39–77.

Sahrawat, K.L. (1983a). Mineralization of soil organic nitrogen under waterlogged conditions in relation to other properties of tropical rice soils. Aust. J. Soil Res. 21, 133–8.

Sahrawat, K.L. (1983b). Nitrogen availability indices for submerged rice soils. Adv. Agron. 36, 415–51.

Saito, M., and Watanabe, I. (1978). Organic matter production in rice field floodwater. Soil Sci. Plant Nutr. 24, 427–40.

Savant, N.K., and De Datta, S.K. (1982). Nitrogen transformations in wetland rice soils. Adv. Agron. 35, 241–302.

Sharma, P.K. and De Datta, S.K. (1985). Puddling influence on soil, rice development, and yield. Soil Sci. Soc. Amer. J. 49, 1451–7.

Shiga, H., and Ventura, W. (1976). Nitrogen supplying ability of paddy soils under field conditions in the Philippines. Soil Sci. Plant Nutr. 22, 387–99.

Shioiri, M. (1948). The effect of drying paddy soils during fallow period. Rep. Agric. Exptl. Stn. 64, 1–24.

Shioiri, M., Aomine, S., Uno, Y., and Harada, T. (1941). The effect of drying of paddy soils. J. Sci. Soil Manure, Japan 15, 331–3.

Shioiri, M., and Harada, T. (1943). Loss of N due to upward movement of ammoniacal nitrogen and its leaching by downward movement. Transformation of nitrogen in flooded soils. J. Sci. Soil Manure, Japan 17, 375–6.

Shioiri, M., and Mitsui, S. (1935). Chemical composition and decomposition in soil of algae and some aquatic plants grown in paddy soils. J. Sci. Soil Manure, Japan 9, 261–7.

Sugihara, S., Konno, T., and Ishii, K. (1986). Kinetics of mineralization of organic nitrogen in soil. Bull. Natl. Inst. Agro-Env. Sci. Jpn 1, 127–66.

Tirol, A.C., Roger, P.A., and Watanabe, I. (1982). Fate of nitrogen from blue-green algae in a flooded rice soil. Soil Sci. Plant Nutr. 28, 559–70.

Vaquer, A. (1984). La production algale dans des rizieres de Camargue pendant la periode de submersion. Ver Internatl. Verein Limnol. 22, 1651–4.

Ventura, W., and Watanabe, I. (1978). Dry season soil conditions and soil nitrogen availability to wet season wetland rice. Soil Sci. Plant Nutr. 24, 535–45.

Ventura, W., and Watanabe, I. (1983). ^{15}N dilution technique for assessing the contribution of nitrogen fixation to rice plant. Soil Sci. Plant Nutr. 29, 123–31.

Ventura, W., and Watanabe, I. (1984). Dynamics of nitrogen availability in lowland rice soils: the role of soil below plow layer and effect of moisture regimes. Philipp. J. Crop Sci. 9, 135–42.

Vlek, P.L.G., and Byrnes, B.H. (1986). The efficacy and loss of fertilizer N in lowland rice. In 'Nitrogen Economy of Flooded Rice Soils'. (Eds S.K. De Datta and W.H. Patrick.) pp. 131–47. (Martinus Nijhoff Publisher: Dordrecht.)

Vlek, P.L.G., and Craswell, E.T. (1979). Effect of nitrogen source and management on ammonia volatilization losses from flooded rice-soil systems. Soil Sci. Soc. Amer. J. 43, 352–8.

Vlek, P.L.G., Stumpe, J.M., and Byrnes, B.H. (1980). Urease activity and inhibition in flooded soil system. Fert. Res. 1, 191–202.

Wada, H., Inubishi, K., and Takai, Y. (1982). Relationship between chlorophyll-type compounds and mineralisable nitrogen. Jpn J. Soil Sci. Plant Nutr. 53, 380–4.

Watanabe, I. (1986). Nitrogen fixation by non-legumes in tropical agriculture with special reference to wetland rice. Plant Soil 30, 343–57.

Watanabe, I., Craswell, E.T., and App, A. (1981a). Nitrogen cycling in wetland rice fields in Southeast and East Asia'. In 'Nitrogen Cycling in Southeast Asian Wet Monsoon Ecosystems'. (Ed. R. Wetselaar.) pp. 4–17. (Aust. Acad. Sci.: Canberra.)

Watanabe, I., and Furusaka, C. (1980). Microbial ecology of flooded soils. Adv. Microb. Ecol. 4, 125–68.

Watanabe, I., and Inubishi, K. (1986). Dynamics of available nitrogen in paddy soils. 1. Changes in available N during rice cultivation and origin of N. Soil Sci. Plant Nutr. 32, 37–50.

Watanabe, I., Padre, B.C.P., and Santiago, S.T. (1981b). Quantitative study on nitrification in flooded rice soil. Soil Sci. Plant Nutr. 27, 373–82.

Watanabe, I., and Roger, P.A. (1984). Nitrogen fixation in wetland rice field. In 'Current Developments in Biological Nitrogen Fixation'. (Eds N.S. Subba Rao) pp. 237–76. (Oxford and IBH: New Delhi.)

Watanabe, I., and Roger, P.A. (1985). Ecology of flooded rice fields. In 'Wetland Soils: Characterization, Classification, and Utilization'. (Intern. Rice Res. Inst.: Manila, Philippines.), pp. 229–43.

Wilson, J.T., Eskew, D.L., and Habte, M.H. (1980). Recovery of nitrogen by rice from blue-green algae added in a flooded soil. Soil Sci. Soc. Amer. J. 44, 1330–1.

Yamagishi, T., Okada, K., Hayashi, T., Kumura, A., and Murata, Y. (1980a). Cycling of carbon in a paddy field and the atmosphere. Jpn J. Crop Sci. 49, 135–45.

Yamagishi, T., Watanabe, J., Okada, K., Hayashi, T., Kumura, A., and Murata, Y. (1980b). Cycling carbon in a paddy field. II. Biomass and gross production of algae. Jpn J. Crop Sci. 49, 146–55.

Yamamoto, T., Kubota, T., and Manabe, H. (1986). Estimation of soil nitrogen mineralization during growth period of rice plant by kinetic method. Jpn J. Soil Sci. Plant Nutr. 57, 487–92.

Yamamuro, S. (1986). Nitrogen cycle in paddy fields (Part 4). The fate of ammonium N by surface application and incorporation in soil and mineralised soil nitrogen on assimilation, immobilisation, denitrification, and absorption by rice plants in semi–ill drained paddy fields. Jpn J. Soil Sci. Plant Nutr. 57, 13–22.

Yoshino, T., and Dei, Y. (1977). Prediction of nitrogen release in paddy soils by means of the concept of effective temperature. J. Central Agric. Exptl. Stn., Konosu 25, 1–62.

Yoshino, T., and Onikura, Y. (1980). Evaluation of nitrogen supplying capacity of paddy soils by an incubation method. J. Central Agric. Exptl. Stn., Konosu 31, 73–86.

Zhu Zhao–liang, Cai Gui–xin, Yu Ying–hua, and Zhang Shao–lin (1984). Nitrogen mineralization of paddy soils in Tai–lake region and the prediction of soil nitrogen supply. Acta Pedologica Sinica 21, 29–36.

NITROGEN MANAGEMENT OF UPLAND CROPS :

FROM CEREALS TO FOOD LEGUMES TO SUGARCANE

R.J.K. Myers

ABSTRACT

Synchrony of the supply of available N in the soil with the demand for N by crops to improve the efficiency of N use is the key to improved N management. Cropping systems in different environments have been examined by simple simulation modelling and absence of synchrony to various degrees has been observed. Cropping systems in the humid tropics are particularly inefficient in their use of soil and fertilizer N because nitrate retention against leaching and denitrification is poor. Some subtropical systems show a similar problem but generally to a lesser extent. Semi-arid cropping systems are more efficient in their use of N because most nitrate that accumulates during fallow or after fertilizer application is retained within the root zone. Asynchrony can occur in semi-arid systems but under these conditions excess nitrate is more stable.

Management procedures to improve N supply and synchrony of supply and demand are considered. These include the use of legumes, fertilizer management, recycling of residues, recycling of animal manures, reduced tillage and adjustment of sowing dates.

INTRODUCTION

This paper considers aspects of N dynamics for a diverse range of soils and crops, over a range of environments, with the emphasis on the tropics. In some cases the paucity of data necessitates looking elsewhere for relevant information. The broadness of the topic dictates that the approach will be general and will be considering three groups of crops, the cereals (grain sorghum, upland rice and corn), the food legumes, and the longer growth cycle carbohydrate producers (cassava and sugarcane). The objective is to find underlying principles of N use such that the common or divergent features between different crops and environments can be identified. This information is then used to describe how the N cycle works for these agricultural systems, and to suggest ways in which N management for crops can be improved.

NITROGEN MANAGEMENT OF CROPS

Management of N is the package of field activities that, inadvertently or by design, interacts to control the supply of available N to the crop. It may do this through control of release of N from soil organic matter, through control of availability of fertilizer N or through the recycling of N from residues. The primary objective is to make N available at an optimum level and time for the crop. A further objective, important in many systems, derives from the susceptibility to loss of plant-available forms of N; that is, to avoid having excess mineral N in the soil when it is not needed by the crop. The major feature of traditional N management for arable crops has been the development of systems whereby excess amounts of nitrate accumulate from mineralization of soil organic matter or by application of fertilizer. This nitrate is stored in the soil and utilized by the crop at a later time. The need to store nitrate in the soil has no doubt contributed to the extensive use of fertile soils in semi-arid regions for upland crop production, since in these soils, nitrate is relatively stable and not rapidly lost by leaching or denitrification. This is not true of the more humid regions (e.g. the humid tropics of South-East Asia and the Amazon, and some humid temperate regions in Europe and the United States) where there have been disappointing yields and inefficient N utilization because of loss of N from the soil. This leads to the important concept of synchrony; that is the matching of the crop's demand for N to the supply of N by the soil and other sources, in both time and space.

Potential yields of arable crops

A major difficulty in the management of crops for efficient use of N is the prediction not only of the potential yield, but also the yield achievable on farms. Yield can be predicted from water use, but considerable variation has been observed in the slopes of yield-water use curves for sorghum (Garrity et al. 1982, R.J.K. Myers, unpublished data), and between water use efficiencies of crops such as sorghum and pearl millet (Kanwar et al. 1984). Angus et al. (1983) examined the water use efficiency for dry matter production of eight dryland crops in the humid tropics and found that transpiration efficiencies of C_4 crops (sorghum and maize), were double those of C_3 crops (peanut, cowpea, mungbean, soybean and rice). French and Schultz (1984) have related wheat yields in temperate Australia to total water use and have described potential yield by a line that follows the upper limits of the data (Fig. 1). They argue that yields below this line were limited by some factor other than water, such as disease, nutrient deficiency, temperature extremes etc. Such variation in yield within a crop species complicates N management because the N requirement of a crop is related to its yield (Fig. 2). Thus N management is complicated by crops failing to achieve their yield potential (Baligar and Bennett 1986).

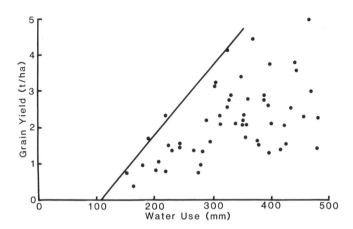

Fig. 1: Relation between grain yield of wheat and water use (soil water plus rainfall). The sloping line indicates the potential yield from water transpired, assuming 110 mm water lost by evaporation (French and Schultz 1984).

Table 1 compares average farm yields of various crops with the yields achievable with a high level of management. Potential yields were estimated using information from a range of sources, and a range of tropical environments, from those with 2000 mm rainfall and a 10 month growing season to those with less than 700 mm rainfall and a 2.5 to 4 month growing season. In some instances, published water use efficiency estimates were used to estimate maximum yields. Since farm yields frequently fall well below the potential, N management practices must be devised to match N supply to the low yields that are actually being obtained. Despite low yields, severe N deficiency is common throughout regions used for crop production in the tropics, even in soils appearing to have adequate levels of total N, and improved N management is necessary. N management is also essential for irrigated crops since, with irrigation, yield increases as does N requirement, but N supply generally does not keep pace and severe N deficiency can result.

Efficiency of uptake of N fertilizer by upland crops

In the tropics, crops grown under both very dry and very wet conditions have poorer utilization of fertilizer N than those grown under less extreme conditions (Table 2). For example, under very wet conditions, an upland rice crop receiving 1310 mm rain recovered only 18% of the applied fertilizer N (Widjang H. Sisworo et al., personal communication), and wet season crops of

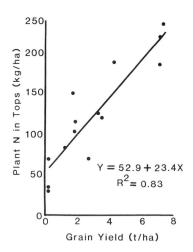

Fig. 2: Relation between plant N in tops and grain yield for grain sorghum (R.J.K. Myers, <u>unpublished data</u>).

Table 1: Average and potential maximum farm yields[1] and estimates of N requirements for crops grown in the humid, sub–humid and semi–arid tropics.

Location	Climate[2]	Crop	Av. farm crop		Potential crop	
			Yield	N req.	Yield	N req.
			t/ha	*kg/ha*	*t/ha*	*kg/ha*
Humid tropics						
Kota Bumi,	1900 mm	Corn	1.2	40	5.0	150
Indonesia	(300 days)	Upland rice	1.2	35	5.0	85
		Cassava	10.0	30	65.0	130
Sub–humid tropics						
Katherine,	960 mm	Sorghum	2.1	110	7.0	210
Australia	(141 days)	Corn	1.9	60	8.0	220
Semi–arid tropics						
Hyderabad,	740 mm	Sorghum	0.7	90	7.5	220
India	(130 days)					
Emerald,	630 mm	Sorghum	1.5	90	6.0	190
Australia	(84 days)	Sunflower	0.6	15	2.4	70
<u>Irrigated</u>	–	Cotton	4.0	165	6.4	250
Various locations		Corn	10.0	230	22.0	400
		Sorghum	8.0	200	17.0	300
		Sugarcane	75	160	120	300

1. Dry matter grain yields, except for cotton (seed plus lint), cassava (fresh tubers) and sugarcane (green weight). 2. Annual rainfall, and days to maturity in brackets.

sorghum at Katherine, N.T. recovered only 12–24% of applied N (Myers 1978, 1983). Better N recoveries of 36% were obtained for corn grown in the dry season (Widjang H. Sisworo et al., personal communication), and of 29–74% of applied N by grain sorghum in a semi-arid environment at Hyderabad, India (Moraghan et al. 1984a).

Table 2: Some examples of nitrogenous fertilizer utilization by arable crops in the tropics.

Crop	Yield[1]	Rainfall[2]	N fert. use	Reference
	t/ha	*mm*	*% of applied*	
Sorghum	1.28	651	24	Myers (1983)
Sorghum	4.36	828	13	Myers (1978)
	3.13	632	12	
Sorghum	4.11–5.33	907	29–56	Moraghan et al.
	3.90–4.91	695	36–74	(1984a,b)
	3.65–6.06	827	47–64	
	6.12–6.57	606	54–67	
Sorghum	2.70	0	24	Myers and Hibberd
	3.64	221	52	(1986)
	5.02	531	63	
	3.39	134	20–39	
	3.23	274	43–45	
Corn/	1.15–1.70/	n.a.	24	W.H. Sisworo
soybean	0.35–0.58			(pers. comm.)
	1.16–1.48/	n.a.	12–21	
	0.60–1.08			
Upland rice	2.32	1310	18	W.H. Sisworo
				(pers. comm.)
Corn	3.01	840	36	W.H. Sisworo
				(pers. comm.)
Sugarcane	78.00	n.a.	22–27	Takahashi (1964)

1. Grain, except sugarcane. 2. During crop growth.

Poor recovery of N in dry environments (Myers and Hibberd 1986) suggests poor uptake of N by the crop. Since this may also suggest an excess supply of N, a partial solution is to apply less N. This would only be possible with better prediction of expected yield. Poor recovery of fertilizer N under wet conditions is more likely due to losses of N via leaching and denitrification. There is scope for reducing these losses by improved N management as demonstrated by the high recoveries obtained in some treatments at Hyderabad (Moraghan et al. 1984a,b).

NITROGEN SUPPLY BY SOILS

Timing of mineralization: synchrony and asynchrony

Arable crops have a period of fallow which may be short where two or more crops are grown per year, or long when only one crop is grown per one or more years. During fallow the soil may be cultivated several times; the soil may be moist some or most of the time; and in the tropics the soil temperatures should be high. Thus nitrate accumulates. The importance of this nitrate build-up is seen with sugarcane where ratoon crops with no fallow are more responsive to N fertilizer than is plant cane (Chapman 1982). On the other hand, nitrate that has accumulated under fallow may not be particularly stable. It is susceptible to leaching from the profile or it can be denitrified. Traditional dryland farming, as developed and fine-tuned in sub-humid and semi-arid areas of the temperate and cool-temperate regions, depends upon the accumulation of available N as nitrate, and this is an efficient management technique because in these environments nitrate losses are small. When such a system is transplanted into the tropics where leaching and/or denitrification losses may be more severe, problems will occur.

A simple simulation of crop growth and N mineralization for several environments (McGill and Myers 1987) is summarized in Figure 3a and b. In this simulation a relative mineralization value is derived from temperature and moisture conditions using known relationships between N mineralization and temperature (Stanford et al. 1973; Myers 1975) and moisture (Stanford and Epstein 1974; Myers et al. 1982). Crop demand for N through time follows the normal sigmoidal growth curve. If mineralization of soil N cannot match this demand, then available N must be added as fertilizer. This simulation seeks to identify problems in N management in particular cropping systems in some humid and semi-arid areas of the world.

Under humid, tropical conditions in Indonesia (Fig. 3a), soil conditions favour mineralization of N for half the year, then the rate declines during the so-called dry season. The simulation was for three crops grown in one year which is a feasible farm practice. Between crops there is some opportunity for mineral N accumulation. The pattern of mineralization coincides well with N demand for the first two crops. The third crop in the cycle is more likely to be N deficient than the other two crops. This is consistent with the local practice of growing a legume at this time. In the semi-arid tropics at Hyderabad, India (Fig. 3b), mineralization will be rapid during the short, warm wet season, when most of the nitrate produced would be used by the first (kharif) crop. Any remaining nitrate would be used by the second (rabi) crop. The simulation suggests that the rabi crop would benefit from N fertilizer although it may be poorly utilized due to dry soil conditions (Finck and Venkateswarlu 1982), or alternatively a legume should be grown in this situation. Both of these simulations demonstrate asynchrony between N uptake requirement and nitrate production in the soil. In the humid tropics, excess soil nitrate may occur between the first and second crops, and

262

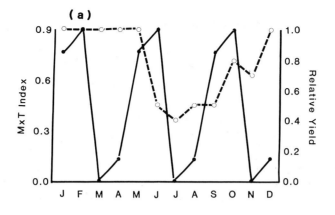

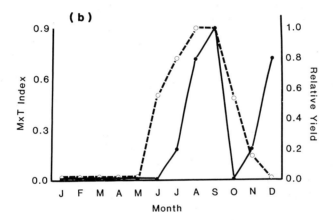

Fig. 3: Moisture x temperature index (o) for N mineralization (see text) compared to relative crop growth rate (yield, ●) for (a) a humid, tropical site in Sumatra, Indonesia with 3 succesive crops, and (b) a semi-arid, tropical site at Hyderabad, India with 2 successive crops (McGill and Myers 1987).

in the semi-arid environment, insufficient N may be mineralized for this second crop.

Other examples of marked asynchrony are demonstrated for summer- and winter-grown grain crops in the subtropics of Australia (Fig. 4). The N requirements of the summer sorghum crop peaks between February and April (Fig. 4a), whereas supply of N in the soil would be expected to peak between October and January. Wheat sown in early winter relies on residual N accumulated during the summer fallow (Fig. 4b). Asynchrony causes nitrate to accumulate in the soil prior to growth. This can be advantageous if the nitrate remains in the root zone to be used by subsequent crops, but undesirable should leaching or denitrification losses occur. In temperate and cool temperate continuous cropping systems (McGill and Myers 1987), synchrony of uptake and formation of nitrate was close, thus minimizing the opportunity for loss of N.

In humid environments, as well as those that are seasonally wet, the opportunity exists for losses of N when N supply by the soil exceeds the current crop demand. Since the best way of reducing losses of N due to leaching or denitrification is not to allow excessive amounts of nitrate to accumulate, cropping systems that more closely synchronize N supply and demand need to be developed.

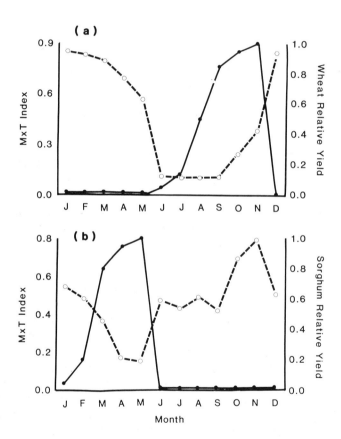

Fig. 4: Moisture x temperature index (o) for N mineralization compared to relative crop growth rate (yield, ●) at sub-tropical sites in Australia for (a) winter crop of wheat and (b) summer crop of sorghum, both near Mundubbera, central Queensland (McGill and Myers 1987).

In food legumes, unless nodulation is ineffective, supply of N from nodules should match demand. This gives them a potential advantage as an

264

alternative crop in environments where asynchrony of supply and demand with non-legumes is severe. For example, growing soybean instead of upland rice in the wet season in the humid tropics would avoid the need for applications of fertilizer N at a time when N is most prone to loss.

Source of mineralized N

The sources of mineral N in soil can be subdivided conveniently into three pools — labile soil organic matter, plant residues and the microbial biomass. Stanford and Smith (1972) define the labile soil organic N pool as being that N which will mineralize during a long-term (up to 5 month) incubation, and have related the rate of release to the size of the pool (N_0), a k factor, and a moisture (θ) and temperature (T) index, viz.

$$dN/dt = N_0.k.\theta.T$$

Crop residues when the C/N ratio is <25 will mineralize N according to a similar relationship. When the C/N ratio is wider, net mineralization is delayed until the ratio falls to 25. Release of N from the microbial biomass may occur similarly after a reduction in its size or a shift in its composition. A shift from low C/N ratio organisms (e.g. bacteria) to higher C/N ratio organisms (e.g. fungi) may result in a release of N from the microbial biomass pool. Commonly though, N is released when microbial cells die and are subsequently decomposed. This occurs continuously in soils though changes in soil environment may promote microbial death and decomposition. Many arable soils are subject to periodic drying and rewetting during which the behaviour of the microbial biomass has been likened to that of a pump (Myers et al. 1987), viz. during favourable moisture conditions the microbial pool recharges, but when dry the cells die, and following rewetting N from dead cells is released.

Rundown of soil organic matter in cropping systems

Two key factors differentiate arable systems from the natural systems from which they were derived. The first is disturbance which stimulates microbial activity and results in more rapid N mineralization as a consequence of improved organism-substrate contact. The second is that a high proportion (in many cases > 50%) of the plant N is removed in harvested product. Both factors contribute to a decline of N availability with intensive cultivation. This fertility decline is particularly evident in tropical lands on infertile Oxisols and Ultisols where shifting cultivation has traditionally been practiced (e.g. Fig. 5). Dalal and Mayer (1986) have recently shown how mineralizable N declines more rapidly than total N in subtropical Vertisols and an Alfisol, confirming earlier observations in both cooler (Campbell and Souster 1982) and warmer (Wood 1985) climates. In contrast, however, there are other observations of a continuation of satisfactory yields after many years of intensive cropping without legumes (Ruschel and Vose 1982). The source of N in these systems deserves further study.

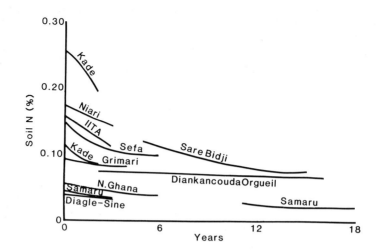

Fig. 5: Changes in soil organic nitrogen with years of cultivation for 10 sites in West Africa (Swift and Sanchez 1984).

LOSS PROCESSES

The main pathways of N loss from arable lands are removal in crop products, leaching, burning of residues, denitrification and erosion. When soil erosion and burning are controlled, the former is of greatest importance and, under some management systems, can account for all of the soil N decline (V.R. Catchpoole, _personal communication_). The other processes, when they occur, exacerbate the situation.

Leaching

This process, being readily measurable, easily modelled, and well-funded for research, has received much attention, though little of it in the tropics. Its impact on options for N management varies between humid and semi-arid environments. In humid environments, nitrate which is leached is lost from the profile. The light-textured Alfisols, Ultisols and Oxisols of the humic tropics allow rapid nitrate movement either downwards or laterally in subsurface horizons. In the semi-arid areas, nitrate accumulates below the root zone; in some situations the leaching losses may be underestimated due to failure to recognize nitrate movement in root channels or cracks (Campbell et al. 1984). In a clay soil, substantial quantities of nitrate can accumulate in the subsoil below the root zone of the annual crops being grown (V.R. Catchpoole, _personal communication_). Accumulation may continue until either it is retrieved by a deeper-rooting plant, or subjected to abnormally wet events whereupon it is ultimately lost by leaching out of the profile or via denitrification. Current research is aimed at searching for deeper-rooting crop cultivars or alternatively avoiding excessive nitrate build-up in the subsoil. The latter may be achieved either through mulching with high C/N materials or by reducing N mineralization through reduced tillage.

Direct measurement of leaching of nitrate from tropical soils is needed so that it is not confused with denitrification loss. There is scope for the use of new techniques in this research such as resin bags placed in the profile to capture nitrate in percolating water.

Denitrification

Cultivated agriculture has produced an environment more favourable for denitrification than occurs in natural systems. Bare fallow increases the likelihood that waterlogging will coincide with nitrate accumulation and the formation of plough pans exacerbates this situation. However, denitrification in cultivated systems can be retarded by the generally low levels of energy-rich material present. Not surprisingly then, N balance studies with ^{15}N on cultivated land often show a deficit which is generally attributed to denitrification and leaching. It is often argued, particularly in semi-arid systems, that because leaching is an unlikely occurrence the major loss pathway is denitrification. It has proved difficult to quantify denitrification in the field, but recent developments (C.J. Smith, in these Proceedings) may resolve this problem.

Ammonia volatilization

Fertilizers and animal urine are the major sources of the ammonia lost from the soil. The principles of the loss process are well-understood, as are those for minimizing loss. Thus ammoniacal fertilizers should be placed below the soil surface, preferably in a band near the seed. Frequently, however, other considerations apply. Often it is desirable to broadcast the fertilizer prior to seeding, or to apply it in solution in irrigation water. It is important to understand the N loss under all these varied conditions so that the net economic benefit of the alternative options can be assessed.

N applied in irrigation water (water run-N) has been investigated in Australia's Murrumbidgee Irrigation Area (Muirhead et al. 1985a). Water-run urea was as effective as banded anhydrous ammonia, and more effective than water-run anhydrous ammonia or nitrate. Urea, being non-polar, is carried into the root zone before being hydrolyzed, whereas ammonium-based fertilizers do not move as far into the root zone and are more susceptible to loss of N as ammonia.

MANAGEMENT TO IMPROVE NITROGEN SUPPLY AND SYNCHRONY

Nitrogen Fixation

Legumes in a rotation often benefit the yield and N content of a subsequent crop through recycling of N from the legume residues or a 'sparing effect' on soil N. The legume does not necessarily add N to the soil. Frequently more N is harvested from legume crops than is fixed and there is no gain in soil N. However, the loss of soil N may have been less than if the legume had not been grown due to nitrate sparing (D.F. Herridge and F.J. Bergersen, in these Proceedings). More work is needed with food legumes but it is doubtful if the benefits to soil N that have been observed with pasture legumes can be equalled by crop legumes.

Fertilizer management

It is puzzling that farmers in many countries tolerate poor efficiency of fertilizer use. They persist with poor agronomy when a little effort could double the efficiency of N uptake and halve fertilizer bills. Several techniques have been discussed by Baligar and Bennett (1986); the basis being to avoid having too much nitrate in the soil at any time through fertilizer application as small split doses, or in a formulation or with methods of application that result in slower nitrification. Alternative chemical or physical formulations (e.g. sulphur-coated urea or 'super-granules') have not yet proven successful in upland systems. Chemical inhibitors also have not yet had a major impact, because of high cost and short-term effectiveness. Urease inhibitors may have some future, since they need to work only for a few days to spread the initial peak of ammonia concentration in the soil solution.

Fertilizer placement is used less than it should be. The technology is not new but a better understanding of the environments where is it most effective is still needed. For example, under low rainfall conditions banded fertilizer was much more effective than N which was broadcast and incorporated by cultivation (Myers and Hibberd 1986), but there was little difference between these methods of application under conditions of higher rainfall.

Banding is well-known as an effective means of applying N to furrow-irrigated crops, but recent work (Wright and Catchpoole 1985) showed that depth of placement is also critically important. The fertilizer was most effective when placed in the ridge below the level of the furrow. When banded above the level of the furrow, fertilizer N moved upwards, nitrite accumulated near the seed and nitrate formed a crust on the ridge. This nitrate was unavailable to the crop due to high soil temperature and rapid drying of the surface soil. Care is also needed with depth of placement in the humid tropics where fertilizer is banded. The recommended method of application of N to upland crops is as a band below the seed, but in the humid tropics this can produce poor results with commonly less than 20% of the applied N recovered by the crop. However, this is with hand placement in shallow furrows in soils with a poor exchange capacity. I believe that deeper placement in a more concentrated band would result in improved recoveries.

Special difficulties arise with irrigated crops such as cotton and sugarcane. These crops often require large applications of N leading to high concentrations of nitrate and the risk of significant losses. When the economic situation is good, farmers are often unconcerned about applying excess amounts of N but this practice certainly needs reconsideration when product prices drop, as has happened in the sugar industry. There is currently an urgent need for research into efficient N management of sugarcane.

Controlled, low-dose application of N in the irrigation water can be an effective technique on corn (Muirhead et al. 1985a,b) to avoid high concentrations of nitrate in the soil and minimize denitrification losses after subsequent irrigations (Freney et al. 1985). Compared with banded N, water-run N will nitrify rapidly, and denitrification is possible following waterlogging.

With sugarcane, it appears to be important to supply N to the crop early in growth. Thus split applications tend to be less successful than single applications (Bieske 1972). However, whether this is because of greater susceptibility to loss from split applications, or to a greater crop demand in the early stages of growth, is not clear.

Better predictions of fertilizer requirement are necessary for most crops. Recent attempts to incorporate plant, weather and soil factors into predictive models (Myers 1984; Godwin and Vlek 1985; McCosker et al. 1986) should lead to improved profitability by avoiding wastage of fertilizer. These models vary in complexity but have in common the philosophy that N requirement is largely determined by the potential yield. Realistic prediction of the latter is thus crucial to their effective use.

Recycling of residues

Crop residues contain N which may be mineralized and therefore become available for crop uptake. However, if their C/N ratio is wide they bring about immobilization of mineral N. Legume residues are usually high in N concentration, have a narrow C/N ratio and generally mineralize N readily. Their potential benefits to subsequent crops depend on the quantity of residues returned and their quality in terms of N, lignin and polyphenol content (Myers and Wood 1987). High polyphenol content in the residue can seriously interfere with N mineralization (Vallis and Jones 1973). Recovery by crops of N applied in legume residues may possibly be higher in the humid tropics than in temperate regions, but evidence is scanty. Whereas a wheat crop recovered only 17% of the N in Medicago littoralis residues in a temperate environment (Ladd and Amato 1986), upland rice recovered 28% of N in cowpea residue in a tropical environment (Widjang H. Sisworo et al., personal communication); however, factors other than climate may have contributed to the difference.

Non-legume residues may contribute N or may immobilize it, depending on their C/N ratio. Immobilized N is not lost and it may later be remineralized. Immobilization prevents it being lost by leaching, gaseous loss or erosion.

The quantity of the residue is also important. In many situations, the quantity of residue is too small to have a major impact on N dynamics, except perhaps with respect to erosion. There exist simple procedures for assessing the likelihood of residues immobilizing or mineralizing significant amounts of N (McGill 1982; D.S. Jenkinson, personal communication). Unfortunately such procedures are not freely available in the literature. They would provide useful tools for extension workers, and it would be desirable for research workers to use them before they embark on expensive field trials. The non-legumes where residues could be important are those that produce large quantities of residue, namely sugarcane, cotton and irrigated cereals. In Australia there is a trend towards trash retention in sugarcane, based on observations that green cane trash contains about 100 kg N/ha and that its retention results in increased soil levels of total N within a single crop cycle (Wood 1986).

Residue management has important socio-economic implications. In many countries, the soil benefits from the residues must be weighed against their alternative traditional uses – as stock feed, as fuel or as construction materials. This is a consideration, not a deterrent to research, particularly in view of findings such as those of Widjang H. Sisworo et al. (personal communication) that cowpea residues can be of almost equal value as, and in percentage utilization terms of more value than, fertilizer N.

Possibly the best niche for recycling of N from residues will be in the humid tropics. One reason is that high temperatures and consistently moist conditions enable a rate of recycling of N higher than in other climates. A second reason is that with fertilizer N the efficiency of utilization can be very poor due to high leaching losses. Against this there are the practical difficulties of growing good legume crops under what are frequently difficult growing conditions.

Manures

In many places animal manures are available but are not returned to the fields. Their value has been well demonstrated (Powell 1986) and the problem is substantially one of extension. Quantity of manure is important. Powell (1986) found that the quantities returned by grazing cattle were insufficient to have any immediate effect on soil fertility. More substantial effects were seen with cattle camps or where manure was transported to the field. In the latter, effectiveness will depend on how much of the urine-N is returned with the dung. If it is high then the N value of the manure from cattle fed crop residues can conceivably be higher than that of the N in the crop residues if they had been left in place, because of the higher proportion of potentially available N in the manure (Myers and Wood 1987). This is despite greater gaseous loss from manures than from crop residues (Kirchmann 1985).

Tillage

One of the main purposes of tillage is weed control. However, an important consequence of tillage is an increased rate of N mineralization, which may exceed immediate crop requirements and result in leaching of nitrate below the root zone (Dalal 1984; V.R. Catchpoole, personal communication). With reduced tillage technology the opportunity now exists for improved N management in such systems. In other systems, reduced tillage may result in increased need for N fertilizer, but where this is used the yield may ultimately be higher (Wells 1984).

Adjustment of sowing dates

A major principle of N management is to have the crop actively growing and taking up N when the soil is most active in making N available. It should then be possible to improve N management by planting the crop at the right time. McGill and Myers (1987) discussed this for several different environments, demonstrating with a simulation model both efficient and potentially inefficient N use depending on the synchrony or asynchrony of the crop and N mineralization which was discussed earlier.

The interaction of water supply with nitrogen

Soil water status is central to the concepts of synchrony of supply and demand for N since it is a major determinant of crop demand and of soil N supply. Additionally water exercises a major control over the rate of gaseous or leaching losses and is the vehicle for N movement from soil to plant. It is thus essential that N-water relationships be properly assessed in dryland cropping systems and included in the simulation models developed as aids to this research.

SOCIO-ECONOMIC CONSIDERATIONS IN RESEARCH

Research on crops important in the tropics has not been balanced. Predominant emphasis has been on the major cash crops such as rice, sorghum, corn, soybean and sugarcane, which are important crops in many countries with a strong applied and basic research capacity. The less widely grown crops and subsistence food crops such as cassava, many food legumes and the millets have received little research attention. Until this situation is rectified it is difficult to prepare a balanced account of N management of crops in the tropics. Even with a well-researched crop such as sugarcane there are special difficulties. Whilst considerable applied research effort has gone into establishing N fertilizer technology, there has been little or no supporting basic research on N dynamics. Further, results with this crop are reported almost exclusively at meetings of specialists in sugarcane resulting in very little interaction with researchers on other crops. In an era when the economics of the cane industry were good, there was over-fertilization with N. The absence of basic knowledge has hindered rapid adjustment of field practices when this situation changed.

CONCLUSIONS

Agronomists have long pursued the goal of optimizing the supply of N to the crop although this has generally meant applying the amount of N required to maximize yield in one application at sowing. Recently attention has turned to the concept of supplying N when it is most required. The concept of synchrony is an important component of the Tropical Soil Biology and Fertility Program under way in various countries. Basic to this concept is the idea of increasing efficiency of N utilization through reducing losses of N, through providing plants with greater uptake capacity, through increasing inputs via fixation of atmospheric N, and through improved efficiency of recycling of N from plant residues. These concepts, together with the end objective of supplying sufficient N at the time that the crop requires it are the goals of scientists seeking to improve N management in the tropics.

REFERENCES

Angus, J.F., Hasegawa, S., Hsiao, T.C., Liboon, S.P., and Zandstra, H.G. (1983). The water balance of post-monsoonal dryland crops. J. Agric. Sci., Camb. 101, 699–710.

Baligar, V.C., and Bennett, O.L. (1986). Outlook on fertility use efficiency in the tropics. Fert. Res. 10, 83–96.

Bieske, G.C. (1972). Split applications of nitrogen fertilizers on ratoon crops. Proc. Queensl. Soc. Sugar Cane Technol., 39th Conf., pp. 73–6.

Campbell, C.A., de Jong, R., and Zentner, R.P. (1984). Effect of cropping, summerfallow and fertilizer nitrogen on nitrate-nitrogen lost by leaching on a brown chernozemic loam. Can. J. Soil Sci. 64, 61–74.

Campbell, C.A., and Souster, W. (1982). Loss of organic matter and potentially mineralizable nitrogen from Saskatchewan soils due to cropping. Can. J. Soil Sci. 62, 651–6.

Chapman, L.C. (1982). Estimating sugar yield responses from N, P and K fertilizers in Queensland. Proc. Aust. Soc. Sugar Cane Technol. pp., 147–53.

Dalal, R.C. (1984). Chronosequential depth distribution of nitrate in cultivated grey and brown clay soils under semi-arid environment. Proc. Nat. Soils Conf., Brisbane, Aust., p. 222.

Dalal, R.C., and Mayer, R.J. (1986). Long-term trends in fertility of soils under continuous cultivation and cereal cropping in southern Queensland. 1. Overall changes in soil properties and trends in winter cereal yields. Aust. J. Soil Res. 24, 265–79.

Finck, A., and Venkateswarlu, J. (1982). Chemical properties and fertility management of vertisols. Trans. 12th Intern. Congr. Soil Sci. II, pp. 61–79.

French, R.J. and Schultz, J.E. (1984). Water use efficiency of wheat in a mediterranean-type environment. 1. The relation between yield, water use and climate. Aust. J. Agric. Res. 35, 743–64.

Freney, J.R., Simpson, J.R., Denmead, O.T., Muirhead, W.A., and Leuning, R. (1985). Transformations and transfers of nitrogen after irrigating a cracking clay soil with a urea solution. Aust. J. Agric. Res. 36, 685–94.

Garrity, D.P., Watts, D.G., Sullivan, C.Y., and Gilley, J.R. (1982). Moisture deficits and grain sorghum performance: Evapotranspiration-yield relationships. Agron. J. 74, 815–20.

Godwin, D.C., and Vlek, P.L.G. (1985). Simulation of nitrogen dynamics in wheat cropping systems. In "Wheat Growth and Modelling", (Eds. W.Day and R.K. Atkin), pp. 311–32. (Plenum Publ. Corp: New York.)

Kanwar, J.S., Rego, T.J., and Seetharama, N. (1984). Fertilizer and water-use efficiency in pearlmillet and sorghum in vertisols and alfisols of semi-arid India. Fert. News 29, 42–52.

Kirchmann, H. (1985). Losses, plant uptake and utilization of manure nitrogen during a production cycle. Acta Agric. Scand. Suppl. 24.

Ladd, J.N., and Amato, M. (1986). The fate of nitrogen from legume and fertilizer sources in soils successively cropped with wheat under field conditions. Soil Biol. Biochem. 18, 417–25.

McCosker, A.N., Hunter, M.N., and Standley, J. (1986). Recommendations for N use in sunflowers. Queensl. Dept. Prim. Ind., Australia RQR86021, pp. 30–1.

McGill, W.B. (1982). N mineralization–immobilization calculations for continuous additions to soil. Mimeo. Prep. Dept. Soil Sci. Univ. of Alberta.

McGill, W.B., and Myers, R.J.K. (1987). Controls on dynamics of soil and fertilizer N. In 'Soil Fertility and Organic Matter as Critical Components of Production Systems'. (Eds. R. Follett, C.V. Cole and J.W.B. Stewart) Soil Sci. Soc. Am. Spec. Publ. No. 19, pp. 71–98 (Amer. Soc. Agron.: Madison.)

Moraghan, J.T., Rego, T.J., Buresh, R.J., Vlek, P.L.G., Burford, J.R., Singh, S., and Sahrawat, K.L. (1984a). Labelled nitrogen fertilizer research with urea in the semi–arid tropics. II. Field studies with a vertisol. Plant Soil 80, 21–33.

Moraghan, J.T., Rego, J.T., and Buresh, R.J. (1984b). Labelled nitrogen fertilizer research with urea in the semi–arid tropics. 3. Field studies on alfisol. Plant Soil 82, 193–203.

Muirhead, W.A., Melhuish, F.M., and White, R.J.G. (1985a). Comparison of several nitrogen fertilizers applied in surface irrigation systems. 1. Crop response. Fert. Res. 6, 97–109.

Muirhead, W.A., Melhuish, F.M., White, R.J.G., and Higgins, M.L. (1985b). Comparison of several nitrogen fertilizers in surface irrigation systems. II. Nitrogen transformations. Fert. Res. 8, 49–65.

Myers, R.J.K. (1975). Temperature effects on ammonification and nitrification in a tropical soil. Soil Biol. Biochem. 7, 83–86.

Myers, R.J.K. (1978). Nitrogen and phosphorus nutrition of dryland grain sorghum at Katherine, Northern Territory. 4. 15–nitrogen studies on nitrogen carrier and method of application. Aust. J. Exp. Agric. Anim. Husb. 19, 481–7.

Myers, R.J.K. (1983). The effect of plant residues on plant uptake and leaching of soil and fertilizer nitrogen in a tropical red earth soil. Fert. Res. 4, 249–60.

Myers, R.J.K. (1984). A simple model for estimating the nitrogen fertilizer requirement of a cereal crop. Fert. Res. 5, 95–108.

Myers, R.J.K., Campbell, C.A., and Weier, K.L. (1982). Quantitative relationship between net nitrogen mineralization and moisture content of soils. Can. J. Soil Sci. 62, 111–24.

Myers, R.J.K., and Hibberd, D.E. (1986). Effect of water supply and fertilizer placement on use of nitrogenous fertilizer by sorghum on clay soils. Proc. 1st Aust. Sorghum Conf., pp. 6.36–6.45.

Myers, R.J.K., McGill, W.B, Vallis, I., and Henzell, E.F. (1987). Nitrogen in grass–dominant, unfertilized pasture systems. Proc. 13th Intern. Congr. Soil Sci., Hamburg August 1986, pp. 761–71.

Myers, R.J.K., and Wood, I.M. (1987). The role of food legumes in the nitrogen cycle of farming systems. In "Food Legume Improvement for Asian Farming Systems". (Ed. E.S. Wallis and D.E. Byth), in press. (Aust. Centre Int. Agric. Res.: Canberra.)

Powell, J.M. (1986). Manure for cropping: a case study from central Nigeria. Expl. Agric. 22, 15–24.

Ruschel, A.P., and Vose, P.B. (1982). Nitrogen cycling in sugarcane. Plant Soil 67, 139–46.

Stanford, G., and Epstein, E. (1974). Nitrogen mineralization–water relations in soils. Soil Sci. Soc. Am. Proc. 38, 103–7.

Stanford, G., Frere, M.H., and Schwaninger, D.H. (1973). Temperature coefficient of soil nitrogen mineralization. Soil Sci. 115, 321–3.

Stanford, G., and Smith, S.J. (1972). Nitrogen mineralilzation potentials of soils. Soil Sci. Soc. Am. Proc. 36, 465–72.

Swift, M.J., and Sanchez, P.A. 1984. Biological management of tropical soil fertility for sustained productivity. Nature and Resources 20, 2–10.

Takahashi, D.T. (1964). N15–nitrogen field studies with sugarcane. Hawaiian Planters' Record 57, 198–222.

Vallis, I., and Jones, R.J. (1973). Net mineralization of nitrogen in leaves and leaf litter of Desmodium intortum and Phaseolus atropurpureus mixed with soil. Soil Biol. Biochem. 5, 391–8.

Wells, K.L. (1984). Nitrogen management in the no–till system. In 'Nitrogen in Crop Production'. (Ed. R.D. Hauck). pp. 535–50 (ASA–CSA–SSSA: Madison).

Wood, A.W. (1985). Soil degradation and management under intensive sugarcane cultivation in North Queensland. Soil Use Manage. 1, 120–6.

Wood, A.W. (1986). Green cane trash management in the Herbert Valley. Preliminary results and research priorities. Proc. Aust. Soc. Sugar Cane Technol. pp., 85–94.

Wright, G.C., and Catchpoole, V.R. (1985). Fate of urea nitrogen applied at planting to grain sorghum grown under sprinkler and furrow irrigation on a cracking clay soil. Aust. J. Agric. Res. 36, 677–84.

THE NITROGEN CYCLE IN PASTURES

K.W. Steele and I. Vallis

ABSTRACT

Awareness that productivity of all pastures not receiving large inputs of fertiliser N is severely restricted by N deficiency, and concerns relating to the impact on human health and the environment of nitrate, ammonia and nitrous oxide originating from pasture, has stimulated research on N cycle processes and control mechanisms in pastoral ecosystems. This paper integrates the various processes of the N cycle which affect the amount and timing of N supply to tropical and temperate pastures, with emphasis on recent developments in Australia and New Zealand.

In high producing pastures, symbiotic N_2-fixation and fertiliser N account for the major input of N. Inputs via asymbiotic fixation or rainfall are only of significance in systems of low productivity. Herbivores play a dominant role in influencing both the magnititude and pathway of N loss from grazed pastures. As stock number per unit area increases there is increased loss of N through retention, transfer, removal in products and via ammonia volatilisation and leaching from animal excreta. However, despite substantial N losses, animal excreta is important in maintaining high pasture productivity at high stocking rates.

Rate limiting steps to the flow of N in pastoral ecosystems and the impact of management are discussed.

INTRODUCTION

Permanent grasslands occur over a wide range of climates and soil types and cover approximately 24% of the world's land surface area (Buringh 1985). Herbage from grasslands comprises a major proportion of forage consumed by herbivores and annual yields range from less than 1 t/ha in many unimproved native pastures in semi-arid and sub-alpine areas to 85 t/ha in irrigated grasslands receiving high fertiliser inputs in the humid tropics (Cooper

1970). The productivity of all pastures not receiving large inputs of N fertiliser is, at least seasonally, limited by N deficiency. Even in highly productive legume-based pastures in temperate regions productivity increases of around 30% may be achieved by elimination of N deficiency (Steele 1982a). Productivity of N-fertilised tropical pastures is commonly 30-100% higher than that of their legume-based counterparts (Myers and Henzell 1985).

In recent years N cycling in grasslands has received much attention from researchers and has been the subject of numerous reviews largely because of (1) the widespread recognition that grassland productivity is severely restricted by N deficiency, (2) concerns relating to the impact on human health and the environment of N lost from pastoral ecosystems as ammonia, nitrous oxide or nitrate, and (3) the recognition that grassland management practices can have large effects on N cycling in pastures affecting productivity and the magnitude and pathways of N loss.

In this paper we integrate the various processes of the N cycle which affect the amount and timing of N supply to tropical and temperate pastures with emphasis on Australian and New Zealand pastures and on developments reported during the 1980's.

AMOUNT AND DISTRIBUTION OF NITROGEN

Representative ranges for N contained in various components of grassland ecosystems are presented in Table 1. Apart from soil organic matter which generally accounts for more than 85% of N in grasslands, plant roots and soil micro-organisms are the largest repositories of N.

From the data summarised in Table 1, Woodmansee et al. (1981) concluded that generally total organic N is higher in moister or colder environments;

Table 1: Representative ranges for nitrogen contained in various components of grasslands (Woodmansee et al. 1981).

Component	Nitrogen content
	$g\ N/m^2$
Tops (live)	1 – 10
Tops (dead)	0 – 2
Litter	<1 – 9
Roots (live)	2 – 10
Roots (dead)	2 – 20
Micro-organisms	2 – 15
Soil organic matter	90 – 1600
Invertebrates	<0.01– 0.3
Large vertebrates	0 – 3

little mineral N is present in most grasslands except under high levels of N fertilisation; the ratio of N in living to dead plant, microbial and animal material is higher in warmer, moister environments because decomposition rates of dead materials are higher; and the amount of N in micro-organisms may be greater in moister environments. A very large overlap in the range of pool sizes for tropical and temperate pastures occurs, but on average smaller pool sizes can be expected in tropical areas because of higher rates of decomposition. Soil organic N is higher in soils containing amorphous clays such as allophane. Rates of exchange of N between pools vary by several orders of magnitude, the productivity of pastures being dependent on the flux of N through highly active components (e.g. microbial biomass, mineral N) and not the size of any component per se.

NITROGEN INPUTS

On soils of low N status, annual symbiotic fixation in newly established legume-dominant temperate pastures may exceed 500 kg N/ha (Steele 1982b). However, in established permanent pastures, symbiotic N_2-fixation decreases to between 100 and 300 kg N/ha and in New Zealand represents approximately 28% of the total N flux through pastures (Hoglund et al. 1979; Hoglund and Brock 1987). Declines in N_2-fixation are due to increases in soil mineral N (Hoglund et al. 1979) and establishment of pest populations which reduce the legume's ability to fix N (Steele et al. 1985). Recent reviews (Vallis and Gardener 1984a; Vallis 1985) indicate that inputs by tropical pasture legumes fall within a similar range to temperate legumes.

Inputs via asymbiotic fixation are usually less than 15 kg N/ha and are only of significance in systems with low productivity (Ball 1979; Weier 1980; Weier et al. 1981; Lambert et al. 1982; Morris et al. 1985). It has been shown, however, that under some conditions Brachiaria humidicola and B. decumbens have obtained 40 and 45 kg N/ha/yr respectively from associative fixation (Boddey and Victoria 1986). Inputs of N through rain and dry deposition range from 0.8 to 22 kg/ha/yr with higher values being in the tropics or near urban industrial areas (Stevenson 1982).

The vast majority of the world's pastures receive no fertiliser N, but considerable differences exist between countries. In Australia an estimated 35,000 tonnes of fertiliser N were used for forage production in 1981-82, compared with an input of 1.2 million tonnes of fixed N by pasture legumes. About two-thirds of this fertiliser N was used on long term pastures and about one-third on forage crops and short-term leys. The dairy industry accounted for 80% of the N fertiliser usage (Myers and Henzell 1985). In recent years an increase in the use of N fertiliser on irrigated pastures sown annually with temperate grasses has been a major cause of increased farm efficiency on dairies in sub-tropical and tropical Australia (Lowe and Hamilton 1986). Recommended rates of N on these irrigated pastures are up to 500 kg/ha/yr (Chopping et al. 1983). In New Zealand, tactical applications of fertiliser N may be used to complement biologically fixed N and overcome short-term feed deficits for stock. However, as in Australia, the total annual input of fertiliser N to grasslands is small (8000 t) relative to biological fixation (>1.1 Mt) (Cullen and Steele 1983).

NITROGEN LOSSES

Processes of N loss have been reviewed in detail elsewhere in these Proceedings, with the exception of removal in animal products. In this section we will restrict our discussion to representative ranges for N losses from pastures.

The major pathways for loss of N from pastures include (1) ammonia emission from soil, plant, animal excreta and fertilisers, (2) biological and chemical denitrification, (3) leaching, (4) retention in animals and animal products, (5) transfer to unproductive areas in the excreta of animals, (6) erosion by wind and and water, and (7) fire (Steele 1987).

Herbivores play a dominant role in influencing both the magnitude and pathway of N loss from grazed pastures. Annual losses vary widely being dependent on stock type, stocking rate, land class, pasture production and climate.

Excluding animal retention, transfer and removal in animal products, the various forms of mineral N (NH_3, NH_4^+, NO_2^-, NO_3^-) are intermediaries in all major pathways of loss, the size and flux of the mineral N pool being a major determining factor in the magnitude of loss (Steele and Brock 1985).

Ammonia volatilisation

The development of techniques to measure NH_3 volatilisation without disturbing the ambient conditions, e.g. micro-meteorological methods (Denmead et al. 1974) has resulted in an improved understanding of NH_3 emission from grasslands. Ammonia emission is larger following surface application of alkaline than neutral or acidic fertilisers. Most of the loss (c. 80%) occurs within two weeks of fertiliser application with losses from tropical pastures fertilised with urea being around 25% (Harper et al. 1983a,b). These studies demonstrated the dynamic nature of NH_3 exchange in pastures. During the two weeks after application there were relatively large effluxes of NH_3 during the day and small influxes at night. After two weeks small net influxes of a short duration were common around sunrise and sunset, and prior to each urea application the daily net flux was into the pasture.

A relatively simple micro-meteorological mass balance method was used by Ryden and McNeill (1984) to determine NH_3 loss from a sward in England that was grazed heavily by yearling steers (12.2/ha) and fertilised with 60 kg N/ha as ammonium nitrate every 28 days. A large diurnal trend in NH_3 flux was observed, high in the day and low at night, with rain having a major depressing effect. The flux during early summer was equivalent to 34% of the fertiliser application but the source of NH_3 (i.e. fertiliser, urine or dung) was not identified.

Other recent research on volatilisation of NH_3 from pastures has used flow-through chambers (Ball et al. 1979; Carran et al. 1982; Vallis et al. 1982; Ball and Keeney 1983; Black et al. 1984; Lockyer 1984; Sherlock and Goh 1984;) and ^{15}N balance methods (Catchpoole et al. 1983; Vallis and Gardener 1984b; Vallis et al. 1985). While there is always some doubt that NH_3 losses measured by chamber techniques accurately reflect those in undisturbed systems (Denmead 1983) they have proved useful for studying the effects of various factors on NH_3 loss. Thus it has been shown that for pastures the percentage losses are similar for medium to high rates of N application in urine (Ball et al. 1979); and losses from urine on an initially dry soil were twice that from wet soil (Carran et al. 1982), although the opposite result was reported by Harper et al. (1983a) for prilled urea fertiliser applied to a tropical pasture. Also, losses from urine or urea fertiliser were much greater when the soil had previously been affected by urine (Black et al. 1984; Sherlock and Goh 1984).

Total losses of N from a tropical pasture after ^{15}N-urea addition were twice the flux of NH_3 from the pasture (Catchpoole et al. 1983). The difference could not be explained by leaching and it was suggested that the cause was probably losses of N as N_2 and N_2O. Only trace amounts of N appear to be lost as NO_2 during volatilisation of NH_3 (Carran et al. 1982; I Vallis, L.A. Harper, V.R. Catchpoole and K.L. Weier, unpublished data).

The average annual loss of urinary N by NH_3 volatilisation in grasslands is probably between 20 and 30% (Ball and Keeney 1983; Simpson and Steele 1983). Ammonia emission from dung on temperate soils is normally much less than from urine (<5%, McDiarmid and Watkin 1972) but may be considerable under hot dry conditions where no dung beetles are active (Gillard 1967). Loss of N directly from plant foliage through volatilisation of NH_3 or amines has been recognised but probably amounts to only a few kg/ha (Wetselaar and Farquhar 1980).

Denitrification

Denitrification may occur through biological and non-biological processes. However, measured losses through the latter have been small (e.g. Galbally and Roy 1978) and the majority of N lost through denitrification can be attributed to biological activity. Relatively few quantitative estimates have been made of the N loss from grasslands through denitrification because of difficulties of measurement under field conditions.

Spatial variability of denitrification, as indicated by N_2O emission, is large with two-fold differences occurring over a few metres (Simpson and Steele 1983). Denmead et al. (1979) measured daily emissions of 217 mg N/m^2 from mown grassland in Australia during spring when soils were warm and wet, and 0.05 mg N/m^2 in winter when soils were cold and dry. Similar rates and patterns of measurement have been reported in New Zealand (Limmer et al. 1982; Limmer and Steele 1983) and Britain (Ryden 1981; Webster and Dowdell 1982). Nitrous oxide absorption by soils may also occur when nitrate concentration in the soil is less than 1 ppm (Ryden 1981).

Nitrous oxide emission following fertiliser application is usually small, accounting for <2% of the N applied (Ryden 1981; Webster and Dowdell 1982) but some large losses have been reported (e.g. O'Hara et al. 1984). Although N_2O emission can occur during nitrification of fertiliser or soil ammonium (Blackmer et al. 1980; Goreau et al. 1980) it is unlikely to be of practical significance in most managed grasslands.

Emission of gaseous N_2 from grasslands may be many times greater than N_2O emission (Ryden 1981; Limmer et al. 1982). The ratio of gaseous N_2 emitted relative to N_2O is reduced by high nitrate concentrations (Blackmer and Bremner 1978), low carbon availability (Focht and Verstraete 1977; Smith and Tiedje 1979), increasing oxygen partial pressure (Krul and Veeningen 1977), low soil pH and low temperature (Nommik 1956).

In both temperate and tropical grassland soils the potential for loss of N through biological denitrification is greatest in the surface layer of mineral soils but can occur from deeper horizons in organic soils (Limmer and Steele 1983; Weier and Doran 1986). Denitrification can occur in aerobic soils if they contain a large population of denitrifying strains of Rhizobium (O'Hara et al. 1983) although the amount of N lost from pasture via rhizobial denitrification is generally small (K.W. Steele, unpublished data).

Leaching losses

Under climatic conditions where, at least seasonally, rainfall exceeds evapotranspiration, substantial loss of N through leaching may occur. Grazing animals have a major effect on the magnitude of this loss. In the United Kingdom the annual loss of nitrate from a grazed grass sward receiving 420 kg N/ha/yr was 5.6 times that lost from a comparable mown sward. The enhanced nitrate leaching in the grazed sward was attributed to the return of N in urine and dung in concentrations which far exceeded the requirements of pasture plants (Ryden et al. 1984). Leaching losses from grazed pastures exceeding 100 kg N/ha/yr have also been measured in New Zealand (Steele et al. 1984) where non-irrigated pastures have been identified as having the greatest impact on the nation's groundwater quality (Burden 1982) and where concentrations of up to 58 mg N/L have been measured in shallow aquifers under intensively grazed grasslands (Baber and Wilson 1972).

Retention in animals and animal products

Henzell and Ross (1973) used published data for the N contents of livestock and their products, apparent digestibility of N in diets with various N concentrations, and the excretion of N to calculate retention of N in animals and animal products. They concluded that typical ranges for the percentage of N retained were:– beef cattle 4–10%, sheep (wethers) 5–13% and dairy cattle 13–28%.

Transfer in excreta

The transfer of N from pasture areas to non–productive areas such as milking sheds, races and gateways depends on the type of stock and management. With dairy cows 10 to 15% of excreta is deposited on unproductive sites (During 1972). Transfer of N also occurs within paddocks especially on hill country. In a 12.5 ha hill country paddock which had been mapped into five strata according to occurrence, colour and vigour of pasture species and the distribution density of obvious urine patches, Gillingham and During (1973) found that the transfer of N between strata varied from a small stock camp which gained 271 kg N/ha/yr to a stratum that was depleted of 73 kg N/ha in the same period.

Surface run–off and erosion

Loss of N in surface run–off and erosion is usually considered to be negligible from grasslands where an adequate plant cover is maintained. Recently a new sensitive method based on concentrations of caesium 137 in the surface soil has been used to show the effects of disturbance of semi–arid native pastures on soil losses by sheet erosion. Over a 17–year period in plots where the trees and shrubs had been thinned, erosion removed about 1 cm more soil than was lost from undisturbed pastures. Where the plots had been grazed by sheep at normal stocking rates the loss over 17 years was 1.5 cm of topsoil (Reece and Campbell 1986).

NITROGEN BALANCES FOR GRAZED PASTURES

To sustain a given level of pasture productivity there is a need to achieve a long–term N balance in the ecosystem, ie:

$$N \text{ inputs } - (N \text{ losses } + \text{ gain in total ecosystem } N) = 0$$

Short–term imbalances in inputs and losses may be masked by changes in the size of the soil organic N pool. Any change in a parameter such as grazing pressure or N input will result in a move towards a new equilibrium level of total ecosystem N. Examples of N balances in humid sub–tropical, monsoonal seasonally dry tropical, and temperate pastures, are shown in Table 2. The amount of N loss from pasture ecosystems generally increases with increasing inputs of N and with increasing productivity. The major pathways for loss of N from pastoral ecosystems are via animals (products, retention, transfer), NH_3 volatilisation and leaching. Partitioning between NH_3 volatilisation and leaching depends on prevailing climatic conditions but is modified by soil and pasture parameters.

Table 2: Nitrogen balances for examples of grazed pastures from Australia and New Zealand (Blank spaces indicate no data available).

Attribute	Australia		New Zealand		
	Samford[1]	Townsville[2]	Woodville[3]	Canterbury[4]	Waikato[5]
Climate	Humid sub-tropical	Monsoonal seasonally dry tropical	Temperate	Temperate	Temperate
Rainfall (mm)	1050	860	1400	771	1600
Stock	Beef cattle	Beef cattle	Sheep	Sheep	Dairy cattle
Stocking rate (SU/ha)	20	5	6	20	24
Pasture	Mixed tropical grasses	Verano stylo	Browntop	Ryegrass white-clover	Ryegrass white-clover
Pasture production (t DM/ha/yr)	c. 10.0	c. 6.0	3.27	11.0	16.5
N inputs (kg/ha/yr)					
Fertiliser	374	0	0	0	0
Rain	4	4	3	10	3
Associative fixation			13		14
Symbiotic N fixation	___	43	17	180	267
TOTAL	378	47	33	190	284
N losses (kg/ha/yr)					
NH_3 emission from:					
fertiliser	70				
urine	20)	10		10	20
faeces	)		4	10	4
plants	)			10	
Denitrification				10	30
Leaching				100	110
Surface runoff			10		
Retention in livestock	8	4		20	8
Livestock products			4		66
Transfer				30	46
Gain in soil	102	33	15	0	0
Unidentified	178	___	___	___	___
TOTAL	378	47	33	190	284

1. Wetselaar & Hutton 1963; Henzell 1972; Catchpoole & Henzell 1975; Catchpoole et al. 1981;
2. C.J. Gardener & I. Vallis unpub. data; Wetselaar & Hutton 1963; C.J. Gardener unpub. data;
Vallis & Gardener 1984a,b; 3. Lambert et al. 1982; 4. Quin 1982 5. Steele 1982b.

RATE LIMITING STEPS TO THE FLOW OF NITROGEN IN THE NITROGEN CYCLE

The productivity of pastures without added fertiliser N is generally well below potential yield despite large amounts of total N present in the soil-plant system. This indicates that productivity is limited by the slow rate of cycling of N in the system into an available form. Pasture deterioration in the brigalow lands of sub-humid central Queensland is influenced more by reduced availability of N than by changes in N reserves (Graham et al. 1981). Inputs of N through symbiotic fixation by legumes and recycling of N within plants are discussed in other chapters, as are losses of N. In this section we concentrate on factors limiting the recycling of herbage N and the supply of mineral N to pasture plants.

Where pasture is burnt or cut for hay the recycling of herbage N is small. Litter production in semi-arid pastures is small (350 kg DM/ha in six months; Christie 1979) compared to temperate (Hunt 1983) and humid sub-tropical pastures (2600 kg DM/ha; Bruce and Ebersohn 1982). In grazed, temperate pastures the N return to the soil in dead leaves varied from 10 to 20% of the annual N uptake in herbage. The rates of N cycling to the litter via leaf senescence peaked at 3.5 kg N/ha/day and rates of N cycling to the soil via litter disappearance ranged up to 140 kg N/ha/yr.

The chemical composition of litter determines the balance between mineralisation and immobilisation of N and the rate of decomposition. Between 10 and 33% of the N in plant litter becomes available to pasture plants within the first year after deposition but in subsequent years <5% per year of the residual litter N is recycled (Vallis 1983). Increasing pasture utilisation effectively reduces cycling through the litter pool, increasing the dependence of plants on mineralisation of stable soil organic matter and animal excreta for their N requirements.

Passage through livestock increases the rate of conversion of N in herbage to mineral forms but at the expense of greater losses from the ecosystem (Vallis and Gardener 1984a). The few studies conducted on tropical pastures (Vallis and Gardener 1984b; Vallis et al. 1985) indicate that the recovery of urinary N by pasture plants may be even lower than the generally 30% or less quoted for temperate pastures in New Zealand (Steele and Brock 1985). However, despite the substantial N losses that occur from animal excreta, particularly urine, animal excreta is important in maintaining high pasture productivity. Productivity is higher on grazed than on comparable cut pastures (Ball 1979). Under extensive agriculture, pasture plants are largely dependent on mineralisation of soil organic matter or N_2-fixation to provide their N requirements for growth. As pasture productivity and animal stocking rate increase, animal excreta contributes proportionally more to the N taken up by pasture plants. For example, on dairy farms in New Zealand rotationally grazed at 4 cows/ha, an average 45% of the grazed area is affected by animal excreta at any one time and this produces 70% of the total dry matter production (Steele and Brock 1985). At high stocking rates, refusal of excreta—affected pasture by stock is minimal.

Roots can also contribute considerable amounts of N to the soil during decomposition but, because of the high C:N ratio, root decay may be associated with net immobilisation of soil mineral N during initial stages of decomposition. In the humid forest zone of the Ivory Coast, Picard (1979) estimated the turnover time for roots of Panicum maximum defoliated at 6 weekly intervals was 0.2 to 0.4 years. The annual input of root dry matter was estimated to be between 9 and 16 tonne/ha. The pattern of N release from decomposing roots appears to be similar to that from herbage, i.e. a rapid release in the first year followed by a slow release (Moore 1974).

Soil micro—organisms are a key component in grassland N cycles because in addition to being mediators of N transformations, micro—organisms act as a small but important reservoir of N and comprise a dynamic component with a rapid rate of turnover. In a typic vitrandept on a high producing dairy farm in the central North Island of New Zealand, microbial N levels fluctuated seasonally and ranged from 240 to 330 kg N/ha. A gradual increase in microbial N levels occurred over autumn and winter indicating an accumulation of N within the biomass, this was followed by a sudden decline in microbial N of about 100 kg N/ha during a short period in spring which coincided with the onset of pasture growth (Sarathchandra and Perrott 1984). Patterns of seasonal change in microbial N differ between years and may be related to temperature and rainfall (Sarathchandra et al. 1987). The soil biomass therefore acts as a sink for mineral N at times of the year when conditions favour an increase in its size or a shift to a population of higher N content, and acts as a source of mineral N when conditions favour a decrease in its size or a shift to a population of lower N content.

The soil fauna, which is usually dominated by earthworms and/or termites, is also important in the cycling of N in pastures. A contrast between temperate and tropical pastures is the general lack of earthworms in the latter. In a review of the influences of earthworms and termites on soil N cycling, Lee (1983) concluded that earthworms are likely to have an important influence on N cycling in humid areas. They may ingest litter at rates exceeding the rate of litter production and return N to the soil as ammonia, urea, mucoproteins and dead tissues. Termites may be of significance in light—textured soils in semi—arid tropical areas. Typically the biomass of termites is only 10% that of earthworms and their specific rate of food intake only 15 to 40%. Further, termites are adapted to conserve N and return little to the ecosystem except through predation. Termites may have a significant role in N cycling where the supply of N is extremely limited, such as in desert ecosystems. The biomass per hectare of grass and litter—feeding termites in a semi—arid environment in North Queensland has been estimated to be 2 — 5 times that of the cattle biomass (Holt and Easey 1984).

MANAGEMENT EFFECTS ON THE NITROGEN CYCLE

It appears inevitable that management practices that increase the rate of N cycling in pastures will automatically increase the rate of N loss. This conclusion assumes that increased uptake of N by pasture plants is accompanied by greater intake and excretion of N by grazing livestock, and that losses of N from excreta–affected areas are thereby increased. If long–term productivity is to be sustained, implementation of management practices to increase the rate of N cycling must be accompanied by practices that increase the inputs of N into the pasture (N fertiliser, increased biological N_2-fixation from legumes) so that the N balance of the system is in equilibrium or positive. It is also necessary to take account of the environmental impact of increasing N losses.

Animal management

The impact of management on the N cycle and productivity of pastures, needs to be considered for both short and long–term effects. For example, increased grazing pressure may give a faster rate of N cycling in the short term (Bromfield and Simpson 1974) but will result in increased N loss and possible changes in botanical composition which may eventually lead to lower productivity. The total amount of N in grassland ecosystems in the western United States increases under light or no grazing, is approximately constant at moderate grazing intensity, but decreases at high grazing intensities (Woodmansee 1979). Net losses of N have also been reported in intensively grazed temperate grasslands in New Zealand (Ball 1982).

In many environments pasture production is very seasonal. In New Zealand up to three–quarters of total annual pasture production occurs during the spring–early summer period which coincides with the time of highest animal production, deposition of excretal N, N_2-fixation and soil microbial activity. Management practices during this period may be expected to have a major impact on annual N budgets (Steele and Brock 1985).

Rotational grazing creates open, low density pastures relative to set stocking. Set stocked pastures, because they are maintained in a more uniform physiological state than rotationally grazed pastures, and are subjected to less intensive return of animal excreta, have lower concentrations of nitrate N in their herbage during most of the year (Brock et al. 1983). Both NH_3 volatilisation and leaching losses can be expected to be lower in denser set–stocked pastures because of the effect of pasture cover on volatilisation loss (Freney et al. 1981) and the better ability of set–stocked pastures to utilise excretal N.

The amount of N cycled through plant litter has a major impact on the size of the soil organic matter pool. Increasing pasture utilisation by animals reduces the amount of litter in pastures and increases the dependence of pasture plants on mineralisation of stable soil organic matter and animal excreta for their N requirements. Where N inputs are insufficient to match losses, a negative N balance may occur (Field and Ball 1982).

The distribution and cycling of N in grazed pastures can be improved by fencing to improve excreta distribution by preventing transfer to non-productive areas and stock camp sites.

Plant factors

Rapid decomposition and release of N occurs from plant roots with a C:N ratio below 20–25, whereas immobilisation takes place at higher C:N ratios (Whitehead 1970). Nitrogen cycling is promoted by shifts in botanical composition of pastures towards legumes since legume tissues have C:N ratios generally <20 while grasses are generally >30. As we noted earlier, however, this increased cycling carries the risk of greater losses of N from the ecosystem. Plant species with high C:N ratios reduce losses of N by causing more of the N to be incorporated into relatively stable organic matter.

Substantial increases in pasture and animal production can be achieved by introduction of legumes suitably-adapted to unimproved pastures. Pasture production increases can also be achieved by identification and removal of limitations to N_2-fixation in established pastures. In 16 <u>Trifolium repens</u>/<u>Lolium</u> <u>perenne</u> pastures in New Zealand, annual N_2-fixation and dry matter production were increased by 57 and 13% respectively by application of insecticide (Steele <u>et al</u>. 1985). Pests and diseases have also proved to be a major problem with tropical pasture legumes. Anthracnose disease in <u>Stylosanthes</u> spp. has caused yield losses of 64–100% in <u>S</u>. <u>guianensis</u> (Anon 1981; Irwin <u>et al</u>. 1986), 53% in <u>S</u>. <u>humilis</u> (Irwin <u>et al</u>. 1986), 26–58% in <u>S</u>. <u>hamata</u> (Lenne and Sonoda 1982), and 78% in <u>S</u>. <u>scabra</u> (Irwin <u>et al</u>. 1986). These losses of yield indicate large reductions in N_2-fixation. Control of rust on <u>Macroptilium</u> <u>atropurpureum</u> increased dry matter yield by 30% (Jones 1982). The <u>Leucaena</u> psyllid has caused considerable damage to stands of leucaena (<u>Leucaena</u> <u>leucochephala</u>) in recent years and strategies for control are under investigation (Bray and Sands 1987).

Soil factors

Surface cultivation of clay soils high in total N but low in available N which have been under a tropical grass pasture may stimulate N cycling. The annual dry matter yield and N content of a stand of green panic (<u>Panicum maximum</u>) was increased from 8 to 10 t/ha and 90 to 120 kg/ha respectively by an annual surface cultivation. Growth of grass was depressed for several weeks after cultivation (because many roots were severed) but was subsequently stimulated (Catchpoole 1984).

In old pastures on soils of low organic N content and where much of the N is contained in pasture roots surface cultivation may not increase N cycling. The yield of dry matter and N in tops of an old buffel grass (<u>Cenchrus</u>

<u>ciliaris</u>) pasture on a sandy loam–surfaced duplex soil low in total N was not increased by surface cultivation. However, when buffel grass was resown after a three month cultivated fallow, pasture dry matter production was increased 57% and N uptake in grass tops 50% (Graham <u>et al</u>. 1985). The N content of the soil biomass in the resown pasture two months after sowing was 55 kg/ha less than in the control, but this difference had almost disappeared by 20 months. Roots were a large reservoir of N in this buffel grass pasture and after two seasons of cummulative growth 89% (207 kg/ha/30 cm depth) of the plant N was in the roots.

Maintenance of an active soil fauna promotes N cycling through litter and organic residues such as animal excreta and removes these materials from the soil surface. The introduction of earthworms, termites and dung beetles to pastures where they are not present should be considered where the organisms are ecologically adapted to the conditions into which they are introduced. In addition to their influence on N cycling, the improvement in soil fertility associated with an active soil fauna is well documented. Further research into factors controlling the N content of the soil microbial biomass is warranted to determine whether the release of N from the microbial biomass can be manipulated to better match plant requirements.

Soil management that improves ammonium retention (e.g. increasing cation exchange capacity by increasing organic matter content) and/or soil structure permitting urine to penetrate to a greater depth, will reduce NH_3 volatilisation losses. Prevention of water–logging by adequate soil drainage will reduce denitrification losses.

Fire

Fire is often used to remove excess, inferior quality, dead pasture herbage at the start of the growing season in extensively managed pastures, and is sometimes used as an aid to the introduction of a legume into the pasture (Mott 1982). In central Queensland burning–off at the start of the growing season reduced the flow of N through litter during the season by 16 kg/ha but had no measurable effect on N uptake by pasture (Filet <u>et al</u>. 1986). It appears that in this situation little N from the decomposing litter was available for plant uptake in the short term. In an improved buffel grass (<u>Cenchrus ciliaris</u>)/Siratro (<u>Macroptilium atropurpureum</u>) pasture a single fire removed 80–120 kg N/ha (80–90% of that present in herbage plus litter). Once again, there was little effect on subsequent pasture growth ('t Mannetje <u>et al</u>. 1983). Regular burning of pastures over many years, however, must deplete the active soil organic matter and ultimately reduce the rate of N cycling.

Shade

Shading, while not a practical management option has the interesting effect of stimulating growth and N uptake by tropical grasses. The effect has been shown for a number of species and locations (Wong and Wilson 1980;

Eriksen and Whitney 1981; Wilson et al. 1986). The cause is uncertain but appears not to be due to differences in associative N_2-fixation, soil moisture or temperature, or to transfer of N from roots to tops (Wilson et al. 1986). The latter authors have pointed out that the better growth of green panic observed under trees could be partly or even largely due to the shade-effect, rather than to 'tree-drip' or redistribution of nutrients from deeper soil layers, and that apparent transfer of symbiotically-fixed N to green panic growing under the leguminous shrub leucaena could contain a component of shade-stimulated N uptake.

CONCLUSION

Increases in the productivity of pastoral ecosystems via manipulation of the N cycle can be achieved by increasing either inputs of N or efficiency of N usage or both. Ideally both should be considered together because inefficient use of N inputs can be costly and environmentally unacceptable. The major limitations to N_2-fixation and the pathways of N loss have been identified in many pastures. Increased N_2-fixation can be achieved through breeding legumes better adapted to their environment. At the present state of knowledge, however, it appears difficult to further reduce the impact of grazing animals on N losses. It may be necessary to alter the excretal pattern of animals or the form of N excreted. Recent advances in understanding the role of micro-organisms as mediators of N transformations and as an important reservoir of N, have highlighted the possibility of altering the seasonal supply of N to pasture plants by manipulation of the soil microbial population.

As management of pastures is intensified losses of N increase, the losses being accentuated by grazing animals. Reducing N losses is necessary to lower production costs and to improve the environmental acceptability of intensively managed pastures.

REFERENCES

Anon (1981). Centro Internacional de Agricultura Tropical (CIAT). Ann. Rep. Trop. Past. Prog., Cali, Colombia.

Baber, H.L., and Wilson, A.T. (1972). Nitrate pollution of groundwater in the Waikato region. Chem. N.Z. 36, 179–83.

Ball, R. (1979). Soil nitrogen relationships in grazed and cut grass–clover systems. PhD Thesis, Massey University, N.Z.

Ball, P.R. (1982). Nitrogen balances in intensively managed pasture systems. In 'Nitrogen Balances in New Zealand Ecosystems'. (Ed. P.W. Gander), pp. 47–66. (DSIR: Palmerston North, N.Z.)

Ball, P.R., and Keeney, D.R. (1983). Nitrogen losses from urine–affected areas of a New Zealand pasture under contrasting seasonal conditions. Proc. XIV Int. Grassl. Cong. Lexington, U.S.A., 1981. pp 342–4.

Ball, P.R., Keeney, D.R., Theobald, P.W., and Nes, P. (1979). Nitrogen balance in urine–affected areas of a New Zealand pasture. Agron. J. 71, 309–14.

Black, A.S., Sherlock, R.R., Smith, N.P., Cameron, K.C., and Goh, K.M. (1984). Effect of previous urine application on ammonia volatilisation from 3 nitrogen fertilisers. N.Z. J. Agric. Res. 27, 413–6.

Blackmer, A.M., and Bremner, J.M. (1978). Inhibitory effect of nitrate on reduction of N_2 by soil microorganisms. Soil Biol. Biochem. 10, 187–91.

Blackmer, A.M., Bremner, J.M., and Schmidt, E.L. (1980). Production of nitrous oxide by ammonia-oxidising chemoautrophic microorganisms in soil. Appl. Environ. Microbiol. 40, 1060-6.

Boddey, R.M., and Victoria, R.L. (1986). Estimation of biological nitrogen fixation associated with Brachiaria and Paspalum grasses using ^{15}N labelled organic matter and fertilizer. Plant Soil 90, 265-92.

Bray, R.A., and Sands, D.P.A. (1987). Arrival of the Leucaena psyllid in Australia: Impact, dispersal, and natural enemies. Leucaena Res. Rep. (Hawaii) 7 (Special Issue), 61-5.

Brock, J.L., Hoglund, J.H., and Fletcher, R.H. (1983). Effects of grazing management on seasonal variation in nitrogen fixation. Proc. XIV Int. Grassl. Cong. Lexington, U.S.A., 1981., pp. 339-41.

Bromfield, S.M., and Simpson, J.R. (1974). Effects of management on soil fertility under pasture. 2. Changes in nutrient availability. Aust. J. Exp. Agric. Anim. Husb. 14, 479-86.

Bruce, R.C., and Ebersohn, J.P. (1982). Litter measurements in two grazed pastures in south east Queensland. Trop. Grassl. 16, 180-5.

Burden, R.J. (1982). Nitrate contamination of New Zealand aquifers: A review. N.Z. J. Sci. 25, 205-20.

Buringh, P. (1985). The land resource for agriculture. Phil. Trans. R. Soc. Lond. B310, 151-9.

Carran, R.A., Ball, P.R., Theobald, P.W., and Collins, M.E.G. (1982). Soil nitrogen balances in urine-affected areas under two moisture regimes in Southland. N.Z. J. Exp. Agric. 10, 377-81.

Catchpoole, V.R. (1984). Cultivating the surface soil to renovate a green panic (Panicum maximum) pasture on a brigalow clay soil in south east Queensland. Trop. Grassl. 18, 96-9.

Catchpoole, V.R., Harper, L.A., and Myers, R.J.K. (1981). Annual losses of ammonia from a grazed pasture fertilised with urea. Proc. XIV Int. Grassl. Congr., Lexington, U.S.A., 1981., pp. 344-7.

Catchpoole, V.R., and Henzell, E.F. (1975). Losses of nitrogen from pastures. C.S.I.R.O., Australia, Div. Trop. Agron. Ann. Rep. 1974-75, p. 82.

Catchpoole, V.R., Oxenham, D.J., and Harper, L.A. (1983). Transformation and recovery of urea applied to a grass pasture in south-eastern Queensland. Aust. J. Exp. Agric. Anim. Husb. 23, 80-6.

Chopping. G.D., Lowe, K.F., and Clarke, L. (1983). Irrigation systems. In 'Dairy Management in the '80s, Focus on Feeding, 1. Seminar sessions and Farm Feeding Workshops'. (Ed. L. Wishart), pp. 109-20. (Queensl. Dep. of Primary Ind.: Brisbane.)

Christie, E.K. (1979). Ecosystem processes in semi-arid grasslands. II Litter production, decomposition and nutrient dynamics. Aust. J. Agric. Res. 30, 29-42.

Cooper, J.P. (1970). Potential production and energy conversion in temperate and tropical grasses. Herb. Abst. 40, 1-15.

Cullen, N.A., and Steele, K.W. (1983). Nitrogen in New Zealand agriculture - A research viewpoint. Proc. Fert. Manuf. Res. Assoc. 8th, Auckland, N.Z., pp. 5-10.

Denmead, O.T. (1983). Micrometeorological methods for measuring gaseous losses of nitrogen in the field. In 'Gaseous Loss of Nitrogen from Plant-Soil Systems'. (Eds J.R. Freney and J.R. Simpson), pp. 133-57. (Martinus Nijhoff/Dr W. Junk: The Hague.)

Denmead, O.T., Freney, J.R., and Simpson, J.R. (1979). Studies on nitrous oxide emission from a grass sward. Soil Sci. Soc. Am. J. 43, 726-8.

Denmead, O.T., Simpson, J.R., and Freney, J.R. (1974). Ammonia flux into the atmosphere from a grazed pasture. Science 185, 609-10.

During, C. (1972). 'Fertilisers and Soils in New Zealand Farming'. 2nd ed. (N.Z. Govt Printer: Wellington.)

Eriksen, F.I., and Whitney, A.S. (1981). Effects of light intensity on growth of some tropical forage species. 1. Interactions of light intensity and nitrogen fertilization on six forage grasses. Agron. J. 73, 427-33.

Field, T.R.O., and Ball, P.R. (1982). Nitrogen balance in an intensively utilised dairy farm system. Proc. N.Z. Grassl. Assoc. 43, 64-9.

Filet, P.G., Whiteman, P.C., and Tothill, J.C. (1986). Nutrient cycling in three grazed Heteropogon contortus management systems. In 'Rangelands; a Resource under Siege'. (Ed P.J. Ross, P.W. Lynch and O.B. Williams), pp. 43-4. (Aust. Acad. Sci.: Canberra.)

Focht, D.D., and Verstraete, W. (1977). Biochemical ecology of nitrification and denitrification. In 'Advances in Microbiol Ecology'. 1. (Ed. M. Alexander), pp. 135-214. (Plenum Press: New York.)

Freney, J.R., Simpson, J.R., and Denmead, O.T. (1981). Ammonia volatilisation. In 'Terrestrial Nitrogen Cycles'. (Eds F.E. Clark and T Roswall), Ecological Bull. 33, pp. 291-302.

Galbally, I.E., and Roy, C.R. (1978). Loss of fixed nitrogen from soil by nitric oxide exhalation. Nature 275, 734-5.

Gillard, P. (1967). Coprophagous beetles in pasture ecosystems. J. Aust. Inst. Ag. Sci. 33, 30–4.

Gillingham, A.G., and During, C. (1973). Pasture production and transfer of fertility within a long–established hill pasture. N.Z. J. Exp. Agric. 1, 227–32.

Goreau, T.J., Kaplan. W.A., Wofsy, S.C., McElroy, M.B., Valois, F.W., and Watson, S.W. (1980). Production of NO_2^- and N_2O by denitrifying bacteria at reduced concentrations of oxygen. Appl. Environ. Microbiol. 40, 526–32.

Graham, T.W.G., Myers, R.J.K., Doran, J.W., Catchpoole, V.R., and Robbins, G.B. (1985). Pasture renovation: the effect of cultivation on the productivity and nitrogen cycling of a Buffel grass (Cenchrus ciliaris) pasture. Proc. XV Int. Grassl. Cong., Kyoto 1985, pp. 640–2.

Harper, L.A., Catchpoole, V.R., Davis, R., and Weier, K.L. (1983a). Ammonia volatilization: soil, plant, and microclimate effects on diurnal and seasonal fluctuations. Agron. J. 75, 212–18.

Harper, L.A., Catchpoole, V.R., and Vallis, I. (1983b). Ammonia loss from fertilizer applied to tropical pastures. In 'Gaseous Loss of Nitrogen from Plant–Soil Systems'. (Eds J.R. Freney and J.R. Simpson), pp. 195–214. (Martinus Nijhoff/Dr W. Junk: The Hague.)

Henzell, E.F. (1972). Loss of nitrogen from a nitrogen–fertilized pasture. J. Aust. Inst. Agric. Sci. 38, 309–10.

Henzell, E.F., and Ross, P.J. (1973). The nitrogen cycle of pasture ecosystems. In 'Chemistry and Biochemistry of Herbage'. (Eds. G.W. Butler and R.W. Bailey). Vol. 2, pp. 227–45. (Academic Press: London.)

Hoglund, J.H., and Brock, J.L. (1987). Nitrogen fixation in managed grasslands. In 'Managed Grasslands'. Ecosystems of the World 17B. (Ed. R.W. Snaydon). pp. 187–96. (Elsevier: Amsterdam.)

Hoglund, J.H., Crush, J.R. Brock, J.L., Ball, R., and Carran, R.A. (1979). Nitrogen fixation in pasture XII. General discussion. N.Z. J. Exp. Agric. 7, 45–71.

Holt, J.A., and Easey, J.F.(1984). Biomass of mound–building termites in a red and yellow earth landscape, north Queensland. Proc. National Soils Conf. Brisbane, Australia. (Aust. Soc. Soil Sci. Inc: Australia), p. 363.

Hunt, W.F. (1983). Nitrogen cycling through senescent leaves and litter in swards of Ruanui and Nui ryegrass with high and low nitrogen inputs. N.Z. J. Agric. Res. 26, 461–71.

Irwin, J.A.G., Cameron, D.F., Davis, R.D., and Lenne, J.M. (1986). Anthracnose problems with Stylosanthes. In 'Proceedings of the Third Australian Conference on Tropical Pastures'. (Eds G.J. Murtagh and R.M. Jones), pp. 38–46. (Trop. Grassl. Soc. Aust.: Brisbane.)

Jones, R.J. (1982). The effect of rust (Uromyces appendiculatus) on the yield and digestibility of Macroptilium atropurpureum cv. Siratro. Trop. Grassl. 16, 130–5.

Krul, J.M., and Veeningen, R. (1977). The synthesis of the dissimulatory nitrate reductase under aerobic conditions in a number of denitrifying bacteria, isolated from activated sludge and drinking water. Water Res. 11, 39–43.

Lambert, M.G., Renton, S.W., and Grant, D.A. (1982). Nitrogen balance studies in some North Island hill pastures. In 'Nitrogen Balances in New Zealand Ecosystems'. (Ed. P.W. Gander), pp. 35–9. (DSIR : Palmerston North, N.Z.)

Lee (1983). The influence of earthworms and termites on soil nitrogen cycling. In 'New Trends in Soil Biology'. (Eds Ph. Lebrun, H.M. Andre, A. De Medts, C. Gregoire–Wibo and G. Wauthy), pp. 35–47. (Dieu–Brichart: Ottignies–Louvain–la–Neuve.)

Lenne, J.M., and Sonoda, R.M. (1982). Effect of anthracnose on yield of the tropical forage legume, Stylosanthes hamata. Phytopathology 72, 207–9.

Limmer, A.W., and Steele, K.W. (1983). Effect of cow urine upon denitrification. Soil Biol. Biochem. 15, 409–12.

Limmer, A.W., Steele, K.W., and Wilson, A.T. (1982). Direct field measurement of N_2 and N_2O evolution from soil. J. Soil Sci. 33, 499–507.

Lockyer, D.R. (1984). A system for the measurement in the field of losses of ammonia through volatilization. J. Sci. Food Agric. 35, 837–48.

Lowe, K.F., and Hamilton, B.A. (1986). Dairy pastures in the Australian tropics and subtropics. In 'Proceedings of the Third Australian Conference on Tropical Pastures'. (Eds G.J. Murtagh and R.M. Jones.), pp. 68–79. (Trop. Grassl. Soc. Aust.: Brisbane.)

Mannetje, L. 't, Cook, S.J., and Wildin, J.H. (1983). The effects of fire on a buffel grass and siratro pasture. Trop. Grassl. 17, 30–9.

McDiarmid, B.N., and Watkin, B.R. (1972). The cattle dung patch. 2. Effect of a dung patch on the chemical status of the soil and ammonia loss from the patch. J. Br. Grassl. Soc. 27, 43–8.

Moore, A.W. (1974). Availability to Rhodes grass (Chloris gayanus) of nitrogen in tops and roots added to soil. Soil Biol. Biochem. 6, 249–55.

Morris, D.R., Zuberer, D.A., and Weaver, R.W. (1985). Nitrogen fixation by intact grass–soil cores using $^{15}N_2$ and acetylene reduction. Soil Biol. Biochem. 17, 87–91.

Mott, J.J. (1982). Fire in improved pastures in northern Australia. Trop. Grassl. 16, 97–100.

Myers, R.J.K., and Henzell, E.F. (1985). Productivity and Economics of legume-based versus nitrogen-fertilised grass-based forage systems in Australia. In 'Forage Legumes for Energy Efficient Animal Production'. (Eds R.F. Barnes, P.R. Ball, R.W. Brougham, G.C. Martin, and D.J. Minson), pp. 40–6. (USDA: Springfield).

Nommik, H. (1956). Investigation on denitrification in soil. Acta. Agric. Scand. 6, 196–228.

O'Hara, G.W., Daniel, R.M., and Steele, K.W. (1983). Effect of oxygen on the synthesis, activity and breakdown of the rhizobium denitrification system. J. Gen. Microbiol. 129, 2405–12.

O'Hara, G.W., Daniel, R.M., Steele, K.W., and Bonish, P.M. (1984). Nitrogen losses from soils caused by rhizobial denitrification. Soil Biol. Biochem. 16, 429–31.

Picard, D. (1979). Evaluation of the organic matter supplied to the soil by the decay of the roots of an intensively managed Panicum maximum sward. Plant Soil 51, 491–501.

Quin, B.F. (1982). The influence of grazing animals on nitrogen balances. In 'Nitrogen Balances in New Zealand Ecosystems'. (Ed. P.W. Gander), pp. 95–102. (DSIR: Palmerston North, N.Z.)

Reece, P.H., and Campbell, B.L. (1986). The use of ^{137}Cs for determining soil erosion differences in a disturbed and non-disturbed semi-arid ecosystem. In 'Rangelands; a Resource under Siege'. (Eds P.J. Joss, P.W. Lynch and O.B. Williams.), pp. 294–5. (Aust. Acad. Sci.: Canberra.)

Ryden, J.C. (1981). N_2 exchange between a grassland soil and the atmosphere. Nature 296, 235–7.

Ryden, J.C., Ball, P.R., and Garwood, E.A. (1984). Nitrate leaching from grassland. Nature 311, 50–4.

Ryden, J.C., and McNeill, J.E. (1984). Application of the micrometeorological mass balance method to the determination of ammonia loss from a grazed sward. J. Sci. Food Agric. 35, 1297–310.

Sarathchandra, S.U., Boase, M.R., Perrott, K.W., and Waller, J.E. (1987). Seasonal changes and the effects of fertilisers on chemical and biochemical characteristics of high producing pastoral soil. II. Nitrogen. N.Z. J. Agric. Res. (In Press).

Sarathchandra, S.U., and Perrott, K.W. (1984). Activities and nutrient contents of the microbial biomass of a yellow brown loam. N. Z. MAF Agric. Res. Div. Ann. Rep., pp. 118–9.

Sherlock, R.R., and Goh, K.M. (1984). Dynamics of ammonia volatilisation from simulated urine patches and aqueous urea applied to pasture. 1. Field experiments. Fert. Res. 5, 181–97.

Simpson, J.R., and Steele, K.W. (1983). Gaseous N exchanges in grazed pastures. In 'Gaseous Loss of Nitrogen from Plant-Soil Systems'. (Eds. J.R. Freney and J.R. Simpson), pp. 215–36. (Martinus Nijhoff/Dr W. Junk: The Hague.)

Smith, M.S., and Tiedje, J.M. (1979). The effect of roots on soil denitrification. Soil Sci. Soc. Am. J. 43, 951–5.

Steele, K.W. (1982a). Nitrogen fixation for pastoral agriculture – biological or industrial? N.Z. Agric. Sci. 16, 118–21.

Steele, K.W. (1982b). Nitrogen in grassland soils. In 'Nitrogen Fertilisers in New Zealand Agriculture'. (Ed. P.B. Lynch), pp. 29–44. (Roy Richards: Auckland, N.Z.)

Steele, K.W. (1987). Nitrogen Losses. In 'Managed Grasslands'. (Ed. R.W. Snaydon), pp. 197–204. (Elsevier: Amsterdam). pp. 197–204.

Steele, K.W., and Brock, J.L. (1985). Nitrogen cycling in 'legume'-based forage production systems in New Zealand. In 'Forage Legumes for Energy Efficient Animal Production'. (Eds R.F. Barnes, P.R. Ball, R.W. Brougham, G.C. Martin, and D.J. Minson), pp. 171–6. (USDA: Springfield.)

Steele, K.W., Judd, M.J., and Shannon, P.W. (1984). Leaching of nitrate and other nutrients from a grazed pasture. N.Z. J. Agric. Res. 27, 5–11.

Steele, K.W., Watson, R.N., Bonish, P.M., Littler, R.A., and Yeates, G.W. (1985). Effect of invertebrates on nitrogen fixation in temperate pastures. Proc. XV Int. Grassl. Cong., Kyoto 1985, pp. 450–1.

Stevenson, F.J. (1982). Origin and distribution of nitrogen in soil. In 'Nitrogen in Agricultural Soils'. (Ed. F.J. Stevenson) Agronomy No.22, pp 1–42.

Vallis, I. (1983). Uptake by grass and transfer to soil of nitrogen from ^{15}N-labelled legume materials applied to a Rhodes grass pasture. Aust. J. Agric. Res. 34, 367–76.

Vallis, I. (1985). Nitrogen cycling in legume-based forage production systems in Australia. In 'Forage Legumes for Energy Efficient Animal Production'. (Eds R.F. Barnes, P.R. Ball, R.W. Broughan, G.C. Martin, and D.J. Minson), pp. 40–6. (USDA: Springfield.)

Vallis, I., and Gardener, C.J. (1984a). Nitrogen inputs into agricultural systems by Stylosanthes. In 'The Biology and Agronomy of Stylosanthes'. (Eds H.M. Stace and L.A. Edye), pp. 359–79. (Academic Press: Sydney.)

Vallis, I., and Gardener, C.J. (1984b). Short-term nitrogen balance in urine-treated areas of pasture on a yellow earth in the subhumid tropics of Queesnland. Aust. J. Exp. Agric. Anim. Husb. 24, 522–8.

Vallis, I., Harper, L.A., Catchpoole, V.R., and Weier, K.L. (1982). Volatilization of ammonia from urine patches in a subtropical pasture. Aust. J. Agric. Res. 33, 97–107.

Vallis, I., Peake, D.C.I., Jones, R.K., and McCowen, R.L. (1985). Fate of urea nitrogen from cattle urine in a pasture–crop sequence in a seasonally dry tropical environment. Aust. J. Agric. Res. 36, 809–17.

Webster, C.P., and Dowdell, R.J. (1982). Nitrous oxide emission from permanent grass swards. J. Sci. Fd. Agric. 33, 227–30.

Weier, K.L. (1980). Nitrogenase activity associated with three tropical grasses growing in undisturbed soil cores. Soil Biol. Biochem. 12, 131–6.

Weier, K.L, and Doran, J.W. (1986). The potential for denitrification in a brigalow clay soil under different management systems. In 'Symposium on Nitrogen Cycling in Agricultural Systems of Temperate Australia'. pp. 327–39. (Aust. Soc. Soil Sci., Riverina Branch: Wagga Wagga, Australia).

Weier, K.L., MacRae, I.C., and Whittle, J. (1981). Seasonal variation in the nitrogenase activity of a Panicum maximum var. trichoglume pasture and identification of associated bacteria. Plant Soil 63, 189–97.

Wetselaar, R., and Farquhar, G.D. 1980. Nitrogen losses from tops of plants. Adv. Agron. 33, 263–302.

Wetselaar, R., and Hutton, J. (1963). The ionic composition of rain water at Katherine, N.T., and its part in the cycling of plant nutrients. Aust. J. Agric. Res. 14, 319–29.

Whitehead, D.C. (1970). 'The Role of Nitrogen in Grassland Productivity'. (Commonwealth Agric. Bur: Farnham Royal, U.K.)

Wilson, J.R., Catchpoole, V.R., and Weier, K.L. (1986). Stimulation of growth and nitrogen uptake by shading a rundown green panic pasture on brigalow clay soil. Trop. Grassl. 20, 134–43.

Wong, C.C., and Wilson, J.R. (1980). The effect of shading on the growth and nitrogen content of green panic and Siratro in pure and mixed swards defoliated at two frequencies. Aust. J. Agric. Res. 31, 269–85.

Woodmansee, R.J. (1979). Factors influencing input and output of nitrogen in grasslands. In 'Perspectives in Grassland Ecology'. (Ed. N. French), pp. 117–34. (Springer-Verlag: Heidelberg.)

Woodmansee, R.G., Vallis, I., and Mott, J.J. (1981). Grassland nitrogen. In 'Terrestial Nitrogen Cycles: Processes, Ecosystem Strategies and Management Impacts'. (Eds F.E. Clark and T. Rosswall.), pp. 443–62. (Swedish Natural Sci. Res. Counc.: Stockholm.)

NITROGEN SUPPLY TO CEREALS IN LEGUME LEY SYSTEMS UNDER PRESSURE

R.L. McCown, A.L. Cogle, A.P. Ockwell, and T.G. Reeves

ABSTRACT

Since the late 1940's the sheep-wheat farming system of southern Australia has relied heavily on annual clover and medic leys fertilized with superphosphate for supply of both forage and soil N. Despite the world-renowned success of this technology, during the 1970's emphasis in these mixed farming systems swung to continuous crop production, with greater usage of N fertilizers. Legume leys declined in area and productivity as inputs were redirected.

Why did farmers move away from an 'ideal' farming system? This paper attempts to answer this question and examines both the economic and ecological consequences of the reduction in use of leys. Emphasis is placed on the mineral N supply, and a simple model is examined for predicting N supply for variable numbers of successive crops as the length of the pasture ley is shortened. The effectiveness of grain legumes as an alternative source of N for cereals is also examined.

An hypothesis is proposed that in their move away from pasture legume leys, farmers were behaving similarly to farmers in the earlier period of rapid pasture expansion, i.e., adapting to changing economic circumstances. Using what is known about the abilities of leys and grain legumes for supplying N for cereals, and the cost/price relativities for the periods of rapid expansion and that of rapid contraction of leys, an economic model (MIDAS) is used to estimate the economically-optimum investment in leys for each period. The results indicate that farmers in both periods were generally making the correct economic decisions.

Combined uncertainties of climate and export markets ensure the persistence of a mixed farming system in this zone, and one in which legume leys will remain an important feature. However their importance will vary as farmers adapt to changed pressures. The long previous period of relative stability seems to have led some agricultural scientists to interpret the phenomenon of the 1970's as the beginning of a reversal of a classic achievement rather than a normal swing of the pendulum (which since 1986 has swung strongly in the other direction).

Lessons from analysis of these historical trends should affect research agendas and assist in appraising prospects for ley farming in other regions of the world.

INTRODUCTION

Ley farming is a form of European mixed farming which features dependence on periods under pasture containing legumes to restore soil productivity between phases of cropping. It is conceptually an intensification of a crop-fallow system (Ruthenberg 1980) achieved by sowing, fertilizing, and managing fallows for greater production of forage and increased rates of soil improvement.

In Australia, the early awareness of the potential benefit of a pasture legume by agricultural scientists (e.g. Farrar 1893 cited by Donald 1965) contributed to the rapid development of a pasture improvement technology, once the presence of a well adapted legume was discovered. Ley farming spread rapidly in the wheat-sheep zone of southern Australia following World War II (Davidson 1981).

Early research in both Britain and Australia compared various ley rotations and continuous crops or pasture. Although leys improved both soil N fertility and soil structure, it was not clear until the mid-1960's that the main crop yield benefits in British ley farming were due to improved N fertility and that the legume was the key to the technology (Cooke 1967). In Australia, the principle role of the legume seems never to have been questioned and research during the last three decades has focused on improving the legume technology.

Recent research on ley farming has been concerned primarily with (1) improvement of the southern Australian Mediterranean system, (Halse and Wolfe 1985), (2) testing the Australian technology in the true Mediterranean region (Chatterton and Chatteron 1984; Springborg 1986), and (3) exploring the technical feasibility of tropical legume ley systems (McCown et al. 1985; Mohamed-Saleem and Otsyina 1986). Ironically, although progress in all three areas can be reported, the recent experience which provides perhaps the most important insight into ley farming and the importance of its N cycle in the future was the swing away from ley farming in southern Australia during the 1970's, and analysis of this phenomenon is the subject of this paper.

In the well-managed Australian ley system prior to this period, a pasture phase of sufficient duration and quality was maintained to (1) provide a supply of N that made periodic cropping profitable, and (2) adequately restore soil structure between crop phases. A major research objective was to identify the most crop-intensive ley rotation consistent with maintenance of soil chemical and physical fertility (e.g. Russell 1960). Recent economic pressures on the ley system raise new research issues; i.e. (1) quantification of the effects of reductions in leys on crop mineral N supply and physical stability and (2) systems of management which compensate for these effects. In this paper we examine the influence of ley length and productivity on soil N, the implications of substitution of grain legumes for pasture leys, the economics of N inputs from leys, and finally, the implications of reduction of leys on ecological stability.

CONSEQUENCES OF REDUCED LEY PRESENCE AND PRODUCTIVITY IN CEREAL ROTATIONS

Effect on soil N accumulation

Although our concern is mainly with mineral N, there are several reasons to begin with effects on soil organic N: organic N level is a determinant of mineral N supply rate, much more is understood about the effects of leys on organic N than on mineral N, and the relationship is relatively simple.

Appreciation of the implications of reducing ley length implies knowledge of the time efficiency of leys in increasing soil organic N levels. Generalizations about absolute rates of soil N accumulation under pasture leys are not very satisfying because accumulation is the net result of accretion and losses, and each is influenced by numerous environmental, nutritional and managerial factors (Clarke and Russell 1977). These authors found the interquartile range of annual increments of soil N accretion from 17 studies to be 39–93 kg/ha/yr.

Fortunately, understanding of the effects on total soil N of reducing leys is more dependent on the _relative_ effects of ley length, and generalization here is more rewarding. Where soil organic N levels are well below the equilibrium levels achieved under continuous pasture, accumulation of organic N progresses at rates that are nearly constant for the first few years; rates diminish as the equilibrium level is approached (Watson 1963; Anon. 1969). Short-term accumulation of soil N is approximately proportional to the length of ley (Mullaly et al. 1967; Barrow 1969; White et al. 1978; Rowland et al. 1980; Holford 1981). This accumulation is highly correlated with cumulative legume herbage production (Watson 1963; Kleinig et al. 1974). The latter relationship is apparently indirect via production and turnover of roots (Watson and Lapins 1964).

The most frequent deviation from the above relationship is lower-than-average N accumulation in the year of pasture establishment due to insufficient numbers of seedlings (Clarke and Russell 1977). While Donald

(1951) found that <u>c</u>. 1500 seedlings/m^2 were sufficient to achieve maximum production of subterranean clover in pure stands, response to much higher densities occurred in the more usual ley pasture situation where grass competition is an important constraint to legume production (Watson <u>et</u> <u>al</u>. 1976). To save costs of sowing the large quantities of seed needed to achieve high populations, reliance is placed on a high degree of hardseededness. Despite considerable progress through genetic selection for increased degrees of hard seed, the problem continues to be cited as a point of serious vulnerability in the annual legume ley system (Carter <u>et</u> <u>al</u>. 1982; Taylor 1985).

    Russell (1981) used data from long-term rotation studies to provide generalized estimates of the even longer-term consequences on soil N of reducing leys. Figure 1 shows the effect of cropping intensity on the equilibrium soil organic N level. All rotations without bare fallows in which the pasture was present in at least three in ten years maintained or increased soil N, whereas when fallow preceded wheat, at least six in ten years of pasture was required. Reduction in equilibrium organic N levels as leys are reduced in rotations is linear over most of the range after a more abrupt decline due to the interruption of continuous pasture with any crop at all.

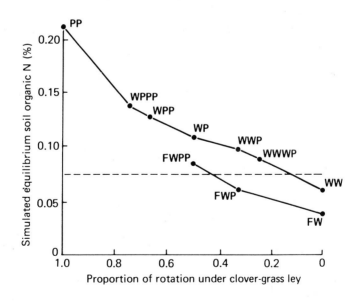

Fig. 1: The relationship between proportion of rotation under a clover-grass ley and the equilibrium soil organic N level (values after 100 years from a starting value of 0.1%) (P = Pasture, W = Wheat, F = Bare Fallow; dashed line approximates the organic N level below which soil structural decline becomes problematic. (Data from Russell 1980).

295

Effect of reducing leys on supply of mineral N to crops

Productive legume leys of modest duration can provide most or all of the mineral N requirements of the first crop and a declining proportion of the requirements of subsequent crops (e.g. Wells 1970; Osborne et al. 1977; Watson et al. 1977; Holford 1980). However, the degree to which the needs of both the first and subsequent crops are met is highly variable. Soil type, botanical composition, the amount of legume and its chemical composition, and management of both soil and herbage all strongly influence the availabilty of legume N from the ley phase to the crop phase (Ladd et al. 1986).

Because of the more dynamic nature of mineral N balance, its characterization requires much more frequent monitoring and sampling of much greater depth of soil profile than organic N. The difficulties and expense of direct measurement of mineral N have resulted in reliance on indirect measures. One of these is N uptake by a nitrophilous crop. This has the merit of integration in time and through the soil profile, but is reduced in value as a bioassay for N when any other factor is limiting, e.g. water (Papastylianou et al. 1981; Strong et al. 1986). While both mineral N balance and test-crop approaches have been used increasingly in recent research on grain legumes (see review by Herridge 1986), little such information exists for pasture leys.

The more prevalent indirect method with pastures has been to relate the amount of a readily mineralizeable N fraction of soil organic matter/plant residues to subsequent mineral N supply (Greenland 1971). The limited value of this approach by 1977 is inferred by Clarke and Russell in the observation that the effects of management and other factors on this fraction "have not yet been elucidated".

More recent research, however, has greatly improved understanding of how a pasture ley influences the mineral N supply via decomposition of residues (summarized by Ladd and Amato 1986). Typical results for mineralization of ^{15}N-labelled medic residues, incorporated in a red brown earth on one occasion, are shown in Figure 2. Although about 40% of residue N mineralized in the first year, it took another seven years to mineralize a further 25%. As yet, no attempt has been made to study the effects of multiple "generations" of legume residues, as would occur under several years of pasture. In the absence of such data, serially combining the decomposition/mineralization function of Figure 2 may serve the purpose of exploring the quantitative effects of reduction in ley duration on N supply to crops in the rotation. Table 1 shows the mineral N available to four crops from two years of an annual legume pasture. It is assumed that each generation of pasture contributes 80 kg N in residues, and that this mineralizes according to Figure 2. For the first crop, 39 kg N is available, 9 from the residues of the first year of pasture, and 30 from the second (most recent) year. The model is expanded to four years pasture followed by four years cereal crop in Figure 3. Two features stand out. Of the mineral N supplied by four years of pasture,

296

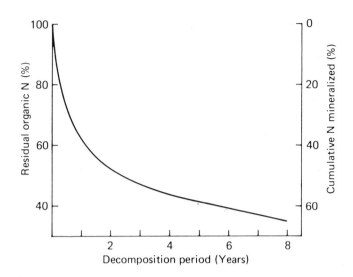

Fig. 2: The decline with time in organic N of soil—incorporated annual legume residues. (Redrawn from Ladd and Amato 1985).

Table 1: Annual increments of mineral N (kg) for four crops (C) from 80 kg organic N in residues of each of two generations of legume pasture (P).

Year:	1	2	3	4	5	6
Rotation:	P	P	C	C	C	C
		30	9	3	2	1
			30	9	3	2
Aggregate Mineral N			39	12	5	3

the most recent pasture legume generation contributes the major share, with contributions from previous generations declining exponentially. Supply to successive crops declines according to the same function. Comparison of the marginal benefits of changing length of ley is aided by presentation as in Figure 4a.

What evidence is there that these simulated patterns of supply of mineral N conform to those from real pasture leys? Holford (1980) reported the effects of three durations of grazed lucerne leys on mineral N supply to four subsequent wheat crops on two soils in northern New South Wales. Uptake of legume N (estimated by subtracting N uptake following continuous wheat from uptake following lucerne leys) in four successive wheat crops on a red brown earth is shown in Figure 4b. The patterns of mineral N uptake are similar to those simulated (Fig. 4a), although the levels were considerably higher.

The marginal benefits in N supply by two additional years of pasture or in N harvest by two additional crops are small (Figs. 4a,b). From the

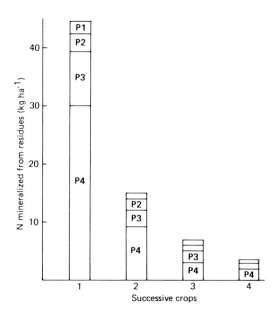

Fig. 3: Simulated mineral N available from annual medic residues for four successive crops. Mineral N from each of four successive years of pasture (P1 – P4) taken from Fig. 2 and combined as per Table 1.

standpoint of N supply for crops, it is clear that frequent short leys offer efficiencies in mineral N supply. However, such benefits of shortening rotations must be weighed against effects on costs of more frequent transition from pasture to crop and back.

There are remarkably few reports of studies which have sufficient duration of pasture leys and crops to describe the N/ley duration function. The results of Rowland et al. (1980) are very similar to Figure 4. In the study of Watson (1963), long leys had a relatively greater benefit than those in Figure 4 but the rapid decline in marginal benefit with leys longer than 1 or 2 years suggested the notion of the "mini–ley" rotation of one or two year pasture – one year crop (Anon. 1969).

In contrast to the results in Figure 3, research on ley rotations at Rutherglen has found the rate of decline in crop yields following long leys is less than after short leys (Bath 1949; Mullaly et al. 1965; Ellington et al. 1979). Since the simple residue mineralization model above does not take into account the effect of legume residues on mineral N supply due to longer term increases in soil organic N or the effect of an associated grass component, reasons for discrepancies regarding long leys are logically sought here. Transitions between residue, "light fraction" (Greenland 1971), and more stable organic matter are gradual, and distinctions not always clear. However, Figure 2 indicates that in year three, about 50% of N from the residue was still in the soil and was declining at about 2.5% per year. A soil with 2000

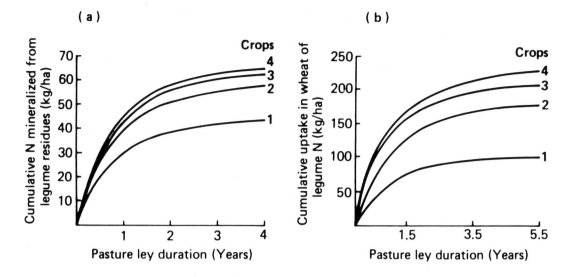

Fig. 4: Cumulative mineral N supply to four successive crops following ley pastures of various duration, (a) soil N mineralized from annual medic residues (plotted from Fig. 3), and (b) uptake by wheat following lucerne ley minus uptake of wheat following wheat. (Data from Holford 1980).

kg of such relatively stable organic N, would supply 50 kg mineral N to a crop. In relation to N from residue mineralization this is a significant amount. However, as far as short-term management is concerned, mineral N from this source would be expected to be less sensitive to changes in the rotation than is that from more recent residues. Further information on the organic N pool dynamics is needed to resolve the discrepancies regarding mineral N contribution by long leys.

Means by which grass in the pasture ley might have an effect is discussed in the next section.

Effects of reduced inputs to leys

Another means of responding to economic pressure to shift resources toward the cropping enterprise is to "neglect" pastures. Reduction of inputs, e.g. fertilizer or management, reduces legume yield and results in proportional reductions in soil N acccumulation (White et al. 1978). Reduced quality of leys, due to lower percentage of legume or lower N concentrations in herbage may be a greater threat to mineral N supply in the cropping phase than reduction in ley length. Ladd et al. (1986) found that within the limits normally experienced in South Australian medic leys, the amount of legume N input did not influence the proportions of N available in the cropping phase. Thus the relationship between amount of legume N input in any given season and later supply of mineral N to crops is linear, and marginal reductions in legume yield are of greater consequence than reduction in ley length.

The problem of reduced legume yield is exacerbated by any substantial presence of grass. Not only does grass competition reduce N_2 fixation and legume production, but pasture residues containing grass have higher C/N and thus a lower rate of mineralization (Ladd and Amato 1985). A further grass-related factor is the soil mineral N "sparing" effect of legumes. Substantially more mineral N is found under pasture legumes than under gramineous crops or pasture, and this "sparing" effect can account for a large proportion of the benefit of the legume to the following cereal crop (Ladd and Amato 1985). Whereas it seems probable that spared N is an important part of the benefit conferred by (weed-free) grain legumes, (see next section) there seems little doubt that the presence of even modest quantities of grass in a pasture would largely negate this component of mineral N supply.

The advent of herbicides which can selectively kill grasses in pasture has prompted research on early winter spraying (Poole 1987). While motivated mainly by the need to control cereal disease hosts and to reduce grass weeds in the following crop, potential benefits to the supply of mineral N are apparent.

GRAIN LEGUMES AS AN ALTERNATIVE SOURCE OF N IN CEREAL ROTATIONS

Grain legumes, mainly field peas, have long been a minor component of the mixed farming system of southern Australia (Wood and Russell 1979). During the 1970's there was a rapid increase in the use of lupins in wheat rotations, made possible by the release of well adapted, low-toxicity cultivars (Halse and Wolfe 1985). This trend plus increased areas of field peas and interest in new grain legumes (e.g. chickpea, faba bean) were spurred by the high returns from crop production during this period and corresponded to the decline in legume pastures in cereal rotations.

In addition to producing grain, these crops benefit mixed livestock-cropping systems in ways similar to pasture leys: they increase yield of subsequent cereal crops and supply a relatively high quality fodder during the dry season. In the papers reviewed by Herridge (1986), cereal yields following grain legumes were 50-80% higher than following cereals. To understand the long term implications of substituting grain legumes for pasture legumes in rotations it is important to understand the differences in their N cycles.

The potential for grain legumes to contribute N to a cereal rotation is greatly diminished by removal of a large proportion of the crop N in grain. This varies among species, but can be as much as 90% (Herridge 1986). In a mixed farming system, where legume haulms provide a valuable dry season fodder, the potential for N contribution to the soil is further reduced. For example, in a study by Papastylianou et al. (1981), of the 120 kg/ha N yield of faba beans, 70 kg N was removed in grain and another 30 kg in haulms, leaving only 20 kg/ha N in legume residues. Assuming that this is later incorporated and that N availability is governed by a mineralization function similar to Figure 2, of this 20 kg N, only about 8 kg N is available to the first crop. Nevertheless, reported benefits of grain legumes in rotations are remarkably large. In the same study, cereal yields following faba beans were no less than when following high–yielding, one–year swards of medic or subterranean clover. All were significantly higher than when following oats. In the next year, cereal yields on the bean treatment were intermediate to those on forage legume and oat treatments. Reeves (1984) found soil mineral N and subsequent wheat yields moderately higher following lupins than following subterranean clover. In the studies reviewed by Herridge (1986), grain legumes provided a benefit to a succeeding cereal crop equivalent to 50–100 kg fertilizer N.

In view of the much lower return of residual N from grain legumes than pasture legumes, how can grain legumes confer such large benefits? In the first place, it now seems clear that most of the soil mineral N immediately following a grain legume crop (in excess of that following a cereal crop) is due to the "sparing" of soil nitrate by the N_2–fixing crop. The average amount of spared nitrate in the six studies cited by Herridge (1986) was 30 kg/ha. Of two studies by Doughton and MacKenzie (1984) where haulms were incorporated immediately following grain harvest, spared nitrate made up a third of the available N following a sufficient period for residue decomposition.

Secondly, legumes in general appear to confer benefits disproportionate to N contribution or sparing. Even with fertilizer N non–limiting the maximum yield obtainable by a cereal following a legume almost always exceeds that following a cereal (De et al. 1983; Rowland et al. 1986). No general explanation can be offered for this phenomenon, but plausible, and intermittently demonstrable, explanations include (1) the legume's effectiveness in breaking cereal disease and pest cycles, (2) the reduction of phytotoxic and allopathic problems, (3) improvement of soil structure, and (4) favourable positioning of N in the soil profile. An example of the last is reported by Strong et al. (1986); they found a greater synchronization and spatial coincidence of nitrate, soil water, and crop roots at crucial crop stages in certain grain legume treatments which had more deeply–positioned legume N. Although deep positioning may not always be an advantage, inferior mineral N efficacy will often be traced to poor coordination in time and space of supply and demand (R.J.K. Myers, in these proceedings):

The generalization that high–yielding grain legumes have little opportunity to cause a net increase to soil N does not necessarily mean that this cannot occur. Determination of the effect of a crop or cropping system on soil organic N requires inclusion not only of the amount of N returned by the crop(s) (R), but the organic N level (N) of the soil as well. This can be expressed as a differential equation (simplified from Clarke and Russell 1977) as:

$$\frac{dN}{dt} = -kN + R \tag{1}$$

where k is the mineralization coefficient. Even though R from grain legumes is normally low, and consequently dN/dt is most often reported as negative, if N and/or k is small enough, dN/dt can be positive. In addition R differs widely between legumes and management practices (Russell 1987).

The Waite Institute's permanent rotation experiment provides a unique opportunity to see the long term effects of a grain legume–cereal rotation on soil N. Most of the soil results of this experiment have been published (Greenland 1971; Clarke and Russell 1977). However those from the only rotation plot which featured a grain legume have not because of concern about an initial total soil N value being lower than most of the experiment (0.138 vs. 0.180%) (J.S. Russell, personal communication). It appears that a valid comparison can be made between this peas–wheat plot and the neighbouring bare fallow plot which had an initial N value of 0.128%. Over a period of 48 years 53% of the organic N in the top 23 cm was lost under bare fallow, while under a peas–wheat rotation 43% was lost. When this effect of peas in retarding decline in organic N is compared with the relative difference between wheat-pasture (WP) and FW (R=0) in Figure 1, it is clear that R from a pea crop was very much less than from a year of pasture. Recent studies indicate that other grain legume species, e.g. lathyrus and lupins, have the potential to contribute more residue N than field peas (Reeves 1984; Strong et al. 1986).

A case of a grain legume increasing N in the above equation is reported by Rowland et al. (1986). This occurred under lupins growing in Western Australia on sandy soils averaging 0.034% total N (in contrast to the 0.138% N in the peas–wheat rotation in the Waite Institute experiment).

Much of what we know about the N cycle of grain legumes has emerged in the past decade. Incomplete as this is, this understanding appears to exceed that of the N cycle of pasture legumes in many respects, in spite of more research over a much longer period on the latter. A few strategically located studies which compare the N cycle of pasture legumes and grain legumes in cereal rotations would greatly clarify differences that might be important over the long term.

THE ECONOMICS OF LEY NITROGEN

The economic origins of the ley system

So far we have considered the farm as an ecosystem within which N cycles. However, interpretation of farmer management of this ecosystem is aided by recognition that the farm system is designed and managed primarily as an economic system and secondarily as an ecosystem. Changes in management are made primarily in response to changed economic circumstances. Agricultural scientists tend to view the rapid, widespread adoption of ley pastures in southern Australia following World War II as a milestone in the evolution of the ideal agroecosystem for this environment. The fact that the motivation to sow pastures was largely economic is often overlooked (Vere and Muir 1986). Improved ecology was an important and fortunate side-effect. The investment returns from sown legume pastures in the 1950's greatly exceeded those from any other form of farm development (Gruen 1959). Davidson (1981) cites Cook and Maleky (1974) that returns on capital investment in pastures in the 1960's were as high as 19%, and raises the question of why larger areas weren't sown.

The primary benefits of sown legume pastures were increased animal carrying capacity and control of skeletonweed (<u>Chondrilla juncea</u>) (Pratley and Rowell 1987). This in no way detracts from their contributions to increased soil N and cereal yields (Donald 1965), and soil conservation (Davidson 1981). The duration of cropping phases in rotations were determined largely by the rate of yield decline following a ley phase; high costs of N fertilizer in Australia deterred extension of the cropping phase by applying N (Donald 1965). During the 1960's this farming system became reknowned world-wide as a rare example of a system that was both economically efficient and ecologically stable.

During the 1970's, there was a trend away from the ley system to more continuous cropping with greater use of N fertilizers. According to Reeves <u>et al</u>. (1984), the area sown to improved pasture increased steadily until 1971, after which pasture area declined. In contrast, the area sown to crops continued to increase, with wheat acreage reaching a record area in 1983-4. Were farmers becoming less responsible managers of their ecosystems, as suggested by some (e.g. Carter <u>et al</u>. 1982; Lloyd-Davies and Myers 1985)? Or were they only continuing to be rational economic managers perceptive of their changing economic environment, as suggested by others (e.g. Davies 1974; Vere and Muir 1986).

The remainder of this section focuses on the economic environments during periods of marked shift in the importance of ley pastures and the N from them. The primary comparison is between the period 1956-60 (rapid expansion of leys)and 1977-81 (expansion of cropping at expense of leys). However a second comparison is made between 1977-81 and 1986 when a marked swing away from cropping back to sheep and leys had begun.

Changes in input cost-product price relativities.

From their study of pasture improvement adoption on the Central and Southern Tablelands of New South Wales, Vere and Muir (1986) concluded that movements in product prices and factor costs were the principal influences accounting for farmers' decisions on areas of ley pastures. Indicators of these influences for Australia generally between the periods of 1956-60, 1977-81 and 1986 are shown in Table 2. The product price ratios (Table 2b) indicate a marked change in price relativities away from the pastoral–based enterprises toward wheat over the period from 1956–60 to 1977–81, and a reversal of that movement between 1977–81 and 1986. Between 1956–60 and 1977–81, the price of urea, needed only for crops, declined relative to the price of wheat, whereas, the price of superphosphate, needed for pastures as well as

Table 2: Product prices, factor costs and changes in factor–product and product–product price relativities over three periods: 1956–60, 1977–81 and 1986.

(a) Unit values of key product and factor variables (in 1986 Australian dollars)

	Unit	1956–60	1977–81	1986
Wheat	$/t	48.00	122.00	115.00
Wool	$/kg	1.14	2.23	3.65
Lamb	$/kg	0.41	1.14	1.20
Urea	$/t	142.50	190.00	280.00
Superphosphate	$/t	12.00	48.00	112.00

(b) Product–product and factor–product price ratios and percentage changes in price relativities.

Ratio	1956–60 (a)	1977–81 (b)	1986 (c)	% (b)/(a)	% (c)/(b)
Wheat/wool	42.11	54.71	31.51	+30	−42
Wheat/lamb	177.07	107.01	95.83	−40	−10
Urea/wheat	2.97	1.56	2.43	−47	+56
Urea/wool	125.00	85.20	76.71	−31	−10
Super/wheat	0.25	0.39	0.97	+57	+150
Super/wool	10.53	21.52	30.65	+104	+43

(c) Alternative cost–price scenarios for estimating the response by farmers to changing economic conditions.

	Unit	Scenario A (Based on 1956–60)	Scenario B (Based on 1977–81)	1986
Wheat	$/t	115.00	115.00	115.00
Wool	$/kg	2.73	2.10	3.65
Lamb	$/kg	0.98	1.07	1.20
Urea	$/t	341.00	179.00	280.00
Superphosphate	$/t	28.70	45.00	112.00

crops, increased relative to the price of wool. However, the increase in the price of superphosphate relative to wool was much greater than the change in relativity between superphosphate and wheat.

It is possible to gain insight to the implications of the differences between the three partial cost–price data sets in Table 2 by deriving for each the mix of crops and pasture leys that is economically optimum. Essential for deriving the optimum crop–pasture mix is a means of comparing the costs and returns from alternative sources of N, and thus we must first examine some of the complexities in estimating the economic contribution of ley N in relation to fertilizer N.

Problems in estimating the economic value of legume N in ley rotations

Functions like those in Figure 4 provide the sort of information required for quantifying mineral N supply from leys, but there is a paucity of specific research information for any given pasture ley/soil situation.

Given the best ley N supply function available and assuming some inputs of N fertilizer, predicting crop yield is complicated by the differential efficacy of biological N compared to fertilizer N. A pragmatic approach has been to lump the ill–defined benefits which are not accounted for by field bio–assays for available N into a 'yield–boost' term (e.g. Pannell and Falconer 1987). The benefit of the legume crop (or ley) to a subsequent cereal crop is expressed as a quantity of fertilizer equivalent N plus a bonus of cereal grain yield.

Problems in valuing biological N have been discussed by Pannell and Falconer (1987). Common to some of these problems is the complexity imposed by the central importance of intermediate products. The pasture ley is both a source of feed for livestock and a source of N for crops. Because neither herbage nor soil N is sold, neither has a market price. In an economic analysis of this system, monetary values of these intermediate products are estimated using both costs of their production and returns from their use as inputs in producing saleable products.

Two concepts which are also crucial to understanding the analysis of the economics of the N cycle are (the law of) diminishing marginal physical returns and opportunity cost. In figure 5a, gross returns increase at a decreasing rate in response to successive increases in the level of the variable input, N. In contrast, since the cost of an additional unit of N is the price of fertilizer N, then marginal cost is constant over the range of the variable input. Profit maximisation occurs at the point where marginal revenue equals marginal cost, i.e. at that point, the additional return realised from a one–unit change in output is equal to the additional cost required to produce that output.

Under economic conditions such as occurred in the 1956–60 period (Table

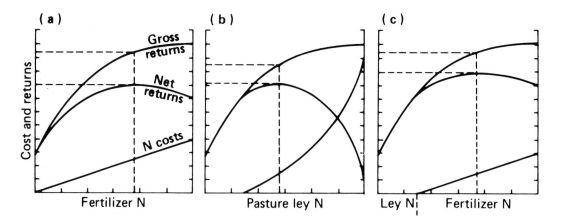

Fig. 5: The economics of mineral N supply from fertilizer and legume leys, (a) fertilizer N with constant unit cost, (b) N from pasture leys with marginal unit costs increasing, and (c) combination of (a) and (b). Product price is a constant. Dashed lines indicate the N rate, yield level, and net returns at the economic optimum (maximum net returns).

2), cropping might be viewed as an opportunistic exploitation of "free" N. However under the cost-price conditions of 1977-81 leys could be an expensive source of N for crops. Leslie (1984) suggested that the cost of ley N might be estimated as the opportunity cost of growing pasture leys when net returns from cropping exceed those from grazing. Fertilizer N has an opportunity cost when returns would have been greater from an alternative use of the capital used to purchase it. Opportunity costs result in both cases from departures from the optimum allocation of resources.

The added complexity when mineral N is supplied by ley pastures rather than by fertilizer is illustrated in Figure 5b. The unit cost of N increases due to rising opportunity costs of keeping land under pasture to accrue N when returns to cropping are high relative to grazing. The income foregone by growing fewer crops exceeds the value of the additional mineral N from the longer pasture leys. The problem is greater with long uninterrupted leys than with frequent short ones (Fig. 4). Nitrogen from leys provides higher net returns than fertilizer at low N levels, but has less scope for increasing net returns by increasing crop output (in response to increase in price) by increasing N inputs. The optimal strategy would probably be some combination of short leys and fertilizer N as illustrated in Figure 5c.

Simulations of optimum mixes of crops and pasture.

In interpreting a move by farmers away from an ecologically stable and sustainable system of management, as occurred during the period of 1977-81, it is useful to have an independent, objective appraisal of what a farmer-entrepreneur <u>should</u> do under such circumstances.

MIDAS (Model of an Integrated Dryland Agricultural System — Morrison et al. 1986) provides a useful tool for analysing crop-pasture rotations with emphasis on N technology (Pannell and Falconer 1987). The model is normative and thus estimates the optimal way in which resources should be combined to maximise incomes. (This approach contrasts with the econometric model of Vere and Muir (1986) which attempts to account for factors influencing what farmers actually did).

The strength of MIDAS lies in its ability to deal with many factors and their interactions, including the complexities discussed earlier. Solution of the single-year equilibrium model is based on a representative farm for average seasonal conditions. The model has been developed to represent farms in the eastern wheatbelt of Western Australia, using weather data from Merridan and is economically specified for 1986 (Falconer 1987). For our purposes here, the average annual rainfall was increased to 500 mm (cf. 310 mm for Merridan) in order to represent the more productive portions of the wheat-sheep zone, and one of the four soil types specified for Merridan was selected as representing "average" productivity.

To estimate how farmers should respond to changing economic conditions, two cost-price scenarios were used to reflect the directional movements (1956-60 and 1977-81) in product-price and factor-cost relativities shown in Table 2; these are compared with the 1986 dollar base of MIDAS. The scenarios "A" and "B" used in the application of MIDAS are shown in Table 2c and may be regarded as representing periods of 'wool favourable relative to wheat' and 'wheat favourable relative to wool', respectively. The 1986 base reflects another period of 'wool favourable relative to wheat'. The model was also run for three scenarios of N technology (i.e ley pasture only, pastures and lupins, and pastures, lupins and N fertilizer). Output includes optimal rotations for two land units and the proportion of the 1500 ha farm occupied by each type of land use. The percentage of arable land on the farm under legume leys was estimated for a complete rotation cycle for each of the three cost-price scenarios (Table 3). With wool favourable relative to wheat (Scenario A), the results predict ley pasture should comprise 83% of total arable land area. Even when lupins and N fertiliser were both available, the model did not include either option in the optimal mix.

When wheat was favourable relative to wool (Scenario B), the model allocated only 24% of the arable area to pasture leys in the optimal configuration (Table 3). The results in Figure 4 suggest that in changing the mix of crops and pastures, a shortening of the ley rotation would result in more optimum use of biological N. If lupins are not grown, the pasture area nearly doubles to 42% but the net income is reduced by 20% (not shown). In the optimum system with lupins, the area of lupins is 22% and the remainder is under wheat.

Using dollar values for 1986 (wool favourable relative to wheat), the model output suggests that farmers should increase their ley pasture area to

Table 3: Percentage of arable land under legume ley pastures (the remainder under crops) to achieve maximum profitability predicted by MIDAS for the three cost–price scenarios in Table 2c.

Other N input options	Predicted optimum percentage of arable land under legume ley pasture		
	Scenario A[1]	Scenario B[2]	1986[1]
Lupins and N fertilizer available[3]	83(0)	24(22)	88(0)
Without lupins	83	42	88
Without lupins or N fertilizer	82	70	87

1. Wool favoured relative to wheat; 2. Wheat favoured relative to wool; 3. Recommended proportional area of lupins shown in brackets.

c. 88% of total arable area.

The results accord with the historical analysis presented earlier, and with the conclusion drawn by Vere and Muir (1986). They support the hypothesis that farmers are rational managers who respond to changes in economic circumstances. The results of MIDAS demonstrate the importance of lupins and N fertilizer in profitably responding to 'wheat favourable to wool' conditions.

RISKS TO ECOLOGICAL STABILITY AS A RESULT OF REDUCED LEYS IN ROTATIONS

Before the adoption of legume ley technology, bare fallowing was essentially the only practice for increasing mineral N supply to cereals which was economic. However, long fallows, which may require as many as ten cultivations for weed control (Poole 1987), leave soil very vulnerable to wind and water erosion and results in low organic matter levels (Fig. 1). The latter lead to problems of hard setting, reduced infiltration, and poor seedling emergence problems on most soils of the zone (Clarke 1986). Thus, whilst the bare fallow practice was the key to profitability, it was seriously degrading the productive resource.

Adoption of pasture ley technology dramatically changed this by providing a much more profitable means of providing N for crops (when returns to animal production were high from sown pastures). Pasture cover reduced erosion and increased soil organic matter. Thus profitability and ecological stability were not in serious conflict — at least not until the 1970's.

Trends away from leys raised concerns by agricultural scientists about adverse ecological consequences. Without being critical of their caution about change that might result in a return to the problems of the bare fallow, we suggest that there is declining need to view pasture leys as central to the prevention of those problems. Firstly, with the advent of conservation tillage technology, there is much less reason to look to pastures to maintain soil stability. Much emphasis in the past has been on maintaining adequate organic matter for a high degree of water stability of aggregates following cultivation. With direct drilling, pore systems which conduct water are left largely intact. Organic matter levels tend to be greater in the soil surface (Dick 1983) where maximum physical stress occurs. Reports of benefits to physical properties and to reduced erosion in the Australian wheat-sheep zone are accumulating (Hamblin 1984; Rowan 1986). Secondly, conservation tillage is increasingly attractive to farmers because of its lower machinery, labour and feed costs. This provides an economic incentive to adopt an ecologically-important practice. Finally, increased cropping at the expense of leys no longer necessitates increased fallow, but rather increased inputs of N. A problem in utilizing the results of the classic long-term rotation studies, e.g. that at the Waite Institute, is that generally N fertilizer was not used, hence cereal yields of crop-intensive rotations were low. They represented low input systems, with emphasis on managing to retard rates of decline in yields and soil organic matter. Systems with higher fertilizer inputs have higher crop yields, more residues and greater contribution to soil organic matter (via equation (1)) (Lucas et al. 1977; Russell 1981).

Trends in the wheat-sheep zone of southern Australia, especially since the early 1970's, indicate a profound change in the ley farming system centred on conservation tillage (Pratley and Cornish 1985). Conservation tillage spurred by cost savings is spreading, but Poole (1987) suggests that optimum systems of tillage and stubble retention for the various regions and soil types will take until the turn of the century to evolve. While to date, use of reduced tillage has often been ad hoc and opportunistic, Pratley and Cornish (1985) suggest that the evolution of new whole-farm systems is required to tap the economic and ecological potential of this technology. The system which is evolving is one which can respond to economic incentives for cropping with reduction in leys in ecologically-sound ways. However, the fact that even in periods of high crop returns 98 percent of southern Australian farms still have important animal production enterprises is evidence that pasture leys will remain an important component of this new system (Poole 1987).

CONCLUSIONS

The alarmist reaction by some agricultural scientists to the swing away from the ley system seems to have resulted from interpretation of this as a reversal of several decades of achievement rather than a swing of the economic pendulum. The subsequent swing back to animal production in response to changed relativities in wheat and wool prices confirms the pendulum interpretation and provides a reminder that farming systems, no matter how elegant their ecological features, must be expected to adapt to changes in the economic environment.

The economic environment is dynamic, thus good economic management avoids options that produce consequences which are not readily reversed. Farmers who disposed not only of their flocks, but of fences and yards, have greatly reduced their ability to respond to subsequent conditions which favour wool production.

When cropping intensification causes serious soil erosion, agricultural scientists should be alarmed. However, it is clear that the role of pasture ley conservationist cannot succeed. Contribution by scientists will depend on acceptance of economics as the primary determinant of change in farming systems and the use of their imagination in identifying ways to assist economically-adaptive change and at the same time minimize adverse ecological consequences (e.g. conservation tillage).

A review can always conclude that more research is needed. Certainly there are enormous gaps in our understanding of the N cycle in the new, flexible, ley system that includes grain legumes and N fertilizer. However, at least as serious are deficiencies in applying existing knowledge to farm management in times of change. Although many farmers are good at adapting, we submit that agricultural science could do a better job in helping the farming community to respond more efficiently. Economic models such as MIDAS have the potential to assist, but are hampered at present by a lack of adequate, appropriate models of mineral N supply. Our need in this paper to improvise the simple model based on residue decomposition points up this important deficiency.

There has been considerable interest in recent years in transferring ley technology from southern Australia to the true Mediterranean region (Chatterton and Chatterton 1984; Springborg 1986) and the concepts to the semi-arid/sub-humid tropics (Jones and Wild 1975; McCown et al. 1985; Mohamed-Saleem et al. 1986). In general the grounds for this have been:
1. Close physical association of crop and animal enterprises with potential benefits from greater integration.
2. Constraints of low soil N for crops and low pasture quality for animals. N fertilizer and improved pastures are not economically feasible, thus possible synergies of legume ley pastures are attractive.
3. Soil physical and/or erosion problems due to overcropping.

What can be learned from analysis of the southern Australian system that might assist in such transfer? What is apparent from the experience of the 1950's and 1960's is that where a well-adapted pasture legume exists and under economic conditions which favour a high proportion of pasture to crops, the system is "ideal". The system during this period is the object of infatuation referred to by Springborg (1986). What has become more clear in subsequent years is that when the profitability of cropping increases relative to livestock production, Australian farmers grow crops at the expense of pasture leys. The system can now be seen as more dynamically linked to the market place than was apparent to many previously. This observation may assist in the interpretation of non—adoption elsewhere.

Springborg (1986) discusses various possible causes of non—adoption in the Middle East and the difficulties of apportioning causality. Two factors suspected to be of major importance are (a) disappointing ("sub—Australian") performance of pasture legumes and (b) insufficient opportunity to benefit financially by increases in livestock production. The Australian experience leaves little doubt that this technology will be adopted only where there is strong incentive and good prospects for increasing profits from livestock production via improved forage supply.

Decline in technical performance of pasture legumes was a primary explanation for decline in leys in southern Australia during the 1970's (Carter et al. 1982). The importance of this relative to that of changed market environment was not clear. The subsequent market reversal (Table 2a) has resulted in a phenomenal swing back to animal production. The degree to which this is being constrained by poor performance of sown legume pastures should help clarify the relative importance of these two factors.

The southern Australian ley system is alive and well. The present version is a more flexible, adaptive one that includes grain legumes, N fertilizer, direct drilling and stubble retention. The mix of these varies both temporally and geographically. Although in the interest of exploring general concepts, this paper does not deal realistically with variation in climate and soils, differentiation in farming systems can be expected to increase as regional research results and farmer experience accrue.

ACKNOWLEDGEMENTS

We are grateful to David Morrison for access to MIDAS and to David Falconer for assistance in partially respecifying the model and conducting the computations for us. The Australian Centre for International Agricultural Research provided salary for Dr Cogle during the period in which this paper was prepared.

REFERENCES

Anonymous. (1969). The mini—ley concept. Rural Res. 67, 24—8.
Barrow, N.J. (1969). The accumulation of soil organic matter under pasture and its effect on soil properties. Aust. J. Exp. Agric. Anim. Husb. 9, 437—44.
Bath, J.G. (1949). The improvement and maintenance of soil fertility by subterranean clover ley rotations at Rutherglen Research Station. British Commonwealth Scientific Official Conference in Agriculture — Melbourne, Australia, July 1949, pp. 1—14.
Carter, E.D., Wolfe, E.C., and Francis, C.M. (1982). Problems of maintaining pastures in the cereal—livestock areas of southern Australia. Proc. 2nd Aust. Agron. Conf., Wagga Wagga 1982, pp. 68—82. (Aust. Soc. of Agron.: Parkville, Vic.).
Chatterton, B., and Chatterton, L. (1984). Alleviating land degradation and increasing cereal and livestock production in North Africa and the Middle East using annual medicago pasture. Agric., Ecosys. Environ. 11, 117—29.

Clarke, A.L. (1986). The impact of agricultural practices on Australian soils – Cultivation. In 'Australian Soils – The Human Impact'. (Eds J.S. Russell and R.F. Isbell) pp. 273–303. (Univ. Queensland Press: Brisbane.)

Clarke A.L., and Russell J.S. (1977). Crop sequential practices. In 'Soil Factors in Crop Production in a Semi–arid Environment'. (Eds J.S. Russell and E.L. Greacen.) pp. 279–300. (Univ. Queensland Press: Brisbane.)

Cooke, G.W. (1967). 'The Control of Soil Fertility'. (Crosby Lockwood & Son Ltd: London.)

Davidson, B.R. (1981). 'European Farming in Australia'. (Elsevier Scientific Publishing Company: Amsterdam.)

Davies, J.S. (1974). The influence of the 1973 Federal Government taxation concession revisions on agriculture. Rev. Marketing Agric. Econ. 42, 202–10.

De, R., Rao, Y.Y., and Ali, W. (1983). Grain and fodder legumes as preceding crops affecting the yield and N economy of rice. J. Agric. Sci., Camb. 101, 463–6.

Dick, W.A. (1983). Organic carbon, nitrogen, and phosphorus concentrations and pH in soil profiles as affected by tillage intensity. Soil Sci. Soc. Am. J. 47, 102–7.

Donald, C.M. (1951). Competition among pasture plants. 1. Intra–specific competition among annual pasture plants. Aust. J. Agric. Res. 2, 355–77.

Donald, C.M. (1965). The progress of Australian agriculture and the role of pastures in environmental change. Farrer Memorial Oration. Aust. J. Sci. 27, 187–98.

Doughton, J.A., and Mackenzie, J. (1984). Comparative effects of black and green gram (mung beans) and grain sorghum on soil mineral nitrogen and subsequent grain sorghum yields on the Eastern Darling Downs. Aust. J. Agric. Anim. Husb. 24, 244–9.

Ellington, A., Reeves, T.G., Boundy, K.A., and Brooke, H.D. (1979). Increasing yields and soil fertility with pasture/wheat/grain–legume rotations and direct–drilling. 49th ANZAAS Congress, Auckland, New Zealand. Section 13, Symposium 16.

Falconer, D.A. (1987). Spreadsheets for MIDAS: Mid–Central Eastern Wheat belt (MCEWN 10D). Western Australia Department of Agriculture, South Perth.

Greenland, D.J. (1971). Changes in the nitrogen status and physical condition of soils under pastures, with special reference to the maintenance of the fertility of Australian soils used for growing wheat. Soils Fertilizers 34, 237–51.

Gruen, F.H. (1959). Pasture improvement – the farmer's economic choice. Aust. J. of Agric. Econ. 3, 19–44.

Halse, N.J., and Wolfe, E.C. (1985). The contribution of science to Australian temperate zone agriculture. 5. The wheat and sheep zone. J. Aust. Inst. of Agric. Sci. 51, 185–95.

Hamblin, A.P. (1984). The effect of tillage on soil surface properties and the water balance of a xeralfic alfisol. Soil Till. Res. 4, 543–59.

Herridge, D.F. (1986). Nitrogen fixation dynamics by rain–fed grain legume crops: Potential for improvement. Proc. 13th Int. Soc. Soil Science Cong., Hamburg pp. 794–804.

Holford, I.C.R. (1980). Effects of duration of grazed lucerne on long–term yields and nitrogen uptake of subsequent wheat. Aust. J. Agric. Res. 31, 239–50.

Holford, I.C.R. (1981). Changes in nitrogen and organic carbon of wheat–growing soils after various periods of grazed lucerne, extended fallowing and continuous wheat. Aust. J. Soil Res. 19, 239–49.

Jones, M.J., and Wild, A. (1975). Soils of the West African Savanna. The maintenance and improvement of their fertility. CAB Tech. Commun. No. 55. (Commonwealth Agricultural Bureaux: England).

Kleinig, C.R., Noble, J.C., and Rixon, A.J. (1974). Herbage production and the accumulation of soil nitrogen under irrigated pastures on the Riverine Plain. Aust. J. of Exp. Agric. Anim. Husb. 14, 49–56.

Ladd, J.N., and Amato, M. (1985). Nitrogen cycling in legume–cereal rotations. In 'Nitrogen Management in Farming Systems in Humid and Subhumid Tropics'. (Eds B.T. Kang and J. van der Heide) pp. 105–27. (Int. Inst. Trop. Agric.: Ibadan).

Ladd, J.N., and Amato, M. (1986). The fate of nitrogen from legume and fertilizer sources in soils successively cropped with wheat under field conditions. Soil Biol. Biochem. 18, 417–25.

Ladd, J.N., Butler, J.H.A., and Amato, M. (1986). Nitrogen fixation by legumes and their role as sources of nitrogen for soil and crop. Biol. Agric. Hort. 3, 269–86.

Leslie, J.K. (1984). Cropping technologies in integrated systems in Central Queensland. In 'Integrated Beef and Crop Production in Central Queensland: Past, Present and Future'. (Ed A. Macqueen) pp. 5.1–5.7. (Aust. Inst. Agric. Sci.: Brisbane.)

Lloyd–Davies, H., and Myers, L.F. (1985). The contribution of science to Australian temperate zone agriculture. 4. Agricultural development in high rainfall southern Australia. J. Aust. Inst. Agric. Sci. 51, 178–84.

Lucas, R.E., Holtman, J.B., and Connor, L.J. (1977). Soil carbon dynamics and cropping practices. In 'Agriculture and Energy'. (Ed W. Lockeretz) pp. 333–51. (Academic Press: New York.)

McCown, R.L., Jones, R.K., and the late Peake, D.C.I. (1985). Evaluation of a no-till, tropical legume ley-farming strategy. In 'Agro-Research for the Semi-Arid Tropics: North-West Australia'. (Ed R.C. Muchow) pp. 450-69. (Univ. Queensland Press: Brisbane).

Mohamed-Saleem, A.M., and Otsyina, R.M. (1986). Grain yields of maize and the nitrogen contribution following Stylosanthes pasture in the Nigerian sub-humid zone. Expl. Agric. 22, 207-14.

Mohamed-Saleem, A.M., Suleiman, H., and Otsyina, R.M. (1986). Fodder Banks: For pastoralists or framers. In 'Potentials of Forage Legumes in Farming Systems of Sub-Saharan Africa'. (Eds I. Haque, S. Jutzi, P.J.H. Neate) pp. 420-37. (International Livestock Centre for Africa: Addis Ababa.)

Morrison, D.A., Kingwell, R.S., and Pannell, D.J. (1986). A mathematical programming model of a crop-livestock farm system. Agric. Syst. 20, 243-68.

Mullaly, J.V., McPherson, J.B., and Bath, J.G. (1965). Effect of length of subterranean clover leys on wheat yield and quality. Proc. Aust. Cereal Agron. Conf., Horsham, Vic., pp. D3-3-4.

Mullaly, J.V., McPherson, J.B., Mann, A.P., and Rooney, D.R. (1967). The effect of length of legume and non-legume leys on gravimetric soil nitrogen at some locations in the Victorian wheat areas. Aust. J. Exp. Agric. Anim. Husb. 7, 568-71.

Osborne, G.J., Batten, G.D., and Kohn, G.D. (1977). Nitrogen and phosphorus requirements of wheat following subterranean clover pasture. Aust. J. Agric. Res. 28, 971-9.

Pannell, D.J., and Falconer, D.A. (1987). The value of nitrogen in a crop-livestock farm system: A bioeconomic modelling approach. In 'Nitrogen Cycling in Agricultural Systems in Temperate Australia'. (Eds P.E. Bacon, J. Evans, R.R. Storrier and A.C. Taylor) pp. 449-66. (Aust. Soc. Soil Sci. Inc: Wagga Wagga).

Papastylianou, I., Puckridge, D.W., and Carter, E.D. (1981). Nitrogen nutrition of cereals in a short-term rotation. 1. Single season treatments as a source of nitrogen for subsequent cereal crops. Aust. J. Agric. Res. 32, 703-12.

Poole, M.L. (1987). Tillage practices for crop production in winter rainfall areas. In 'Tillage — New Directions in Australian Agriculture'. (Eds P.S. Cornish and J.E. Pratley.) pp. 24-47. (Inkata Press: Melbourne.)

Pratley, J.E., and Cornish, P.S. (1985). Conservation Farming — A crop establishment alternative or a whole-farm system? In 'Crop and Pasture Production — Science and Practice.' Proc. 3rd Aust. Agron. Conf., Hobart, Tasmania, 1985. pp. 95-111. (Aust. Soc. Agron.: Hobart.)

Pratley, J.E., and Rowell, D.L. (1987). From the first fleet — Evolution of Australian farming systems. In 'Tillage — New Directions in Australian Agriculture.' (Eds P.S. Cornish and J.E. Pratley.) pp. 2-23. (Inkata Press: Melbourne.)

Reeves, T.G. (1984). Lupins in crop rotations. Proc. 3rd Inter. Lupin Conf, La Rochelle, France, June 1984. pp. 207-26.

Reeves, T.G., Mears, P.T., and Ockwell, A.P. (1984). Report of a Working Party on Pasture/Crop/ Animal Systems in Temperate and Sub-Tropical Australia — II. Statistical and Biological Data, Victorian Dept. of Agric., Rutherglen.

Rowan, J.N. (1986). Conservation practices. In 'Australian Soils — The Human Impact.' (Eds J.S. Russell and R.F. Isbell) pp. 397-414. (Univ. Queensland Press: Brisbane.)

Rowland, I.C., Halse, N.J., and Fitzpatrick, E.N. (1980). The role of legumes in Australian dryland agriculture. Proc. Inter. Cong. Dryland Farming, Adelaide, 1980, pp. 676-98. (Department of Agriculture: Adelaide S.A.).

Rowland, I.C., Mason, M.G., and Hamblin, J. (1986). Effects of lupins on soil fertility. Proc. 4th Inter. Lupin Conf., Geraldton, Western Australia, August 1986, pp. 96-111.

Russell, J.S. (1960). Soil fertility changes in the long-term experimental plots at Kybybolite, South Australia. 1. Changes in pH, total nitrogen, organic carbon, and bulk density. Aust. J. Agric. Res. 11, 902-25.

Russell, J.S. (1980). Nitrogen in dryland agriculture. Proc. Inter. Cong. on Dryland Farming, Adelaide, S.A. 1980, pp. 201-226.

Russell, J.S. (1981). Models of long term soil organic nitrogen change. In 'Simulation of Nitrogen Behaviour of Soil-Plant Systems'. (Eds M.J. Frissel and J.A. van Veen) pp. 222-32. (Centre Agric. Publ. Document.: Wageningen.)

Russell, J.S. (1987). Concepts of nitrogen cycling in agricultural systems in Australia. In 'Nitrogen Cycling in Agricultural Systems in Temperate Australia'. (Eds P.E. Bacon, J. Evans, R.R. Storrier and A.C. Taylor) pp. 1-13. (Aust. Soc. Soil Sci. Inc.: Wagga Wagga).

Ruthenberg, H. (1980). 'Farming Systems in the Tropics', 3rd Ed. (Clarendon Press: Oxford).

Springborg, R. (1986). Impediments to the transfer of Australian dry land agricultural technology to the Middle East. Agric. Ecosys. Environ. 17, 229-51.

Strong, W.M., Harbison, J., Nielsen, R.G.H., Hall, B.D., and Best, E.K. (1986). Nitrogen availability in a Darling Downs soil following cereal, oilseed and grain legume crops. 2. Effects of residual soil nitrogen and fertiliser nitrogen on subsequent wheat crops. Aust. J. Exp. Agric. 26, 353-9.

Taylor, G.B. (1985). Effect of tillage practices on the fate of hard seed of subterranean clover in a ley farming system. Aust. J. Exp. Agric. Anim. Husb. 25, 568–73.

Vere, D.T., and Muir, A.M. (1986). Pasture improvement adoption in south–eastern New South Wales. Rev. Marketing Agric. Econ. 54, 19–32.

Watson, E.R. (1963). The influence of subterranean clover pastures on soil fertility. 1. Short–term effects. Aust. J. Agric. Res. 14, 796–807.

Watson, E.R., and Lapins, P. (1964). The influence of subterranean clover pastures on soil fertility. 2. The effects of certain management systems. Aust. J. Agric. Res. 15, 885–94.

Watson, E.R., Lapins, P, Arnold, G.W., Barron, R.J.W., and Anderson, G.W. (1977). Effect on cereal crop and sheep production of two rotations in a ley farming system in the south–west of Western Australia. Aust. J. Exp. Agric. Anim. Husb 17, 1011–9.

Watson, E.R., Lapins, P., and Barron, R.J.W. (1976). Effects of initial clover seeding rate and length of ley on pasture production, soil nitrogen and crop yields in a ley farming system. Aust. J. Exp. Agric. Anim. Husb. 16, 484–90.

Wells, G.J. (1970). Skeleton weed (Chondrilla juncea) in the Victorian Mallee. 2. Effect of legumes on soil fertility, subsequent wheat crop and weed population. Aust. J. Exp. Agric. Anim. Husb. 10, 622–9.

White, D.H., Elliott, B.R., Sharkey, M.J., and Reeves, T.G. (1978). Efficiency of land–use systems involving crops and pastures. J. Aust. Inst. Agric. Sci. 44, 21–7.

Wood, I.M., and Russell, J.S. (1979). Grain legumes. In 'Australian Field Crops Vol 2: Tropical Cereals, Oilseeds, Grain Legumes and Other Crops'. Eds J.V. Lovett and Elec Lazenby). pp. 232–62. (Angus and Robertson: Sydney).

NITROGEN CYCLING IN TROPICAL EVERGREEN TREE CROP ECOSYSTEMS

R.A. Stephenson and R.J. Raison

ABSTRACT

The nitrogen requirements of tropical tree crops are high, and they are usually heavily fertilised to achieve increased production. Tree crops are buffered against fluctuations in the external supply of available N because of storage and cycling within the tree. This allows some flexibility in N management. Moreover, N reserves within the orchard/plantation ecosystem are often considerable, and inputs may be large where legume cover crops and/or shade trees are grown. Efforts to enhance and exploit these N resources are rare, and increased research in this area is justified. In addition to fixing N, leguminous shade trees may cycle N from deeper soil layers. N-rich legume litter decomposes rapidly in tropical environments and synchronisation of N input from litter with periods of high N demand by crops during flowering and fruit formation is thought to be important in traditional farming systems.

Current knowledge of pools and fluxes of N in tropical tree crop systems is insufficient to draw up a complete N budget. The data are highly variable, due to differences in climate, soils and cultural techniques. Optimisation of N cycling may need radical reappraisal of spatial and perhaps temporal planting patterns of crop and cover to minimise competition between them. The search for superior N_2 fixing cover crops and/or shade trees should continue, with particular emphasis on shade tolerant legumes for use in mature orchards.

INTRODUCTION

Numerous tree crops, many important as cash crops for export and as staple foods, are found in the tropics and subtropics. These include tea, coffee, cocoa, coconuts, oil palm, citrus, mango, date, fig, avocado and various spices and nuts (Anon. 1984). Fruit trees are often grown in combination with other crops and animals (Nair 1985); such agroforestry systems provide both food and useful wood products (King 1979). Not only are food-producing trees important to the economy in the tropics, their incorporation in farming systems provides greater ecological stability and efficient use of nutrients (Clarke 1976 ; King 1979 ; Huxley 1986). Plant nutrients, especially nitrogen (N) undergo constant and rapid cycling within the soil and plant compartments of well managed tree-based ecosystems, with minimal losses (Nair 1985). It is vital that management systems be developed to elicit maximum efficiency of production from these crops.

Nitrogen is the most common nutrient limiting growth and production of tree crops, and supplementary sources are often used to increase yields (Childers 1975 ; Reuther 1977 ; Kramer and Kozlowski 1979 ; Khanna and Nair 1980). To satisfy this N requirement with a minimal input of fertiliser requires an ecosystems approach (see Nair 1979). This seldom occurs, largely because of the complexity of N cycling within tree crop ecosystems and our poor understanding of it. Apart from complexity due to cultural manipulations, most tree crops have distinct vegetative and reproductive stages which influence their N requirements and affect cycling within the ecosystem. The major factors influencing the N nutrition of tree crops are considered here and these demonstrate the importance of an ecosystem approach.

GENERAL FEATURES OF NITROGEN CYCLING IN TREE CROP ECOSYSTEMS

Numerous reviews of N cycling in natural tree ecosystems are available (e.g. Date 1973 ; Heal et al. 1982 ; Pereira 1982) but few relate to the tropics. Despite luxuriant growth and great diversity of species in tropical rainforests, soil nutrient reserves are often low (Bartholomew 1973 ; Janzen 1973 ; Richards 1973 ; Sanchez 1976 ; Nair 1985), and the forest exhibits a reasonably closed nutrient cycle. Once the system is disturbed, productivity can drop rapidly, and for viable agro-ecosystems to be maintained, inputs of fertilisers are usually required (Ruthenberg 1971). However, agroforestry systems can be developed and managed so that they approach the closed nutrient systems which they replaced (Huxley 1980, 1986). Clarke (1976) suggested that it would be ecologically more sound to encourage cultivation of trees in the tropics rather than annual crops. Tree crop ecosystems will have lower biomass and probably less capacity for roots to recycle leached N than is the case for rainforests. It may be possible to balance greater leaching losses of N by increased N_2 fixation from compatible leguminous cover crops (Agamuthu and Broughton 1981, 1985).

Much of the information on tropical tree ecosystems comes from Malaysia on rubber (which is outside the scope of this review but see Broughton 1977) and oil palm (eg. Agamuthu and Broughton 1981, 1985). There are also studies on coffee and cocoa in South America (Aranguren et al. 1982a,b ; Bornemisza 1982 ; Roskoski 1982 ; Roskoski et al. 1982 ; Santana and Cabala-Rosand 1982). Robertson and Rosswall (1986) prepared a comprehensive N budget for the West

African region, which emphasises the importance of N fluxes through tree crops. Manguiat et al. (1981) concluded that there was still insufficient data to assemble complete N budgets for any of the tropical areas.

Despite the small absolute amount of N contained in tree crops, the cycling of N within them will be of considerable economic importance. The challenge is to exploit these tree ecosystems by developing efficient farming systems and cultural practices.

EFFECTS OF TREE GROWTH PHASE ON NITROGEN REQUIREMENTS AND CYCLING

Young tree crops have a large year round requirement for N from the soil (Hansen 1980), especially during canopy development. Moreover, young trees are susceptible to weed and cover crop competition (Baxter and Newman 1971). Cultural manipulation is therefore required to avoid this and ensure adequate N for unimpeded growth. After the canopy stabilises, N cycling e.g. as a result of leaf litter decomposition or retranslocation prior to leaf shed, progressively provides a greater proportion of the annual N needs of the stand.

Once fruiting commences, the N requirements change; flowering, fruit set and early fruit growth (e.g. Taylor 1969) all have a high N requirement. Productive trees will have an additional N requirement equivalent to that removed in the annual crop. Manipulation of N supply also has to be considered in relation to maintaining the long-term balance between vegetative and reproductive growth.

Fruit quality is also importantly related to N status. Figure 1 indicates the complexity of yield and quality responses of orange fruits to N status. Thus the management of tree crop N requirements will often be more demanding than that for most agricultural crops.

NITROGEN POOLS, CYCLES AND BUDGETS

Soil N reserves

Tropical soils vary widely in their reserves of N; from a few tonnes to more than 50 t/ha in the surface 100 cm (Bartholomew 1972 ; Sanchez 1976 ; Edwards and Grubb 1982). Almost all soil N is bound in organic matter and rates of N supply depend largely on rates of net mineralisation. Only a small fraction of total soil N is mineralisable, and management should aim to maintain or increase this. Assessment of management induced changes in N-supplying capacity can be made using in situ measurement of N mineralisation (Raison et al. 1987). Some tropical soils also fix significant amounts of ammonium-N (see Young and Aldag 1982).

Mineralisation of N is usually high in the moist tropics throughout the year. Where aridity limits mineralisation in wet-dry tropical environments, enhanced mineralisation follows re-wetting of the soil. This may predispose leaching (Date 1973 ; Janzen 1973) especially in systems where soil exchange capacity is low or vegetation is shallow rooted.

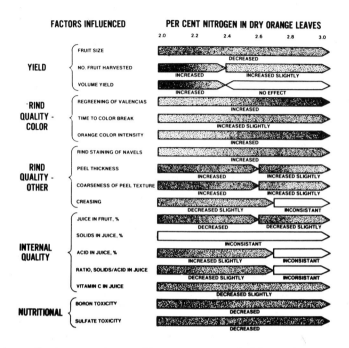

Fig. 1: Effects of increasing nitrogen on yield and quality of oranges. The greater the intensity of stippling, the greater the effect on the factor. [From Embleton et al. (1975)].

Nitrogen is often cycled in large quantities in natural, closed tropical forest ecosystems (Melillo and Gosz 1983) and is probably not the major limiting factor to growth (Richards 1973 ; Vitousek 1984). Clearing usually leads to N losses and agricultural crops which have a high N requirement may not obtain sufficient N for optimum crop growth (Bartholomew 1972), particularly during active growth phases (Date 1973). The shortfall in N supply is sometimes boosted either by the inclusion of a suitable legume cover crop, fertiliser applications or other cultural inputs.

N in tree biomass

Quantities of N held in the biomass of tropical tree crops are variable and generally low (Table 1). There is a need for further data of this type for high yielding crops. The highest total values recorded are similar in magnitude to those for some temperate forests (Melillo 1981) and legume crops (Ralph 1986) but are smaller than for other forests (Gosz 1981 ; Heal et al. 1982). Values are much smaller than those for tropical rainforests (Nye and Greenland 1960 ; Tanner 1985) but are generally higher than the N immobilised in annual crops such as wheat (Ralph 1986). Tropical food trees are not deciduous and the annual return of organic matter and N in leaf litter is less than for temperate deciduous forests (Heal et al. 1982). The foliage contains a large proportion of the total N of palms (40 to 50%) and regular pruning of oil palm leaves at harvest may therefore enhance recycling of N in litter.

Table 1: Nitrogen pools within tropical food trees. [Note that in coffee and cocoa plantations, shade trees may contain much more N than the food crop (e.g. Roskoski et al. 1982).]

Crop	Leaves	Wood	Roots	Inflor./fruit	Total	Reference
	- - - - - - - - - - - - *kg N/ha* - - - - - - - - - - - -					
Citrus	34	26	15	10	85	5, 12
Cocoa	11–45	54–149	43–88	1–20 (beans) 100 (dry pods)	109–302	2,4,12,13,14
Coconut	163	58	21	91	333	10
Coffee	62	2–158	9–249	5–90	20–491	1,3,6,7,12,13
Oil Palm	193	183	82	30–67	488	8, 9, 11, 15

1. Aranguren et al. (1982a); 2. Aranguren et al. (1982b); 3. Bornemisza (1982); 4. Boyer (1973); 5. Cameron & Compton (1945); 6. Carvajal (1959); 7. Cooil & Fukunaga (1959); 8. Corley et al. (1976); 9. Georgi (1931); 10. Khanna & Nair (1980); 11. Rees & Tinker (1963); 12. Robertson & Rosswall (1986); 13. Roskoski et al. (1982); 14. Santana & Cabala-Rosand (1982); 15. Tinker & Smilde (1963).

Incomplete recovery of fine roots which often contain large amounts of N (Heal et al. 1982) may account for some differences within crop estimates. The small amounts of N in oil palm roots probably reflect their shallow rooting habit. In contrast, almost half the N in coffee trees was recovered from the roots. Taylor and van den Ende (1969) found that N in roots of dormant peach trees was the most sensitive indicator of N status of the trees. Little is known about such aspects of tropical evergreen trees and more attention should be given to measuring the dynamics of fine roots and associated N turnover.

Most of the N in orange trees occurs in leaves (50%) and bark (Cameron and Compton 1945). This contrasts to the situation in tropical rainforest where a relatively small proportion is contained in leaves of trees (9%) and shrubs (16%), most of their N being in the massive bole and in branches (Singh and Pandey 1981 ; Grubb and Edwards 1982). In peach, on the other hand, most N is stored in roots (Taylor 1967a,b ; Taylor and van den Ende 1970).

There is a need for further data on partitioning of annual N uptake within fruit trees. Up to 25% of tree biomass N is removed each year in the harvested product. Unless this is replaced by an external input, productivity will decline rapidly. Management should aim to conserve site N and balance inputs and outputs.

Cycling of N within trees

The N cycle of tree crops is influenced by the capacity of roots to absorb N throughout the year and for roots, stems and foliage to accumulate and subsequently redistribute stored N (e.g. Kramer and Kozlowski 1979 ; Titus and Kang 1982 ; Weinbaum and Muraoka 1986). Most information on the dynamics of N within fruit trees comes from deciduous crops. These undergo substantial changes: major export of N from leaves prior to abscission in autumn, subsequent storage of N reserves over winter, and remobilisation in spring to support new vegetative and early fruit growth. The literature has been reviewed by Taylor (1967a), Tromp (1970), and more recently by Titus and Kang (1982) and Habib (1984). With evergreen trees of the tropics and sub-tropics, storage and recycling patterns of N differ due to their multiple flushing characteristics, e.g. in citrus (Sato 1961) and in macadamia (Macadamia integrifolia) (Stephenson et al. 1986 a and b). More frequent fluctuations of smaller amplitude are usual (Tromp 1970). Nevertheless, seasonal fluctuations in N levels do occur and may influence phenological cycling of the tree and yield of fruit.

In evergreen trees changes in levels of stored N due to remobilisation from senescing leaves will obviously be less important than in deciduous crops, except perhaps when leaf drop occurs at critical stages. In south east Queensland, for example, macadamia often has noticeable leaf drop in spring, coinciding with flowering and nut set. Developing nuts could benefit from enhanced N status following mobilisation of N from these leaves. In apples, up to half pre-abscission leaf N was relocated for storage in woody tissues (Titus and Kang 1982). This represents a considerable N resource. In general, mobile reserves of N are important in the nutrition of tree crops.

Natural inputs of N

Natural N inputs are relatively minor, but often variable (see Table 3 later). Estimates of non-symbiotic N_2 fixation in tree crop plantations vary from 2-40 kg N/ha/yr but most are less than 10 kg N/ha/yr. Suwanarit et al. (1981) considered that levels as high as 40 kg N/ha/yr (Tinker 1976) are unlikely to be achieved in the field. Many estimates are based on laboratory studies and their applicability to field environments is uncertain. N_2 fixed by epiphylls is an insignificant source of N for coffee (Roskoski 1982). Presumably the same applies to other tropical tree crops.

N in harvested products

Significant amounts of N are removed with the harvested crop (Table 2),

Table 2: Total annual yield and quantities of N removed in the harvested portion of tree crops

Crop	Yield	Plant part	N removed	Reference
	kg/ha		*kg/ha*	
Apple	110,500	fruit	45	Murneek 1942 (247 trees/ha)
Citrus	17,000	fruit	20	Robertson & Rosswall 1986
Cocoa	1,400	dry beans	32	Chew & Khoo 1980
"	1,120	"	10	Ng & Thamboo 1967a
"	1,000	"	20–24	Santana & Cabala–Rosand 1982
"	636	"	45	Aranguren et al. 1982b
"	600	"	13	Cooke 1972
"	262	"	13	Robertson & Rosswall 1986
Coconut	2,900	nuts	40	Chew & Khoo 1980
"	2,389	dry copra	16	Ng & Thamboo 1967a
"	1,400	"	62	Cooke 1972
"	269	nuts	36	Georgi & Teik 1932
"	–	–	56–91	Murray 1977
"	–	–	120	Khanna & Nair 1980
Coffee	1,120	beans	19–26	Dierendonck 1963
"	1,500	dry berries	38	Cooke 1972
"	800	"	5	Robertson & Rosswall 1986
"	700	"	20	Aranguren et al. 1982a
"	–	"	30–60	Roskoski 1982
"	–	beans	17–80	Bornemisza 1982
Durian	6,720	fruit	16	Ng & Thamboo 1967b
Oil palm	26,676	fresh fruit	77	Chew & Khoo 1980
"	10,000	bunch	29	Ng & Thamboo 1967a
"	2,500	oil	162	Cooke 1972
Rambutan	6,720	fruit	13	Ng & Thamboo 1967a
Tea	1,071	dry leaf	25	Ng & Thamboo 1967a

although the rates are low when compared with some agricultural crops (Nair 1985). There may also be a significant amount of N associated with related harvested organs such as male inflorescences in oil palm (11 kg/ha, Ng and Thamboo 1967a).

The amount of N removed with the harvested product varies with species (Table 2) from 5 kg N/ha/yr for coffee berries up to 162 kg N/ ha/yr for oil palm oil. The 45 kg N/ha removed by a relatively productive apple crop is included for comparison.

The N composition of fruit components vary (Georgi and Teik 1932 ; Ng and Thamboo 1967a,b). Where the crop undergoes preliminary on-farm processing, there may be scope for returning some of the N in residues to the orchard. Aranguren et al. (1982b) estimated that 20 kg N/ha remain in pod residues after cocoa beans have been extracted. Coffee pulp residues contain 8-40 kg N/ha (Bornemisza 1982). For such quantities of N, the return of residues to the field is likely to be economically feasible. Residues are traditionally returned to the field in Venezeula (Aranguren 1982a,b), and is becoming common in Costa Rica (Bornemisza 1982). The quantity of N in harvest residues is much less for other tree crops and it is seldom economic to return these to the plantation (e.g. Bornemisza 1982).

Other N fluxes

Examples of the relative magnitude of inputs and internal transfers of N in tropical tree crops are given in Table 3. Crops vary markedly in the magnitude and relationship between the N fluxes. With cocoa, N transfers in leaf litter dominate, whereas with coffee they are relatively small. N removed in citrus fruits (Table 2) exceeds that in litter fall (Table 3). For coconut, both these transfers of N are substantial. With oil palm, litter, prunings associated with harvest and the harvested portion contain similar amounts of N. In most cases, transfer of N in litter fall from tree crops is less than that for tropical rainforests where quantities can exceed 200 kg N/ha/yr (Ayanaba and Dart 1977 ; Vitousek 1984).

Legumes (either cover crops or shade trees) grown in association with food crops may cycle more N in litterfall than is in the crop itself (Roskoski et al. 1982 ; Agamuthu and Broughton 1985). Most of the N deposited in litter is rapidly mineralised. The addition of litter from leguminous cover crops increased mineralisation rates by two to three times in a rubber plantation (Pushparajah 1981). In addition, a significant internal transfer of N usually occurs prior to leaf abscission. Together these provide a substantial amount of the annual N requirement of crop trees and if peaks of N transfer to the soil by litter and its subsequent mineralisation are synchronised with periods of high N demand by crop plants, as suggested by Aranguren et al. (1982a) for coffee in Venezuela, the value of recycling is considerably enhanced.

Leaching of N beyond the root zone can be high under tropical agriculture (Keeney 1982), but there are few reports on leaching under food tree crops. In a fertilised oil palm/cover crop system, Agamuthu and Broughton (1985) measured leaching losses of 41 kg N/ha/yr. Santana and Cabala-Rosand (1982), however, concluded that leaching of N from a cocoa

Table 3: Fluxes of N in plantations of tropical food trees, excluding the economic harvest portion.

	Coconut	Oil palm	Cocoa	Coffee	Citrus
	– – – – – – – – – *kg N/ha/year* – – – – – – – – –
Recycled or lost
Harvest residues	0.4		4 , 20	8–40 ,2 ,3	n.a.
Non-woody litterfall	69 , 33	69	37 ,5 ,30–60	2 , 28	14
Woody litterfall	n.a.	n.a.	5–6	0.2	2
Pruning	n.a.	72	n.a.	1–5	n.a.
Leaching		41			
Inputs
Symbiotic N_2 fixation (legume shade trees/ cover crop)		150			40
Non-symbiotic fixation		40			0.4–0.6
Rain	2	21	17	3 , 10	
Fertiliser N	175	136		100–300	75–125

1. Robertson & Rosswall (1986); 2. Agamuthu et al. (1980); 3. Agamathu & Broughton (1985); 4. Tinker (1976); 5. Corley et al. (1976); 6. Santana & Cabala–Rosand (1982); 7. Aranguren et al. (1982b); 8. Aranguren et al. (1982a); 9. Bornemisza (1982); 10. Roskoski (1982); 11. Khanna & Nair (1980); 12. Embleton et al. (1973); na = not applicable, trace

plantation in Brazil was unimportant. It is possible to minimise losses using various cultural practices, e.g. Nair (1985) proposes incorporating compatible species into tree crop ecosystems in such a way that root systems fully exploit the available soil volume without resulting in serious competition.

No estimates of gaseous losses of N from tropical plantations were found in the literature. Losses of N by ammonia volatilisation from surface–applied N fertilisers can be high (Acquaye and Cunningham 1965). Erosion losses of N vary from 0.5 to 103 kg N/ha/yr, the latter occurring on steep cultivated slopes (Greenland 1977). There is a need to manage ground cover so as to minimise soil erosion which can result in rapid decline of soil fertility (Raison 1984).

Little is known of the contribution of death and decay of fine roots to the N cycle in tree crops. Heal et al. (1982) draw attention to data which indicate that input of N into the forest floor and soil from this source may be much greater than that which occurs through above–ground litter. Nye and Greenland (1960) suggested that root decay makes an important contribution to N cycling in tropical rain forest fallows. Observations of root flushing patterns of sub-tropical fruit trees in Queensland (A.W. Whiley <u>personal communication</u>) suggest that this contribution to the N cycle may apply to tree crops generally, although confirmation is needed. Roots also support a substantial microbial population in the rhizosphere (Nair 1985) which enhances nutrient cycling.

Apart from the influence of phenological and seasonal cycling on N reserves in the tree, these are often manipulated by routine horticultural practices such as pruning (Taylor and Ferree 1986) and fruit thinning (Jentsch and Eaton 1982). These practices will only effect the overall N budget if prunings are removed from the orchard.

MANAGEMENT TO ENHANCE NITROGEN SUPPLY

Horticultural production is characterised by a greater degree of tree and environment manipulation than occurs in most agricultural and forestry crops. Management should, among other things, be designed to maintain a satisfactory root environment conducive to good tree performance. An idealised N cycle for a tree food crop plantation is shown in Figure 2. This is characterised by high N fixation, low losses and efficient N cycling.

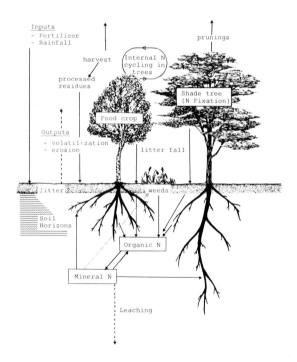

Fig 2: Idealised N cycle for a tree food crop plantation incorporating deep rooted leguminous shade trees.

Manipulation of rooting patterns

Managers often attempt to enhance N cycling in tree crop ecosystems by manipulating tree rooting patterns (Soong and Yap 1976 ; Agamuthu et al. 1981; Agamuthu and Broughton 1986), together with those of associated cover crops (Chandapillai 1968). Gray (1969) found that root development of rubber trees was 37% more prolific after growing for 10 years in a plot in which a legume was previously grown. Agamuthu and Broughton (1985, 1986) attribute superior tree growth and yield of oil palm to stimulation of a more extensive and effective root system when established in association with a legume cover crop.

Trees can contribute directly to efficient nutrient cycling by returning N from lower soil horizons to the topsoil in leaf litter (Nair 1985 ; Reid and Wilson 1985). Nair (1985) proposed that planting patterns of trees, covers and other associated plants be designed to optimise root function and N cycling in the ecosystem. Such an approach requires a radical redesign of tree crop farming systems. Jordan and Opoku (1966) proposed the establishment of widely spaced tree crop plantations to fully exploit the potential contribution of leguminous covers. This approach is worthy of serious consideration, even if yield is slightly reduced.

Use of legume cover crops

N input

Tropical legumes, because of their high N fixation capacity, can make a significant contribution to the N economy of tree crop ecosystems (Agamuthu et al. 1980 ; Haynes 1980 ; Nair 1985). The potential value of legume cover crops and/or shade trees grown in association with tree food crops is demonstrated in Table 3. Few data are available and many cultural problems must be overcome before such gains can be achieved. Symbiotic N_2 fixation can replace N removed at harvest and other N losses, thus achieving an essentially closed N system (Tables 2 and 3). N_2 fixation is high under ideal tropical conditions; e.g. Vigna and Calopogonium can fix 300 kg N/ha in 12-14 weeks, and fixation of 100 kg N ha/yr or more is not unusual (de Leon 1928 ; Vincent 1974 ; Nutman 1976 ; Greenland 1977 ; Anon. 1979).

To obtain the maximum benefit for tree crops from legume covers, conditions for efficient fixation must be optimised within constraints imposed by the trees. Despite this, use of legume covers with tropical plantation crops has become widespread (Evans et al. 1983), resulting in improved tree growth and production.

Competition

Sometimes cover crops reduce tree growth and yield (Acquaye and Smith 1965 ; Pacheco et al. 1973). The uncertainty of deriving benefits from cover cropping and hence resistance to adopting this practice is often due to competition between the tree crop and the covers, particularly for available N (Jordan and Opoku 1966 ; Haynes 1980) and water. The degree of competition is often directly related to the vigour of the cover, particularly at critical phenological stages of trees (Haynes 1980). Detrimental effects are often temporary and, once the tree canopy begins to close, cover crop species tend to be shaded out (Anon. 1936 ; Gray 1969). Once this occurs, N accumulated in the legume biomass over many years (Broughton 1977) becomes available to trees and this is often reflected in higher growth rates and yield (Broughton 1977 ; Agamuthu et al. 1981). Several authors (Mainstone 1963 ; Gray 1969 ; Soong and Yap 1976) have encouraged the identification and introduction of shade—tolerant legumes capable of extending the period of N fixation and production of N—rich litter.

The conflicting requirements of vigorous growth and extensive cover by legumes on the one hand, and the minimisation of competition between trees and covers on the other are resolved by maintaining bare areas around the trees by hand weeding, herbicides or mulching (Jacks et al. 1955 ; Broughton 1976). This results in greater exploitation of the surface layers of soil by tree roots (Atkinson and White 1976), and improved tree growth and production (Baxter and Newman 1971). Removal of cover competition by cultivation is considered to be detrimental (Joffe 1955 ; Wallwork 1976 ; Haynes 1980).

Grazing animals

The introduction of grazing animals under plantation crops is widespread (Thomas 1978 ; Reid and Wilson 1985) but the use of improved legumes is a recent innovation (Whiteman 1980). Grazing animals can enhance N cycling via the N inputs in urine and dung (Joffe 1955). Although Whiteman (1980) and Watson (1983) cite examples of coconut yields being enhanced by animals grazing on undersown improved pasture, presumably due to increased nutrient (particularly N) cycling, Watson (1983) was unable to demonstrate this on fertile soils in the Solomon Islands. Reports of damage to developing tree crops by cattle are common (Samuel 1974 ; Whiteman 1980) and there is insufficient light under mature oil palm, and cocoa for all but the most shade—tolerant legumes to persist. Consequently, grazing of improved legume pastures has been largely restricted to mature coconut plantations (Thomas 1978; Whiteman 1980).

Shade trees

Leguminous shade trees may contribute substantial amounts of N to the understorey crop trees through litterfall. They are an important component of the N cycle in many tropical areas (Oladokun 1980 ; Thiagalingam 1983 ; Bourke 1984 ; Nair 1985), although for maximum yield crops are often grown without shade and with increased fertiliser applications (Willey 1975).

N fertilisation

Fruit tree crops generally have a high requirement for N (Khanna and Nair 1980 ; Titus and Kang 1982 ; Page 1984 ; Kato 1986). Despite the inputs of N by cover crops, N fertilisers are used to supplement these. Wetselaar et al. (1981) cited predictions of large increases in fertiliser use in the tropics generally. Attempts have been made to use leaf analysis as a means of predicting the adequacy of the nutrient status of a wide range of tree crops for growth and yield (Leece 1976 ; Smith 1986). Data on N and other elements for a range of fruits and nuts including avocado, banana, citrus, coffee, macadamia, mango, oil palm, papaw and guava are summarised by Robinson (1986). Such information can guide N fertilisation, but must be combined with more detailed information on N requirements by the crop, and N inputs from cycling, before quantities of fertiliser and timing of application can be judged.

Broughton (1976) found that fertiliser alone did not produce oil palm yields as high as those attained with legume covers. Fertilisers were, in fact, more efficiently used in legume plots. Where cover crops are grown, they may utilize much of the N fertiliser applied and eventually recycle it through litter. Bould and Jarrett (1962) obtained the best tree growth and production using cover crops, and were only able to demonstrate a yield response when N fertiliser was applied to trees having a grass sward as understorey. Baxter and Newman (1971) however, recommended applying fertilisers to the bare strip only so the cover would not compete with trees for N.

Tree crops differ from agricultural crops in that there is often a considerable time lag (a year or more) between N application and tree growth and yield responses (Cameron and Compton 1945 ; Taylor 1967a ; Weinbaum and Muraoka 1986). In addition, the application strategy (timing x amount) can influence the response (Ludders 1980). This may be related to the storage capacity and recycling of N within trees which was discussed earlier.

Fertiliser N inputs usually exceed the amounts of N cycled annually in plantations and are several-fold greater than the amounts of N removed in harvested products. There appears to be considerable scope for reducing the reliance of tree crops on fertiliser N and for increasing efficiency with which fertiliser is used.

CONCLUSIONS

Tree crop growth, particularly in the tropics, is commonly limited by N supply. Unfortunately, information on N cycling in tropical tree crops is far from complete. Despite optimism that symbiotically fixed N might supply the tree's N requirements, this has not yet been shown. Enhanced oil palm yields after legume covers died out, however, were attributed to recycled N promoting improved tree root development, the advantages of which persisted for many years. This important breakthrough highlights the need for long term studies on tree root systems and their capacity to exploit soil and fertiliser N. The cumulative benefits of leguminous cover crops appear to be considerable.

Tree root growth is under-researched generally. Greater emphasis is needed on the effects of recycled N from legume litter on tree root growth, including lateral and vertical spread, density and timing of root growth flushes. The contribution of fine root decay to N cycling also needs to be determined. Extensive sampling using soil coring under trees and use of in-field root observation facilities may provide valuable insights into N cycling/rooting relationships.

Cyclic growth patterns of many tropical and sub-tropical tree crops are often reflected in N cycling between various organs within the tree. Much more data is needed on the amount (and timing) of N required for optimum growth and production, on N immobilised in the tree biomass, and on N cycling within the tree.

The enhancement of N inputs by biological N_2 fixation by an associated legume cover crop should be retained as an objective for tropical tree crops research. The search for a productive shade-tolerant legume which is capable of continuous input of N to mature orchards and plantations should continue.

REFERENCES

Anon. (1936). 'Use of Leguminous Plants in Tropical Countries as Green Manure, as Cover and as Shade'. (Inter. Inst. Agric:Rome.)

Anon. (1979). 'Tropical Legumes : Resources for the Future'. (National Academy of Sciences: Washington D.C.)

Anon. (1984). 'FAO Production Yearbook'. Vol. 37. (FAO: Rome)

Acquaye, D.K., and Cunningham, R.K. (1965). Losses of nitrogen by ammonia volatilization from surface-fertilized tropical forest soil. Trop. Agric. (Trin.) 42, 281-92.

Acquaye, D.K., and Smith, R.W. (1965). Effect of ground covers and fertilizers on establishment and yield of cocoa on clear felled land in Ghana. Expl. Agric. 1, 131-9.

Agamuthu, P., and Broughton, W.J. (1981). Nitrogen cycling in a legume-oil palm ecosystem in Malaysia. In 'Nitrogen Cycling in South-East Asian Wet Monsoonal Ecosystems'. (Eds R. Wetselaar, J.R. Simpson and T. Rosswall), pp. 113-8. (Aust. Acad. Sci: Canberra.)

Agamuthu, P., and Broughton, W.J. (1985). Nutrient cycling within the developing oil palm-legume ecosystem. Agric. Ecosys. Envir. 13, 111-23.

Agamuthu, P., and Broughton, W.J. (1986). Factors affecting the development of the rooting system in young oil palms (Elaeis guineensis JACQ.). Agric. Ecosys. Envir. 17, 173-80.

Agamuthu, P., Chan, Y.K., Jesinger, R., Khoo, K.M., and Broughton, W.J. (1980). Effect of diphenyl ether pre-emergence herbicides on legume cover establishment under oil palm (Elaeis guineensis JACQ.) Agro-Ecosystems 6, 193-208.

Agamuthu, P., Chan, Y.K., Jesinger, R., Khoo, K.M., and Broughton, W.J. (1981). Effect of differently managed legumes on the early development of oil palms (Elaeis guineensis JACQ.) Agro-Ecosystems 6, 315-23.

Aranguren, J., Escalante, G., and Herrera, R. (1982a). Nitrogen cycle of tropical perennial crops under shade trees. I. Coffee. Plant Soil 67, 247-58.

Aranguren, J., Escalante, G., and Herrera, R. (1982b). Nitrogen cycle of tropical perennial crops under shade trees. II. Cacao. Plant Soil 67, 259-69.

Atkinson, D., and White, G.C. (1976). Soil management with herbicides - The response of soils and plants. Proc. Brit. Crop Protec. Conf. - Weeds. 3, 873-83.

Ayanaba, A., and Dart, P.J. (1977). 'Biological Nitrogen Fixation in Farming Systems of the Tropics'. (John Wiley and Sons: New York.)

Bartholomew, W.V. (1972). Soil nitrogen and organic matter. <u>In</u>: 'Soils of the Humid Tropics'. (Nat. Acad. Sci.: Washington D.C.), pp. 63–81.

Bartholomew, W.V. (1973). Soil nitrogen in the tropics. N.C. Agric. Exp. Stn. Tech. Bull. 219, 68–89.

Baxter, P., and Newman, B.J. (1971). Orchard and soil management trials. II. Effect of herbicides and nitrogen on growth and yield of young apple trees in permanent pasture. <u>Aust. J. Exp. Agric. Anim. Husb.</u> 11, 105–12.

Bornemisza, E. (1982). Nitrogen cycling in coffee plantations. <u>Plant Soil</u> 67, 241–6.

Bould, C., and Jarrett, R.M. (1962). The effect of cover crops and NPK fertilizers on growth, crop yield and leaf nutrient status of young dessert apple trees. <u>J. Hort. Sci.</u> 37, 58–82.

Bourke, R.M. (1984). Food, coffee and casuarina: An agroforestry system from the Papua New Guinea Highlands. <u>Agrofor. Syst.</u> 2: 273–9.

Boyer, J. (1973). Cycles de la matière organique et des elements mineraux dans une cacaoyere Camerounaise. <u>Cafe, Cacao, The</u> 17, 3–23.

Broughton, W.J. (1976). Effect of various covers on the performance of <u>Elaeis guineensis</u> (JACQ.) on different soils. <u>In</u> 'Proceedings of the International Oil Palm Conference, Kuala Lumpur, 1976'. (Eds D.A. Earp and W. Newall) pp. 501–25. (Incorp. Soc. Planters: Kuala Lumpur.)

Broughton, W.J. (1977). Effect of various covers on soil fertility under <u>Hevea brasiliensis</u> Muell. Arg. and on growth of the tree. <u>Agro-Ecosystems.</u> 3, 147–70.

Cameron, S.H., and Compton, O.C. (1945). Nitrogen in bearing orange trees. <u>Proc. Am. Soc. Hort. Sci.</u> 46, 60–8.

Carvajal, J.F. (1959). Mineral nutrition and crop requirements of the coffee crop. MAG Tech. Bull. 9, 1–16.

Chandapillai, M.M. (1968). Studies of root systems of some cover plants. <u>J. Rubb. Res. Inst. Malays.</u> 20: 117–29.

Chew, P.S., and Koo, K.T. (1980). Nutrient responses of perennial tree crops on coastal clay soil in Peninsular Malaysia. <u>In</u> 'Proceedings of the Conference on Classification and Management of Tropical Soils, Kuala Lumpur, 1977'. (Ed. K.T. Joseph). pp. 446–56. (Malaysian Society of Soil Science : Kuala Lumpur.)

Childers, N.F. (1975). 'Modern Fruit Science'. 6th Ed. (Rutgers University: N.J.)

Clarke, W.C. (1976). The maintenance of agriculture and human habitats within the tropical forest ecosystem. <u>In</u> 'Report of Symposium on Ecological Effects of Increasing Human Activities on Tropical and Subtropical Forest Ecosystems'. pp. 103–14. (Aust. Gov. Publ. Serv.:Canberra.)

Cooil, B.J., and Fukunaga, E.T. (1959). Mineral nutrition of coffee. II. Intensive fertilizer application and its effects. <u>In</u> 'Progress in Coffee Production Techniques'. pp. 91–95. (Interamerican Inst. Agric. Sci. : Costa Rica.)

Cooke, G.W. (1972). 'Fertilizing for Maximum Yield'. (Crosby, Lockwood, Staples:London.)

Corley, R.H.V., Hardon, J.J., and Wood, B.J. (1976). 'Oil Palm Research'. (Elsevier: Amsterdam.)
Date, R.A. (1973). Nitrogen, a major limitation in the productivity of natural communities. <u>Soil Biol. Biochem.</u> 5, 5–18.

Dierendonck, F.J.E. van (1963). Coffee production and fertilizer treatment. <u>Outlook Agric.</u> 4, 13–21.

Edwards, P.J., and Grubb, P.J. (1982). Studies on mineral cycling in a montane rainforest in New Guinea. IV. Soil characteristics and the division of mineral elements between vegetation and soil J. Ecol. 70, 649–66.

Embleton, T.W., Jones, W.W., and Platt, R.G. (1975). Plant nutrition and citrus fruit crop quality and yield. <u>HortSci.</u> 10, 48–50.

Embleton, T.W., Reitz, H.J., and Jones, W.W. (1973). Citrus fertilization. <u>In</u> 'The Citrus Industry.' Vol 3. (Eds L.D. Batchelor and H.J. Webber) pp. 122–82. (Div. Agric. Sci., University of California: Berkley.)

Evans, D.O., Yost, R.S., and Lundeen, G.W. (1983). 'A Selected and Annotated Bibliography of Tropical Green Manures and Legume Covers'. (Hawaiian Inst. Trop. Agric. Human Res. University of Hawaii Research Extension Series 028:Honolulu.)

Georgi, C.D.V. (1931). The removal of plant nutrients in oil palm cultivation. <u>Malayan Agric. J.</u> 19, 484.

Georgi, C.D.V., and Teik, G.L. (1932). The removal of plant nutrients in coconut cultivation. <u>Malayan Agric. J.</u> 20, 358–64.

Gosz, J.R. (1981). Nitrogen cycling in coniferous ecosystems. Ecol. Bull. (Stockholm) 33: 405–26.

Gray, B.S. (1969). Ground covers and performance. <u>J. Rubb. Res. Inst. Malays.</u> 21, 107–12.

Greenland, D.J. (1977). Contribution of micro-organisms to the nitrogen status of tropical soils. <u>In</u> 'Biological Nitrogen Fixation in Farming Systems of the Tropics'. (Eds A. Ayanaba and P.J. Dart) pp. 13–25. (John Wiley and Sons: N.Y.)

Grubb, P.J., and Edwards, P.J. (1982). Studies of mineral cycling in a montane rainforest in New Guinea. III. The distribution of mineral elements in the above-ground material. J. Ecol. 70, 623-48.

Habib, R. (1984). The formation of nitrogen reserves in fruit trees: a review. Fruits 39, 623-35.

Hansen, P. (1980). Yield components and fruit development in 'Golden Delicious' apples as affected by the timing of nitrogen supply. Scient. Hortic. 12, 243-57.

Haynes, R.J. (1980). Influence of soil management practice on the orchard agro-ecosystem. Agro-Ecosystems. 6, 3-32.

Heal, O.W., Swift, M.J., and Anderson, J.M. (1982). Nitrogen cycling in United Kingdom forests: the relevance of basic ecological research. Phil. Trans. R. Soc. Lond. B296, 427-44.

Huxley, P.A. (1980). The need for agroforestry and special considerations regarding field research. In 'Nuclear Techniques in the Development of Management Practices for Multiple Cropping Systems'. (IAEA Tech. Doc. 235: Vienna.)

Huxley, P.A. (1986). The prediction of biological productivity and sustainability of tree-crop mixtures. Trop. Agric. (Trin.) 63, 68-70.

Jacks, G.V., Brind, W.D., and Smith, G.A. (1955). Mulching. Commonw. Bur. Soil Sci. Tech. Commun. No. 49.

Janzen, D. (1973). Tropical Agro-ecosystems. Science 182, 1212-9.

Jentsch, D.W. and Eaton, G.W. (1982). Nitrogen fertilizer and fruit removal effects upon leaf mineral content in apple trees. Scient. Hortic. 18, 49-56.

Joffe, J.S. (1955). Green manuring viewed by a pedologist. Adv. Agron. 7, 141-87.

Jordan, D., and Opoku, A.A. (1966). The effect of selected soil covers on the establishment of cocoa. Trop. Agric. (Trin.) 43, 155-66.

Kato, T. (1986). Nitrogen metabolism and utilization in citrus. Hortic. Rev. 8: 181-216.

Keeney, D.R. (1982). Nitrogen management for maximum efficiency and minimum pollution. In 'N in agricultural soils'. (ed. F.J. Stevenson), pp. 605-49 (Amer. Soc. Agron.:Madison.)

Khanna, P.K., and Nair, P.K.R. (1980). Evaluation of fertilizer practices for coconuts under pure and mixed cropping systems in the West Coast of India based on a systems-analysis approach. In 'Proceedings of the Conference on Classification and Management of Tropical Soils, 1977'. (Ed. K.T. Joseph) pp. 457-66. (Malaysian Soc. Soil Sci.: Kuala Lumpur.)

King, K.F.S. (1979). Agroforestry and the utilization of the fragile ecosystems. For. Ecol. and Manag. 2, 161-8.

Kramer, P.J., and Kozlowski, T.T. (1979). 'Physiology of Woody Plants' (Academic Press: N.Y.)

Leece, D.R. (1976). Diagnosis of nutritional disorders of fruit trees by leaf and soil analyses and biochemical indices. J. Aust. Inst. Agric. Sci. 42, 3-19.

Leon, J., de (1928). Cover crops, fertilizer and top working experiments with manderin trees at the Tanauan Citrus Experiment Station from 1923 to 1927. Philippine Agric. Rev. 21, 173-82.

Ludders, P. (1980). Effects of time and amount of nutrient additives on nutrient status and distribution and on fruit quality. In 'Mineral Nutrition of Fruit Trees'. (Eds D. Atkinson, J.R. Jackson, R.D. Sharples and W.M. Waller) pp. 165-72. (Butterworths: London.)

Mainstone, B.J. (1963). Residual effects of ground cover and of continuity of N fertilizers. Emp. J. Exp. Agric. 31, 213-25.

Manguiat, I.J., Agamuthu, P., Kawana, A., Majid, N.M., Pushparajah, E., Riswan, J., Rosswall, T., Singh, K.P., Soetanto, S., Srivastava, P.B.L., Suivanarit, P., Thayib, M.H. (1981). Nitrogen balance in forests and plantation crops. In 'Nitrogen Cycling in South-East Asian Wet Monsoonal Ecosystems'. (Eds R. Wetselaar, J.R. Simpson and T. Rosswall). pp. 203-9. (Aust. Acad. Sci.: Canberra.)

Melillo, J.M. (1981). Nitrogen cycling in deciduous forests. Ecol. Bull. (Stockholm) 33: 427-42

Melillo, J.M., and Gosz, J.R. (1983). Interactions of biogeochemical cycles in forest ecosystems. In 'The Major Biogeochemical Cycles and Their Interactions'. (Eds B. Bolin and R.B. Cook) pp. 177-222. (John Wiley and Sons: New York.)

Murneek, A.E. (1942). Quantitative distribution of nitrogen and carbohydrates in apple trees. Mo. Agric. Exp. Stn. Res. Bull. 348, 1-28.

Murray, D.B. (1977). Coconut palm. In 'Ecophysiology of Tropical Crops'. (Eds P. de T. Alvin and T.T. Kozlowksi) pp. 383-407. (Academic Press: N.Y.)

Nair, P.K.R. (1979). Intensive multiple cropping with coconuts in India: Principles, Programmes and Prospects. J. Agron. Crop Sci. Suppl. 6.

Nair, P.K.R. (1985). Fruit trees in tropical agroforesty systems. Nairobi, Kenya, ICRAF/EWC Working Paper No. 32.

Ng, S.K., and Thamboo, S. (1967a). Nutrient contents of oil palms in Malaya. I. Nutrients required for reproduction: Fruit bunches and male inflorescence. Malay. Agric. J. 46, 3-45.

Ng, S.K., and Thamboo, S. (1967b). Nutrient removal studies on Malayan fruits: Durian and Rambutan. Malay. Agric. J. 46, 164-82.

Nutman, P. (1976). 'Symbiotic Nitrogen Fixation in Plants'. (Cambridge Univ. Press:London.)

Nye, P.H., and Greenland, D.J. (1960). The soil under shifting cultivation. Commonw. Bur. Soils Tech. Comm. No. 51

Oladokun, M.A.O. (1980). Legume cover crops, organic mulch and associated soil conditions and plant nutrient content for establishing Tuillou coffee. HortSci. 15, 305-6.

Pacheco, E.B., Santos, H.L. dos, Teixeira, S.L., Chandinalli, L.R., and Feldmann, R. de O. (1973). Soil management of corrado vegetation area under citrus and its influence on plant growth and fruit yield. Pesq. Agropec. Bras. Agron. 8, 109-13.

Page, P.E. (1984). 'Tropical Tree Fruits for Australia', (Queensl. Dep. Prim. Indu.:Brisbane.)

Pereira, J. (1982). Nitrogen cycling in South American savannas. Plant Soil. 67, 293-304.

Pushparajah, E. (1981). Nitrogen cycle in rubber (Hevea) cultivation. In 'Nitrogen Cycling in South-East Asian Wet Monsoonal Ecosystems'. (Eds R. Wetselaar, J.R. Simpson and T. Rosswall) pp. 101-8. (Austral. Acad. Sci.: Canberra.)

Raison, R.J. (1984). Potential adverse effects of forest operations on the fertility of soils supporting fast growing plantations. In 'Site and Productivity of Fast Growing Plantations'. (Eds D.C. Grey, A.P.G. Shonau and C.J. Schutz). pp 457-72. (I.U.F.R.O: Pretoria/Pietermaritzburg.)

Raison, R.J., Connell, M.J. and Khanna, P.K. (1987). Methodology for studying fluxes of soil mineral-N in situ. Soil Biol. Biochem. 19, 521-30.

Ralph, W. (1986). Nitrogen and carbon cycling through crop and pasture. Rur. Res. 130, 28-31.

Rees, A.R., and Tinker, P.B.H. (1963). Dry-matter production and nutrient content of plantation oil palms in Nigeria. Plant Soil 19, 19-32.

Reid, R., and Wilson, G. (1985). 'Agroforestry in Australia and New Zealand'. (Goddard and Dobson Publishers: Box Hill.)

Reuther, W. (1977). Citrus. In 'Ecophysiology of Tropical Crops'. (Eds P. de T. Alvim and T.T. Kozlowski) pp. 409-39. (Academic Press: N.Y.)

Richards, P. (1973). The tropical rainforest. Scient. Amer. 229, 58-67.

Robertson, G.P., and Rosswall, T. (1986). Nitrogen in West Africa: The regional cycle. Ecol. Monog. 56, 43-72.

Robinson, J.B. (1986). Fruits, Vines and Nuts. In 'Plant Analysis: an Interpretation Manual'. (Eds D.J. Reuter and J.B. Robinson) pp. 120-147. (Inka Press: Melbourne.)

Roskoski, J.P. (1982). Nitrogen fixation in a Mexican coffee plantation. Plant Soil. 67, 283-91.

Roskoski, J.P., Bornemisza, E., Aranguren, J., Escalante, G., and Santana, M.B.M. (1982). Report of the work group on coffee and cacao plantations. Plant Soil. 67, 403-7.

Ruthenberg, H. (1971). 'Farming Systems in the Tropics'. (Clarendon Press: London.)

Samuel, C. (1974). Cattle in oil palm. 1. The effects of an integrated grazing system. Planter 50: 201-12.

Sanchez, P.A. (1976). 'Properties and Management of Soils in the Tropics'. (Wiley: New York.) Santana, M.B.M., and Cabala-Rosand, P. (1982). Dynamics of nitrogen in a shaded cacao plantation. Plant Soil. 67, 271-81.

Sato, K. (1961). The effects of growth and fruiting on leaf composition of citrus trees. In 'Plant Analysis and Fertilizer Problems'. (Ed. W. Reuther) pp. 400-8. (Amer. Inst. Biol. Sci.: Washington D.C.)

Singh, K.P., and Pandey, O.N. (1981). Cycling of nitrogen in a tropical deciduous forest. In 'Nitrogen Cycling in South-East Asian Wet Monsoonal Ecosystems'. (Eds R. Wetselaar, J.R. Simpson and T. Rosswall), pp. 123-130. (Aust. Acad. Sci.: Canberra.)

Smith, F.W. (1986). Interpretation of Plant Analysis: concepts and principles. In 'Plant Analysis: an Interpretation Manual'. (Eds D.J. Reuter and J.B. Robinson) pp. 1-12. (Inka Press: Melbourne.)

Soong, N.K. and Yap, W.C. (1976). Effect of cover management on physical properties of rubber-growing soils. J. Rubb. Res. Inst. Malays. 24, 145-59.

Stephenson, R.A., Cull, B.W., and Stock, J. (1986a). Vegetative flushing patterns of macadamia trees in South East Queensland. Scient. Hortic. 30, 53-62.

Stephenson, R.A., Cull, B.W., Mayer, D.G., Price, G., and Stock, J. (1986b). Seasonal patterns of macadamia leaf nutrient levels in South East Queensland. Scient. Hortic. 30, 63-71.

Suwanarit, P., Chantanao, A., and Srimahasongkham, S. (1981). Nitrogen fixation by Azotobacter in tropical soils. In 'Nitrogen Cycling in South-East Asian Wet Monsoonal Ecosystems'. (Eds R. Wetselaar, J.R. Simpson and T. Rosswall) pp. 26-8. (Aust. Acad. Sci:Canberra.)

Tanner, E.V.J. (1985). Jamaican montane forests : nutrient capital and cost of growth. J. Ecol. 73, 553-68.

Taylor, B.H., and Ferree, D.C. (1986). The influence of summer pruning and fruit cropping on the carbohydrate, nitrogen and nutrient composition of apple trees. J. Am. Soc. Hort. Sci. 111, 342-6.

Taylor, B.K. (1967a). Storage and mobilization of nitrogen in fruit trees: a review. J. Aust. Inst. Agric. Sci. 33, 23-9.

Taylor, B.K. (1967b). The nitrogen nutrition of the peach tree. I. Seasonal changes in nitrogenous constituents in mature trees. Aust. J. Biol. Sci. 20, 379—87.

Taylor, B.K. (1969). The role of nutrition in fruit set and fruit growth. J. Aust. Inst. Agric. Sci. 35, 168—74.

Taylor, B.K., and van den Ende, B. (1969). The nitrogen nutrition of the peach tree. IV. Storage and mobilization of nitrogen in mature trees. Aust. J. Agric. Res. 20, 869—81.

Taylor, B.K., and van den Ende, B. (1970). The nitrogen nutrition of the peach tree. VI. Influence of autumn nitrogen applications on the accumulation of nitrogen, carbohydrate and macroelements in one—year—old peach trees. Aust. J. Agric. Res. 21, 693—8.

Thiagalingam, K. (1983). The role of casuarina in agroforestry. In 'Casuarina Ecology Management and Utilisation'. (Eds S.J. Midgley, J.W. Turnbull and R.D. Johnston) pp. 175—79. (CSIRO:Melbourne.)

Thomas, D. (1978). Pastures and livestock under tree crops in the humid tropics: Trop. Agric., (Trin.) 55, 39—44.

Tinker, P.B.H. (1976). Soil requirements of the oil palm. In 'Oil Palm Research'. (Eds R.H.V. Corley, J.J. Hardon and B.T. Wood) pp. 165—81. (Elsevier:Amsterdam.)

Tinker, P.B.H., and Smilde, K.W. (1963). Dry matter production and nutrient content of plantation oil palms in Nigeria. Plant Soil. 19, 350—63.

Titus, J.S., and Kang, S.M. (1982). Nitrogen metabolism, translocation and recycling in apple trees. Hortic. Rev. 4, 204—46.

Tromp, J. (1970). Storage and mobilization of nitrogenous compounds in apple trees with special reference to arginine. In 'Physiology of Tree Crops'. (Eds L.C. Luckwill and C.V. Cutting) pp. 143—59. (Academic Press:London.)

Vincent, J.M. (1974). Root nodule symbioses with Rhizobium. In 'The Biology of Nitrogen Fixation'. (Ed. A. Quispel). pp. 265—341. (North Holland Publishing Co:Amsterdam.)

Vitousek, P.M. (1984). Litter fall, nutrient cycling and nutrient limitation in tropical forests. Ecology 65, 285—98.

Wallwork, J.A. (1976). 'The Distribution and Diversity of Soil Fauna'. (Academic Press: London.)

Watson, S.E. (1983). 'The Productivity of Pastures on Open Plains and under Coconuts in the Solomon Islands' (Unpublished PhD Thesis, University of Queensland).

Weinbaum, S.A., and Muraoka, T.T. (1986). Nitrogen redistribution from almond foliage and pericarp to the almond embryo. J. Am. Soc. Hort. Sci. 111, 224—8.

Wetselaar, R., Denmead, O.T., and Galbally, I.E. (1981). Environmental problems associated with terrestrial nitrogen transformations in agrosystems in the wet monsoonal tropics. In 'Nitrogen cycling in South—East Asian Wet Monsoonal Ecosystems'. (Eds R. Wetselaar, J.R. Simpson and T. Rosswall) pp. 157—64. (Aust. Acad. Sci.:Canberra.)

Whiteman, P.C. (1980). 'Tropical Pasture Science' (Oxford University Press:New York.)

Willey, R.W. (1975). The use of shade in coffee, cocoa and tea. Hortic. Abstr. 45, 791—8.

Young, J.L., and Aldag, R.W. (1982). Inorganic forms of nitrogen in soil. In 'Nitrogen in Agricultural Soils'. (ed. F.J. Stevenson). pp. 43—66. (Amer. Soc. Agron. Inc: Madison.)

NITROGEN CYCLING IN MULTIPLE CROPPING SYSTEMS

B.T. Kang

ABSTRACT

Nitrogen is the key nutrient for sustaining or increasing food production in the tropics. Multiple cropping systems, particularly intercropping, sequential cropping and rotations, have benefited from inclusion of legumes in the production sequence. Multiple cropping systems are more efficient in using native, biologically fixed, and fertilizer N.

The N benefit to non-legumes in multiple cropping with legumes is mostly derived from N left in crop residue. The amount of N transferred from grain legumes is usually low because most of the N is removed in the grain harvest. There are some indications that in intercropping systems, N excretions from the legumes to the associated crop may take place. The N contribution is higher from sole than from intercropped grain legumes. Sole cropped grain legumes can contribute between 40 to 70 kg N/ha to the succeeding crop. Inclusion of woody legumes in an alley cropping system can help maintain soil productivity with less N input.

INTRODUCTION

Nitrogen is the most important plant nutrient for food crop production in the tropics. Because of its high demand for growth and its high mobility in the soil, N is usually also the most commonly exhausted nutrient in the soil.

Although natural N supply processes are important, they seldom provide an abundance of N for long to sustain production on the same land (Bartholomew 1972). In the traditional slash–and–burn cultivation system widely practised in the humid and subhumid tropics, available soil N supply exhausted during the short cropping cycle is mainly restored during the subsequent fallow period through biological N_2 fixation. However, rapid population increases during the last few decades have resulted in longer use of the farm lands. The shorter fallow periods as a consequence are inadequate to regenerate soil fertility, thus leading to lower crop yields. Although yields can be augmented by fertilizer use, this practice is only a partial solution for maintaining the productivity of the upland soils widely distributed in the tropics (Bache and Heathcote 1969; Kang and Juo 1983). On these poorly buffered soils injudicious N application with continuous cropping can lead to rapid soil acidification and eventual decline in crop yield (Bache 1965; Bache and Heathcote 1969; Le Mare 1972; Nnadi and Arora 1985). There is a need for a different approach to N management for upland food crop production on these soils.

Multiple cropping is widely practised throughout the humid and subhumid tropics (Dalrymple 1971; Beets 1982) and is mainly associated with traditional production systems. Farmers for various reasons have learned to better exploit the environment to achieve higher total crop production by using crop combinations. Although multiple cropping systems have been practised for centuries, it is only recently that there has been growing interest to study them scientifically (Okigbo and Greenland 1976). Information on fertilizer use is also limited (Oelsligle et al. 1976). This paper reviews some aspects of multiple cropping systems to provide a better basis for developing integrated N management for sustained food crop production in the tropics.

MULTIPLE CROPPING SYSTEMS

Types

Multiple cropping is defined as the growing of more than one crop on the same piece of land during one calendar year (Andrews and Kassam 1976). This includes intercropping (growing of two or more crops simultaneously), relay intercropping (growing two or more crops simultaneously during part of the life cycle of each), and sequential cropping (growing two or more crops in sequence). Intercropping can include row intercropping (crops are planted in alternate rows), mixed intercropping (crops grown with no distinct row arrangement) or strip intercropping (growing two or more crops simultaneously in strips).

Mixed intercropping and relay cropping are the most common types of multiple cropping systems practised by resource-poor farmers in the tropics, particularly in Africa (Dalrymple 1971; Andrews and Kassam 1976; Okigbo and Greenland 1976; Beets 1982; Francis 1986). In the humid tropics, farmers also practise multi-storey cropping in home gardens. This is an association of tall and short woody (cash) perennials with a subsistence crop with no particular arrangement (Michon 1983).

Other intercropping systems include live mulch and alley cropping systems. The live mulch system consists of growing food crops such as maize in an established field of low-growing leguminous cover crops (Akobundu 1980). In alley cropping, food crops are grown between hedgerows of woody species that are periodically pruned to prevent shading the associated crops (Kang et

334

al. 1981).

Advantages and management of multiple cropping systems

The various systems have been developed over centuries in the different regions of the tropics and are closely related to prevailing local and socio-economic constraints and conditions (Beets 1982; Steiner 1982). Intercropping is generally more prevalent and complex in the forest than in the savanna areas, mainly because of the longer growing season resulting from higher moisture availability (Okigbo and Greenland 1976; Steiner 1982; Kang 1983). In areas of intensive subsistence agriculture and high population pressure as in parts of Asia, labor is abundant and multiple cropping is the most logical way to produce crops (Beets 1982).

Many reasons are cited for the popularity of intercropping among small farm holders: (1) better resource utilization; (2) higher yield stability; (3) higher returns, and (4) nutritionally more balanced produce (Andrews 1972; Norman 1972; Willey and Osiru 1972).

The success of a multicrop system will depend largely on proper management to minimize the competitive pressure of one crop on the other. Particularly important are the spatial and temporal arrangements of the crops grown in association. Productive combinations involve crops with different maturity durations or different statures (Willey 1979; De 1980).

NITROGEN FERTILIZER USE IN MULTIPLE CROPPING SYSTEMS

A marked advantage of a multiple crop is that crop species with different rooting patterns can better exploit together the limited nutrient resources, especially N in the soil and assure a reasonable production, whereas sole crops would give only marginal yields (Baldy 1963; Trenbath 1976; Willey 1979; Beet 1982).

With the increase in cropping intensity in various parts of the tropics, fertilizer is needed to maintain stable yields. However, fertilizers are costly and of limited availability in developing countries, and thus need to be used judiciously and efficiently in production systems. Although many of the intercropping systems show greater advantages under limited nutrient conditions, they can also benefit from fertilizer use because they provide greater yield stability than sole crops due to higher total yields (Rao and Willey 1980). De (1980) mentioned that in legume/cereal intercropping, a change in the N management for the cereal crop can make the system more viable and productive. Basal application of low N fertilizer rates and other nutrients would meet the early requirements of the cereal crop which at the same time acts as starter fertilizer for the legume. Top dressing of N thereafter close to the cereal rows benefits the cereal crop without affecting the productivity of the legumes. Kurtz et al. (1952) clearly demonstrated the importance of an abundance of N and moisture in reducing competition between intercrops.

Efficient fertilizer use in multiple cropping systems, depends on the knowledge of the nutrient requirement and nutrient uptake patterns of the associated crops, so that fertilizers can be applied in a way that will benefit the responsive crop(s) without depressing the yield of the non-responsive crop(s) (Oelsligle et al. 1976; Steiner 1982).

NITROGEN UPTAKE AND TRANSFER IN INTERCROPPING SYSTEMS

Cereal crops grown in mixtures will compete for available N, although combined they make more effective use of this N (Kassam and Stockinger 1973). Legume/non-legume mixtures are believed to be more advantageous than cereal mixtures, because legumes can fix N_2 symbiotically, so that competition for N is minimized. In addition, legumes can contribute to the N requirement of the non-legume. As illustrated in Figure 1, there are a number of possible pathways by which N can enter the non-legume and legume components. The non-legume can obtain N from (1) native soil sources, (2) fertilizers, and (3) legumes. Henzell and Vallis (1977) discussed the various processes involved in N transfer from legumes to other crops. They concluded that the processes are probably the same whether the non-legume is grown at the same time in mixtures or in rotation at a later stage, although the amount of N transferred may be different.

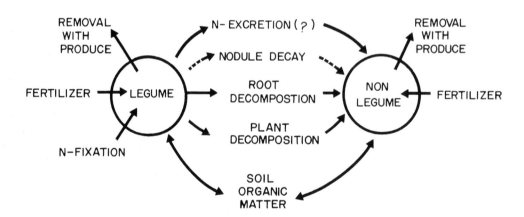

Fig. 1: Nitrogen cycling in multiple cropping systems involving legumes and non-legumes.

Legume/non-legume intercropping

Intercropping pulses and cereals is widely practised by traditional farmers in the sub-humid and semi-arid tropics. One or two pulses such as black gram, cowpeas, chickpeas, groundnuts, mung bean, phaseolus bean, pigeon peas and soybeans are combined with cereals such as maize, millet and sorghum (Norman 1972; De 1980; Reddy and Willey 1982; Faris et al. 1983). Pulses are also intercropped with crops such as cassava (Leihner 1983); and sugarcane (Yadav 1981). Among these crop combinations, N management in cereal-legume mixtures, has been intensively investigated in recent years.

Multiple cropping with legumes can provide two types of benefits for N supply (Reddy et al. 1983):

(i) Current transfer of N from the legume during the life of the intercropped species.
(ii) Residual supply, in which N fixed by the legume grown as a relay intercrop or in sequential cropping, is available after senescence of the legume and the decomposition of its residues.

Under conditions where N may be limiting, legumes may compete for N with the associated crop. Fortunately, however, legumes are usually weaker competitors for mineral N than, for example, grasses (Henzell and Vallis 1977). Legumes also utilize mineral N in preference to forming nodules and fixing N_2 (Allos and Bartholomew 1959). There is however, still some uncertainty about the extent of the N contribution from intercropped legumes.

Some researchers believe that N fixed in the root nodules can be transferred by excretion to the associated non-legume crop. Virtanen et al. (1937) demonstrated in greenhouse experiments that under certain circumstances grain legumes excreted part of the N fixed in the root nodules. However, as expressed by Eaglesham et al. (1981), this process of N excretion is difficult to demonstrate and the conditions conducive to significant N transfer are also difficult to define. There are, however, some field results which indicate possible N excretion. Agboola and Fayemi (1972) observed that maize intercropped with green gram did not respond to N fertilization in low N status soil. Remison (1978) made a similar observation for maize grown in association with cowpea in western Nigeria. De (1980) and Bandyopadhay and De (1986) reported large increases in maize and sorghum yields and N uptake when intercropped with grain legumes. Eaglesham et al. (1981) showed that maize grown in association with cowpeas in western Nigeria showed significantly higher N uptake than maize grown alone. Results of a ^{15}N experiment indicated possible N excretion by the intercropped cowpea which benefited the maize only under conditions of low soil mineral N but this effect could not be demonstrated where mineral N was plentiful (Eaglesham et al. 1981). Recent ^{15}N studies by Patra et al. (1986) support this observation; they observed that 78% of the total N uptake by maize (22.2 kg N/ha) was of atmospheric origin and was obtained by transfer of fixed N from cowpea grown in association with the maize. Inoculation of the cowpea with Rhizobium increased both the amount of N fixed by the legume crop and N transferred to the intercropped maize.

Henzell and Vallis (1977) attributed some of the growth benefits of non-legumes when intercropped with legumes, to the sparing effect of the legumes on the availability of N for the associated crop, since nodulated legumes usually take up N less rapidly from the soil than the non-nodulated plants. For the same planted area, a cereal intercrop will have access to a larger pool of N than when sole cropped. Depletion of available soil N by the intercropped cereal would also tend to enhance the rate of N fixation by the legumes.

Intercropping studies conducted with cereals and legumes have given mixed results on the N benefit to the cereal. Waghmare and Singh (1984) observed large increases in sorghum yield and N uptake when intercropped with green gram, grain cowpea and particularly fodder cowpea, but showed little benefit when intercropped with groundnut and soybeans. The advantage of fodder cowpea was that it was harvested at 55 days after planting and left biologically fixed N available in the soil for the sorghum, which was harvested later. Groundnut and cowpeas which matured later than the sorghum and which translocated much of their fixed N_2 to their grain, contributed little to the sorghum. Pandy and Pendleton (1986) observed improvement of maize yield when part of the intercropped soybean crop was ploughed under 42 days after planting. They estimated the N contribution from the soybean green manure at about 28 kg N/ha. Most results (Wahua and Miller 1978; Ahmed and Gunasena 1979; Yadav 1981; Ahmed and Rao 1982; Nambiar et al. 1983; Reddy et al. 1983) showed that: (1) with no N applied, the yields of sole and intercropped cereal were similar, this gives little support to the belief that grain legumes in intercropping might benefit the associated cereal during the growing season; (2) sole and intercropped cereal responded similarly to applied N; (3) intercropping tended to reduce the yield and N uptake of legumes particularly with increased rates of N fertilizers; and (4) land equivalent ratios decreased with increased N rates. Ahmed and Rao (1982) also showed that the yield advantage of maize/soybean intercropping varied with N fertilizer level, ranging from 64% at zero N application to 42% at 100% of the recommended N rate. They concluded that intercropping seems to be especially useful for small subsistence farmers who use little or no fertilizers.

The large decrease in yield of intercropped legumes at high N doses is mainly attributed to shading (Wahua and Miller 1978; Ahmed and Gunasena 1979; Yadav 1981; Nambiar et al. 1983). Shading also caused less nodulation in the legumes (Wahua and Miller 1978; Yadav 1981; Nambiar et al. 1983). Graham and Rosas (1980) however, did not detect any decrease in the N_2 fixation of climbing beans intercropped with maize. This was probably because seasonal N_2 fixation rates had peaked before competition for light and nutrients from maize affected the bean crop. Neves et al. (1982) clearly demonstrated in screenhouse trials that shading of cowpea reduced photosynthesis, and through this affected N_2 fixation. The lack of photosynthates due to shading also explained the reduction in seed yield of the legume.

In intercropping studies with cassava and cowpeas, Leihner (1983) showed positive response of tuber yield of sole cassava to application of 50 kg N/ha. Root yield declined at higher N rates. However, tuber yield of cassava intercropped with cowpea increased with higher N rates. Cowpea yield was not affected by intercropping or N application. In this case, N fertilization benefited the intercrops.

Non—legume intercropping

Kassam and Stockinger (1973) have determined the N uptake in a pearl millet/sorghum intercrop. The more aggressive millet utilized 80% of total available N, leaving only 20% for the intercropped sorghum. Combined, they removed 182 kg N/ha, which was more than the 132 kg N/ha applied as fertilizer, indicating the high foraging efficiency of the mixture. Continuous cereal—cereal intercropping can therefore exhaust soil N without adequate inputs of fertilizer N.

In maize/cassava intercropping systems, N application mostly benefits the maize crop. On high N soil, N application reduced cassava root yields because of early shading by the maize. However on low N soil, N application benefited both maize and cassava crops (Kang and Wilson 1981; Van der Heide et al. 1985). On an acid Ultisol, maize and cassava showed a combined high recovery of fertilizer N applied, with the cassava showing a higher recovery percentage than maize (Van der Heide et al. 1985). The higher N recovery by the cassava is attributed to its deeper rooting system which allows better utilization of available N from deeper soil layers.

NITROGEN IN RESIDUES OF GRAIN LEGUME CROPS

Grain legumes fix large amounts of N_2, thus it is generally assumed that they will enrich the soil N. Nutman (1976) gave the following figures for N fixed by commonly grown pulses: cowpeas, 73–354 kg N/ha; groundnuts, 72–124 kg N/ha; pigeon peas, 168–208 kg N/ha; and soybeans, 55–168 kg N/ha. However, pulses have a high N harvest index and much of this N may be removed in the grain. Pate and Minchin (1980) estimated the N harvest indices of cowpeas, groundnuts, and soybeans as 61%, 80% and 75% respectively. In addition to N removal with the grains, vegetative parts may also be removed for feeding animals or at harvest, consequently the legume crop may return only a small amount of the total fixed N as a residue to the soil. To illustrate the amount of N returned as a residue from a grain legume crop, some examples are discussed.

Pigeon peas

This crop, widely grown in the tropics particularly on the Indian subcontinent, has a high N uptake which varies greatly with growth duration and dry matter yield (Sheldrake and Narayanan 1979; Rao et al. 1981). Sheldrake and Narayanan (1979) reported values for medium duration cultivars of 89 to 113 kg N/ha on a Vertisol and 79 kg N/ha on an Alfisol. If the seeds and plant parts are removed at harvest, they estimated the amount of N contribution to the soil at 40 kg N/ha arising from fallen leaves (32–36 kg N/ha) and from the root system (c. 10 kg N/ha). With only seed removal at harvest, the N contribution increased to 56–70 kg N/ha. A more detailed study of N uptake by 11 cultivars of pigeon peas grown on an Alfisol was reported by Rao et al. (1981). They showed large varietal differences in N uptake and N partition between different plant parts (Table 1). The N_2 fixation of medium and late maturing cultivars was estimated to be 13–55 kg N/ha. Their data also confirmed the results of Sheldrake and Narayanan (1979), that considerable amounts of N were lost as fallen leaves, ranging from 9–28 kg N/ha, which represents between 12–26% of total N uptake. Nitrogen uptake in pods and seeds ranged from 33–56 kg N/ha, amounting to 25–63% of total N uptake. Assuming the return to the soil of all non-seed material, the crop can contribute 26–101 kg N/ha. Since, in most pigeon pea growing areas, plant parts will also be removed at harvest for firewood and/or fodder, the amount of N contribution by medium and late maturing cultivars in crop residue amounted to only 21–36 kg N/ha, which is much lower than the figures reported by Sheldrake and Narayanan (1979).

Table 1: N uptake in different plant parts of pigeon pea cultivars grown on an Alfisol, and comparison with sorghum grown on the same soil as a monoculture (Rao et al. 1981)

Cultivar[1]	Pod with seed[2]	Plant top	Fallen plant parts	Root and nodule	Total in residue	Total uptake	Balance against sorghum
			— kg N/ha —				
Prabhat	43.4(63)	11.6	11.4	2.7	25.7	69.1	+ 4.4
Prant A–3	44.3(62)	14.9	8.5	3.9	27.3	71.6	+ 6.9
T–21	55.7(52)	13.1	16.5	4.6	52.2	107.9	+ 43.2
Upas–120	46.1(50)	24.9	15.5	5.3	45.7	91.8	+ 27.1
BDN–1	52.3(44)	37.2	24.6	4.1	65.9	118.2	+ 53.5
No 148	53.2(44)	40.9	17.6	8.1	66.6	119.8	+ 55.1
JA 275	33.7(43)	19.0	17.7	7.5	44.2	77.9	+ 13.2
ICP–7035	33.4(33)	34.7	21.0	11.9	67.6	101.0	+ 36.3
ICP–7065	49.0(45)	23.0	28.1	7.6	58.7	107.7	+ 43.0
T–7	33.3(25)	64.2	21.3	15.3	100.8	134.1	+ 69.4
NP (WR)–15	34.0(30)	54.1	14.7	11.5	80.3	114.3	+ 49.6
Sorghum	9.5	51.6	0	3.6		64.7	

1. Days to maturity for cultivars respectively; 115, 115, 130, 125, 130, 150, 170, 170, 175, 215, 240 and 175 days; 2. In brackets is N harvest index = % of N in seed of total N uptake.

Cowpeas

Cowpea grown mostly as an intercrop in West Africa, usually has low yields in traditional farming systems. With good management the crop can give a good dry matter and N yield. Eaglesham et al. (1982), showed a total N uptake of cowpea cultivars ranging from 73 to 118 kg N/ha. Observations by the author in southern Nigeria also showed that the early maturing cowpea cv. Prima grown on sandy soil takes up c. 112 kg N/ha at maturity. The N yield in seeds plus pods was about 60%, which is in close agreement with results reported by Pate and Minchin (1980). Eaglesham et al. (1977) reported the N partitioning at harvest of cowpea cv. K 2809 grown in the screenhouse as 66.7% of the total N in pod walls and seeds, 30.3% in the plant top (foliage, stem, branches and peduncle) and only 2.9% in the roots and nodules.

Eaglesham et al. (1982) also studied the N balance of four cowpea cultivars in field trials on a N depleted soil. With applications of 25 kg N/ha to the crop, they observed large differences in N accrual among the cultivars (Table 2). Comparing the amount of fixed N and N in seeds, only two of the cultivars, Ife Brown and TVU 1190, gave appreciable positive N balances of + 24 and + 52 kg N/ha. Although the crop residue can contribute between 30 to 85 kg N/ha, much of this N was derived from mineral N uptake. Since only a small portion of N is present in roots and nodules (Eaglesham et al. 1977), removing the seeds plus plant tops at harvest will usually result in soil N depletion.

Table 2: N uptake and balances of four cowpea cultivars grown in low N status soil (adapted from Eaglesham et al. 1982)

Cultivar	Mineral N uptake	Fixed N	Total N uptake	N in seed[1]	N in residue	N balance
			— kg N/ha —			
TVU 4552	27	49	76	46(66)	30	+3
ER-1	32	50	82	48(59)	34	+2
Ife Brown	25	81	106	57(57)	49	+24
TVU 1190	33	101	134	49(37)	85	+52

1. In brackets is N harvest index = % of N in seed of total N uptake

Soybeans

Soybean is widely grown in the Asian and Latin American tropics. It is the crop most studied in terms of N_2 fixation. In North America, N_2 fixation was between 15 and 263 kg N/ha (LaRue and Patterson 1981), with an average value of N_2 fixation of 75 kg N/ha (at an average seed yield of 1471 kg/ha), this is about 50% of the whole plant N. This N_2 fixation figure can also be expected in the tropics as similar yields are attainable (Franco 1985). Hammond et al. (1957) studied N uptake and partitioning in soybean cv. Richland grown on two soil types (Table 3). They observed large differences in N uptake by the crop grown on the two soil types. The N harvest index of 80% in the seed was higher than the 75% reported by Pate and Minchin (1980). With only seed removal, the crop residue contributed between 22 to 41 kg N/ha to the soil. Eaglesham et al. (1982) estimated in a field study, that soybean cv. N 59-5253 grown in a low N status soil took up a total of 216 kg N/ha, of which 188 kg N/ha was derived from N_2 fixation. With a high N harvest index of 92% in the seed, only a paltry 12 kg N/ha was left as a residue after grain harvest.

Table 3: N uptake and N content in various plant organs of mature soybean cv. Richland grown on two Iowan soils (adapted from Hammond et al. 1957)

Soil series	N in leaves[1]	N in stem and roots	N in pods	N in seed[2]	Total N uptake	N in residue
	— — — — — — — — — — *kg N/ha* — — — — — — — — — —					
Webster silt loam	25	8	8	164(80)	205	41
Clarion loam	13	6	4	96(80)	119	22

1. Includes N in fallen leaves; 2. In brackets is N harvest index = % of N in seed of total N uptake

NITROGEN CONTRIBUTION FROM GRAIN LEGUMES TO SUCCEEDING CROPS

Extensive evidence indicates that grain legumes in a cropping sequence increase soil N content and yield of the succeeding non-legume crop (Wild 1972; Jones 1974; Saxena and Yadav 1975; Giri and De 1980; Rao et al. 1981; Kang 1983). However, there are also reported instances (see Dart and Krantz 1977) when they have not contributed to soil N. Eaglesham et al. (1982) have postulated that at intermediate levels of soil N soybeans may even deplete soil N because of inhibition of nodulation and N_2 fixation.

The N contribution from grain legumes to crops grown in relay cropping, sequential cropping or in rotation, will largely depend on the amount of N returned to the soil as crop residue. The potential availability of N in the residue may, however, be variable. Root fragments of legumes and nodules decomposed rapidly in the soil (Chulan and Waid 1981), but Henzell and Vallis (1977) reported that residues containing less than 1.5% N do not mineralize in a short time. In addition, drying and application of plant residues to the soil surface will also slow down decomposition and mineralization (Henzell and Vallis 1977; Read et al. 1985).

Wild (1972) in northern Nigeria found that soil N status was improved following groundnuts. Jones (1974) also found that groundnuts enhanced soil N status, which substantially benefited the yield of the succeeding maize crop, even when the maize received fertilizer N. Despite the removal of the above ground parts of the legume crops at harvest, maize yield was higher following a single groundnut crop than after N-fertilized sorghum or cotton crops or after a cowpea crop. According to Giri and De (1980), pearl millet yield was increased following groundnuts and cowpeas grown for a full season. They estimated the legumes to contribute about 60 kg N/ha to the cereal crop. Improvement in soil N status in the top 30 cm following sole crops of groundnuts and cowpeas of 70 and 50 kg N/ha respectively was also reported by Nnadi (1980). On the other hand, intercropping of the cowpea, soybeans and maize did not increase available N over that of maize alone.

Searle et al. (1981) showed higher soil N status following groundnuts and soybeans grown as sole crops or when intercropped with maize. Nitrogen uptake by a succeeding wheat crop, which is a good measure of residual N, showed large differences. With no N applied to the wheat, at 19 weeks after planting N uptake was higher after groundnut (54 kg N/ha) than after soybean (19 kg N/ha) or maize (12 kg N/ha). Waghmare and Singh (1984) also reported an increase in wheat yield when grown after sorghum intercropped with fodder and grain cowpeas or with groundnut as compared to sole sorghum. But a preceding intercrop of sorghum with soybean did not prove better than sole sorghum for the wheat crop. Intercropping with fodder cowpea (harvested 55 days after sowing) resulted in the highest savings of fertilizer N for the wheat crop. This was followed by intercrops of sorghum with groundnuts, grain cowpeas and green gram. Significant increases in wheat yield were also reported when grown after black gram (110%), green gram (108%) and soybean (44%), compared with those after pigeon pea (Narwall et al. 1983). The preceding crops of green gram and black gram were estimated to reduce the N requirement of the wheat crop by 30 to 60 kg N/ha compared with only 30 kg N/ha after pigeon pea or soybean. De (1980) studied the effect on subsequent wheat yield of preceding maize or sorghum intercropped with groundnut, mungbean, cowpea and black gram. Intercropping black gram either with maize or sorghum benefited wheat yield irrespective of rates of N applied to the wheat crop. An increase in wheat yield was also observed following intercropping of sorghum and mung. There was no benefit from intercropping sorghum and maize with either cowpea or groundnut.

A large residual N effect was observed following sole cropped pigeon pea. According to Rao et al. (1983), sole pigeon pea increased grain yield of the succeeding maize crop by 57% over fallow treatment, but intercropped pigeon pea had little residual effect. Based on the N fertilizer requirement of maize crops (following either fallow, sorghum grown alone, or a sorghum/pigeon pea intercrop) to attain similar yields to that following a sole crop of pigeon pea, the N contribution of sole pigeon pea to the maize crop was estimated at between 38 to 49 kg N/ha. Yadav (1981) reported that sugarcane yield increased by an average of 43% and 40% following sole pigeon pea and a pigeon pea intercrop with maize than when grown after sole maize. According to Ruschel and Vose (1982), sugarcane grown in rotation with soybean could benefit from between 21 to 63 kg N/ha from the legume.

NITROGEN CONTRIBUTION FROM A LEGUMINOUS LIVE MULCH SYSTEM

Akobundu (1980) observed that maize yields after five seasons of continuous cropping without N fertilizer were superior when grown in succeeding years in live mulch plots of Psophocarpus palustris and Centrosema pubescens. In the live mulch plots, maize showed little response to N fertilizer, while in a no-tillage plot the crop responded to applications of over 60 kg N/ha. According to Mulongoy and Akobundu (1985), this benefit of live mulch is only observed in well established plots, while in newly established plots, they reported strong competition for nutrients between the live mulch cover crops and the maize. They estimated the average N contribution from a well-established Psophocarpus plot at about 31 kg N/ha, which is less than 30% of the N content of the live mulch biomass. Arachis repens as a live mulch contributed the least amount of N. The better N contribution from well-established plots was attributed to the higher amount of organic matter that had accumulated under the mulch. This system, which provides good soil cover for soil conservation, also has the potential for steady addition of organic matter and for better soil fertility maintenance even under continuous cropping in the humid tropics.

NITROGEN CONTRIBUTION IN THE ALLEY CROPPING SYSTEM

The food crops grown in the alleys formed by hedgerows of woody species which are kept pruned during the cropping period, can benefit from nutrients recycled by the hedgerows. If the hedgerows are legume species then the alley crop can also benefit from symbiotically fixed N_2 (Kang et al. 1981). On non-acid soils in the humid tropics, inclusion of leguminous woody species such as Leucaena leucocephala and Gliricidia sepium in alley cropping systems has contributed substantial amounts of N to the production system. They can reduce the requirement for fertilizer N for the associated crops (Kang et al. 1981, 1984). To illustrate the amount of N contribution, data for 1986 for a maize crop in a well established trial are presented in Table 4. The woody hedgerows were established on an Alfisol in 1981, spaced 4m apart and pruned to 75 cm height. Plots were sequentially cropped annually with maize followed by cowpeas.

The woody legumes had high N yield compared to non-legumes (Table 4). Intercropping with the woody species increased N uptake by the maize, particularly when grown with the leguminous species which contributed in excess of 40 kg N/ha. The positive contribution from the non-legume species of 13 to 19 kg N/ha may be due to nutrient recycling, since all four woody species have a deep root system. The sequentially cropped cowpea after maize may also contribute N to the system. Kang et al. (1985) also showed that maize and Leucaena have different root feeding zones, with Leucaena extracting moisture from deeper soil layers. Alley cropping systems have good potential for sustained food crop production with low N input, for upland soils.

Table 4: Nitrogen yield from hedgerow prunings during one maize crop, N uptake by the alley cropped maize and estimated N gain from the hedgerows to the system (B.T. Kang, unpublished data)

Woody hedgerow	N yield from prunings[1]	N uptake by maize	Estimated N gain	Maize grain yield
	- - - - - - - - *kg N/ha* - - - - - - -			*kg/ha*
Control	–	26.2	–	1632
Non-legumes:				
Acioa barterii	24.5	38.8	12.6	2588
Alchornea cordifolia	62.0	44.9	18.7	2557
Legumes:				
Gliricidia sepium	127.8	68.6	42.4	3349
Leucaena leucocephala	231.1	68.1	41.9	3210

1. Not including N removed with wood harvested

CONCLUSIONS

Multiple cropping systems have given traditional resource-poor farmers in the tropics an efficient way of utilizing soil N, and of exploiting biologically fixed N_2 provided by legume/non-legume intercropping systems. Although much agronomic data have been generated in the past three decades on the N benefits from multicropping systems, basic information is still needed on how best to maximize N contribution and N transfer from the legumes to non-legumes and also on methods of N fertilization to increase efficiency of N use. Information on the effects of timing and placement of N fertilizers, soil moisture dynamics, and spatial root distribution on N utilization in multiple cropping systems is inadequate and deserves more attention.

Transfer of N from grain legumes to non-legumes in intercropping or sequential cropping systems is generally low, particularly when in addition to the grain, other plant parts are also removed at harvest. Even less N contribution is expected with the increasing use of legume genotypes with N grain harvest indices as high as 80 - 95% (Henzell and Vallis 1977).

Since more N appears to be transferred in sequential or rotational systems than in intercropping systems with annual legumes, better rotational systems need to be developed. Better methods for assessing the N contribution from legumes in rotational systems are needed. Comparing soil N status or N uptake by a succeeding crop for plots previously planted with legumes or cereals tends to exaggerate the N contribution of the legume crop, since plots under preceding cereals tend to be more exhausted of N.

Shrub and tree legumes have great potential as alternative N sources to fertilizer N for resource-poor farmers. Unlike green manure crops, woody species have many uses in the farmer's household so that acceptability by traditional farmers of systems such as alley cropping with woody hedgerows is less of a problem. Research in the tropics on N cycling in woody species and food crop intercropping systems and multistorey cropping systems needs to be expanded.

REFERENCES

Agboola, A.A., and Fayemi, A.A. (1972). Fixation and excretion of nitrogen by tropical legumes. Agron. J. 64, 409-12.

Ahmed, S., and Gunasena, H.P.M. (1979). N utilization and economics of some intercropped systems in tropical countries. Trop. Agric. 56, 115-23.

Ahmed, S., and Rao, M.R. (1982). Performance of maize-soybean intercrop combination in the tropics: results of a multi-location study. Field Crops Res. 5, 147-61.

Akobundu, I.O. (1980). Live mulch: a new approach to weed control and crop production in the tropics. Proc. British Crop Protection Conf. Weeds 1980, pp 377-82.

Allos, H.F., and Bartholomew, W.V. (1959). Replacement of symbiotic fixation by available nitrogen. Soil Sci. 87, 61-7.

Andrews, D.J. (1972). Intercropping with sorghum in Nigeria. Expl. Agric. 8, 139-50.

Andrews, D.J., and Kassam, A.H. (1976). The importance of multiple cropping in increasing world food supplies. In 'Multiple Cropping'. (Eds R.I. Papendic, P.A. Sanchez, and G.B. Triplett). pp. 1-10. (ASA Spec. Publ. 27: Madison.)

Bache, B.W., (1965). The harmful effects of ammonium sulphate on fine sandy soils at Samaru. Proc. 2nd FAO Meeting on Soil Fertility and Fertilizer Use in West Africa, Dakar, Senegal I, pp. 15-20.

Bache, B.W., and Heathcote, R.G. (1969). Long-term effects of fertilizers and manure on soil and leaves of cotton in Nigeria. Expl. Agric. 5, 241-7.

Baldy, C.H. (1963). Cultures associees et productivite de l'eau. Ann. Agron. 14, 489-534.

Bandyopadhay, S.K., and De, R. (1986). N relationship in legume non legume association grown in an intercropping system. Fert. Res. 10, 73-82.

Bartholomew, W.V. (1972). Soil nitrogen and organic matter. In 'Soils of the Humid Tropics'. pp. 63-81. (Nat. Acad. Sci.: Washington D.C.)

Beets, W.C. (1982). 'Multiple Cropping and Tropical Farming Systems'. (Westview Press Inc.: Colorado.)

Chulan, A., and Waid, J.S. (1981). Loss of nitrogen from decomposing nodules and roots of the tropical legume Centrosema pubescens to soil. In 'Nitrogen Cycling in South-East Asian Wet Monsoonal Ecosystems'. (Eds R. Wetselaar, J.R. Simpson and T. Rosswall). pp. 150-3. (Australian Academy of Science: Canberra.)

Dalrymple, G.D. (1971). 'Survey of Multiple Cropping in Less Developed Nations'. (U.S. Agency for Int. Development: Washington D.C.)

Dart, P.J., and Krantz, B.A. (1977). Legumes in the semi-arid tropics. In 'Exploiting the Legume-Rhizobium Symbiosis in Tropical Agriculture'. (Eds J.M. Vincent, A.S. Whitney and J. Bose). pp. 119-54. (Univ. of Hawaii: Hawaii.)

De, R. (1980). Role of legumes in intercropping systems. In 'Nuclear Techniques in the Development of Management Practices for Multiple Cropping Systems'. pp. 73-84, (IAEA: Vienna.)

Eaglesham, A.R.J., Ayanaba, A., Rao, V.R., and Eskew, D.L. (1981). Improving the nitrogen nutrition of maize by intercropping with cowpea. Soil Biol. Biochem. 13: 169-71.

Eaglesham, A.R.J., Ayanaba, A., Rao, V.R., and Eskew, D.L. (1982). Mineral nitrogen effects on Vigna unguiculata and soybean Glycine max crops in a Nigerian soil II. Amounts of nitrogen fixed and accrual to the soil. Plant Soil 68, 183-92.

Eaglesham, A.R.J., Minchin, F.R., Summerfield, R.J., Dart, P.J., Huxley, A.P., and Day, J.M. (1977). Nitrogen nutrition of cowpea (Vigna unguiculata) 3. Distribution of nitrogen within effectively nodulated plants. Expl. Agric. 13, 369-80.

Faris, M.A., Burity, H.A., Dos Reis, O.V., and Mafra, R.C. (1983). Intercropping of sorghum or maize with cowpeas, and common beans under two fertility regimes in northeastern Brazil. Expl. Agric. 19, 251-61.

Francis, C.A. (1986). Introduction: Distribution and importance of multiple cropping. In 'Multiple Cropping Systems'. (Ed. C.A. Francis). pp. 1-19. (Macmillan Publ. Comp.: New York.)

Franco, A.A. (1985). Contribution of biologically fixed nitrogen to foodcrop production in Brazil. In 'Nitrogen Management in Farming Systems in Humid and Subhumid Tropics'. (Eds B.T. Kang and J. van der Heide). pp. 147–66. (Inst. Soil Fertility: Haren.)

Giri, G., and De, R. (1980). Effect of preceding grain legumes on growth and nitrogen uptake of dryland pearl millet. Plant Soil 56, 459–64.

Graham, P.H., and Rosas, J.C. (1980). Plant and nodule development and nitrogen fixation in climbing cultivars of Phaseolus vulgaris L. grown in monoculture or associated with Zea mays, L. J. Agric. Sci. 90, 311–7.

Hammond, L.C., Black, C.A., and Norman, A.G. (1957). Nutrient uptake by soybeans on two Iowa soils. Iowa State Agricultural Experiment. Stat. Res. Bull. 384, pp. 463–512.

Henzell, E.F., and Vallis, I. (1977). Transfer of nitrogen between legumes and other crops. In 'Biological Nitrogen Fixation in Farming Systems in the Tropics'. (Eds A. Ayanaba and P. Dart). pp. 73–88. (J. Wiley & Sons: Chichester.)

Jones, M.J. (1974). Effects of previous crop on yield and nitrogen response of maize at Samaru, Nigeria. Expl. Agric. 10, 273–80.

Kang, B.T. (1983). Fertilizer use in multiple cropping systems in Nigeria, Tanzania and Senegal. In 'Fertilizer Use under Multiple Cropping Systems'. pp. 36–45. (FAO Fertilizer and Plant Nutrient Bull. 5: Rome.)

Kang, B.T., Grimme, H., and Lawson, T.L. (1985). Alley cropping sequentially cropped maize and cowpea with Leucaena on a sandy soil in southern Nigeria. Plant Soil 85, 267–77.

Kang, B.T., and Juo, A.S.R. (1983). Management of low activity clay soils in tropical Africa. In 'Proc. Fourth International Soil Classification Workshop, Rwanda'. Part I (Eds F.H. Beinroth and H. Eswaran). pp. 450–70. (ABOS–AGCD: Brussels.)

Kang, B.T., and Wilson, G.F. (1981). Effect of maize plant population and nitrogen application on maize–cassava intercrop. In 'Tropical Root Crops: Research Strategies for the 1980s'. (Eds E.R. Terry, K.A. Oduro and F. Caveness). pp. 129–33. (IDRC: Ottawa.)

Kang, B.T., Wilson, G.F., and Lawson, T.L. (1984). Alley cropping a stable alternative to shifting cultivation. Int. Inst. Tropical Agriculture, Ibadan, Nigeria.

Kang, B.T., Wilson, G.F., and Sipkens, L. (1981). Alley cropping maize (Zea mays L.) and Leucaena (Leucaena leucocephala Lam) in southern Nigeria. Plant Soil 63, 165–79.

Kassam, A.H., and Stockinger, K.R. (1973). Growth and nitrogen uptake of sorghum and millet mixed cropping. Samaru Agric. Newsl. 15, 28–33.

Kurtz, T., Melsted, S.W., and Bray, R. (1952). The importance of nitrogen and water in reducing competition between intercrops and corn. Agron. J. 44, 13–7.

LaRue, T.A., and Patterson, T.G. (1981). How much nitrogen do legumes fix? In 'Advances of Agronomy'. (Ed N.C. Brady). pp. 15–38. (Academic Press Inc.: N.Y.)

Leihner, D. (1983). Management and evaluation of intercropping systems with cassava. Centr. Int. Agric. Tropical (CIAT), Cali, Columbia.

Le Mare, P.H. (1972). A long term experiment on soil fertility and cotton yield in Tanzania. Expl. Agric. 8, 299–310.

Michon, G. (1983). Village forest gardens in West Java. In 'Plant Research and Agroforestry'. (Ed. P.A. Huxley). pp. 13–24, (Int. Council for Res. in Agroforestry: Nairobi.)

Mulongoy, K., and Akobundu, I.O. (1985). Nitrogen uptake in live mulch systems. In 'Nitrogen Management in Farming Systems in Humid and Subhumid Tropics'. (Eds B.T. Kang and J. van der Heide). pp. 285–90. (Inst. for Soil Fertility: Haren.)

Nambiar, P.T.C., Rao, M.R., Reddy, M.S., Floyd, C.N., Dart, P.J., and Willey, R.W. (1983). Effect of intercropping on nodulation and N_2-fixation by groundnut. Expl. Agric. 19, 79–86.

Narwall, S.S., Malik, D.S., and Malik, R.S. (1983). Studies in multiple cropping I. Effects of preceding grain legumes on the nitrogen requirement of wheat. Expl. Agric. 19, 143–51.

Neves, Maria, C.P., Summerfield, R.J., and Michin, F.R. (1982). Effects of complete leaf shading during the late reproductive period on carbon and nitrogen distribution and seed production by nodule–dependent cowpea (Vigna unguiculata) plants. Trop. Agric. (Trinidad.) 59, 248–53.

Nnadi, L.A. (1980). Nitrogen economy in selected farming systems of the savanna region. In 'Nitrogen Cycling in West African Ecosystems'. pp. 345–51. (SCOPE/UNEP Royal Swedish Academy of Sciences: Stockholm.)

Nnadi, L.A. and Arora, Y. (1985). Effect of nitrogen sources on crop yield and soil properties in the savanna. In 'Nitrogen Management in Farming Systems in Humid and Subhumid Tropics'. (Eds B.T. Kang and J. van der Heide). pp. 223–34. (Inst. for Soil Fertility: Haren.)

Norman, D.H. (1972). Mixed cropping in northern Nigeria III. Mixtures of cereals. Expl. Agric. 15, 41–8.

Nutman, T.S. (1976). 'IBP Field Experiments on Nitrogen Fixation in Peanuts'. (Cambridge Univ. Press: Cambridge, Massachusetts, USA.)

Oelsligle, D.D., McCollum, R.E., and Kang, B.T. (1976). Soil fertility management in tropical multiple cropping. In 'Multiple Cropping'. (Eds R.I. Papendick, P.A. Sanchez and G.B. Triplett). pp. 275–92. (ASA Spec. Publ. 27: Madison.)

Okigbo, B.N., and Greenland, D.J. (1976). Intercropping systems in tropical Africa. In 'Multiple Cropping'. (Eds R.I. Papendick, P.A. Sanchez, and G.B. Triplett). pp. 63–101. (ASA Spec. Publ. 27: Madison.)

Pandy, R.K., and Pendleton, J.W. (1986). Soyabeans as green manure in maize intercropping system. Expl. Agric. 22, 179–85.

Pate, J.S., and Minchin, F.R. (1980). Comparative studies of carbon and nitrogen nutrition of selected grain legumes. In 'Advances in Legume Science'. (Eds R.J. Summerfield and A.H. Bunting). pp. 105–25. (Royal Botanic Gardens: Kew.)

Patra, D.C., Sachdev, M.S., and Subbiah, B.V. (1986). ^{15}N studies on the transfer of legume fixed nitrogen to associated cereals in intercropping systems. Biol. Fertil. Soils 2, 165–71.

Rao, J.V.D.K., Dart, P.J., Matsumoto, T., and Day, J.M. (1981). Nitrogen fixation by pigeonpea. In 'Proc. Int. Workshop on Pigeon Peas I'. (Ed. Y.L. Nene). pp. 190–9, (ICRISAT: Patancheru, India.)

Rao, J.V.D.K., Dart, P.J., and Sastry, P.V.S. (1983). Residual effect of pigeon pea (Cajanus cajan) on yield and nitrogen response of maize. Expl. Agric. 19: 131–41.

Rao, M.R., and Willey, R.W. (1980). Evaluation of yield stability in intercropping studies on sorghum/pigeon pea. Expl. Agric. 16, 105–16.

Read, M.D., Kang, B.T., and Wilson, G.F. (1985). Use of Leucaena leucocephala (Lam De wit) leaves as a nitrogen source for crop production. Fert. Res.: 8, 107–16.

Reddy, M.S., Rego, T.J., Burford, J.R., and Willey, R.W. (1983). Fertilizer management in multiple cropping systems with particular reference to ICRISAT's experience. In 'Fertilizer Use Under Multiple Cropping Systems'. pp. 46–55. (FAO Fertilizer and Plant Nutrients Bull. 5: Rome.)

Reddy, M.S., and Willey, R.W. (1982). Improved cropping systems for the deep Vertisols of the India semi-arid tropics. Expl. Agric. 18, 277–87.

Remison, S.U. (1978). Neighbour effects between maize and cowpea at various levels of N and P. Expl. Agric. 14, 205–12.

Ruschel, A.P., and Vose, P.B. (1982). Nitrogen cycling in sugarcane. Plant Soil 67, 139–46.

Saxena, M.C., and Yadav, D.S. (1975). Multiple cropping with short duration pulses. Indian J. Genet. Plant Breed. 35, 194–208.

Searle, P.G.E., Comudom, Y., and Shedden, D.C. (1981). Effect of maize + legume intercropping systems and fertilizer nitrogen on crop yields and residual nitrogen. Field Crops Res. 4, 133–45.

Sheldrake, A.R., and Narayanan, A. (1979). Growth, development and nutrient uptake in pigeonpeas (Cajanus cajan). J. Agric. Sci. 92, 513–26.

Steiner, K.G. (1982). 'Intercropping in Tropical Smallholder Agriculture with Special Reference to West Africa'. (German Agency for Technical Cooperation, (GTZ): Eschborn.)

Trenbath, B.R. (1976). Plant interactions in mixed crop communities. In 'Multiple Cropping'. (Eds R. Papendick, P.A. Sanchez and G.B. Triplett). pp. 129–69. (ASA Spec. Publ. 27: Madison.)

Van der Heide, J. van der Kruijs, A.C.B.M., Kang, B.T., and Vlek, P.L.G. (1985). Nitrogen management in multiple cropping systems. In 'Nitrogen Management in Farming Systems in Humid and Subhumid Tropics'. (Eds B.T. Kang and J. van der Heide). pp. 291–306. (Inst. for Soil Fertility: Haren.)

Virtanen, A.I., Von Hausen, S., and Lamie, T. (1937). Investigations on the root nodule bacteria of leguminous plants XX. Excretion of nitrogen in associated cultures of legumes and non-legumes. J. Agric. Sci. 27, 584–610.

Waghmare, A.B., and Singh, S.P. (1984). Sorghum-legume intercropping and the effects of nitrogen fertilization.1. Yield and nitrogen uptake by crops, Expl. Agric. 20, 251–9.

Wahua, T.A.T., and Miller, D.A. (1978). Effects of intercropping on soybean N_2-fixation and plant composition of associated sorghum and soyabeans. Agron. J. 70, 292–5.

Wild, A. (1972). Mineralization of soil nitrogen at a savanna site in Nigeria. Expl. Agric. 8, 91–7.

Willey, R.W. (1979). Intercropping. Its importance and research needs. Part I. Competition and yield advantages. Field Crop Abst. 32, 1–10.

Willey, R.W., and Osiru, D.S.O. (1972). Studies on mixtures of maize and beans (Phaseolus vulgaris) with particular reference to plant population. J. Agric. Sci. 79, 517–29.

Yadav, R.L. (1981). Intercropping pigeonpea to conserve fertilizer nitrogen in maize and produce residual effects on sugarcane. Expl. Agric. 17, 311–15.

PART IV

ADVANCES IN NITROGEN METHODOLOGY

MEASUREMENT OF NITROGEN FIXATION IN THE FIELD

S.F. Ledgard and M.B. Peoples

ABSTRACT

The principal techniques for measuring biological N_2 fixation in the field in agricultural or natural ecosystems provide either short-term or time-integrated determinations. Short-term estimates of symbiotic activity may be obtained from acetylene reduction assays or by analysis of the N-solutes transported in the xylem stream leaving the roots of legumes. Time-averaged values for the proportion of plant N derived from atmospheric N_2 have been calculated using N-difference methods, or the dilution of isotopically enriched (naturally or by addition of ^{15}N-labelled N) soil N by N_2 fixed from air.

All but the N-solute technique are suitable for use in areas of research on biological N_2 fixation other than nodulated legumes. The four methods are discussed and their advantages and limitations highlighted. It is concluded that careful appraisal of all techniques is necessary before selecting the method(s) most applicable to the system under study.

INTRODUCTION

It is essential that the fixation of atmospheric N_2 by living organisms be reliably determined in the field if researchers are to assess the effect of different management and cultural practices on the overall N-balance of agricultural and natural systems. This paper discusses primarily the measurement of N_2 fixation in legumes; however, most of the techniques described have also been used in studies of N_2 fixation by free-living microorganisms and non-legumes (Boddey 1987). We examine only those methodologies applicable to the field and confine discussion to recent findings. The use of $^{15}N_2$ (or $^{13}N_2$) gas will not be considered. Although exposure of plant roots to a $^{15}N_2$ – enriched atmosphere will provide definitive proof of N_2 fixation, the requirement for repeated measurements, the provision of suitable gas enclosures and the high cost of $^{15}N_2$ discourages its field use (Boddey 1987).

The advantages and limitations of various methods are discussed with emphasis on the newer procedures involving measurement of translocated products of N_2 fixation and the use of ^{15}N.

ACETYLENE REDUCTION ASSAY (ARA)

The enzyme nitrogenase not only reduces atmospheric N_2 to NH_3, but also catalyses the reduction of acetylene to ethylene. The ARA commonly involves the enclosure of detached nodules, severed root systems or whole plants in a closed vessel containing 10% acetylene; the accumulation of ethylene in the gas phase being determined by gas chromatography (Turner and Gibson 1980). The assay provides an instantaneous determination of N_2 fixation under the prevailing conditions. Long term estimates require a series of measurements to cover diurnal, daily and seasonal variations in N_2 fixation. The high sensitivity of the ARA makes it a particularly useful diagnostic tool for the detection of nitrogenase activity in non-legumes and free-living microorganisms as well as legumes. It has been widely used, due to its rapidity, simplicity and relatively low (equipment and resource) cost. However, there are several limitations of the ARA which may prevent it accurately quantifying nitrogenase activity in vivo.

Inherent problems

(i) A major limitation is the need to calibrate the rates of ethylene production with the actual rates of N_2 fixation. Although the theoretical acetylene/N_2 conversion ratio is 3:1 (provided all electrons transferred are allocated to these substrates), measured values range from 1.5:1 to 25:1. Since acetylene reduction does not relate directly to N_2 fixation, but only measures the electron flux through nitrogenase, the number of electrons used in H_2 production influence the acetylene reduction measurements obtained. Various factors affect electron allocation to H_2, and it has been demonstrated that the ranking of treatments (e.g. Rhizobium strains) based on their ability to reduce acetylene may be invalid if appropriate conversion ratios are not determined (Witty and Minchin 1987).

(ii) In some nodulated legumes there is a substantial decline in nitrogenase activity within a few minutes of the start of the assay. The non-linear rates of ethylene accumulation, attributed to acetylene-induced changes in nodule gas exchange, may underestimate nitrogenase activity by as much as 50% (Witty and Minchin 1987).

Procedural limitations

(i) Apart from the requirement to recover the entire nodule population of legumes to attain estimates of N_2 fixation per plant, the removal of the shoot, and washing or shaking of the roots to remove soil, can have a detrimental effect on nitrogenase activity, as will nodule detachment (Witty and Minchin 1987). To avoid disturbance of N_2-fixing systems in the field, in situ methods have been devised to measure acetylene reduction in plant rhizospheres, soil and aquatic environments (Denison et al. 1983; Grant 1986).

(ii) At low nitrogenase activities in non-legume studies there can be an overestimation of acetylene reduction rates due to endogenous ethylene production by agents other than the organism of interest (Boddey 1987).

Despite these technical limitations, ARA remains one of the most commonly utilized methodologies in N_2 fixation research. Although it has been suggested that the procedure is incapable of even comparative estimates of nitrogenase activity in field-grown legumes, the results obtained may still represent a useful indication of symbiotic potential provided the interpretative difficulties of the technique are appreciated and the data are used with caution (Turner and Gibson 1980; Witty and Minchin 1987).

N-SOLUTE METHOD

The products of N_2 assimilation in legume nodules are exported to the shoot through the xylem. Nodulated legumes are often placed into two broad groups (Table 1) depending on whether they transport fixed N as ureides (allantoin and allantoic acid) or amides (asparagine and glutamine). Legumes

Table 1: Nitrogenous solutes of xylem sap of effectively nodulated legumes[1].

Major constituents as amides	Major constituents as ureides
Acacia alata, A. extensa	Albizia lophantha
A. insauvis, A. pulchella	Cajanus cajan
Arachis hypogea	Calopogonium caeruleum
Cicer arientinum	Cyamopsis tetragonoloba
Lathyrus cicera, L. sativus	Glycine max
Lens culinaris, L. esculenta	Hardenbergia spp.
Lotus corniculatus	Kennedia spp.
Lupinus albus, L. angustifolius	Lablab purpureus
L. cosentinii, L. mutabilis,	Macroptilium atropurpureum
Medicago minima, M. sativa	Macrotyloma uniflorum
Pisum arvense, P. sativum	Pueraria javanica, P. phaseoloides
Trifolium pratense, T. subterraneum,	Phaseolus vulgaris, P. lunatus
T. repens	Psophocarpus tetragonolobus
Vicia calacarata, V. ervilia	Vigna angularis, V. mungo, V. radiata,
V. faba, V. sativa	V. triloba, V. unguiculata, V. umbellata

1. Information derived from Pate and Atkins 1983; Peoples et al. 1986; 1987b; Hansen and Pate 1987; and M.B. Peoples and D.F. Herridge, unpublished data.

in soil absorb inorganic N (predominantly nitrate) and this also passes to the shoot in the xylem either as free nitrate or as organic products of nitrate reduction in the root (Fig. 1). Since the bulk of the nitrogenous solutes in xylem sap represent current products of N uptake and assimilation it is possible to devise an assay system based on sap analyses to assess plant reliance on symbiosis in the presence of nitrate; provided there are substantial differences in sap composition between N_2–fixing and nitrate–fed plants.

Nitrate reductase activity in the roots of ureide–exporters is characteristically low (Peoples et al. 1987b) so that most of the nitrate absorbed from the soil passes unreduced to the shoot (Fig. 1A). Decreasing the dependence of nodulated plants on N_2 fixation with nitrate results in a progressive decline in the proportion of organic N present as ureides and a marked increase in xylem nitrate content (Fig. 2). The relative abundance of ureides (Fig. 2C) can be used as a quantitative measure of the N_2–fixing status for many species in this group of legumes (Table 1).

The response of amide–producing legumes (Table 1) is less well defined. A proportion of incoming nitrate is commonly reduced in the roots (Fig. 1B), but the extent of this varies between species and with the external nitrate supply (Peoples et al. 1987b). White lupin, faba bean and tree legumes such

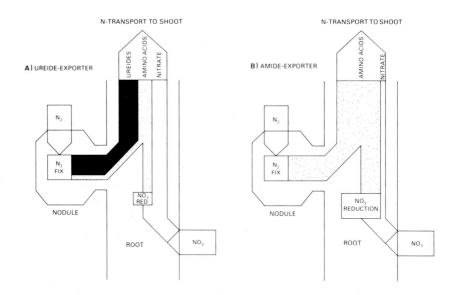

Fig. 1: Diagrammatic representation of the pathways of nitrogen transport from the nodulated root systems of (A) ureide–exporting, and (B) amide–exporting legumes relying on both symbiotic N_2 fixation and nitrate uptake from the soil. The areas indicating nitrate reduction are proportional to the relative extents of nitrate metabolism occurring within the roots of species from each class of legume.

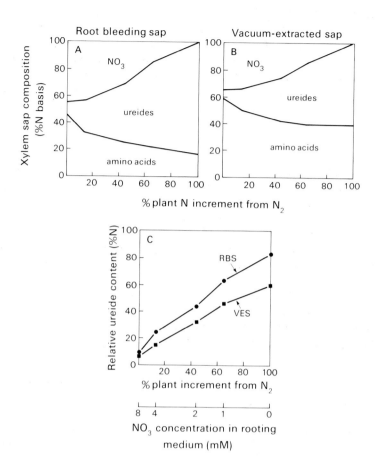

Fig. 2: Changes in the composition of xylem sap collected as (A) root-bleeding exudate, or (B) vacuum-extracted from stems of nodulated soybean fed a range of constantly maintained levels of nitrate, and (C) the relationship between the abundance of ureides and plant dependence upon N_2 fixation. Relative ureide contents of root-bleeding sap (RBS) and vacuum-extracted sap (VES) are expressed as a proportion of total sap N (ureide-N + α amino-N + nitrate-N). The N content of solutes were calculated from the data of Herridge (1984) for soybean as described by Peoples et al. (1987a).

as Acacia appear to have a high capacity for nitrate reduction in their roots, and do not greatly change the spectrum of products exported from the roots with altered N-nutrition (Hansen and Pate 1987; Peoples et al. 1987b). By contrast, in peanut, chickpea, lentil and pea sufficient nitrate escapes the reductase system of the roots to change the relative proportions of nitrate and organic N in xylem sap in response to increased plant reliance on soil N (Fig. 3A; Peoples et al. 1987b). Although there are major shifts in the relative importance of asparagine and/or glutamine in xylem exudates in response to changes in symbiotic dependence (e.g. Fig. 3A), the specialised equipment needed to quantify individual amino compounds prohibits the routine analysis of large numbers of samples. Of the two N-components which can be determined by simple colorimetric assay (i.e. nitrate and total α-amino NH_2), levels of nitrate respond more readily to alterations in N-nutrition. Compositional relationships based on nitrate are not as desirable as those relying primarily upon N_2 fixation products, because the presence of nitrate depends upon a correlation between its concentration, rate of uptake, and the extent of storage and reduction within the root. Nonetheless, the proportion of N-solutes as nitrate in xylem exudate is a reasonable indicator of relative dependence on soil-N in peanut (Fig. 3B) and other amide-exporters (Peoples et al. 1987b).

It is possible to collect xylem sap from decapitated root stumps in glasshouse grown plants or in the field in the humid tropics (Peoples et al. 1987a); but difficulties have been experienced in recovering xylem exudate from unwatered field-grown legumes in Australia and the United States (see Herridge 1982). Pre-watering field plots may assist plants to bleed, but this may not always be possible or convenient and does not guarantee success. To overcome this limitation, techniques have been devised to examine legume N-solutes under almost any environmental condition. Although there may be quantitative differences from root-bleeding exudates in absolute concentration or relative composition, relationships between N-solutes and symbiotic

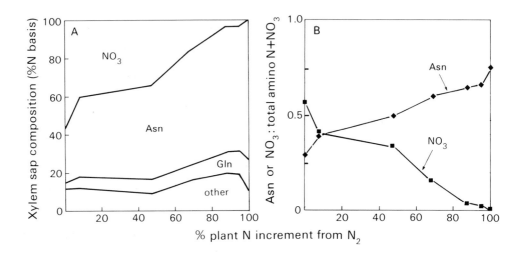

Fig. 3: Changes in (A) the xylem exudate composition of nodulated peanut fed a range of constantly maintained levels of nitrate, and (B) the relationships between compositional features (N/N ratios) and plant dependence on N_2 fixation. Data derived from Peoples et al. (1986). Asn = asparagine, Gln = glutamine, other = those amino acids listed in Peoples et al. (1986).

dependence have been prepared for sap extracted from freshly harvested shoot segments by mild vacuum (Fig. 2) and for the soluble-N components of aqueous extracts of plant tissues (Herridge 1982; Peoples et al. 1987b).

Advantages of N-solute analysis

(i) The collection of xylem exudate is simple and the analysis of its N-components (i.e. ureides, total α-amino N, and nitrate) can be done by colorimetric assays in a test tube. Thus expensive or sophisticated equipment is not needed and many analyses may be performed daily.

(ii) It is not necessary to dig roots from the soil and recover nodules.

(iii) It is not necessarily a totally destructive technique; sufficient sap can often be collected from stem segments or laterals to enable complete analysis.

Possible difficulties

(i) The form of transport N must be known for the species investigated.

(ii) Physiological, environmental and nutritional variables must be considered before glasshouse-derived relationships can be applied to field-grown crops.

(iii) A number of factors which may influence the accuracy of the ureide bioassay in soybean have been considered (Peoples et al. 1987a), but only the effect of plant age, and delays in sampling xylem contents by vacuum-extraction appear to be important. The change in xylem ureide content with stage of physiological development has been overcome by developing age-dependent correlations (Herridge 1984). Vacuum-extraction of sap within 10 min. of stem decapitation minimizes time related changes in N-solutes. Less detailed information is available on possible sources of error in species other than soybean; however, it seems that age-related changes in compositional relationships may not exist in all legumes (Peoples et al. 1987b).

(iv) The major disadvantage of the N-solute technique is that it is a point rather than time-integrated assay. To estimate seasonal fixation, repeated measurements must be combined with sequential sampling from a crop for dry matter and total N content. Only then can total inputs of N be determined for short intervals of time and partitioned between symbiotic and soil derived N. Studies using ureide relationships in soybean in this way have shown close agreement with methods using acetylene reduction (Patterson and LaRue 1983), soil nitrate depletion and N-difference (Herridge et al. 1984) methods, and ^{15}N natural abundance (D.F. Herridge, F.J. Bergersen, and M.B. Peoples, unpublished data).

N–DIFFERENCE METHOD

With this method, the amount of N fixed by a legume is estimated from the difference in N yield between the legume and a non N_2-fixing plant grown in the same soil. It is a relatively simple procedure, commonly used when only total N analysis is available. A major assumption is that the legume and reference plant assimilate the same amount of soil N. Even when the supposed 'ideal' reference plant (a non-nodulating isoline of the legume) is used, results may be incorrect due to differences in soil N uptake because of differences in root morphologies (Boddey et al. 1984). Comparisons with several other procedures for measuring N_2 fixation indicate that erroneous estimates may be obtained using the N–difference method (Ruschel et al. 1979).

^{15}N ISOTOPE DILUTION TECHNIQUES

The isotope ^{15}N occurs in atmospheric N_2 at a constant abundance of 0.3663 atoms %. The proportions of plant N derived from N_2 and from soil N can be calculated, if the isotopic abundances in these two sources of N are sufficiently different, by comparing the ^{15}N enrichment of nodulated legumes to that of non N_2-fixing reference plants. In many cases the very small natural difference in abundance of ^{15}N between soil N and N_2 can be used, provided a suitably precise mass spectrometer is available. More usually, the difference between soil N and N_2 is increased by incorporation in the soil of small amounts of fertilizer-N highly enriched with ^{15}N.

^{15}N enrichment

There has been a large increase in use of methods involving artificial ^{15}N enrichment of soil N (Chalk 1985). The percentage of legume N fixed from atmospheric N_2 (P) is estimated from the following equation:

$$P = 100 \left[1 - \left[\frac{\text{atoms\% } ^{15}\text{N}(\textit{legume N}) - \text{atoms\% } ^{15}\text{N}(\textit{air } N_2)}{\text{atoms\% } ^{15}\text{N}(\textit{soil-derived N}) - \text{atoms\% } ^{15}\text{N}(\textit{air } N_2)} \right] \right] \quad (1)$$

where the atoms % ^{15}N in air N_2 is 0.3663. The atoms % ^{15}N of soil-derived N in the legume is commonly estimated from the atoms % ^{15}N of a non N_2-fixing reference plant grown in the same soil over the same period as the N_2-fixing legume.

A major assumption is that the legume and reference plant absorb the same relative amounts of N from added ^{15}N and soil N. Its main advantage is that the method provides a 'time-averaged' estimate of P which is the integral of any changes in P that may have occurred during the measurement period. The latter can be determined, if desired, by taking a series of samplings. For example (Table 2), the initial high levels of inorganic soil N after sowing a white clover/ryegrass pasture resulted in almost all clover N being derived from the soil; substantial amounts of clover N were not fixed until levels of inorganic soil N were reduced. The estimate of P is independent of yield, although it is necessary to measure dry matter and N yield to determine the amount of N fixed. Apart from the high cost of a mass spectrometer and of ^{15}N-labelled material there are a number of other possible limitations.

Table 2: Effect of time after sowing a white clover (_Trifolium repens_)/ryegrass (_Lolium_ _perenne_) pasture on plant N accumulation, percentage of clover N fixed from atmospheric N_2 (P), and inorganic soil N. [Six week fallow before sowing 6 April 1986 into a sandy loam soil (0.56% N in 0–75 mm depth) near Hamilton, New Zealand].

	Days after sowing			
	47	98	144	170
Grass N (kg/ha)	11	41	83	100
Clover N (kg/ha)	1	6	8	13
P (%)	4	24	59	68
Standard error of P	5	4	5	5
Inorganic soil N (µg/g)				
0–75 mm depth	41	13	8	6
75–200 mm depth	41	5	3	4

(i) It is important that addition of ^{15}N-labelled material does not affect N_2 fixation. The effect of N on N_2 fixation is well established (e.g. Table 2) and therefore low rates of N application should be used (e.g. < 2 kg N/ha). If a carbon source is being used to immobilise added ^{15}N or if ^{15}N-labelled organic matter is being added, it is also important that there is not a stimulation of N_2 fixation due to labelling treatment effect on soil N availability (Chalk 1985).

(ii) Transfer of fixed N from legume to reference plant is a factor of importance where the legume and reference plant are grown together, and would lead to N_2 fixation by the legume being underestimated. In practice, N transfer appears to be a relatively slow process and the effects on estimating P can be limited by restricting the measurement period. Alternatively it can be compensated for by measuring it using the techniques of Vallis et al. (1967) or Ledgard et al. (1985a).

(iii) N_2 fixation associated with the reference plant can occur and lead to an underestimate of P. In temperate climates, N_2 fixation associated with non-legumes appears to be insignificant; but in tropical environments, N_2 fixation associated with some non-legumes can supply up to 40% of the plant's

N (Boddey 1987). It is important to check reference plants for such activity (e.g. using acetylene reduction) or to compare several different reference plants.

(iv) The nitrogen uptake characteristics of the reference plant may not match those of the legume. This is the most important factor affecting the estimate of P, which can vary substantially depending on the reference plants used (Witty 1983; Ledgard et al. 1985c, d). These differences occur because the legume and reference plant differ in the ratio (R) of N assimilated from added ^{15}N to N assimilated from indigenous soil N due to; (1) the legume and reference plant differing in their uptake of N from different soil depths where there are differences in the isotopic composition of plant–available soil N, or (2) the plants differing in their pattern of N assimilation with time in association with time–related changes in the isotopic composition of plant–available N. Variations in the ^{15}N concentration of soil N with depth occur because ^{15}N–labelled compounds are commonly applied to the soil surface to avoid disturbance of the soil. Such differences in ^{15}N distribution with soil depth can significantly affect estimates of P (Table 3). However, in studies where large discrepancies (e.g. negative P values) have occurred, associated measurements indicate that temporal changes in ^{15}N enrichment generally have more effect on estimating P than differences in ^{15}N content with soil depth (Witty and Ritz 1984; Ledgard et al. 1985d). Thus, it is important to choose a reference plant with a seasonal pattern of N assimilation similar to that of the legume. Where plants are imperfectly matched, the error in estimating P will depend on the rate of change in ^{15}N concentration of plant–available soil N with time. This is largely affected by the form of ^{15}N material used.

Table 3: Influence of differences in distribution of added ^{15}N on estimates of the percentage of clover N fixed from atmospheric N_2 ($\underline{P}$).

	P	S.E.D.
1. [15]N – nitrate applied to soil surface with 2 mm water	64.4	3.8
[15]N – nitrate applied to soil surface with 10 mm water	80.6	
2. [15]N – ammonium applied to soil surface with 2 mm water	24.4	5.0
[15]N – ammonium mixed in 0–15 cm soil depth	68.4	

1. Trifolium subterraneum/Lolium rigidum, established pasture (Ledgard et al. 1985c); 2. T. repens/L. perenne, newly cultivated and sown. (S.F. Ledgard, unpublished data).

Form of added [15]N

    Witty and Ritz (1984) showed the advantage in using slow–release forms of ^{15}N for reducing the temporal decline in enrichment of plant–available soil N.

Their estimates of $\underline{P}$ for soybean, using maize as the poorly-matched reference plant, were −9% using KNO_3 and 51% using slow-releasing oxamide. An obvious slow-release form is ^{15}N-labelled plant material, which can be obtained from previous ^{15}N experiments (Boddey et al. 1984). However, incorporation of organic matter may affect the availability of soil N. This will be unimportant where crop residues are returned to soil as a normal management practice but otherwise it may give erroneous results. This procedure is likely to be least useful where plants are already established (e.g. perennial pastures) and where soil disturbance should be avoided.

The most commonly used forms of ^{15}N are inorganic N compounds. Steele and Littler (1987) found that ^{15}N-enriched $NaNO_3$ sprayed onto the surface of soil growing perennial pasture gave estimates of $\underline{P}$ consistently higher (by 2–12%) than those using $(NH_4)_2SO_4$ or urea. The difference disappeared by the third harvest and they recommended using infrequent applications of ^{15}N and delaying measurements until several harvests after application. Other field studies have shown no effect on $\underline{P}$ from using different forms of inorganic ^{15}N (Rennie 1986). The choice of ^{15}N-labelled material should be determined by experimental factors such as management practices, availability of ^{15}N-labelled materials and the duration of the experiment.

Procedures for checking suitability of reference plants

The only definitive check is to estimate whether the legume and reference plant assimilate added ^{15}N and indigenous soil N in the same ratio ($\underline{R}$). Wagner and Zapata (1982) estimated $\underline{R}$ indirectly by measuring the relative uptake of labelled and indigenous soil sulphur. They found that the ratio of uptake of labelled and soil sulphur was similar for various legume/non-legume combinations but some plants differed in their relative absorption of N and sulphur. Ledgard et al. (1985b) described a regression approach for assessing $\underline{R}$. This involved two or more treatments with the same rate of N addition but at different ^{15}N concentrations, and a natural ^{15}N abundance treatment. Application to several legume/grass associations established that the reference plant can induce real changes in $\underline{P}$ by influencing the soil N status in the association, and/or cause erroneous estimates of $\underline{P}$ using the ^{15}N enrichment procedure.

Natural ^{15}N abundance

Almost all N transformations in the soil result in isotopic fractionation. The net effect is a small increase in the ^{15}N abundance of soil N compared with atmospheric N_2 (Shearer and Kohl 1986). In looking at small differences in ^{15}N concentration the term $\delta^{15}N$ (‰) is commonly used;

$$\delta^{15}N = 1000 \, (R_{sample} - R_{standard})/R_{standard}$$

where R = mass 29/mass 28, and the standard is commonly atmospheric N_2. An estimate of $\underline{P}$ is obtained using the following equation (synonymous with equation 1):

$$ P = \left[\frac{\delta^{15}N(\textit{available soil N}) \;-\; \delta^{15}N(\textit{legume})}{\delta^{15}N(\textit{available soil N}) \;-\; B} \right] \times 100 \qquad (2) $$

where $\delta^{15}N(\textit{available soil N})$ is commonly obtained from a non N_2-fixing reference plant grown in the same soil as the legume, and B is the $\delta^{15}N$ of fixed N_2 in the legume (determined by analysis of nodulated legumes grown in N-free media).

This method has been used to estimate N_2 fixation in legumes, non-legumes and free-living organisms in agricultural systems and natural ecosystems, including marine, forest and desert environments (Shearer and Kohl 1986). Several studies have shown results similar to those from ^{15}N enrichment techniques and with similar precision (Table 4). Natural ^{15}N abundance has proven useful in demonstrating the effects of soil moisture stress (Domenach and Corman 1985), rhizobial strain (Rennie and Kemp 1983) and paddock history (Bergersen et al. 1985) on N_2 fixation by crop legumes. It gives an integrated estimate of P over time and can be applied to established experiments because no pretreatment is necessary.

Table 4: Estimates of the percentage (P) of nitrogen fixed by field-grown pasture and crop legumes using natural ^{15}N abundance and ^{15}N enrichment methods.

Legume	Reference plant	Natural ^{15}N abundance		^{15}N-enrichment	
		P	S.E.	P	S.E.
		%		%	
Pastures legumes:					
Subterranean clover[1]	Ryegrass	93	3	95	1
Subterranean clover[2]	Ryegrass	85	3	70	4
Subterranean clover[2]	Phalaris	86	4	50	5
Lucerne[2]	Ryegrass	81	7	88	3
Lucerne[2]	Phalaris	64	6	70	5
Crop legumes:					
Soybean[3]	Ryegrass	64	–	58	
Soybean[3]	Non-nodulated soybean	65	–	62	–
Field bean[4]	Uninoculated field bean	56	8	50	5
Lupin[5]	Wheat	81	4	85	4

1. Bergersen and Turner 1983; 2. Ledgard et al. 1985c; 3. Domenach et al 1979; 4. Rennie 1986; 5. Evans et al 1987.

Although the principles of the technique are similar to those of ^{15}N enrichment studies, the main limitations are quite different (Shearer and Kohl 1986). An isotope ratio mass spectrometer capable of accurately measuring differences of 0.1‰ (about 0.00004 atoms % ^{15}N) is needed and sample preparation requires great care to avoid isotopic fractionation and contamination from ^{15}N-enriched material. It is preferable to have the $\delta^{15}N$ of plant-available soil N above 6‰ because the accuracy in estimating P decreases markedly at values below this (Fig. 4). Low soil ^{15}N abundances have been detected in intensively-grazed pastures (Steele 1983) and considerable spatial variability found in ^{15}N values of inorganic soil N in a study on shrub legumes in a dry forest environment (Hansen and Pate 1987). However, it is more usual to find relatively uniform ^{15}N enrichments in soil (Shearer and Kohl 1986), with $\delta^{15}N$ values for soil-derived N commonly above 8‰ in cropping soils (Rennie and Kemp 1983; Bergersen et al. 1985), thereby enabling reliable and accurate estimates of P.

Use of the natural ^{15}N abundance method assumes that isotopic fractionation during N_2 fixation is nil or a known constant value. Numerous studies have measured significant isotopic fractionation and have established a value for B, viz. $\delta^{15}N$ *(fixed N_2 in legume)* – $\delta^{15}N$ *(atmospheric N_2)*, for use in equation 2. Field studies of lucerne grown with ryegrass (Ledgard et al. 1985b) produced estimates of P of 61% or 81% using a B value of 0‰ or

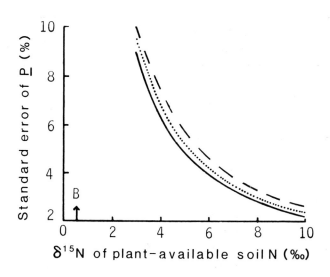

Fig. 4: Effect of $\delta^{15}N$ of plant available soil N on the error in estimating the percentage (P) of legume N fixed. P values of 90 (– – –), 75 (·····), or 50 (———)% were used. Data are for subterranean clover using a B value of 0.6 (S.E. ± 0.1)‰, and typical values from field studies for S.E. of $\delta^{15}N$ (clover and grass) = ± 0.2‰. (Ledgard et al. 1985c).

Table 5: Summary of the advantages and limitations of methods for estimating N_2 fixation by legumes in the field.

Advantages	Limitations

Acetylene reduction method
: relatively inexpensive	: indirect (must establish conversion factor for N_2 fixation, which may vary)
: simple	
: rapid	: short–term estimate
: very sensitive	: acetylene may inhibit nitrogenase activity
	: unless using an in situ assay, incomplete recovery or damage to nodules can cause considerable error

N–Solute method
: inexpensive	: indirect (must establish calibration with plant N_2–fixing status)
: simple	
: analyses done by simple colorimetric assays	: short–term estimate
: need not be totally destructive	: cannot be used for interspecific comparisons without calibrating each species
: may be done on an individual plant basis	: calibrations may be influenced by development stage in some species
: assesses plant dependence on atmospheric N_2 and soil N	: cannot be used on some amide–producers

N difference method
: direct	: requires suitable non–N_2–fixing reference plant
: relatively simple	
: adjusts for soil – derived N	: legume and reference plant must absorb the same amount of soil N

Isotope dilution techniques
: direct	: requires suitable non–N_2–fixing reference plant
: give time–integrated estimate of % N fixed	
: assesses plant dependence on atmospheric N_2 and soil N	

(a) ^{15}N enrichment method
: potentially accurate	: expensive
	: requires addition of ^{15}N–labelled compound
	: legume and reference plant must absorb the same relative amounts of N from the soil and added N
	: ^{15}N enrichment of plant–available soil N can change with depth and time

(b) Natural ^{15}N abundance method
: no ^{15}N addition required	: relatively expensive
: $\delta^{15}N$ of plant–available soil N is relatively constant with depth and time	: insensitive if $\delta^{15}N$ (soil) nears $\delta^{15}N$ (air)
	: field variability may be large in some cases
	: may have to allow for isotopic fractionation during N_2 fixation

0.97‰ (established in a pot experiment) respectively. In some legumes (e.g. Lupinus spp.), rhizobial strain can influence the $\delta^{15}N$ of shoots of plants grown with atmospheric N_2 as their sole source of N, but due to an inverse effect on the $\delta^{15}N$ of nodules, there is no effect on the value for whole plants (Bergersen et al. 1986). Recent studies (S.F. Ledgard, unpublished data) produced no difference in B value (shoots or whole plants) for T. repens and T. pratense grown with Rhizobium leguminosarum biovar. trifolii strain 2668 (the strain used for seed-inoculation in New Zealand) or with field isolates of rhizobia even under nutritional (Mo, P) or environmental (temperature, moisture) stress, illustrating that a value for B can be established in a pot trial and applied to field studies.

Isotopic fractionation during uptake of soil N was examined for a large number of legumes and non-legumes by Mariotti et al. (1980) and found to be insignificant. Thus, there appears to be no need to allow for this in estimating P using natural ^{15}N abundance.

Studies by Domenach et al. (1979) and Ledgard et al. (1985c) found that the use of different reference plants did not influence estimates of P. Since ^{15}N abundance of plant-available soil N is generally uniform with soil depth (Ledgard et al. 1984) and may show little temporal variation (Bergersen et al. 1985; Ledgard et al. 1985c), the major limitation of ^{15}N enrichment techniques (i.e. choice of reference plant) appears to be relatively unimportant with the natural ^{15}N abundance method.

CONCLUSIONS

The methodology of measuring N_2 fixation has advanced considerably in recent years, particularly in defining the limitations of the current techniques (summarised in Table 5). These limitations must be recognised to obtain accurate estimates of N_2 fixation in the field, and procedures must be used to either reduce their effect or to check on their relative importance in the calculation of symbiotic activity. Ideally, two or more methods should be used and results related to soil N measurements and plant nodulation. Studies on N_2 fixation should also be associated with research on N transfer or effects on subsequent plant growth if an overall estimate of the value of fixed N_2 is to be obtained. Finally, methodology should be chosen in light of the overall objectives of the study. Sophisticated techniques may not be appropriate when only qualitative or comparative determinations are required.

ACKNOWLEDGEMENTS

MBP gratefully acknowledges the Australian Centre for International Agricultural Research for financial support (Project No. 8305 - Development and evaluation of methods of measuring biological nitrogen fixation). The authors also wish to thank staff of the CSIRO Division of Plant Industry word-processing centre for their assistance in preparing the manuscript.

REFERENCES

Bergersen, F.J., and Turner, G.L. (1983). An evaluation of ^{15}N methods for estimating nitrogen fixation in a subterranean clover–perennial ryegrass sward. Aust. J. Agric. Res. 34, 391–401.

Bergersen, F.J., Turner, G.L., Amarger, N., Mariotti, F., and Mariotti, A. (1986). Strain of Rhizobium lupini determines natural abundance of ^{15}N in root nodules of Lupinus spp. Soil Biol. Biochem. 18, 97–101.

Bergersen, F.J., Turner, G.L., Gault, R.R., Chase, D.L., and Brockwell, J. (1985). The natural abundance of ^{15}N in an irrigated soybean crop and its use for the calculation of nitrogen fixation. Aust. J. Agric. Res. 36, 411–23.

Boddey, R.M. (1987). Methods for quantification of biological nitrogen fixation with Gramineae. CRC Critical Reviews, (CRC Pres: Boca Raton) (In press).

Boddey, R.M., Chalk, P.M., Victoria, R.L., and Matsui, E. (1984). Nitrogen fixation by nodulated soybean under tropical field conditions estimated by the ^{15}N isotope dilution technique. Soil Biol. Biochem. 16, 583–8.

Chalk, P.M. (1985). Estimation of N_2 fixation by isotope dilution: an appraisal of techniques involving ^{15}N enrichment and their application. Soil Biol. Biochem. 17, 389–410.

Denison, R.F., Sinclair, T.R., Zobel, R.W., Johnson, W.M., and Drake, G.M. (1983). A non-destructive field assay for soybean nitrogen fixation by acetylene reduction. Plant Soil 70, 173–82.

Domenach, A.M., Chalamet, A., and Pachiaudi, C. (1979). Estimation de la fixation d'azote par le Soja à l'aide de deux methodes d'analyses isotopiques. C.R. Acad. Sci. Ser. D289, 291–3.

Domenach, A.M., and Corman, A. (1985). Use of ^{15}N natural abundance method for the study of symbiotic fixation of field-grown soybeans. Soil Sci. Plant Nutr. 31, 311–21.

Evans, J., Turner, G.L., O'Connor, G.E., and Bergersen, F.J. (1987). Nitrogen fixation and accretion of soil nitrogen by field-grown lupins (Lupinus angustifolius) Field Crop Res. (In press).

Grant, I.F. (1986). Modification and field testing of a portable gas chromatograph for in situ acetylene reduction activity assay of tropical soils. Plant Soil 95, 435–9.

Hansen, A.P., and Pate, J.S. (1987). Evaluation of the ^{15}N natural abundance method and xylem sap analysis for assessing N_2 fixation of understorey legumes in jarrah (Eucalyptus marginata) forest in S.W. Australia. J. Exp. Bot. (In press).

Herridge, D.F. (1982). Relative abundance of ureides and nitrate in plant tissues of soybean as a quantitative assay of nitrogen fixation. Plant Physiol. 70, 1–5.

Herridge, D.F. (1984). Effects of nitrate and plant development on the abundance of nitrogenous solutes in root-bleeding and vacuum-extracted exudates in soybean. Crop Sci. 25, 173–9.

Herridge, D.F., Roughley, R.J., and Brockwell, J. (1984). Effect of rhizobia and soil nitrate on the establishment and functioning of the soybean symbiosis in the field. Aust. J. Agric. Res. 35, 149–61.

Ledgard, S.F., Freney, J.R., and Simpson, J.R. (1984). Variations in natural abundance of ^{15}N in the profiles of some Australian pasture soils. Aust. J. Soil Res. 22, 155–64.

Ledgard, S.F., Freney, J.R., and Simpson, J.R. (1985a). Assessing nitrogen transfer from legumes to associated grasses. Soil Biol. Biochem. 17, 575–7.

Ledgard, S.F., Morton, R., Freney, J.R., Bergersen, F.J., and Simpson, J.R. (1985b). Assessment of the relative uptake of added and indigenous soil nitrogen by nodulated legumes and reference plants in the ^{15}N dilution measurement of N_2 fixation. 1. Derivation of method. Soil Biol. Biochem. 17, 317–21.

Ledgard, S.F., Simpson, J.R., Freney, J.R., and Bergersen, F.J. (1985c). Field evaluation of ^{15}N techniques for estimating nitrogen fixation in legume–grass associations. Aust. J. Agric. Res. 36, 247–58.

Ledgard, S.F., Simpson, J.R., Freney, J.R., and Bergersen, F.J. (1985d). Effect of reference plant on estimation of nitrogen fixation by subterranean clover using ^{15}N methods. Aust. J. Agric. Res. 36, 663–76.

Mariotti, A., Mariotti, F., Amarger, N., Pizelle, G., Ngambi, J.M., Champigny, M.L., and Moyse, A. (1980). Fractionnements isotopiques de l'azote lors des processus d'absorption des nitrates de fixation de l'azote atmospherique par les plants. Physiol. Vegetale 18, 161–81.

Pate, J.S., and Atkins, C.A. (1983). Nitrogen uptake, transport and utilization. In 'Nitrogen Fixation. 3. Legumes.' (Ed. W.J. Broughton). pp. 245–98. (Clarendon Press: Oxford.)

Patterson, T.G., and LaRue, T.A. (1983). N_2 fixation (C_2H_2) and ureide content of soybeans: Ureides as an index of fixation. Crop Sci. 23, 825–31.

Peoples, M.B., Bergersen, F.J., Herridge, D.F., Sudin, M.N., Wahab, F.A., Kewi, C. and Moris, N. (1987a). Estimation of nitrogen fixation in legumes in the tropics by xylem sap analysis. Proc. Biotechnology of Nitrogen Fixation in the Tropics, Universiti Pertanian Malaysia (In press).

Peoples, M.B., Pate, J.S., Atkins, C.A. and Bergersen, F.J. (1986). Nitrogen nutrition and xylem sap composition of peanut (Arachis hypogaea L. cv Virginia Bunch). Plant Physiol. 82, 946—51.

Peoples, M.B., Sudin, M.N. and Herridge, D.F. (1987b). Translocation of nitrogenous compounds in symbiotic and nitrate—fed amide—exporting legumes. J. Exp. Bot. 38, 567—79.

Rennie, R.J. (1986). Comparison of methods of enriching a soil with nitrogen—15 to estimate nitrogen fixation by isotope dilution. Agron. J. 78, 158—63.

Rennie, R.J., and Kemp, G.A. (1983). N_2—fixation in field beans quantified by ^{15}N isotope dilution. 1. Effect of strain of Rhizobium phaseoli. Agron. J. 75, 640—4.

Ruschel, A.P., Vose, P.B., Victoria, R.L., and Salati, E. (1979). Comparison of isotope techniques and non—nodulating isolines to study the effect of ammonium fertilization on dinitrogen fixation in soybean, Glycine max. Plant Soil 53, 513—25.

Shearer, G., and Kohl, D.H. (1986). Review: N_2—fixation in field settings: estimations based on natural ^{15}N abundance. Aust. J. Plant Physiol. 13, 699—756.

Steele, K.W. (1983). Quantitative measuremments of N turnover in pasture systems with particular reference to the role of ^{15}N. In 'Nuclear Techniques in Improving Pasture Management'. pp. 17—35. (IAEA: Vienna).

Steele, K.W., and Littler, R.A. (1987). Field evaluation of some factors affecting estimation of nitrogen fixation in pastures by ^{15}N isotope dilution. Aust. J. Agric. Res. 38, 153—61.

Turner, G.L., and Gibson, A.H. (1980). Measurement of nitrogen fixation by indirect means. In 'Methods for Evaluating Biological Nitrogen Fixation' (Ed. F.J. Bergersen). pp. 111—38 (J.Wiley & Sons: Chichester.)

Vallis, I., Haydock, K.P., Ross, P.J. and Henzell, E.F. (1967). Isotopic studies on the uptake of nitrogen by pasture plants. III. The uptake of small additions of ^{15}N—labelled fertilizer by Rhodes grass and Townsville lucerne. Aust. J. Agric. Res. 18, 865—77.

Wagner, G.H., and Zapata, F. (1982). Field evaluation of reference crops in the study of nitrogen fixation by legumes using isotope techniques. Agron. J. 74, 607—12.

Witty, J.F. (1983). Estimating N_2—fixation in the field using ^{15}N—labelled fertilizer: Some problems and solutions. Soil Biol. Biochem. 15, 631—9.

Witty, J.F., and Ritz, K. (1984). Slow release ^{15}N fertilizer formulations to measure N_2—fixation by isotope dilution. Soil Biol. Biochem. 16, 657—61.

367

DETERMINATION OF MICROBIAL BIOMASS CARBON AND NITROGEN IN SOIL

D.S. Jenkinson

ABSTRACT

Methods for measuring the quantities of C and N held in the soil microbial biomass are reviewed, with particular attention to the fumigation-incubation method, the fumigation-extraction method, the substrate-induced respiration method and the adenosine triphosphate method. The advantages and disadvantages of the different methods are discussed, as is their intercalibration. A new value (0.57) is proposed for k_N, the factor used to calculate microbial biomass N from the size of the flush caused by fumigation in the FI method. Ways of preparing and storing soil for biomass measurements are considered.

INTRODUCTION

Before examining current methods for measuring microbial biomass in soil, it is well to consider why such measurements are needed. Although only accounting for some 1–3% of the soil organic carbon (Jenkinson and Ladd 1981), the microbial biomass is both the agent of biochemical change in soil <u>and</u> a repository of plant nutrients that is more labile than the bulk of the soil organic matter. Biomass measurements can reveal changes brought about by soil management long before such changes can be detected in total organic carbon or N (Powlson and Jenkinson 1981; Powlson <u>et al</u>. 1987). They can also serve as a sensitive indicator of toxicity – for example, heavy metals from contaminated sewage sludge caused a decline in soil microbial biomass (Brookes and McGrath 1984). Measurements of biomass are also useful because any self-respecting model for the turnover of organic matter in soil contains at least one biomass compartment. Agreement between the amount of biomass predicted by a particular model and that actually measured provides a useful test of the adequacy of the model.

It is also well to be clear about what biomass measurements do not provide. They are essentially 'standing crop' measurements, not measurements of microbial activity such as respiration, heat production, dehydrogenase activity or mineralization of N (Nannipieri 1984; Van Veen et al. 1984). The capacity of the microbial biomass to process N and release it in a form available to plants is given by $N_m = p \, B_N$, where N_m is the N processed each year by the biomass, B_N is the N held in the microbial biomass (as kg N/ha, to a specified depth) and p is a rate constant, defined as the quantity of N processed by the biomass each year, expressed as a fraction of that in the biomass. In a group of soils, biomass C (and N) can correlate closely with release of mineral N under specified conditions (e.g. Carter and Rennie 1982; Sarathandra et al. 1984) but presumably this is merely because the different soils in the group all have similar values of p.

This paper reviews methods developed over the last 10 years for measuring C and N in microbial biomass. The adenosine triphosphate method, although not strictly a way of measuring either biomass C or N, will also be considered, because of the light it throws on the other methods. Biomass measurement by direct microscopic observation will not be considered here for lack of space. Although too tedious for routine use, it is the most direct method for measuring soil biomass and can be a useful tool on occasion (Jenkinson et al. 1976; West et al. 1986c). It also provides a unique view of the complexity of the soil biomass.

Another question not addressed here is the separation of the biomass into 'active' and 'dormant' compartments, other than to point out that such a distinction is largely meaningless in the soil context. In the energy-limited world of the soil, the arrival of substrate will allow a particular organism that follows a 'K' strategy (Andrews and Harris 1986) to go through a short phase of rapid growth and division, and then shut down metabolic activity as much as possible and wait for the next input of substrate. Other organisms (Winogradski's autochthonous population or in more modern terms 'r' strategists — see Gerson and Chet 1981) may grow and divide steadily over long periods, perhaps using widely distributed but highly recalcitrant substrates such as the humic substances. 'Active' and 'dormant' compartments are meaningless for this part of the population.

MEASUREMENT OF MICROBIAL BIOMASS CARBON IN SOIL

Fumigation incubation (FI)

This method and its variants have been widely used; it is sometimes referred to as the Chloroform Fumigation Incubation Method or CFIM (Voroney and Paul 1984), but is perhaps better described as the 'FI' method, as the technique is not limited to the use of chloroform as fumigant. Biomass C (B_C) is given by:

$$B_C = F_C \, / \, k_C \qquad (1)$$

where k_C is the fraction of the killed biomass mineralized to CO_2 when the fumigated soil is incubated under standard conditions. F_C, the flush of decomposition brought about by fumigation, is taken as the difference between the CO_2–C evolved by fumigated soil and that from an unfumigated soil incubated under identical conditions – the 'control'.

In the original method (Jenkinson and Powlson 1976b), moist soil is exposed to ethanol–free chloroform for 24h, the fumigant removed by repeated evacuation, the soil inoculated with unfumigated soil and then incubated at 25°C for 10 days at 55% of its water holding capacity. Chloroform is commonly used as fumigant, because it kills most of the population (Shields et al. 1974; Lynch and Panting 1980; Chaussod et al. 1986) and is easy to remove from the soil, but other fumigants have also been used: carbon disulphide (Kudeyarov and Jenkinson 1976) and methyl bromide (Powlson and Jenkinson 1976). As purchased, chloroform always contains ethanol as stabilizer, which must be removed if erroneously high measurements are to be avoided. The literature contains a few values for biomass C by the fumigation method that seem unreasonably high, more than 10% of soil organic C in biomass for example, and this may well have been the result of incomplete removal of ethanol.

Evolution of CO_2 can be measured by sorption in alkali followed by titrimetric (Jenkinson and Powlson 1976b), or colorimetric (Chaussod et al. 1986) determination. Less satisfactorily, CO_2 can be allowed to accumulated in the gas phase and measured by gas chromatography (Sparling 1981; Chaussod and Nicolardot 1982; Schnürer et al. 1985). Martens (1987) showed that problems can arise if CO_2 is allowed to accumulate in soil incubations, particularly with neutral and alkaline soils (see also Jenkinson and Powlson 1976b). In such soils accumulation of HCO_3^- in the soil solution can lead to low results if only the gas phase is analysed for CO_2.

It is now usual to incubate at 50% water holding capacity (c. −0.01 MPa) rather than the 55% originally specified: this change has little effect on the biomass measurement but loss of N by denitrification is less likely to occur (D.S. Powlson, personal communication). Shorter incubation periods (7 days) at a higher temperature (28°C) have been used with satisfactory results (Chaussod and Nicolardot 1982), as have lower temperatures (22°C; Anderson and Domsch 1978a). The value of k_C is usually found by adding a known quantity of microbial C to a soil, fumigating and measuring the proportion of the added microbial C mineralized under the standard incubation conditions. Values of k_C thus obtained from organisms grown in vitro are then assumed (without much evidence) to be applicable to the soil population. The most used values for k_C are 0.45 (Jenkinson and Ladd 1981) and 0.411 (Anderson and Domsch 1978a). As the first is derived from incubations at 25°C and the second at 22°C there is probably little real difference between them. Fig. 1 shows values of k_C taken (or calculated) from the literature; the mean k_C for soils of pH 5.0 and above is 0.46 ± 0.046 (S.D.), the results for each soil being weighted according to the number of organisms tested in it. There are

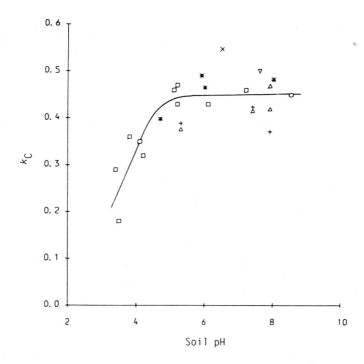

Fig. 1: The relationship between soil pH and the fraction of killed biomass C mineralized to CO_2 (k_C). Results from Nicolardot et al. (1984) (each point being a mean for 2 bacteria, 1 actinomycete and 2 fungi),△; Vance et al. (1987a) (means for 4 bacteria, 4 fungi),□; Nicolardot et al. (1986a) (means for 4 bacteria), +; Anderson and Domsch (1978a) (means for 12 bacteria and 14 fungi), *; Jenkinson, (1976) (mean for 7 bacteria, 1 actinomycete, 2 yeasts, 2 fungi and 1 nematode),▽; Voroney (1983) (mean for 4 bacteria and 10 fungi), +; Adams and Laughlin (1981) (mean for 5 bacteria, 1 actinomycete and 4 fungi), x; Amato and Ladd (1987) (mean for 3 bacteria and 2 fungi), o. All results normalised to a 10 day incubation period at 25°C, using the relationship between temperature and reaction rate in Jenkinson et al. (1987).

indications (Nicolardot et al. 1984) that k_C may be slightly greater in sandy soils than in soils of heavier texture but the effect is small and normalization of the data in Fig. 1 to a common clay content did not alter the mean k_C for soils of pH 5 and above, although the spread of values was reduced. For practical purposes, k_C can be taken to be 0.45 for incubations at 25°C in soils of pH 5.0–8.5, regardless of texture.

The major problem with the FI method is the choice of control. A control is necessary because not all the CO_2 evolved from a fumigated soil comes from the killed organisms. This can be seen from experiments on soils containing a biomass heavily labelled with ^{14}C in a matrix of less heavily-labelled dead organic matter (Jenkinson 1966; Voroney 1983). The CO_2 evolved immediately after fumigation of such a soil is very heavily labelled but after a few days the specific activity falls to a value similar to that in the corresponding unfumigated 'control' soil. The question at issue is how to allow for this background 'basal respiration', or, in other words, which is the best control to use.

Once the flush is over, both fumigated and unfumigated soils respire at much the same rate for long periods (although the rates are rarely exactly the same; Jenkinson and Powlson 1976a; Ross et al. 1980b; Chaussod and Nicolardot 1982; Chaussod et al. 1986). The large complex natural population present in unfumigated soil is thus no more effective in mineralizing CO_2 than the much smaller specialized population (largely bacterial: see Ridge 1976) found in fumigated soil, suggesting that, although basal respiration is carried out by micro-organisms, they do not govern the rate, which is set by independent physical or chemical processes (see Shen et al. 1987 for a discussion of this). The flush is usually taken as the difference between CO_2 evolved by the fumigated and unfumigated soils in 10 days, (Jenkinson and Powlson 1976b), but other controls have been proposed: the CO_2 evolved by the unfumigated soil over the 10–20 day period (Jenkinson and Powlson 1976b); the CO_2 evolved by the unfumigated soil over a (variable) period when the fumigated soil is respiring faster than the control (Ross et al. 1980b); the CO_2 evolved by the fumigated soil over the 7–14 day period, the flush being measured during the first 7 days (Chaussod and Nicolardot 1982). None of these controls are entirely satisfactory, particularly with freshly sampled soils. Another approach (Voroney and Paul 1984) has been to use no control. CO_2 evolved from non-biomass organic matter then inflates the calculated value for biomass C and it is not surprising that this procedure gives 'biomasses' that are often twice those obtained when a control is deducted (Schnurer et al. 1985). My current procedure is to pre-incubate soils for 7–10 days over soda lime to allow the effects of sampling to subside and then fumigate, taking F_C to be [(CO_2–C from fumigated soil, 0–10 days) – (CO_2–C from unfumigated soil, 0–10 days).

The original FI method was devised for well-drained agricultural soils and is unsuitable for strongly acid (pH <4.5) soils, for soils that have just had large additions of fresh substrate and for waterlogged soils. In strongly acid soils, the FI method gives lower results than the ATP method or direct microscopy for three reasons. Inoculation is much more important with acid soils (Chapman 1987; Vance et al. 1987a) than with near neutral soils, in which it is rarely critical (Powlson and Jenkinson 1976; Martens 1985). The flush of decomposition in strongly acid soils is preceded by a very long lag phase unless a large inoculum is used. Secondly, there is evidence (Fig. 1) that k_C declines sharply with pH in soils of pH <5. Thirdly, in strongly acid soils the unfumigated control respires more actively than the fumigated soil once the flush is over, so that an unfumigated 'control' overestimates basal respiration in the fumigated soil (Powlson and Jenkinson 1976; Sparling and Williams 1986). Presumably, in strongly acid soils, the specialized bacterial population dominant in fumigated soil is not as effective in metabolising soil organic matter (and for that matter, killed micro-organisms, see Fig. 1) as the intact natural population. Indeed by simply omitting the control and using the same inoculation technique and value of k_C that are used for near-neutral soils, it is possible to obtain biomass measurements in strongly acid soils by the FI method that are consistent with measurements by direct microscopy or from ATP content (Vance et al. 1987b). The pH below which a control should not be subtracted is about 4.2.

The FI method gives unacceptable results in soils that have recently received large additions of substrate (Jenkinson and Powlson 1976b; Sparling et al. 1981; Martens 1985). Again this is because of the control problem. If no control is used, the CO_2 from both basal respiration and that part of the substrate decomposed by the population in the fumigated soil counts as 'microbial biomass'. If an unfumigated soil is taken as the control, correct results will only be obtained if the proportion of substrate decomposed is the same in fumigated and unfumigated soil. This is a reasonable assumption for a simple substrate like glucose, which is effectively decomposed both by the specialized bacterial population in fumigated soil and by the natural population in unfumigated soil, although Martens (1985) showed that glucose decomposes a little faster in fumigated soil. However, it is not true for a complex substrate such as plant material which is decomposed more effectively in unfumigated than in fumigated soil, making the measured biomass values too small, or even negative (Martens 1985). None of the variants of the FI method currently available are suitable for biomass measurements during the early weeks after addition of fresh substrate, unless the fresh addition is negligibly small compared to the biomass.

In the FI method, the killed biomass is measured by allowing it to decompose under carefully controlled aerobic conditions and collecting the CO_2 diffusing from the soil. In waterlogged soil, CO_2 and CH_4 are produced under conditions that restrict diffusion of gases, so that the FI method cannot be used in its unmodified form.

Fumigation extraction (FE)

Voroney (1983) pointed out that there was a close correlation between biomass C and C rendered extractable to K_2SO_4 solution by fumigation. Vance et al. (1987c) showed that B_C could be calculated from the expression $B_C = 2.64\ E_C$, where E_C is C extracted by 0.5 $\underline{M}$ K_2SO_4 from fumigated soil, less that extracted from unfumigated soil, and 2.64 is an empirically-determined proportionality factor. This new method sidesteps the control problem in the FI method and can be used over the whole range of soil pH (Vance et al. 1987c). It may also be applicable to waterlogged soils (K. Inubushi, personal communication) and to soils that have recently received substrate, although this has not been tested. One problem, common to all extraction methods, is that the soluble organic matter liberated by fumigation will partition between soil and extractant in a way that may not be the same in all soils, particularly if the soils differ in clay content. In the FE method for biomass P (Brookes et al. 1982), this is allowed for by measuring the recovery of a spike of PO_4: so far no comparable correction has been made for C, if indeed this turns out to be necessary.

Substrate induced respiration (SIR)

Anderson and Domsch (1978b) introduced a method for measuring biomass in which the soil was amended with an excess of substrate and the respiration rate measured shortly afterwards, before the biomass had time to proliferate. The quantity of substrate added must be more than enough to saturate the respiration capacity of the soil population in the particular soil under examination. They found a close linear relationship between respiration rates thus measured and soil microbial biomass by the FI method and used this relationship to calibrate the SIR method. Martens (1987) also found a close correlation (r=0.98) between biomass as measured by the two methods but, in contrast, Sparling (1981) found no correlation. This discrepancy may have arisen because some of Sparling's soils were highly organic and acidic, soils for which the FI method is unreliable. Martens (1987) also pointed out that Sparling allowed CO_2 to accumulate during incubation, a technique which can lead to low results with certain soils.

Glucose is the most often-used substrate in the SIR method, although not all the soil microbial biomass responds to glucose amendment under the conditions of the SIR measurement. Thus Smith et al. (1985) found that more CO_2 was evolved from glucose supplemented with nutrient broth than from glucose alone and showed that this extra CO_2 did not come from the glucose. Presumably substrates in the broth were metabolised by a section of the population that did not respond to glucose. An assumption implicit in the SIR method for biomass C is that a constant fraction of the microbial biomass responds to glucose in a similar way in different soils.

Difficulties arise because the pattern of response of the soil microbial population to glucose differs between soils (Anderson and Domsch 1978b). Some glucose-amended soils respire at a steady rate for several hours before the rate rises as a result of cell division, in others the rate falls initially and then rises (e.g. Smith et al. 1985), whereas in a third group of soils the rate increases steadily from the beginning. Anderson and Domsch used the respiration rate measured 1 hour after glucose addition, except in soils showing an initial fall, for which they used the minimum rate.

West and Sparling (1986) modified the SIR method by adding glucose to the soil in solution rather than as the solid; the CO_2 evolved from the suspension was then measured over the 0.5-2.5 hour period. West et al. (1986c) calibrated this modified method against the FI method and found an almost identical relationship to that established by Anderson and Domsch (1978b). However, Sparling et al. (1986) found no significant relationship between biomass measured by the modified SIR method and by a modified FI method in a group of New Zealand soils, many of them highly organic. This lack of agreement could indicate that the assumptions implicit in either (or both) of the methods are not valid in those particular soils; certainly two different methods of measuring biomass should give similar values when applied to the same soil.

SIR measurements must be calibrated against other ways of estimating biomass, if they are to be used as a measure of microbial biomass in soil. More work is needed on the calibration, particularly in view of the discrepancies recently observed (Sparling et al. 1986; Martens 1987).

ATP as a measure of microbial biomass

The use of adenosine 5'-triphosphate (ATP) as a measure of microbial biomass in soil was reviewed by Jenkinson and Ladd (1981). Much progress has been made since then, particularly in improving techniques for measuring soil ATP (Ahmed et al. 1982; Tate and Jenkinson 1982; Eiland 1983; Verstraete et al. 1983; Maire 1984; Webster et al. 1984; West et al. 1986b). The importance of using acid extractants that rapidly inactivate ATPases (Lundin and Thore 1975) is now generally recognized. When neutral extractants such as $NaHCO_3/CHCl_3$ are used (Paul and Johnson 1977) some of the ATP originally present in the biomass is degraded by ATPases to ADP and AMP during extraction (Brookes et al. 1987). Acidic extractants such as H_2SO_4 or trichloracetic acid (TCA) deactivate ATPases rapidly, so that there is less opportunity for ATP hydrolysis. Thus, although the neutral $NaHCO_3/CHCl_3$ reagent and the acidic TCA reagent both extract similar quantities of total adenine nucleotides (ATP, ADP and AMP) from soil (Brookes et al. 1987), the proportion of ATP is lower with the neutral extractant. This is why the ATP content of soils is lower as measured by $NaHCO_3/CHCl_3$ extraction than as measured using TCA (Jenkinson et al. 1979; Verstraete et al. 1983). A neutral extractant has been used to measure the ATP in 'active' biomass (Verstraete et al. 1983) but whether the fraction of the total ATP thus measured does indeed indicate 'active' biomass or merely the use of an inefficient technique for extracting ATP remains to be seen.

If ATP is to serve as a measure of microbial biomass it must be present in the soil biomass in constant proportion and absent from all other soil constituents (Jenkinson and Ladd, 1981). There is now considerable evidence that the ATP content of the soil biomass is sufficiently constant, under carefully specified conditions, for it to be a useful if rough measure of soil biomass. Fig. 2 shows the ATP content of the soil microbial biomass, as measured in a wide range of Australian, Danish, English and New Zealand soils from arable, grassland and woodland sites. Taken as a whole, the ATP content of the biomass is 11.7 $\pm$ 0.29 (SE) μmol/g biomass C, corresponding to 5.39 μmol ATP/g dry biomass, if dry biomass is taken to contain 46% C. Note that Fig. 2 only contains data from experiments on soils that had been incubated aerobically at 20–25°C for at least a few days before measurements were made: ATP measurements with non–acid extractants are also excluded.

375

The quantity of ATP in a microbial cell is set by the balance between its formation and decomposition. Recent work has thrown some light on the factors that govern ATP levels in soil. When soils are air dried, ATP levels drop

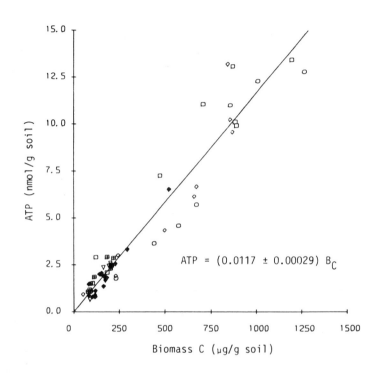

Fig. 2: The relationship between ATP content and biomass C in soil. Results from Eiland (1983), excluding soils not given a pre-incubation at 25°C,▽; West et al. (1986c), soils given a 7 day preincubation only,□; Oades and Jenkinson (1979), excluding two soils of pH 4.8 and using ATP x 0.81 to allow for the use of crude luciferase – see Tate and Jenkinson (1982),◇; Jenkinson et al. (1979), again using ATP x 0.81, o; Brookes and McGrath (1984),⊞; Eiland (1985), page 88, preincubated soils only,◈. These results were selected from those in the literature on the basis that (1), ATP was determined by acid extraction (2), the soils had had an aerobic (i.e. excluding soils incubated at >60% WHC) pre-incubation of at least a few days at 25°C before ATP and biomass C were measured and (3) that biomass C was measured by the FI method, recalculated using k_C = 0.45 where necessary.

sharply (Ahmed et al. 1982; Brookes et al. 1983; Sparling et al. 1986; West et al., 1986c), largely because of dephosphorylation to ADP and AMP (Brookes et al. 1983). On rewetting, ATP levels rise within minutes (Ahmed et al. 1982; Brookes et al. 1983), far too quickly for the increase to be caused by the synthesis of new microbial cells. Waterlogging decreases the concentration of ATP in soil (Tate and Jenkinson 1982). Likewise, change of temperature can alter the concentration of ATP in the soil (Tate and Jenkinson, 1982), in all probability because of a change in the ATP concentration of the soil microbial biomass.

Soils must first be pre-incubated under strictly controlled conditions if biomass is to be calculated from ATP content: the water content during this pre-incubation _must_ be such that the soil does not contain anaerobic microsites. Although living plant root fragments contain ATP, these die during pre-incubation and the ATP they contain disappears rapidly (Sparling et al. 1985), so that pre-incubated soils only contain microbial ATP.

MEASUREMENT OF MICROBIAL BIOMASS NITROGEN IN SOIL

Fumigation incubation (FI)

Fumigation causes a large but transient increase in the mineralization of N (for reference to early work see Jenkinson and Powlson 1976a) and much effort has been put into relating the N thus released to the N originally in the biomass. In principle, biomass N (B_N) can be calculated in the same way as biomass C, from the expression

$$B_N = F_N / k_N \qquad (2)$$

where F_N is the flush of N mineralized after fumigation and k_N is the fraction of the N in the killed biomass that is mineralized under the particular conditions used. As with C, F_N is taken as the N mineralized by the fumigated soil incubated under standard conditions, less that by a 'control'. The control is usually taken to be the N mineralized by an unfumigated soil incubated under the same conditions as the fumigated soil (Jenkinson and Powlson 1976a). Nicolardot and Chaussod's (1986) alternative approach, in which F_N is taken to be NH_4–N mineralized by the fumigated soil over the 0–7 day period, less that mineralized by the same fumigated soil over the 7–14 day period, has much to recommend it, particularly as it is unnecessary to measure changes in NO_3–N, since newly-fumigated soils do not nitrify. Nicolardot and Chaussod (1986) also developed a more sophisticated way of establishing the 'control' by mathematical analysis of the curve for cumulative mineralization of N in fumigated soil. The ratio of N mineralized by fumigated soil to N mineralized by control is much greater than the corresponding ratio for C (Ayanaba et al. 1976; Voroney and Paul 1984), so that the errors introduced by an inappropriate control are less important with N than with C.

The principal problem in calculating biomass N by equation (2) lies in establishing k_N. Most organisms contain a fairly constant proportion of C, usually 45 $\pm$ 5% on a dry matter basis, irrespective of growth medium, and it is likely that the same holds for the soil population (Jenkinson 1976). However, the N content of micro-organisms varies widely, falling sharply if they are grown in a medium deficient in N. Voroney (1983) found an inverse relationship between k_N and C/N ratio in a collection of organisms with C/N ratios ranging from 3.7 to 11.3 (see also Nicolardot et al. 1986a). This is why k_N cannot be obtained as simply as k_C, merely by growing organisms in vitro, adding them to soil, fumigating and observing how much of their N is mineralized to NH_4 under standardized conditions. The C/N ratio of the organisms added must be similar to that of the native population if a valid result is to be obtained.

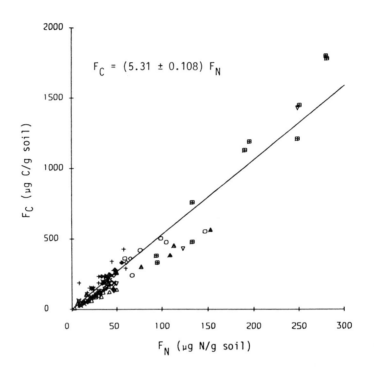

Fig. 3: The relationship between C flush (F_C) and N flush (F_N) in 104 fumigated soils. Results from Ayanaba et al. (1976),△; Powlson and Jenkinson (1976), excluding one soil of pH 3.9,▽; Ross et al. (1980b), excluding 3 soils of pH <4.7, o; Carter and Rennie (1982), whole profile samples only,◇; Voroney (1983),; Adams and Adams (1983), excluding 6 soils of pH <4.7,⊞; Woods and Schuman (1986), excluding 3 soils on parent material,; Carter (1986), x; Nicolardot et al. (1986b), excluding one soil of pH 4.8, +.

There are however indications that the C/N ratio of the soil microbial biomass is relatively constant in soils that do not contain large quantities of freshly-added plant material of wide C/N ratio. Fig. 3 shows that the ratio F_C/F_N is 5.31 ± 0.108 (SE) in a large group of soils from different countries. If the C/N ratio of the population varied widely from soil to soil, this relative constancy would not be observed. In the extreme case, a biomass of very wide C/N ratio would not release any N, making the F_C/F_N ratio infinite. Fungi of wide C/N ratio can indeed immobilize N during the initial stages of decomposition (Jenkinson 1976).

A second line of argument lends support to the hypothesis that the soil microbial biomass has a relatively constant C/N ratio. In most soils, organisms are bathed in a dilute solution of inorganic N and it is unusual for additions of inorganic N to stimulate respiration and increase the evolution of CO_2 (Jansson 1958). In such soils, microbial activity is limited by lack of carbonaceous substrate, rather than by lack of N. The obvious exceptions are soils that have just received (or already contain) large quantities of straw or other plant debris of wide C/N ratio. Such soils apart, soil organisms live in an environment where their C/N ratio is governed more by their physiology than by deficiencies in the external N supply.

Accepting the argument that the C/N ratio of the native soil biomass is relatively constant, k_N can be calculated. If $B_C/B_N = \beta$ and $F_C/F_N = \alpha$, it follows from equations (1) and (2) that

$$k_N = \frac{\beta k_C}{\alpha} \qquad (3)$$

Shen et al. (1984) and Anderson and Domsch (1980) assembled data suggesting that β, the C/N ratio of the soil microbial biomass, is about 6.7. If α, the ratio (C flush)/(N flush), is taken as 5.31 (Fig. 3), k_C as 0.45 and β as 6.7, a value of 0.57 is obtained for k_N. A more sophisticated treatment is given by Shen et al. (1984), in which allowance is made for immobilization of N during the flush. They used a value of 4.45 for α, leading to a value of 0.68 for k_N. The value of 5.31 (Fig. 3) is preferable for α, being based on a wider range of soils. A weakness of this approach is that the value of 6.7 used for β is based on data from organisms grown in vitro: techniques are needed for extracting the microbial biomass from soil so that β can be measured directly.

In this treatment, k_N is greater than k_C because β is greater than α. The C/N ratio of whole microbial cells is normally greater than that of their cytoplasm, so that the cytoplasm contains a higher proportion of the cell N than of the cell C. Thus Nicolardot (1986) found that the C/N ratio of intact Aspergillus flavus cells was 7.6, but 5.6 in the cytoplasmic fraction. He also showed that the cytoplasmic fraction was more rapidly decomposed than the killed cells as a whole, particularly in fumigated soil. This is why β is greater than α: the N mineralized during the flush comes preferentially from the rapidly-decomposing and N-rich cytoplasm. It is implicit in the use of equation (2) that the proportion of cytoplasm in the biomass does not vary greatly from soil to soil.

The value of k_N thus calculated is not very different from those measured using organisms grown in vitro, provided that organisms of C/N ratio

> 6.7 are excluded. Several workers have determined k_N by adding organisms to soil, fumigating, incubating and measuring the N mineralized. Jenkinson (1976) obtained a mean value of 0.54 for 9 organisms (omitting 3 of C/N ratio > 6.7); Amato and Ladd (1987) a value of 0.62 for 3 organisms (2 omitted) and Voroney (1983) a mean of 0.49 for 6 organisms, (8 omitted), giving a weighted mean from 18 organisms of 0.54.

Immobilization can also be a problem in measuring biomass N by the FI method but may not be as serious as sometimes thought. Thus Nicolardot et al. (1986b), using ^{15}N techniques, found that the ratio of gross mineralization of N to net mineralization of N was 1.06, 1.04 and 1.10 in three fumigated soils incubated for 7 days. A similar value (1.07) was obtained by Shen et al. (1984), who also showed that little more N was immobilized by fumigated soil than by unfumigated soil, even though the fumigated soil evolved more than twice as much CO_2 as the fumigated control. There are two reasons why relatively little N is immobilized during the flush: firstly, the recolonising population is much smaller than the killed population (Jenkinson and Powlson 1976a; Martens 1985), and secondly, the C/N ratio of the killed population is probably similar to that of the recolonising population, so that the N requirements of the recolonizers are largely met from the N in the killed organisms.

The argument, that $B_N = F_N/0.57$, is of course restricted to soils that do not contain large amounts of undecomposed plant debris of wide C/N ratio. If the F_C/F_N ratio of a soil lies much to the right of the regression line in Fig. 3, say >7.5, neither biomass C nor biomass N should be measured by the FI method.

Biomass N can be measured in waterlogged soils by a modification of the FI method (Inubushi et al. 1984). In this modification, a thin layer of wet soil is exposed to $CHCl_3$, the $CHCl_3$ removed, the soil incubated under anaerobic conditions and the NH_4–N released by fumigation then measured.

Fumigation extraction (FE)

Brookes et al. (1985a;b) found a close linear relationship between total N rendered extractable by fumigation and the flush of N, as measured by FI in a wide range of soils. Biomass N can be calculated from the relationship $B_N = 2.22\ E_N$ where E_N is N extracted by $0.5\underline{M}$ K_2SO_4 from fumigated soil less that extracted from unfumigated soil and 2.22 is an empirically–determined proportionality factor for a 24 hour exposure to $CHCl_3$. The value of 2.22 is based on $k_N = 0.57$; if k_N is taken to be 0.68 (see above), the proportionality factor becomes 1.85, in accord with the original publication (Brookes et al. 1985b).

Amato and Ladd (1987) found that the ninhydrin–reactive N (mainly α–amino acid N and NH_4–N) solubilized by chloroform fumigation correlated closely with biomass C (as measured by the FI method) in a group of 25 soils. If the chloroform was removed and the soils incubated, as in the FI method, the α-amino acid N was converted to NH_4-N. Biomass N can be calculated from $B_N = m$ $E_{\alpha N}$ where m is a proportionality factor (tentatively set at 3.1) and $E_{\alpha N}$ is the ninhydrin-reactive N released from soils that have been fumigated but not incubated.

These new direct extraction methods for biomass N show great promise, and like the direct extraction method for biomass C, may be useful where the conventional FI method breaks down, e.g. in soils that are actively immobilizing N. However, as yet they have only been calibrated against the FI method and therefore depend on the validity of biomass measurements by this method.

SAMPLING AND PREPARING SOILS FOR BIOMASS MEASUREMENTS

The traditional way of preparing soil for chemical analysis, by air drying, sieving and grinding, will not do for the delicate and labile microbial biomass.

Soils are usually homogenised by coarse sieving before measuring biomass. Lynch and Panting (1980) found that coarsely-sieved soil contained much less biomass (as measured by the FI method) than intact soil cores. However, Jenkinson and Powlson (1980) found that coarse sieving (6.25 mm) had no effect on biomass measurements and attributed Lynch and Panting's (1980) findings to incomplete fumigation of the sieved soil, which had been sieved and then packed tightly into its original volume before fumigation. Ross et al. (1985) likewise found that coarse (6 mm) sieving had no effect on biomass C measurements provided the soils were not very wet when sieved. It would be surprising if coarse sieving could cause much loss of microbial biomass in well-structured topsoils: indeed fine grinding only killed about a quarter of the biomass (Powlson 1980).

Soils are usually at the wrong water content for biomass measurements when taken from the field. Most of the present methods for measuring biomass are critically influenced by water content. If too wet, soils must first be dried down to the desired water content in such a way that no part becomes air dry during the process; if too dry, they must be wetted to the desired water content. Soils that are air-dry when collected from the field pose particularly severe problems. If air-dry for a long time, they may well contain an accumulation of dead plant material that has not had an opportunity to decompose. Such soils in effect contain 'freshly added' substrate and the FI method should not be used on them. The rewetting of air- dried soil not only allows organisms killed during the drying process to decompose but also releases substrate over and above that in the killed organisms (Jenkinson 1966; Bottner 1985; Chaussod et al. 1986; Shen et al. 1987).

When a soil is sampled and prepared for biomass measurement respiration is stimulated, even if the water content is not altered. The reasons are ill understood but include, (1) the killing of organisms by mechanical disruption (although probably a minor effect unless the soil is finely ground), (2) the exposure of previously inaccessible substrates, again probably a minor effect with coarse sieving, (3) the decomposition of root fragments detached during sampling and sieving, and (4) complex shifts in balance between competing sections of the soil population, brought about by changes in the environment, one of these changes being isolation from living roots.

There is much to be said for giving soil a 7–10 day pre– incubation under carefully standardized conditions before measuring biomass. This allows the effects of disturbance to subside and living fragments of roots to die (Sparling et al. 1985). Problems caused in the FI method by abiotic evolution of CO_2 from calcareous soils can be avoided if soils are pre–incubated in the presence of alkali to keep the concentration of CO_2 in the gas phase low (Jenkinson and Powlson 1976b; see also Martens 1987).

The biomass present in a soil that has received a pre–incubation is not of course the same as that in the soil immediately before it was collected from the field. However there are reasons for believing that the difference is not large, except with soils sampled air–dry or that have just received fresh substrate. Tate and Jenkinson (1982) and West et al. (1986a) showed that, although soil ATP increased significantly during the first day of a pre–incubation, it changed little during the following week. Although biomass C, as measured by the FI method, appeared to decrease during the pre– incubation, there was little change in biomass N or P (West et al. 1986a), suggesting that the decline in biomass C was an artifact.

Prolonged storage of soil prior to biomass measurement is best avoided (Ross et al. 1980a). If this is not possible, soil can be kept just above $0^{\circ}C$ (Ross et al. 1980a; Federle and White 1982; Chaussod et al. 1986), or stored at $-15^{\circ}C$: in both cases a pre– incubation should be given before measuring biomass. Some loss of biomass occurs when soils are frozen and thawed and this loss increases with length of storage at $-15^{\circ}C$ (Jenkinson and Powlson 1976a).

CONCLUSIONS

We are now beginning to understand the weaknesses and strengths of the different methods for measuring microbial biomass in soil and the situations where a particular method should or should not be used. However, we are still far from having a method of universal applicability, if indeed such a method can be developed. Present methods are at their most reliable (and best calibrated) in situations where the biomass is not in rapid change, for example when comparing biomass in soils taken from different ecosystems or from contrasting agricultural systems. They are at their least reliable in situations where the biomass is in rapid change, for example in air dry soils that have just been rewetted, in laboratory studies of the decomposition of organic matter in soil, or immediately after large quantities of plant residue have been ploughed under. In such situations it is wise to use a battery of methods, rather than put all the eggs in one basket.

ACKNOWLEDGEMENTS

I thank Prof. K.H. Domsch for some of the data used in Fig. 1; Miss L.C. Parry for statistical work; Dr P.C. Brookes, Dr K. Inubushi and Dr D.S. Powlson for helpful comments.

REFERENCES

Adams, T.McM., and Adams, S.N. (1983). The effects of liming and soil pH on carbon and nitrogen contained in the soil biomass. J. Agric. Sci., Camb. 101, 553–8.

Adams, T.McM., and Laughlin, R.J. (1981). The effects of agronomy on the carbon and nitrogen contained in the soil biomass. J. Agric. Sci., Camb. 97, 319–27.

Ahmed, M., Oades, J.M., and Ladd, J.N. (1982). Determination of ATP in soil: effect of soil treatments. Soil Biol. Biochem. 14, 273–9.

Amato, M., and Ladd, J.N. (1987). Effects of fumigation on soil enzyme activity and extraction of organic N from microbial biomass. Soil Biol. Biochem. 19, in the press.

Anderson, J.P.E., and Domsch, K.H. (1978a). Mineralization of bacteria and fungi in chloroform fumigated soils. Soil Biol. Biochem. 10, 207–13.

Anderson, J.P.E., and Domsch, K.H. (1978b). A physiological method for the quantitative measurement of microbial biomass in soils. Soil Biol. Biochem. 10, 215–21.

Anderson, J.P.E., and Domsch, K.H. (1980). Quantities of plant nutrients in the microbial biomass of selected soils. Soil Sci. 130, 211–16.

Andrews, J.H., and Harris, R.F. (1986). r and K-selection and microbial ecology. In: 'Advances in Microbiological Ecology'. (Ed. K.C. Marshall) pp.99–144. (Plenum Press: New York.)

Ayanaba, A., Tuckwell, S.B., and Jenkinson, D.S. (1976). The effects of clearing and cropping on the organic reserves and biomass of tropical forest soils. Soil Biol. Biochem. 8, 519– 25.

Bottner, P. (1985). Response of microbial biomass to alternate moist and dry conditions in a soil incubated with $^{14}C-$ and $^{15}N-$labelled plant material. Soil Biol. Biochem. 17, 329– 37.

Brookes, P.C., Kragt, J.F., Powlson, D.S., and Jenkinson, D.S. (1985a). Chloroform fumigation and the release of soil nitrogen. Soil Biol. Biochem. 17, 831–5.

Brookes, P.C., Landman, A., Pruden, G., and Jenkinson, D.S. (1985b). Chloroform fumigation and the release of soil nitrogen: a rapid direct extraction method to measure microbial biomass nitrogen in soil. Soil Biol. Biochem. 17, 837–42.

Brookes, P.C., and McGrath, S.P. (1984). Effects of metal toxicity on the size of the soil microbial biomass. J. Soil Sci. 35, 341–6.

Brookes, P.C., Newcombe, A.D., and Jenkinson, D.S. (1987). Adenylate energy charge measurements in soil. Soil Biol. Biochem. 19, 211–7.

Brookes, P.C., Powlson, D.S., and Jenkinson, D.S. (1982). Measurement of microbial biomasss phosphorus in soil. Soil Biol. Biochem. 14, 319–29.

Brookes, P.C., Tate, K.R., and Jenkinson, D.S. (1983). The adenylate energy charge of the soil microbial biomass. Soil Biol. Biochem. 15, 9–16.

Carter, M.R. (1986). Microbial biomass and mineralizable nitrogen in solonetzic soils. Soil Biol. Biochem. 18, 531–7.

Carter, M.R., and Rennie, D.A. (1982). Changes in soil quality under zero tillage farming systems: distribution of microbial biomass and mineralizable C and N potentials. Can. J. Soil Sci. 62, 587–97.

Chapman, S.J. (1987). Inoculation in the fumigation method for soil biomass determination. Soil Biol. Biochem. 19, 83–87.

Chaussod, R., and Nicolardot, B. (1982). Mesure de la biomasse microbienne dans les sols cultive s. I. Rev. Ecol. Biol. Sol. 19, 501–12.

Chaussod, R., Nicolardot, B., and Catroux, G. (1986). Mesure en routine de la biomasse microbienne des sols par la methode de fumigation au chloroforme. Sci. du Sol, 201–211.

Eiland, F. (1983). A simple method for quantitative determination of ATP in soil. Soil Biol. Biochem. 15, 665–70.

Eiland, F. (1985). Determination of adenosine triphosphate (ATP) and adenylate energy charge (AEC)in soil and use of adenine nucleotides as measures of soil microbial biomass and activity. Thesis, Royal Veterinary and Agricultural University, Copenhagen.

Federle, T.W., and White, D.C. (1982). Preservation of estuarine sediments for lipid analysis of biomass and community structure of microbiota. Appl. Environ. Microbiol. 44, 1166-69.

Gerson, U., and Chet, I. (1981). Are allochthonous and autochthonous soil microorganisms r and K-selected? Rev. Ecol. Biol. Sol. 18, 285-9.

Inubushi, K., Wada, H., and Takai, Y. (1984). Determination of microbial biomass nitrogen in submerged soil. Soil Sci. Plant Nutr. 30, 455-9.

Jansson, S.L. (1958) Tracer studies on nitrogen transformation in soil with special attention to mineralization immobilization relationships. Ann. Roy. Agric. Coll. Sweden 24, 101- 361.

Jenkinson, D.S. (1966). Studies in the decomposition of plant material in soil. II. J. Soil Sci. 17, 280-302.

Jenkinson, D.S. (1976). The effect of biocidal treatments on metabolism in soil. IV. Soil Biol. Biochem. 8, 203-8.

Jenkinson, D.S., Davidson, S.A., and Powlson, D.S. (1979). Adenosine triphosphate and microbial biomass in soil. Soil Biol. Biochem. 11, 521-7.

Jenkinson, D.S., Hart, P.B.S., Rayner, J.H., and Parry, L.C. (1987). Measuring the turnover of organic matter in long-term experiments at Rothamsted. Intecol Bulletin, in the press.

Jenkinson, D.S., and Ladd, J.N. (1981). Microbial biomass in soil: measurement and turnover. In 'Soil Biochemistry'. Vol. 5 (Eds E.A. Paul and J.N. Ladd) pp.415-71. (Marcel Dekker: New York).

Jenkinson, D.S., and Powlson, D.S. (1976a). The effects of biocidal treatments on metabolism in soil. I. Soil Biol. Biochem. 8, 167-77.

Jenkinson, D.S., and Powlson, D.S. (1976b). The effects of biocidal treatment on metabolism in soil. V. Soil Biol. Biochem. 8, 209-13.

Jenkinson, D.S., and Powlson, D.S. (1980). Measurements of microbial biomass in intact cores and in sieved soil. Soil Biol. Biochem. 12, 579-81.

Jenkinson, D.S., Powlson, D.S., and Wedderburn, R.W.M. (1976). The effects of biocidal treatments on metabolism in soil. III. Soil Biol. Biochem. 8, 189-202.

Kudeyarov, V.N., and Jenkinson, D.S. (1976). The effects of biocidal treatments on metabolism in soil. VI. Soil Biol. Biochem. 8, 375-8.

Lundin, A., and Thore, A. (1975). Comparison of methods for extraction of bacterial adenosine nucleotides determined by firefly assay. Appl. Microbiol. 30, 713-21.

Lynch, J.M., and Panting, L.M. (1980). Cultivation and the soil biomass. Soil Biol. Biochem. 12, 29-33.

Maire, N. (1984). Extraction de l'adenosine triphosphate dans les soils. Soil Biol. Biochem. 16, 361-6.

Martens, R. (1985). Limitations in the application of the fumigation technique for biomass estimations in amended soils. Soil Biol. Biochem. 17, 57-63.

Martens, R. (1987). Estimation of microbial biomass in soil by the respiration method: importance of soil pH and flushing methods for the measurement of respired CO_2. Soil Biol. Biochem. 19, 77-81.

Nannipieri, P. (1984). Microbial biomass and activity measurements in soil: ecological significance. In 'Current Perspectives in Microbial Ecology'. (Eds M.J. Klug and C.A. Reddy) pp.515-521. (American Society for Microbiology: Washington.)

Nicolardot, B. (1986). Etude comparée de la mineralisation de differentes fractions cellulaires d'Aspergillus flavus dans un sol fumigé ou non au chloroforme. C.R. Acad. Sci. Paris 303, III, 489-94.

Nicolardot, B., and Chaussod, R. (1986). Mesure de la biomasse microbienne dans les sols cultivés. III. Rev. Ecol. Biol. Sol. 23, 233-47.

Nicolardot, B., Chaussod, R., and Catroux, G. (1984). Décomposition de corps microbiens dans des sols fumigés au chloroforme: effets du type de sol et de microorganisme. Soil Biol. Biochem. 16, 453-8.

Nicolardot, B., Guiraud, G., Chaussod, R., and Catroux, G. (1986a). Minéralisation dans le sol de matériaux microbiens marqués au carbone 14 et a l'azote 15: quantification de l'azote de la biomasse microbienne. Soil Biol. Biochem. 18, 263-73.

Nicolardot, B., Guiraud, G., Chaussod, R., and Catroux, G. (1986b). Dynamique de la biomasse microbienne et minéralisation réorganisation de l'azote dans les sols cultivés. Role compartimental de la biomasse microbienne. Rapport Final Contrat 84330. INRA/Ministere de l'Environment et du Cadre de Vie. Dijon, 37 p.

Oades, J.M., and Jenkinson, D.S. (1979). Adenosine triphosphate content of the soil microbial biomass. Soil Biol. Biochem. 11, 201-4.

Paul, E.A., and Johnson, R.L. (1977). Microscopic counting and adenosine 5'-triphosphate measurement in determining microbial growth in soils. Appl. Environ. Microbiol. 34, 263-9.

Powlson, D.S. (1980). The effects of grinding on microbial and non-microbial organic matter in soil. J. Soil Sci. 31, 77-85.

Powlson, D.S., Brookes, P.C., Christensen, B.T. (1987). Measurement of microbial biomass provides an early indication of changes in total soil organic matter due to straw incorporation. Soil Biol. Biochem. 19, in the press.

Powlson, D.S., and Jenkinson, D.S. (1976). The effects of biocidal treatments on metabolism in soil. II. Soil Biol. Biochem. 8, 179—88.

Powlson, D.S., and Jenkinson, D.S. (1981). A comparison of the organic matter, biomass, adenosine triphosphate and mineralizable nitrogen contents of ploughed and direct-drilled soils. J. Agric. Sci., Camb. 97, 713—21.

Ridge, E.H. (1976). Studies on soil fumigation II. Soil Biol. Biochem. 8, 249—53.

Ross, D.J., Speir, T.W., Tate, K.R., and Orchard, V.A. (1985). Effects of sieving on estimations of microbial biomass, and carbon and nitrogen mineralization, in soil under pasture. Aust. J. Soil. Res. 23, 319—24.

Ross, D.J., Tate, K.R., Cairns, A., and Meyrick, K.F. (1980a). Influence of storage on soil microbial biomass estimated by three biochemical procedures. Soil Biol. Biochem. 12, 369—74.

Ross, D.J., Tate, K.R., Cairns, A., and Pansier, E.A. (1980b). Microbial biomass estimations in soils from tussock grasslands by three biochemical procedures. Soil Biol. Biochem. 12, 375—83.

Sarathchandra, S.U., Perrott, K.W., and Upsdell, M.P. (1984). Microbiological and biochemical characteristics of a range of New Zealand soils under established pasture. Soil Biol. Biochem. 16, 177—83.

Schnürer, J., Clarholm, M., and Rosswall, T. (1985). Microbial biomass and activity in an agricultural soil with different organic matter contents. Soil Biol. Biochem. 17, 611—8.

Shen, S.M., Brookes, P.C., and Jenkinson, D.S. (1987). Soil respiration and the measurement of microbial biomass carbon by the fumigation technique in fresh and air-dried soil. Soil Biol. Biochem. 19, 153—8.

Shen, S.M., Pruden, G., and Jenkinson, D.S. (1984). Mineralization and immobilization of nitrogen in fumigated soil and the measurement of microbial biomass nitrogen. Soil Biol. Biochem. 16, 437—44.

Shields, J.A., Paul, E.A., and Lowe, W.E. (1974). Factors influencing the stability of labelled microbial materials in soils. Soil Biol. Biochem. 6, 31—7.

Smith, J.L., McNeal, B.L., and Cheng, H.H. (1985). Estimation of soil microbial biomass: an analysis of the respiratory response of soils. Soil Biol. Biochem. 17, 11—6.

Sparling, G.P. (1981). Microcalorimetry and other methods to assess biomass and activity in soils. Soil Biol. Biochem. 13, 93—8.

Sparling, G.P., Ord, B.G., and Vaughan, D. (1981). Changes in microbial biomass and activity in soils amended with phenolic acids. Soil Biol. Biochem. 13, 455—60.

Sparling, G.P., Speir, T.W., and Whale, K.N. (1986). Changes in microbial biomass C, ATP content, soil phospho-monoesterase and phospho-diesterase activity following air-drying of soils. Soil Biol. Biochem. 18, 363—70.

Sparling, G.P., West, A.W., and Whale, K.N. (1985). Interference from plant roots in the estimation of soil microbial ATP, C, N and P. Soil Biol. Biochem. 17, 275—8.

Sparling, G.P., and Williams, B.L. (1986). Microbial biomass in organic soils: estimation of biomass C and effect of glucose or cellulose amendments on the amounts of N and P released by fumigation. Soil Biol. Biochem. 18, 507—13.

Tate, K.R., and Jenkinson, D.S. (1982). Adenosine triphosphate measurements in soil: an improved method. Soil Biol. Biochem. 14, 331—5.

Vance, E.D., Brookes, P.C., and Jenkinson, D.S. (1987a). Microbial biomass measurements in forest soils: determination of k_C values and tests of hypotheses to explain the failure of the chloroform fumigation-incubation method in acid soils. Soil Biol. Biochem. 19, (in the press).

Vance, E.D., Brookes, P.C., and Jenkinson, D.S. (1987b). Microbial biomass in forest soils: the use of the chloroform fumigation-incubation method in strongly acid soils. Soil Biol. Biochem. 19, (in the press).

Vance, E.D., Brookes, P.C., and Jenkinson, D.S. (1987c). An extraction method for measuring soil microbial biomass C. Soil Biol. Biochem. 19, (in the press).

Van Veen, J.A., Ladd, J.N., and Frissel, M.J. (1984). Modelling C and N turnover through the microbial biomass in soil. Plant Soil 76, 257—74.

Verstraete, W., Van de Werf, H., Kucnerowicz, F., Ilaiwi, M., Verstraeten, L.M.J., and Vlassak, K. (1983). Specific measurement of soil microbial ATP. Soil Biol. Biochem. 15, 391—6.

Voroney, R.P. (1983). Decomposition of crop residues. PhD Thesis, University of Saskatchewan.

Voroney, R.P., and Paul, E.A. (1984). Determination of k_C and k_N in situ for calibration of the chloroform fumigation-incubation method. Soil Biol. Biochem. 16, 9—14.

Webster, J.J., Hampton, G.J., and Leach, F.R. (1984). ATP in soil: a new extractant and extraction procedure. Soil Biol. Biochem. 16, 335—42.

West, A.W., Ross, D.J., and Cowling, J.C. (1986a). Changes in microbial C, N, P and ATP contents, numbers and respiration on storage of soil. Soil Biol. Biochem. 18, 141–8.

West, A.W., and Sparling, G.P. (1986). Modifications to the substrate–induced respiration method to permit measurement of microbial biomass in soils of differing water contents. J. Microbiol. Methods 5, 177–89.

West, A.W., Sparling, G.P., Cowling, J.C., Tate, K.R., and Reynolds, J. (1986b). Luciferase preparation, assay conditions and calculated extraction efficiency as factors in the estimation of soil ATP content. Biol. Fert. Soils 2, 205–11.

West, A.W., Sparling, G.P., and Grant, W.D. (1986c). Correlation between four methods to estimate total microbial biomass in stored, air–dried and glucose amended soils. Soil Biol. Biochem. 18, 569–76.

Woods, L.E., and Schuman, G.E. (1986). Influence of soil organic matter concentrations on carbon and nitrogen activity. Soil Sci. Soc. Am. J., 50, 1241–4.

DENITRIFICATION IN THE FIELD

C.J. Smith

ABSTRACT

Several methods are compared for their suitability to quantify N loss by direct measurement of gaseous products of denitrification appropriate for the field. The methods either involve (i) the use of highly labelled fertilizer to increase the ^{15}N enrichment of the NO_3^- pool and measurement of $^{15}N_2$, (ii) replacement of atmospheric N_2 with a $^{15}N_2/He/O_2$ mixture and monitoring isotopic dilution of $^{15}N_2$ by denitrification of unlabelled NO_3^-, or (iii) the use of acetylene to inhibit N_2O reduction to N_2 so that total denitrification losses can be measured as N_2O. Determining N losses due to denitrification by direct field measurements remains a problem, especially in systems with low fertility and where denitrification rate is substrate limited. Nevertheless, the available methodology can be used effectively to expand current knowledge of denitrification in agricultural and natural ecosystems.

INTRODUCTION

Interest in denitrification exists because (a) it is a major mechanism of loss of soil N resulting in decreased fertilizer efficiency, (b) it contributes significantly to N_2 in the atmosphere (CAST 1976; Crutzen and Ehhalt 1977), and (c) it is a major step in the global nitrogen cycle (Knowles 1982). In situ measurement of denitrification is essential to isolate factors in agricultural and non-agricultural ecosystems which can minimize the loss of N. It is important to study denitrification in relation to nitrification to get a better understanding of N transformation in the soil. Unfortunately techniques to measure denitrification which rely on the addition of NO_3^- artificially increase denitrification above rates one would observe with the slower, natural nitrification/ denitrification sequence. This problem is exacerbated in non polluted flooded soils and sediments where NO_3^- concentrations are generally low and close coupling exists between the two processes (Jenkins and Kemp 1984).

Several recent reviews deal with the microbiology and physiology of denitrification and methods available for its study (Focht 1978; Rolston 1978; Payne 1981; Knowles 1982). This paper does not cover all aspects of the subject but draws attention to recent developments in methods for the direct measurement of gaseous products of denitrification in the field.

^{15}N ISOTOPES

In order to quantify N_2 production due to denitrification against the large background of N_2 in ambient air, ^{15}N has been used to distinguish the source of the N. This is achieved by measuring concentrations of ^{15}N-labelled N_2 and N_2O evolved from small plots of soil treated with highly enriched ^{15}N-labelled NO_3^- (Rolston et al. 1978; Ryden et al. 1981; Siegel et al. 1982) or by monitoring isotopic dilution of a ^{15}N enriched N_2 atmosphere from unamended soil (Limmer et al. 1982).

Advances in combined gas chromatography – mass spectrometry, not only separate the gases before introduction into the mass spectrometer, but enable identification of compounds (Focht 1978; Focht et al. 1979, 1980). The technology has been used to study denitrification under laboratory conditions and will not be covered in this paper.

^{15}N-labelled N_2 evolution

Recent work by Siegel et al. (1982) has led to the development of mass spectrometer procedures for determination of N_2 and N_2O evolution from soil treated with ^{15}N-labelled fertilizer. These procedures utilize a double-collector instrument equipped with dual inlet systems, but they can also be conducted on a triple-collector instrument (Mulvaney 1984). In all cases the ratio difference (e.g. for double-collector $^{30}N_2/$ ($^{28}N_2+^{29}N_2$) or $^{29}N_2/^{28}N_2$; and for the triple-collector $^{29}N_2/^{28}N_2$ or $^{30}N_2/^{28}N_2$) is measured. The ratios are used to calculate the mole fraction of ^{15}N in the NO_3^- pool from which the N_2 or N_2O are derived and a fraction proportional to the amount of ^{15}N-labelled N_2 or N_2O in the atmospheric sample analyzed. Equations used are based on assumptions that (i) the NO_3^- being denitrified is isotopically uniform, and (ii) the N_2 evolved does not contribute to a significant increase in background N_2 in the gas being analysed. Approximations were originally made to simplify the solution of the equations (Mulvaney and Kurtz 1982; Siegel et al. 1982), however with microcomputers calculations can now be made with fewer approximations (Mulvaney and Boast 1986).

The field procedure involves insertion of brass cylinders (9.14 cm i.d.) 14 cm into soil, and enclosing a small volume of air (400 cm^3) using a rubber stopper. A novel feature is a closable inner chamber attached to the rubber stopper. The inner chamber is in contact with the remainder of the enclosed air space in the brass cylinder during the incubation period, but at sampling it can be isolated and a gas sample taken without inducing a pressure drop at the soil surface (Siegel et al. 1982). The equivalent of 100 kg/N ha as KNO_3 with an enrichment of 63.3 atom% ^{15}N is applied to soil with two hour gas sampling times. Samples are taken daily, however flexibility exists in frequency of sampling and total length of the sampling time.

A precise measurement of the $^{30}N_2/(^{28}N_2+^{29}N_2)$ ratio is needed. This ratio is small (1.34×10^{-5}) at natural abundance and is subject to interferences from NO, N_2O and O_2. However, it is because of this extremely low abundance that $^{30}N_2$ concentrations change rapidly by denitrification of the highly enriched substrate. The first and essential step to using this technique is the elimination of the interfering gases. Complete exclusion of O_2 is necessary since it reacts with N forms in the mass spectrometer to form NO and thus contaminate the $^{30}N_2$ peak, causing overestimation of the N_2 flux. Siegel et al. (1982) did this by passing the gas sample successively through two liquid N_2 cold traps with a high capacity O_2 trap between them, then a commercial O_2 trap and a high efficiency cold trap (see Fig. 2 in Siegel et al. 1982).

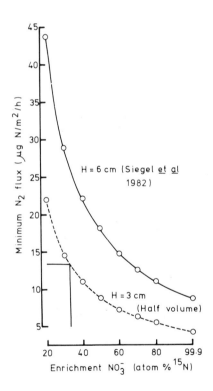

Fig. 1: Effect of NO_3^- enrichment and height (volume) of the chamber on minimum detectable N_2 flux calculated using the procedure of Siegel et al. (1982).

Nitrous oxide is determined by dilution of ^{15}N-labelled N_2O produced by denitrification with a known amount of unenriched N_2 followed by reduction of the N_2O to N_2 in a heated tube containing reduced copper (Mulvaney and Kurtz 1982). The procedure permits N_2O analysis of an atmospheric sample previously analysed for ^{15}N-labelled N_2 by the method of Siegel et al. (1982). Both procedures require a sophisticated inlet system for the mass spectrometer.

Precision of the ion ratio measurement determines the minimum N_2 flux that can be quantified with confidence. Using the equations and approximations outlined by Siegel et al. (1982) and assuming analytical precision does not change with NO_3^- enrichment, the minimum detectable N_2 evolution is inversely related to the NO_3^- enrichment and is directly proportional to the internal volume of the chamber (Fig. 1). Minimum detectable N_2 flux decreases from 44 to 9 $\mu gN/m^2/h$ as the mole fraction of ^{15}N in the NO_3^- undergoing denitrification increases from 0.20 to 0.999. The reason for improved sensitivity is that the mole fraction of $^{30}N_2$ increases at the expense of $^{29}N_2$ as the enrichment of the NO_3^- is increased, reaching 100 at 100 atom% [15]N. However, at the highest enrichment the fraction of $^{29}N_2$ approaches zero so that computation of its mole fraction is subject to large errors.

Field experiments with 2 hour collection times following the addition of 100 kgN/ha at 63.3 atom % [15]N and 10 mm of water, allowed the detection of 4 ng N_2/cm^2, which corresponds to a detection limit of 5 g N_2/ha/day. This method generally represents one approach to the direct measurement of N_2 evolution from soil, but the scope of its application is limited by:

(i) the cost of highly enriched ^{15}N-labelled fertilizer; and
(ii) only the loss of N_2 derived from fertilizer can be measured.

The method cannot be applied to non-agricultural ecosystems or sites of low fertility without technical enhancement of the indigenous NO_3^- pool. It is doubtful whether the N_2 evolution reflects the natural rate of in situ denitrification, as denitrification exhibits Michaelis-Menton kinetics with respect to NO_3^- concentration (Focht 1978; Ryzhova 1979). Addition of ^{15}N-labelled NO_3^- may consequently induce artifically high rates of denitrification. The proposed method does have the advantage of enabling non-destructive monitoring of the soil NO_3^- isotopic composition while denitrification is occurring.

Isotopic dilution of ^{15}N enriched N_2 atmosphere

The second ^{15}N technique is capable of measuring in situ N_2 evolution from unamended soil. Ambient concentration of N_2 is reduced, and a small volume of ^{15}N-labelled N_2 is introduced into the headspace gas and becomes diluted as unlabelled N_2 is evolved (Limmer et al. 1982).

A double-walled gas lysimeter is inserted into the soil to enable N_2 to be purged from the system (using helium or argon containing the desired O_2

concentration to maintain the existing O_2 concentration within the soil profile) and to introduce $^{15}N_2$. A continuous flow and positive pressure maintained at the base of the lysimeter prevents the inward diffusion of soil gases. It is necessary to monitor the soil atmosphere in order to maintain the existing O_2 concentration within the soil profile.

The headspace of the lysimeter is purged for 20 min with He/O_2 gas at 100 cm^3/min and a pressure of 2 cm H_2O. The soil core is then flushed with $^{15}N_2/He/O_2$ at the same flow and pressure conditions. After 10 min., the $^{15}N_2/He/O_2$ valve is closed and He/O_2 gas flow directed to the lysimeter base-flow ports (see Fig. 3. in Limmer et al. 1982). Samples of the headspace gas are taken 10 and 70 min after commencement of the base-flow.

Dinitrogen enrichment in the atmosphere is monitored on a VG Micromass 602C mass spectrometer. Due to the limited amplification of the major beam (m/e 28) the minimum concentration for which double-beam ^{15}N determinations can be made is approximately 5000 ppm (v/v). However, increasing N_2 to 12000 ppm (v/v) almost doubles the precision of the ^{15}N determination (Limmer et al. 1982). Nitrous oxide is also determined by using the instrument on single beam and scanning for the principal fragmentation products. It is necessary to remove CO_2 as it will interfere with N_2O analysis. Limmer et al. (1982) used Carbsorb to remove >95 percent of the CO_2.

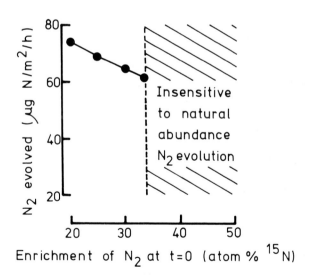

Fig. 2: Effect of ^{15}N enrichment of the atmosphere on sensitivity of N_2 evolution [Calculated according to Limmer et al. 1982 with N_2 = 10,000 ppm].

391

This technique has been used on silt loam soils and has the advantage of not requiring the addition of ^{15}N-labelled NO_3^-. The overall sensitivity of the technique depends on the $^{15}N_2$ enrichment and the ability to precisely and accurately measure the $^{15}N_2$ dilution which is attributed to denitrification of unlabelled substrate. However, it is a requirement of the Micromass double-beam facility that the m/e 29 (minor) does not exceed m/e 28 (major). The two m/e signals are equal at 33.3 atom % ^{15}N which is the maximum usable gas enrichment. At enrichments above 33.3 atom % ^{15}N, m/e 29 and m/e 30 become the major and minor peaks respectively, and the technique becomes insensitive to the evolution of N_2 at natural abundance.

As the m/e ratios 28/29 are increased from 1.0 the minimum detectable limit of N_2 evolution increases from 62 to 74 $\mu gN/m^2/h$ (Fig. 2). These evolution rates are similar to the minimum rates detectable using the previously discussed method proposed by Siegel et al. (1982).

Failure to remove all the background N_2 is a major limitation. All the N_2 evolution may not be attributable to denitrification as some may come from release of N_2 dissolved in the soil solution, or N_2 within the soil atmosphere. Consequently the technique overestimates the denitrification rate because it is not possible to distinguish between N_2 from these different sources. It may be possible to overcome this problem by repeated incubation and purging cycles. Seitzinger et al. (1980) utilized repeated incubation to reduce the background concentrations of N_2 in the atmosphere and dissolved in water (413 $\mu mol/Kg$ at $22^\circ C$ and $31^\circ/oo$), enabling direct measurement of denitrification from coastal marine sediment cores maintained in the laboratory at in situ conditions.

The two ^{15}N methods discussed require either (i) a relatively sophisticated inlet system for the mass spectrometer (Siegel et al. 1982) or (ii) the establishment and maintenance of subambient N_2 concentrations (Limmer et al. 1982). Exclusion of O_2 is required since it may react with N molecules in the mass spectrometer to form NO (mass 30) and contaminate the m/e 30 peak due to $^{15}N_2$.

Non-random N_2 isotopic distribution

The following section discusses a recent method for measuring ^{15}N content of a gas mixture with non-random N_2 isotopic distribution (Craswell et al. 1985). N_2 gas evolved during denitrification of ^{15}N enriched fertilizer does not achieve a random N isotope distribution, thus the mixture of $^{15}N_2$ and atmospheric N_2 cannot be analyzed by simple measurement of the m/e 28 and 29 ratios. The procedure involves use of a high voltage arc passed between tungsten electrodes sealed in a glass vessel to fix some of the N_2 into the NO_x pool. The NO_x is absorbed and oxidized by acidic potassium permanganate and is reduced to NH_4^+ (monatomic species) by the reduced iron procedure (Tedesco and Keeney 1972). The NH_4^+ is distilled by routine steam distillation procedures and the $(NH_4)_2SO_4$ – boric acid salt is analysed for ^{15}N content on a Micromass 622 dual collector mass spectrometer. The method is precise, accurate, and avoids many of the problems associated with direct ^{15}N analysis of gas samples with non–random N isotopic distribution. Craswell et al. (1985) outlined the method of gas analysis and indicated that the method is applicable to measurement of denitrification losses from ^{15}N–enriched fertilizers. However the method can also be used for measurement of total denitrification losses from fertilized soils.

The precision of the mass spectrometer used for ^{15}N analysis will determine the minimum detectable N_2 evolution. With the Micromass 622 used by Craswell et al. (1985) the ^{15}N enrichment of compressed air analysed by the procedure is 0.3647 $\pm$ 0.0006 atom % [15]N. Consequently in these calculations I assumed that the minimum ^{15}N enrichment that can be determined with confidence is 0.002 atom % [15]N above background (i.e. 0.3680 atom % [15]N). Enrichment of the NO_3^- pool, has been varied from 10 to 80 atom % [15]N and a linear N_2 evolution rate has been assumed during the cover period. Equations used for calculation of the total denitrification rate are:

$$N_{2\,den.} \quad atom\%^{15}N_{NO3} + N_{2\,atm.} \quad 0.366 = N_{2\,mix.} \quad 0.368 \tag{1}$$

$$N_{2\,den.} \quad + N_{2\,atm.} \quad = N_{2\,mix.} \tag{2}$$

where $N_{2\ den.}$ is N_2 evolved from denitrification of NO_3^- at known atom $\%^{15}N$; $N_{2\ atm.}$ is N_2 within cover at zero time with enrichment of 0.366 atom $\%^{15}N$ and $N_{2\ mix.}$ is total amount N_2 in cover at t_1 with enrichment of 0.368 atom $\%^{15}N$.

In order to calculate total denitrification it is essential to know the isotopic composition of the N substrate as it will be isotopically diluted by natural N transformations occurring in the soil. Minimum N_2 flux (denitrification) decreases from 14 to 1 mgN_2–N/m^2 in the cover period as the enrichment of the NO_3^- pool is increased. Sensitivity can be increased by reducing the cover volume (height) (Fig. 3). It is important to use a mass spectrometer with high precision, apply highly enriched fertilizer to maximize ^{15}N enrichment in the NO_3^- pool, and accurately measure the isotopic composition of the NO_3^-. Data presented based on these assumptions indicates that this procedure lacks some sensitivity for the measurement of field denitrification when compared with the direct measurement of $^{30}N_2$ as proposed by Siegel et al. (1982).

INHIBITION OF NITROUS OXIDE REDUCTION BY ACETYLENE

An alternative technique for measuring denitrification in the field involves the use of acetylene (C_2H_2) to inhibit the reduction of N_2O to N_2. Ryden et al. (1979) were among the first of a number of workers (Chan and Knowles 1979; Rolston et al. 1982; Ryden 1982; Ryden and Dawson 1982; Mosier et al. 1985; Duxbury and McConnaughey 1986; Terry et al. 1986) to use this technique to quantify denitrification in agricultural systems. Areas of concern with the technique are inhibition of nitrification (Walter et al.

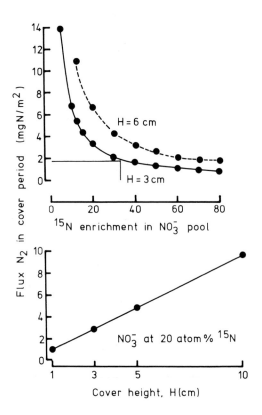

Fig. 3: Minimum N_2 evolution detectable using the technique outlined by Craswell et al. (1985).

1979; Berg et al. 1982) and the potential decomposition of C_2H_2 by soil microorganisms (Watanabe and de Guzman 1980; Yeomans and Beauchamp 1982; Terry and Duxbury 1985; Terry et al. 1986). Metabolism of C_2H_2 could increase N_2O emissions by providing an energy source for denitrification or it could reduce N_2O emission by decreasing the partial pressure of C_2H_2 below that required to inhibit N_2O reduction.

Ryden (1982) confirmed that the C_2H_2 technique cannot be used to measure denitrification when limited by inhibition of nitrification during the denitrification measurement. This is a serious limitation in permanently flooded soil and sediments where NO_3^- concentrations are generally low and close coupling exists between nitrification and denitrification. Inhibition of nitrification will also lead to a decrease in O_2 consumption which may reduce the denitrification rate.

The technique offers potential for short term denitrification measurement, however the rates are likely to be more sensitive to errors caused by gas diffusion in soil. It cannot be used where nitrification and denitrification occur concomitantly and where NO_3^- limits the denitrification rate. These problems can be overcome by rotating field plots, and limiting the time soil is exposed to C_2H_2. The criteria for selection of a new site should be the differential NO_3^- levels in C_2H_2-treated and untreated areas.

The method is relatively quick and easy to perform and suitable instrumentation is generally found in most laboratories.

The methodology will only be given in general terms since the nature of the ecosystem to be investigated governs the selection of appropriate methods. The technique can be divided into two phases: the introduction of C_2H_2, and the measurement of N_2O.

Acetylene is introduced to establish concentrations of 1.0% in the soil or sediment atmosphere by:

(i) radial diffusion of C_2H_2 from multiple C_2H_2 -supply probes inserted into the soil to various depths (Ryden et al. 1979; Hallmark and Terry 1985; Mosier et al. 1985; McConnaughey and Duxbury 1986). The time required to establish the necessary C_2H_2 concentration will vary with soil porosity and this emphasizes the need for inclusion of a system whereby C_2H_2 concentration can be monitored (Duxbury and McConnaughey 1986).

(ii) saturating water with C_2H_2 or dissolving the desired volume of pure C_2H_2 into the water (Chan and Knowles 1979; Terry et al. 1986). It is possible to use irrigation water to carry C_2H_2 into the soil. It is necessary to establish that the C_2H_2 concentration within the soil is sufficient to inhibit N_2O reductase activity, and to monitor the persistence of the C_2H_2.

Nitrous oxide evolution is measured at locations with and without C_2H_2 treatments and, after suitable intervals, the amount of N_2 evolved is calculated by difference in apparent N_2O fluxes. A range of techniques have been used for N_2O measurements.

These include the use of a static chamber, collecting gas samples with time to monitor N_2O concentration using a gas chromatograph equipped with an electron capture detector (Duxbury and McConnaughey 1986; McConnaughey and Duxbury 1986; Terry et al. 1986). A linear increase in N_2O concentration will only occur when the minimum measurable increase in N_2O within the static chamber has negligible effect on concentration gradient. Hutchinson and Mosier (1981) proposed an equation for N_2O evolution which overcomes this problem of decreasing N_2O concentration gradient with time.

Ryden et al. (1979) used a 5A molecular sieve adsorption trap for N_2O collection in a continuous flow of air through the soil chamber. Traps were changed at 3 h intervals, N_2O was displaced from the molecular sieve and quantified using a gas chromatograph. It is essential that the chamber design avoids creation of small negative pressure changes within the chamber (Denmead

1979). Such changes may induce mass flow of gas from the soil and lead to erroneous flux measurements.

Mosier et al. (1985) have further developed the analytical system to enable continuous monitoring of N_2O evolution from open systems. The procedure is an extension of the open circulation system developed for measuring N_2O emission from soils in the field (Denmead 1979). Nitrous oxide concentration is measured with an infrared gas analyzer fitted with an N_2O measuring cell. Minimum detectable evolution of N_2O (denitrification) is estimated to be 16–24 $\mu gN/m^2/h$ based on flow rate 20 L/h and having the analytical capability to detect a 10% flux contribution to the N_2O adsorbed from air drawn through the cover (Ryden et al. 1979). However, the use of an infrared gas analyzer offers potential for improvement in sensitivity. Minimum N_2O discrimination is less than 2 $ngN_2O–N/m^2/s$, or 7 $\mu gN/m^2/h$ at flow rate 1 L/min (Denmead 1979). Mosier et al. (1985) using infrared gas analysis and a flow rate of 4.5 L/min reported a detection limit of 20 $\pm$ 10 $ngN/m^2/s$, i.e. 72$\pm$36 $\mu gN/m^2/h$ with a N_2O concentration difference of 88 ppb (v/v).

Static chambers (Duxbury and McConnaughey 1986; Terry et al. 1986) have the advantage of simplicity in design, and correction factors can be used based on the diffusion equation to compensate for errors introduced by increasing gas concentrations. Chamber height can be varied to improve the sensitivity of N_2O measurement (Mathias et al. 1980). Again chamber design must be modified in relation to the N_2O flux and the limit of detection required by the researcher to give the desired sensitivity in the measurement of denitrification.

RELEASE OF DENITRIFICATION GASES

Considerable attention has been given to methods of measuring denitrification gas fluxes from soil but little has been given to the dynamics of gas production and subsequent diffusion to the soil surface. Release of denitrification gases at the soil surface is greatly controlled by diffusion, and any attempt to measure N_2 or N_2O evolution must consider this controlling mechanism. Jury et al. (1982) have developed a two dimensional model of gas production and diffusion through soil to study the efficacy of open and closed chambers for measuring N_2O production. The model clearly illustrates many of the faults with the various measurement techniques and suggests that in many cases the measurements do not relate well to production rates because gas concentration is still increasing in the soil. Due to the large lag caused by gaseous diffusion, accurate information will only be obtained from an in situ denitrification event, irrespective of the measurement technique, if all gas produced is trapped. This may require frequent or continuous monitoring for up to several weeks after gaseous production ceases if soil water content remains high.

Direct measurements of denitrification in the field are essential to our understanding of this process in agricultural and natural ecosystems. I feel that all the current methods for doing this have some limitations which introduce uncertainties in the results. The methods are also limited to use on very small areas or at a single micro-location.

The sensitivity of the three ^{15}N methods using a cover height of 5 cm, ^{15}N enrichment of 33.33 atom% ^{15}N and 1 hour incubation are 3,000, 62, and 44 gN/m^2/h for the Craswell et al., Limmer et al., and Siegel et al. methods, respectively. The Siegal et al. method is the most sensitive and a N$_2$ evolution rate of 5 gN/ha/d can be measured with confidence. However, their method requires a sophisticated inlet system for the mass spectrometer and its use is restricted to sites with high fertility. Under low fertility conditions the Limmer et al. method has an advantage when precautions are taken to remove the background N$_2$.

The alternative method is to use acetylene to inhibit N$_2$O reduction to N$_2$ so that total denitrification losses are measured as N$_2$O. Unfortunately the acetylene interferes with nitrification, and incomplete blockage, gas solubilities and flux calculations may also introduce errors into the results.

REFERENCES

Berg, P., Klemedtsson, L., and Rosswall, T. (1982). Inhibitory effect of low partial pressure of acetylene on nitrification. Soil Biol. Biochem. 14, 301-3.

CAST (1976). Effect of increased nitrogen fixation on stratospheric ozone. Counc. Agric. Sci. Tech. Rep. No. 53, Iowa State Univ., Ames.

Chan, Y., and Knowles, R. (1979). Measurement of denitrification in two freshwater sediments by an in-situ acetylene inhibition method. Appl. Environ. Microb. 37, 1067-72.

Craswell, E.T., Byrnes, B.H., Holt, L.S., Austin, E.R., Fillery, I.R.P., and Strong, W.M. (1985). Nitrogen-15 determination of non randomly distributed dinitrogen in air. Soil Sci. Soc. Am. J. 49, 664-8.

Crutzen, J.P., and Ehhalt, D.H., (1977). Effects of nitrogen fertilizers and combustion on the stratospheric ozone layer. Ambio 6, 112-7.

Denmead, O.T. (1979). Chamber systems for measuring nitrous oxide emission from soils in the field. Soil Sci. Soc. Am. J. 43, 89-95.

Duxbury, J.M., and McConnaughey, P.K. (1986). Effect of fertilizer source on denitrification and nitrous oxide emissions in a maize field. Soil Sci. Am. J. 50, 644-8.

Focht, D.D., (1978). Methods for analysis of denitrification in soils. In 'Nitrogen in the Environment' Vol.2. (Eds D.R. Nielsen and J.G. MacDonald) pp. 433-90. (Academic Press Inc.: New York.)

Focht, D.D., Stolzy, L.H., and Meek, B.D. (1979). Sequential reduction of nitrate and nitrous oxide under field conditions as brought about by organic amendments and irrigation management Soil Biochem. 11, 37-46.

Focht, D.D., Valoras, N., and Letey, J. (1980). Use of interfaced gas chromatography-mass spectrometry for detection of concurrent mineralization and denitrification in soil. J. Environ. Qual. 9, 218-23.

Hallmark, S.L., and Terry, R.E. (1985). Field measurement of denitrification in irrigated soils. Soil Sci. 140, 35-44.

Hutchinson, G.L., and Mosier, A.R. (1981). Improved soil cover method for field measurement of nitrous oxide fluxes. Soil Sci. Soc. Am. J. 45, 311-6.

Jenkins, M.C., and Kemp, W.M. (1984). The coupling of nitrification and denitrification in two estuarine sediments. Limnol. Oceanogr. 29, 609-19.

Jury, W.A., Letey, J., and Collins, T. (1982). Analysis of chamber methods used for measuring nitrous oxide production in the field. Soil Sci. Soc. Am. J. 46, 250-6.

Knowles, R. (1982). Denitrification. Microb. Rev. 46, 43–70.

Limmer, A.W., Steele, K.W., and Wilson, A.T. (1982). Direct field measurement of N_2 and N_2O evolution from soil. J. Soil Sci. 33, 499–507.

McConnaughey, P.K., and Duxbury, J.M. (1986). Introduction of acetylene into soil for measurement of denitrification. Soil Sci. Soc. Am. J. 50, 260–3.

Matthias, A.D., Blackmer A.M., and Bremner, J.M. (1980). A simple chamber technique for field measurement of emission of nitrous oxide from soils. J. Environ. Qual. 9, 251–6.

Mosier, A.R., Melhuish, F.M., and Meyer, W.S. (1985). Direct measurement of denitrification using acetylene blockage and infrared gas analysis in a root zone lysimeter. In 'Root Zone Limitations to Crop Production on Clay Soils'. (eds W.A. Muirhead and E. Humphreys) pp. 101–15. (CSIRO:Melbourne.)

Mulvaney, R.L., (1984). Determination of ^{15}N–labeled dinitrogen and nitrous oxide with triple-collector mass spectrometers. Soil Sci. Soc. Am. J. 48, 690–2.

Mulvaney, R.L., and Boast, C.W. (1986). Equations for determination of nitrogen–15 labeled dinitrogen and nitrous oxide by mass spectrometry. Soil Sci. Soc. Am. J. 50, 360–3.

Mulvaney, R.L., and Kurtz, L.T. (1982). A new method for determination of ^{15}N–labeled nitrous oxide. Soil Sci. Soc. Am. J. 46, 1178–84.

Payne, W.J. (1981). 'Denitrification'. (Wiley: New York.)

Rolston, D.E. (1978). Application of gaseous–diffusion theory to measurement of denitrification. In 'Nitrogen in the Environment'. (Eds D.R. Nielson and J.G. MacDonald) Vol. 1, pp. 309–35. (Academic Press, Inc.: New York.)

Rolston, D.E., Hoffman, D.L., and Toy, D.W. (1978). Field measurement of denitrification. I. Flux of N_2 and N_2O. Soil Sci. Soc. Am. J. 42, 863–9.

Rolston, D.E., Sharplay, A.N. Toy, D.W., and Broadbent, F.E. (1982). Field measurement of denitrification: III. Rates during irrigation cycles. Soil Sci. Soc. Am. J. 46, 289–96.

Ryden, J.C. (1982). Effects of acetylene on nitrification and denitrification in two soils during incubation with ammonium nitrate. J. Soil Sci. 33, 263–70.

Ryden, J.C., and Dawson, K.P. (1982). Evaluation of acetylene–inhibition technique for the measurement of denitrification in grassland soils. J. Sci. Food Agric. 33, 1197–206.

Ryden, J.C., Lund, L.J. Focht, D.D., and Letey, J. (1979). Direct measurement of denitrification loss from soils. II. Development and application of field methods. Soil Sci. Soc. Am. J. 43, 110–8.

Ryden, J.C., Lund, L.J., and Whaley, S.A. (1981). Direct measurement of gaseous nitrogen losses from an effluent irrigation area. J. Water Pollut. Control Fed. 53, 1677–82.

Ryzhova, I.M. (1979). Effect of nitrate concentration on the rate of soil denitrification. Soviet Soil Sci. 2, 168–71.

Seitzinger, S., Nixon, S. Pilson, M.E.Q., and Burke. S. (1980). Denitrification and N_2O production in near–shore marine sediments. Geolchimica Cosmo Chimica Acta. 44, 1853–60.

Siegel, R.S., Hauck, R.D., and Kurtz, L.T. (1982). Determination of [30]N_2 and application to measurement of N_2 evolution during denitrification. Soil Sci. Soc. Am. J. 46, 68–74.

Tedesco, M.J., and Keeney, D.R. (1972). Determination of (nitrate+nitrite)–N in alkaline permanganate solutions. Commun. Soil Sci. Plant Anal. 3, 339–44.

Terry, R.E., and Duxbury, J.M. (1985). Acetylene decomposition in soils. Soil Sci. Soc. Am. J. 49, 90–4.

Terry, R.E., Jellen, E.N., and Breakwell, D.P. (1986). Effect of irrigation methods and acetylene exposure on field denitrification measurements. Soil Sci. Soc. Am. J. 50, 115–20.

Walter, H.M., Keeney, D.R., and Fillery, I.R. (1979). Inhibition of nitrification by acetylene. Soil Sci. Soc. Am. J. 43, 195–6.

Watanabe, I., and de Guzman, M.R. (1980). Effect of nitrate on acetylene disappearance from anaerobic soil. Soil Biol. Biochem. 12, 193–4.

Yeomans, J.C., and Beauchamp, E.G. (1982). Acetylene as a possible substrate in the denitrification process. Can. J. Soil Sci. 62, 139–44.

LABORATORY TECHNIQUES FOR DETERMINATION

OF DIFFERENT FORMS OF NITROGEN

J.M. Bremner and J.C. Yeomans

ABSTRACT

The following topics are discussed: Kjeldahl methods and automated nitrogen analyzers for determination of total nitrogen; steam distillation, ion electrode, colorimetric, and ion chromatography methods for determination of ammonium, nitrite, and nitrate; estimation of non-exchangeable (fixed) ammonium in soil; colorimetric and enzymatic methods for determination of urea in soil; estimation of amino acids and hexosamines in soil hydrolysates; and methods for determination of gaseous forms of nitrogen.

INTRODUCTION

Most of the progress made during the past 30 years in research on nitrogen (N) cycling in agricultural ecosystems has resulted from advances in methodology for such research. This paper reviews advances in laboratory techniques for determination of different forms of N. Space limitations have made it impossible to discuss some of these advances in any detail, but detailed discussions can be found in articles and reviews cited.

TOTAL NITROGEN

Kjeldahl methods

Methods proposed for determination of total N in soil have been discussed in recent reviews by Nelson and Sommers (1980), Bremner and Hauck (1982), and Bremner and Mulvaney (1982). The methods recommended in the revised American Society of Agronomy (ASA) monograph on 'Methods of Soil Analysis' (Bremner and Mulvaney 1982) are essentially semimicro modifications of a macro-Kjeldahl procedure described by Bremner (1965a) in the first edition of this monograph, and they include procedures for quantitative recovery of nitrate-N, nitrite-N, and non-exchangeable (fixed) ammonium-N (Table 1). In the Kjeldahl methods most commonly used for soil analysis, organic-N in the sample is converted to NH_4^+-N by digestion with H_2SO_4 containing K_2SO_4 (to raise the temperature of digestion) and $CuSO_4$ and Se (to promote oxidation of organic matter) (usually 1 g K_2SO_4, 0.1 g $CuSO_4$, and 0.01 g Se per 3 mL H_2SO_4), and the NH_4^+-N thus produced is determined by estimation of the NH_3 liberated by distillation of the digest with alkali (Bremner 1965a).

Table 1: Pretreatments used in modifications of the Kjeldahl method.

Pretreatment	Purpose	References
Water	To improve N recovery in analysis of clay soils	Bal 1925; Bremner 1960; Bremner & Mulvaney 1982; Moraghan et al. 1983
HF and HCl	To recover non-exchangeable (fixed) NH_4^+-N	Keeney & Bremner 1967a; Bremner & Mulvaney 1982; Cooper 1986
$KMnO_4$ and H_2SO_4, then reduced Fe	To recover NO_2^--N and NO_3^--N	Bremner 1965a; Bremner & Mulvaney 1982
Salicylic acid and H_2SO_4, then $Na_2S_2O_3$ or Zn dust	To recover NO_2^--N and NO_3^--N	Bremner 1965a; Bremner & Mulvaney 1982
Aqueous solution of $Na_2S_2O_3$	To recover NO_2^--N and NO_3^--N	Dalal et al. 1984
Devarda alloy and H_2SO_4	To recover NO_3^--N	Liao 1981
Zn, Cr(III)K sulfate and H_2SO_4	To recover NO_3^--N	Pruden et al. 1985

Recent developments in Kjeldahl analysis include the introduction of heated aluminum blocks for Kjeldahl digestion, the development of rapid distillation methods of determining NH_4^+ in digests obtained when digestion is performed with a block digester, the use of NH_3 electrodes for direct determination of NH_4^+ in Kjeldahl digests, and the introduction of the Kjel–Foss Automatic Analyzer, an automated instrument for rapid Kjeldahl analysis of agricultural and food products.

Aluminum block digesters with internal electric heaters and temperature controls are marketed by several companies, and those supplied by Tecator, Inc. (Herndon, Virginia) and Technicon Instruments Corp. (Tarrytown, New York) have been evaluated for Kjeldahl analysis of soils (Schuman et al. 1973; Nelson and Sommers 1980). These digesters have significant advantages over previous digesters because they permit better temperature control during digestion, require less attention and fume hood space, and allow simultaneous digestion of 40 samples.

Tecator, Inc. has developed distillation systems that permit rapid determination of NH_4^+ in Kjeldahl digests when digestion is performed with their 40–tube block digester. These systems have the important advantage that they do not require transfer of the digest for NH_4^+ analysis, but their use has been limited by their high cost. Bremner and Breitenbeck (1983) described a simple and inexpensive steam distillation apparatus that permits direct distillation of NH_4^+ from the Pyrex tubes used for Kjeldahl digestion in commercial 40–tube block digesters, and showed that this apparatus allows rapid and precise determination of NH_4^+ in semimicro–Kjeldahl analysis of soils and plant materials using these digesters.

Ammonia electrodes with an internal reference electrode are now available from several companies in the United States and Britain. Their attraction for total N analysis is that they allow direct analysis of Kjeldahl digests for NH_4^+ and thereby eliminate the need for the customary distillation step before NH_4^+ analysis. Bremner and Tabatabai (1972) showed that the highly specific Model 95–10 NH_3 electrode developed by Orion Research, Inc. (Cambridge, Massachusetts) was satisfactory for direct determination of NH_4^+ in Kjeldahl digests of soils. Gallaher et al. (1976) have described a semiautomated Kjeldahl procedure for determination of total N in soils and plant materials in which an NH_3 electrode is used for direct determination of NH_4^+ in digests from an aluminum block digester.

The Kjel–Foss Automatic Analyzer for rapid Kjeldahl analysis was developed by Foss Electric of Hillerod, Denmark and is marketed in the United States by Foss America, Inc., Fishkill, NY. It uses H_2O_2 to reduce the time required for conversion of total N to NH_4^+ by the Kjeldahl digestion technique, and it permits 120 to 180 analyses per day. This analyzer has proved satisfactory for determination of total N in plants, animal feeds, and fertilizers (Wall et al. 1975; Noel 1976; Larson and Peterson 1979; Bjarno 1980), and it clearly deserves evaluation for routine determination of total N in soil.

A significant proportion of the total N in some soils (particularly subsoils) is as NH_4^+ trapped in the lattices of clay minerals. This fixed (non–exchangeable) NH_4^+ is usually recovered satisfactorily by Kjeldahl procedures commonly used for total N analysis of soils, but there is evidence that certain subsoils contain fixed NH_4^+–N that is not recovered quantitatively by these procedures (Bremner and Mulvaney 1982). Total N in such subsoils can be determined by pretreating the soil sample with HF–HCl solution to decompose clay minerals before Kjeldahl digestion (Table 1).

Bal (1925) found that the total N values obtained in Kjeldahl analysis of some Indian soils with a high % clay were increased significantly when the soils were treated with water before digestion, and that the effect of this pretreatment was markedly greater with the clay fractions of these soils than with the silt or fine silt fractions. From these and other observations, Bal concluded that these soils contained material that cemented the soil particles together and protected organic matter inside the particles from the action of H_2SO_4, and that this cementing material was not readily soluble in concentrated H_2SO_4 but was easily dissolved by dilute H_2SO_4. Walkley (1935) found that ball–milling of such soils before total N analysis had an even greater effect than treatment with water and deduced that the lower N values obtained with these soils when the treatment with water was omitted were not due to the presence of cementing materials insoluble in concentrated H_2SO_4 but to failure of the soil crumbs to disperse in this acid.

Since work during 1956–1960 showed that a considerable proportion of the N in some soils (particularly clay soils and subsoils) is in the form of NH_4^+ trapped in the lattices of clay minerals, the observations by Bal and Walkley suggested that their clay soils contained NH_4^+–N (and possibly organic N) within the lattices of clay minerals which was not recovered by the Kjeldahl method unless the soils were first ball–milled to destroy clay lattices or were treated with water to permit expansion of these lattices during digestion with H_2SO_4. However, Bremner (1959, 1960) and Bremner and Harada (1959) found that a pretreatment with water had no effect on total N values obtained when soils containing substantial amounts of clay and clay–fixed NH_4^+–N were analyzed by the macro–Kjeldahl method described by Bremner (1965a) and later work by Bremner and Mulvaney (1982) showed that a pretreatment with water did not increase the total N values obtained in Kjeldahl analysis of diverse Iowa soils by a semi–microversion of this method and tended to slightly decrease the values obtained by this method with soils containing more than about 3% of their total N as NO_3^-. However, Moraghan et al. (1983) found that a pretreatment with water led to a significant increase in the value obtained in total N analysis of an Indian Vertisol soil (51% clay) by macro– and semimicro–Kjeldahl methods. They also found that the effect of the water pretreatment was greatly reduced when the soil was ground to pass a 100–mesh screen instead of a 16–mesh screen before Kjeldahl analysis and that it appeared to be associated with incomplete digestion of organic N rather than incomplete recovery of fixed NH_4^+–N.

The Kjeldahl methods commonly used for total N analysis of soils do not give quantitative recovery of NO_3^-–N or NO_2^-–N and must be modified to obtain complete recovery of these forms of N. The modifications most commonly used are the permanganate–reduced Fe and salicylic acid–thiosulfate ($Na_2S_2O_3$) modifications (Table 1). In the permanganate–reduced Fe modification, the sample is treated before Kjeldahl digestion with $KMnO_4$ and H_2SO_4 to oxidize NO_2^- to NO_3^-, and then with reduced Fe to reduce NO_3^- to NH_4^+, the reduction being effected by nascent hydrogen generated by reaction of the Fe with the H_2SO_4 used in the $KMnO_4$ pretreatment. In the salicylic acid–thiosulfate modification, the sample is treated with salicylic acid dissolved in concentrated H_2SO_4, and the nitro compounds formed by reaction of salicylic acid with NO_3^- in acid medium are reduced to the corresponding amino compounds by heating the mixture with $Na_2S_2O_3$ before Kjeldahl digestion (Table 1). Dalal et al. (1984) recently confirmed older work by Cheng and Bremner (1964) indicating that it is not necessary to add salicylic acid to obtain quantitative recovery of NO_3^- and NO_2^- by the salicylic acid–$Na_2S_2O_3$ modification of the Kjeldahl method, and they have described a modification of the Kjeldahl method in which only $Na_2S_2O_3$ is used to obtain quantitative recovery of NO_2^- and NO_3^- (Table 1). Liao (1981) has described a modified Kjeldahl method for soil analysis that involves the use of Devarda alloy and H_2SO_4 for reduction of NO_3^- to NH_4^+ before Kjeldahl digestion (Table 1), and Pruden et al. (1985) have recently described a modified Kjeldahl method that involves the use of Cr(II) for reduction of NO_3^- to NH_4^+ (Table 1). In the latter method, the sample is mixed with Zn powder and treated with an acidified (H_2SO_4) solution of $Cr(III)K(SO_4)_2$, and the mixture is allowed to stand at room temperature for 2 hours. The usual Kjeldahl digestion reagents are then added and digestion is performed on a block digester. This method has the attraction that the pretreatment does not involve heating, and it gives quantitative recovery of NO_3^- added to soil or plant material. However, like the method of Liao (1981), it does not give quantitative recovery of NO_2^-–N (recovery of NO_2^- is about 80%).

Automated N analyzers

In the classical Dumas combustion method of determining total N, the sample is heated with CuO at a high temperature (usually above 600°C) in a stream of purified CO_2, and the gases liberated are led over hot Cu to reduce nitrogen oxides [mainly N_2O] to N_2, and then over CuO to convert CO to CO_2. The N_2–CO_2 mixture is collected in a nitrometer containing concentrated alkali, which absorbs the CO_2, and the volume of N_2 gas is measured. Early work showed that, although Dumas and Kjeldahl methods gave similar total N values with mineral soils, Dumas methods gave considerably higher values with soils containing substantial amounts of organic matter (Dyck and McKibbin 1935; Bremner and Shaw 1958). Subsequent work showed that the higher values obtained by the Dumas method with such soils were due to incomplete combustion, which resulted in methane (CH_4) being formed and measured with the N_2 produced by combustion (Stewart et al. 1963, 1964). Stewart et al. (1964) and Keeney and Bremner (1967a) showed that when Coleman Model 29 and Model 29A Nitrogen Analyzers (automated Dumas combustion instruments) were modified to improve combustion and eliminate interference by CH_4, the results obtained when they were used for total N analysis of mineral and organic soils agreed closely with those obtained by Kjeldahl analysis. Keeney and Bremner (1967a) also showed that, when modified to improve combustion, the Coleman Model 29A analyzer gave nearly quantitative (96–98%) recovery of NO_3^--N and NO_2^--N added to soils, but did not give complete recovery of total N in two subsoils containing substantial amounts of non–exchangeable (fixed) NH_4^+-N.

The Coleman Model 29 and Model 29A Nitrogen Analyzers did not gain acceptance for soil analysis because they were difficult and costly to operate and allowed only 30–40 analyses per day. Other automated N analyzers have been developed in the United States, Italy, Britain, and Japan, but most of these have been designed and used for analysis of pure compounds with high N contents (Colombo and Giazzi 1982; Kirsten and Hesselius 1983; Kirsten and Jansson 1986), and they do not appear suited for total N analysis of soils, which usually contain less than 0.5% N. It seems likely, however, that the Leco UO–14SP Nitrogen Determinator could be adapted satisfactorily for soil analysis, because there is evidence (Wong and Kemp 1977) that this instrument is satisfactory for total N analysis of sediments, and the problems in total N analysis of sediments are similar to those encountered in soil analysis. There also is evidence that the Leco CHN–600 Elemental Analyzer can be used for total N analysis of soil (Sheldrick 1986). This instrument utilizes gas chromatography for determination of the N_2 produced by combustion of the sample in O_2 at $950^{\circ}C$. Other automated instruments that merit consideration for total N analysis of soils are the Leco TN–15 Nitrogen Determinator and the Leco TC–36 Nitrogen/Oxygen Analyzer. Total N analysis by these instruments involves conversion of total N to N_2 by fusing the sample in a graphite crucible at high temperature ($2500–3000^{\circ}C$) in an inert (He) atmosphere and subsequent determination of N_2 by a gas chromatographic procedure. Two new automated N Analyzers, the Leco FP–228 and Carlo Erba NA–1500 instruments, are recommended by their manufacturers for total N analysis of soils. The Leco FP–228 instrument is similar in principle to the Leco CHN–600 analyzer, but it determines only total N and permits more rapid N analysis (160 sec) than the Leco CHN–600 analyzer. The Carlo Erba NA–1500 instrument also utilizes gas chromatography for determination of N_2, but it involves flash combustion of the sample in O_2 at $1800–2000^{\circ}C$. The Leco instruments are manufactured by Laboratory Equipment Corporation, St. Joseph, Missouri. The Carlo Erba instruments are manufactured by Carlo Erba Strumentazione, Milano, Italy.

Dalal and Henry (1986) have recently proposed the use of near infrared diffuse reflectance (NIR) spectrophotometry for simultaneous estimation of total N, moisture, and organic carbon in air–dried soils. This technique allows rapid, nondestructive estimation of total N in soil, but it seems useful only for routine analysis of finely ground soils that have a limited range in color and contain moderate amounts of organic matter (0.3–2.5% C).

Bremner and Hauck (1982) and Keeney and Nelson (1982) recently reviewed the methods available for determination of inorganic forms of N in soils and concluded that the best methods for most research on N transformations in soils are the steam distillation methods of determining exchangeable NH_4^+, NO_3^- and NO_2^- (Bremner and Keeney 1966), the KOBr–HF method of determining non–exchangeable (fixed) NH_4^+ (Silva and Bremner 1966), and the modified Griess–Ilosvay colorimetric method of determining NO_2^- (see Bremner 1965b).

The methods for exchangeable NH_4^+, NO_3^-, and NO_2^- proposed by Bremner and Keeney (1966) involve extraction of the soil sample with 2$\underline{M}$ KCl and analysis of the extract by steam distillation techniques involving the use of MgO and Devarda alloy for reduction of NO_3^- and NO_2^- to NH_3 and of sulfamic acid for destruction of NO_2^- ($HNO_2 + NH_2SO_3H \longrightarrow N_2 + H_2SO_4 + H_2O$) (Table 2). These methods have been used extensively during the past 20 years, because besides being rapid, simple, and precise, they are applicable to colored extracts, require only one extraction of the soil sample, permit isotope–ratio analysis of inorganic forms of N, and are not subject to interference by common soil constituents. These important advantages are not shared by colorimetric methods of determining inorganic forms of N in soils, and only one such method (the Griess–Ilosvay method of determining NO_2^-) has gained significant acceptance for soil analysis during the past 20 years.

Keeney and Nelson (1982) discussed problems in the use of colorimetric methods for determination of NH_4^+ and NO_3^- in soils and recommended methods designed to minimize these problems. In these procedures, NH_4^+, NO_2^-, and NO_3^- are extracted by 2$\underline{M}$ KCl and determined by direct analysis of the extract by indophenol blue, Griess–Ilosvay, and Cd reduction methods, respectively (Table 3). These methods are similar to procedures now commonly used in automated analysis schemes for determination of inorganic forms of N in natural waters. The methods for NH_4^+ and NO_3^- have significant advantages over colorimetric methods previously proposed for analysis of soils for these forms of N. The method for determination of NH_4^+ by the indophenol blue reaction is a modification of a method proposed by Kempers (1974). The method for determination of NO_3^- involves use of a copperized Cd column for reduction of NO_3^- to NO_2^- and is similar to methods described by Henriksen and Selmer–Olsen (1970) and Jackson et al. (1975).

Work to evaluate various methods proposed for determination of non–exchangeable (fixed) NH_4^+ in soils (Bremner 1959; Nelson and Bremner 1966; Bremner et al. 1967) has shown that several of these methods have serious defects and has indicated that the method of Silva and Bremner (1966) is the best currently available. In this method, the soil sample is treated with alkaline potassium hypobromite (KOBr–KOH) solution to convert exchangeable NH_4^+ and organic N to N_2. The residue from this treatment is washed with 0.5$\underline{M}$ KCl and treated with HF–HCl solution to decompose clay minerals and release fixed NH_4^+. The NH_4^+ released by the HF–HCl treatment is then determined by collection and titration of the NH_3 liberated by steam distillation of the soil–acid mixture with KOH (Table 2).

Table 2: Steam distillation methods for determination of inorganic forms of N in soil (Bremner and Keeney 1966; Silva and Bremner 1966).

Form of N	Method[1]
Ammonium–N	Steam distillation of soil extract with MgO
Nitrate–N	Steam distillation of soil extract with MgO and Devarda alloy after destruction of nitrite with sulfamic acid and removal of ammonium by steam distillation with MgO[2]
(Ammonium + nitrate)–N	Steam distillation of soil extract with MgO and Devarda alloy after destruction of nitrite with sulfamic acid[2]
(Nitrate + nitrite)–N	Steam distillation of soil extract with MgO and Devarda alloy after removal of ammonium by steam distillation with MgO
(Ammonium + nitrate + nitrite)–N	Steam distillation of soil extract with MgO and Devarda alloy
Non–exchangeable (fixed) ammonium–N	Steam distillation of soil with KOH after removal of exchangeable ammonium and labile organic N compounds by KOBr–KOH solution and treatment of residue with HF–HCl solution to decompose clay minerals

1. In each method, ammonia liberated by steam distillation is collected in boric acid–indicator solution and determined by titration with $0.005\underline{N}$ H_2SO_4.
2. If nitrite is absent, the sulfamic acid treatment is omitted.

Attempts to use ion electrodes for determination of NH_4^+, NO_3^-, and NO_2^- in soil indicate that only the highly specific NH_3–sensing electrodes developed for determination of NH_4^+ have any attraction for soil analysis. These NH_3 electrodes respond to the activity of NH_3 in solutions made alkaline (pH>11) to convert NH_4^+ to NH_3, and they can be used in conjunction with any sensitive pH–mV meter. Banwart et al. (1972) demonstrated that the Orion Model 95–10 NH_3 electrode permits direct determination of NH_4^+ in soil extracts and water samples, and others have shown that NH_3 electrodes supplied by Orion Research Inc. or by Electronic Instruments Ltd. (Chertsey, Surrey, England) can be used for determination of NH_4^+ in fresh and saline waters, sewage effluents, and acid extracts of fresh silage (Thomas and Booth 1973; Beckett and Wilson 1974; Byrne and McCormack 1978). Besides being highly specific (only volatile amines interfere), these NH_3 electrodes have the advantage that they are not subject to significant interference by substances known to be present in soils. In contrast, the NO_3^- electrodes currently available are not highly specific, and they are subject to interference by anions commonly found in soil extracts. Numerous studies to evaluate NO_3^- electrodes and reduce interference by these anions have been reported (Keeney and Nelson 1982), but many of these studies have given conflicting results, and none of the NO_3^- electrode methods thus far described seems likely to be satisfactory for all soils. As noted by Keeney and Nelson (1982), the NO_3^- electrodes currently available require continual restandardization and lack the sensitivity needed for satisfactory analysis of many soil extracts. These criticisms also apply to the NO_2^- (NO_x) electrode developed by Orion Research, Inc., which responds to NO_x (NO, NO_2, N_2O_3, N_2O_4) produced by acidification of solutions containing NO_2^-. This electrode gives satisfactory recovery of NO_2^- added to soil extracts (Tabatabai 1974), but is much less sensitive than current colorimetric methods of determining NO_2^-, and it is not suited for determination of the small amounts of NO_2^- normally present in soils, plants, or natural waters.

Siegel (1980) and Orion Research, Inc. (1986) have recently described methods for determination of NO_3^- in soil extracts in which an Orion NH_3 electrode is used to determine NH_3 produced through reduction of NO_3^- by Devarda alloy or Ti(III)Cl$_3$ under alkaline conditions (Table 3). If NO_2^- is present, it is also reduced to NH_3 and measured as NO_3^- in these methods.

The Griess–Ilosvay colorimetric methods commonly used for determination of NO_2^- in soil extracts are well suited for determination of small amounts of NO_2^-, but they require aliquot dilution for analysis of extracts containing substantial amounts of NO_2^-. Puttanna and Prakasa Rao (1986) have recently described a modified Griess–Ilosvay procedure for analysis of extracts containing up to 75 µg NO_2^-–N/mL.

Recent work has led to development of methods of determining NO_3^- in soil extracts that involve the use of Griess–Ilosvay colorimetric methods for determination of NO_2^- formed through reduction of NO_3^- by copperized cadmium (Keeney and Nelson 1982), microbial NO_3^- reductase (Rice et al. 1984), hydrazine (Markus et al. 1985), or Ti(III)Cl$_3$ (Al–Wehaid and Townshend 1986) (Table 3). These methods are more sensitive than colorimetric procedures previously proposed for determination of NO_3^- in soil extracts.

The separation and determination of NO_3^-, NO_2^-, and other anions is now easily carried out by ion chromatography. Suppressed ion chromatography (SIC) for the simultaneous determination of soluble anions and cations was first described by Small et al. (1975). Dick and Tabatabai (1979) showed that SIC (Dionex Model 10 Ion Chromatograph) could be utilized for determination of NO_3^- in soil extracts and that the results obtained by this method agreed

Table 3: Methods for determination of inorganic forms of N in soil extracts.

Form of N	Method	References
Ammonium	1. NH_3 electrode	Banwart et al. 1972
	2. Colorimetric (indophenol blue method)	Keeney & Nelson 1982; Markus et al. 1985; Wang & Oien 1986
	3. Ion chromatography	Nieto & Frankenberger 1985b (SCIC)
Nitrite	1. NO_x electrode	Tabatabai 1974
	2. Colorimetric (modified Griess–Ilosvay method)	Bremner 1965b; Keeney & Nelson 1982; Puttanna & Prakasa Rao 1986
	3. Ion chromatography	Nieto & Frankenberger 1985a (SCIC)
Nitrate	1. NO_3^- electrode	Bremner et al. 1968; Oien & Selmer-Olsen 1969; Keeney & Nelson 1982
	2. NH_3 electrode [electrode is used to determine NH_3 formed by reduction of NO_3^- by Devarda alloy (A) or $Ti(III)Cl_3$ (B) under alkaline conditions]	Siegel 1980 (A); Orion Research, Inc. 1986 (B)
	3. Colorimetric [modified Griess–Ilosvay method is used to determine NO_2^- formed by reduction of NO_3^- by copperized cadmium (C), microbial NO_3^- reductase (D), hydrazine (E), or $Ti(III)Cl_3$ (F)]	Keeney & Nelson 1982 (C); Rice et al. 1984 (D); Markus et al. 1985 (E); Al–Wehaid & Townshend 1986 (F)
	4. Ion chromatography	Dick & Tabatabai 1979 (SIC); Nieto & Frankenberger 1985a (SCIC); Barak & Chen 1987 (SCIC)

SCIC, single column ion chromatography; SIC, suppressed ion chromatography.

closely with those obtained by the steam distillation technique. They also showed that the SIC method gave quantitative recovery of NO_3^- added to soils and that NO_2^- did not interfere with NO_3^- analysis by this method. Single column ion chromatography (SCIC) has the advantages that it does not require the suppressor column needed with SIC and allows faster analysis and better resolution than SIC. Separation is by a low–capacity anion–exchange resin, and anions are quantified by conductimetric methods (Karlson and Frankenberger 1986). Nieto and Frankenberger (1985a) and Barak and Chen (1987) found that SCIC permitted rapid, sensitive, precise, and simultaneous determination of NO_3^-, NO_2^- and other anions in soil extracts. Nieto and Frankenberger (1985b) also showed that SCIC can be used for determination of NH_4^+ in soil extracts and that the results of NH_4^+ analysis by this method agreed closely with those obtained by the steam distillation technique.

Norman and Stucki (1981) have described an ultraviolet spectrophotometric difference method for determination of NO_3^- in soil extracts. This method measures the UV absorbance of the soil extract at 210 nm before and after addition of Raney nickel catalyst and acid to reduce NO_3^- to nonabsorbing species, NO_3^- being calculated from the difference between the two absorbance measurements. If NO_2^- is present, it is removed with sulfamic acid before absorbance measurements. Norman et al. (1985) subsequently described a simpler method for determining NO_3^- in soil extracts which is also based on measurement of UV absorbance at 210 nm but allows for interference by non-nitrate species by subtraction of an empirically-determined multiple of the absorbance of the soil extract at 270 nm from its absorbance at 210 nm. The results by this method agreed closely with those obtained by the steam distillation technique (Norman et al. 1985).

ORGANIC FORMS OF NITROGEN

The methods currently available for determination of organic forms of N in soils have been discussed in a recent article by Stevenson (1982). They involve hydrolysis of the nitrogenous organic material in soil by hot mineral acid (usually 6N HCl) and analysis of the hydrolysate by chromatographic, steam distillation, or colorimetric techniques. Ion-exchange chromatography has been used to identify and estimate amino acids and hexosamines in soil hydrolysates (Sowden et al. 1977; Stevenson 1982), and both ion-exchange chromatography and ion-exclusion chromatography have been used to identify and estimate purines and pyrimidines in hydrolysates of soils or soil fractions (Cortez and Schnitzer 1979). These techniques have been little used, however, and most research to characterize organic N in soils and estimate amino acid-N and hexosamine-N has been performed by N fractionation techniques similar to that described by Bremner (1965c), which involves the use of rapid steam distillation techniques to estimate amino acid-N, hexosamine-N, and NH_4^+-N released by acid hydrolysis of soil.

Although only a trace amount of the organic N in most soils is in the form of urea, research to develop methods for determination of urea in soil has been stimulated by urea's increasing importance as a N fertilizer in world agriculture. The methods now available for this analysis have been discussed in a recent article by Bremner (1982). The most sensitive and precise of these methods are the colorimetric procedure described by Douglas and Bremner (1970) and the two enzymatic procedures described by Keeney and Bremner (1967b). The procedure of Douglas and Bremner (1970) involves extraction of the soil sample with 2M KCl containing a urease inhibitor (phenylmercuric acetate) and analysis of the extract by a colorimetric method based on the reaction of urea with diacetyl monoxime and thiosemicarbazide in the presence of H_3PO_4 and H_2SO_4. Douglas et al. (1978) have described an automated modification of this method, and Mulvaney and Bremner (1979) have described a modification to eliminate problems caused by impurities in some batches of H_3PO_4. The two enzymatic procedures proposed by Keeney and Bremner (1967b) use urease to convert urea–N to NH_4^+–N. In one, the soil sample is extracted with $2\underline{M}$ KCl, and an aliquot of the extract is treated with K phosphate buffer (pH 8) and urease. In the other, the soil sample is treated directly with K phosphate buffer (pH 8) and urease. In both, the NH_4^+ produced is separated by a rapid steam distillation technique and determined by titration of the distillate with standard acid. Besides being specific and precise, these enzymatic procedures have the important advantage that they permit isotope ratio analysis of urea–N in ^{15}N–tracer studies of urea transformations in soils.

GASEOUS FORMS OF NITROGEN

Bremner and Blackmer (1982), Bremner and Hauck (1982), and Smith (1983) have recently reviewed the methods available for determination of N_2, N_2O, NO, NO_2, NH_3, and volatile amines in research on N transformations in soils. Comparison with an early review by Cheng and Bremner (1965) shows that there have been striking advances in methodology for determination of gaseous forms of N in soil research during the past 20 years and that these advances have largely resulted from the introduction of gas chromatographic (GC) methods of analysis. Gas chromatographic methods are applicable both to investigations of the soil atmosphere as it occurs in the field and to laboratory studies of emission and uptake of nitrogenous gases by soils.

Bremner and Blackmer (1982) discussed GC methods available for determination of N_2, N_2O, and NO in laboratory research on denitrification and other processes leading to formation of these gases in soils and drew attention to the important advantages of the method of Blackmer and Bremner (1977), which involves use of an ultrasonic detector and two columns of Porapak Q at different temperatures. This method permits determination of O_2, Ar, and CO_2 as well as N_2, N_2O, and NO, and it has a combination of features (sensitivity, specificity, durability, versatility, etc.) that make it well suited for research on denitrification in soils or routine determination of the major constituents of soil atmospheres. It should be noted, however, that none of the GC methods thus far proposed for determination of NO in soil atmospheres allows determination of NO in the presence of large amounts of N_2 or appreciable amounts of O_2 or Ar and that no GC method has been developed for determination of NO_2 in soil atmospheres. Several GC methods have been proposed for detection or estimation of NO_2, but most have been designed for analysis of simple gaseous mixtures containing high concentrations of NO_2, and none permit determination of small amounts of NO_2 in gaseous mixtures as complex as soil atmospheres. Moreover, it has been found that NO_2 can vitiate GC determination of NO by reacting with materials commonly used to pack GC columns for N gas analysis (Bremner and Blackmer 1982).

Several workers have used alkaline or acidic $KMnO_4$ solutions to trap NO and NO_2 evolved from soils incubated in closed systems, and have determined

(NO + NO$_2$)–N by analysis of these solutions for (NO$_2^-$ + NO$_3^-$)–N or NO$_2^-$–N (see Bremner and Blackmer 1982). Galbally and Roy (1978) measured emissions of NO and NO$_2$ from soil by a chamber technique involving the use of a highly sensitive chemiluminescent detector developed for estimation of NO and NO$_x$ (NO + NO$_2$) in surface air. Similar detectors have been used extensively for NO and NO$_x$ analysis in air pollution research during the past decade. They originated from research by Fontijn et al. (1970) leading to development of a detector that utilizes the chemiluminescent reaction of NO with ozone (O$_3$) for determination of small amounts of NO in air. Subsequent research has increased the sensitivity of this method of determining NO and resulted in development of methods of converting NO$_2$ to NO for determination of small amounts of NO + NO$_2$ in air by chemiluminescent detectors (Bremner and Blackmer 1982). These chemiluminescent techniques are subject to several interferences, but they are much more sensitive than other methods proposed for determination of NO and NO$_x$ (they can detect <1 ppb of NO in air), and there seems little doubt that they will prove valuable for measuring emissions of NO or NO$_2$ from soils.

Many GC techniques have been proposed for detection of NH$_3$ and volatile amines, but most have been designed for analysis of solutions of these compounds, and no techniques for direct GC analysis of soil atmospheres for NH$_3$ and volatile amines have been reported. However, GC methods have been developed for NH$_4^+$ and amine analysis of acidic solutions used to sorb NH$_3$ and volatile amines from air, and these methods may prove useful for analysis of acidic solutions used to sorb and estimate NH$_3$ and volatile amines evolved from soils in closed systems (see Bremner and Blackmer 1982; Bremner and Hauck 1982).

Bremner and Douglas (1971) described a simple method of determining NH$_3$ evolved from soils in laboratory incubation experiments which has been used extensively for laboratory studies of NH$_3$ volatilization from soils treated with urea and other N fertilizers. In this method, an acidic solution (0.25$\underline{M}$ H$_2$SO$_4$) is used to trap NH$_3$ evolved during incubation and then analysed for NH$_4^+$ by a distillation–titration technique. Volatile amines can interfere (Elliott et al. 1971), but this is not normally a problem in soil research.

Several workers (e.g. Aneja et al. 1978; Baumgartner et al. 1979) have described sensitive and rapid methods for NH$_3$ analysis of air involving use of a chemiluminescent detector for determination of NO produced through oxidation of NH$_3$ by a thermal catalytic converter. These methods have not been evaluated for research to assess emissions of NH$_3$ from soils, but they seem promising for such research.

REFERENCES

Al–Wehaid, A., and Townshend, A. (1986). Spectrophotometric flow–injection determination of nitrate based on reduction with titanium(III) chloride. Anal. Chim. Acta 186, 289–94.

Aneja, V.P., Stahel, E.P., Rogers, H.H., Witherspoon, A.M., and Heck, W.W. (1978). Calibration and performance of a thermal converter in continuous atmospheric monitoring of ammonia. Anal. Chem. 50, 1705–8.

Bal, D.V. (1925). The determination of nitrogen in heavy clay soils. J. Agric. Sci. 15, 454–9.

Banwart, W.L., Tabatabai, M.A., and Bremner, J.M. (1972). Determination of ammonium in soil extracts and water samples by an ammonia electrode. Commun. Soil Sci. Plant Anal. 3, 449–58.

Barak, P., and Chen, Y. (1987). Three-minute analysis of chloride, nitrate, and sulfate by single column anion chromatography. Soil Sci. Soc. Am. J. 51, 257–8.

Baumgartner, R., McClenny, W.A., and Stevens, R.K. (1979). Optimized chemiluminescence system for measuring atmospheric ammonia. EPA-600/2–79–028. (U.S. Environmental Protection Agency: Research Triangle, NC).

Beckett, M.J., and Wilson, A.L. (1974). The manual determination of ammonia in fresh waters using an ammonia–sensitive membrane electrode. Water Resour. Res. 8, 333–40.

Bjarno, O.-C. (1980). Kjel-Foss automatic analysis using an antimony–based catalyst: Collaborative study. J. Assoc. Off. Anal. Chem. 63, 657–63.

Blackmer, A.M., and Bremner, J.M. (1977). Gas chromatographic analysis of soil atmospheres. Soil Sci. Soc. Am. J. 41, 908–12.

Bremner, J.M. (1959). Determination of fixed ammonium in soil. J. Agr. Sci. 52, 147–60.

Bremner, J.M. (1960). Determination of nitrogen in soils by the Kjeldahl method. J. Agr. Sci. 55, 11–33.

Bremner, J.M. (1965a). Total nitrogen. In 'Methods of Soil Analysis, Part 2'. (Eds C.A. Black et al.) pp. 1149–78. (American Society of Agronomy: Madison, WI.)

Bremner, J.M. (1965b). Inorganic forms of nitrogen. In 'Methods of Soil Analysis, Part 2'. (Eds C.A. Black et al.) pp. 1179–237. (American Society of Agronomy: Madison, WI.)

Bremner, J.M. (1965c). Organic forms of nitrogen. In 'Methods of Soil Analysis, Part 2'. (Eds C.A. Black et al.) pp. 1238–55. (American Society of Agronomy: Madison, WI.)

Bremner, J.M. (1982). Urea. In 'Methods of Soil Analysis, Part 2, 2nd edition'. (Eds A.L. Page et al.) pp. 699–709. (American Society of Agronomy: Madison, WI.)

Bremner, J.M., and Blackmer, A.M. (1982). Composition of soil atmospheres. In 'Methods of Soil Analysis, Part 2, 2nd edition'. (Eds A.L. Page et al.) pp. 873–901. (American Society of Agronomy: Madison, WI.)

Bremner, J.M., and Breitenbeck, G.A. (1983). A simple method for determination of ammonium in semimicro-Kjeldahl analysis of soils and plant materials using a block digester. Commun. Soil Sci. Plant Anal. 14, 905–13.

Bremner, J.M., Bundy, L.G., and Agarwal, A.S. (1968). Use of a selective ion electrode for determination of nitrate in soils. Anal. Letter 1, 837–44.

Bremner, J.M., and Douglas, L.A. (1971). Decomposition of urea phosphate in soils. Soil Sci. Soc. Am. Proc. 35, 575–8.

Bremner, J.M., and Harada, T. (1959). Release of ammonium and organic matter from soil by hydrofluoric acid and effect of hydrofluoric acid treatment on extraction of soil organic matter by neutral and alkaline reagents. J. Agr. Sci. 52, 137–46.

Bremner, J.M., and Hauck, R.D. (1982). Advances in methodology for research on nitrogen transformations in soils. In 'Nitrogen in Agricultural Soils'. (Ed F.J. Stevenson.) pp. 467–502. (American Society of Agronomy: Madison, WI.)

Bremner, J.M., and Keeney, D.R. (1966). Determination and isotope-ratio analysis of different forms of nitrogen in soils: 3. Exchangeable ammonium, nitrate, and nitrite by extraction distillation methods. Soil Sci. Soc. Am. Proc. 30, 577–82.

Bremner, J.M., and Mulvaney, C.S. (1982). Nitrogen — Total. In 'Methods of Soil Analysis, Part 2, 2nd edition'. (Eds A.L. Page et al.) pp. 595–624. (American Society of Agronomy: Madison, WI.)

Bremner, J.M., Nelson, D.W., and Silva, J.A. (1967). Comparison and evaluation of methods of determining fixed ammonium in soils. Soil Sci. Soc. Am. Proc. 31, 466–72.

Bremner, J.M., and Shaw, K. (1958). Denitrification in soil I. Methods of investigation. J. Agric. Sci. 51, 22–39.

Bremner, J.M., and Tabatabai, M.A. (1972). Use of an ammonia electrode for determination of ammonium in Kjeldahl analysis of soils. Commun. Soil Sci. Plant Anal. 3, 159–65.

Byrne, E., and McCormack, S. (1978). Determination of ammonium nitrogen in silage samples by an ammonia electrode. Commun. Soil Sci. Plant Anal. 9, 667–84.

Cheng, H.H., and Bremner, J.M. (1964). Use of the salicylic acid-thiosulfate modification of the Kjeldahl method for determination of total nitrogen in soils. Agron. Abstr. p. 21.

Cheng, H.H., and Bremner, J.M. (1965). Gaseous forms of nitrogen. In 'Methods of Soil Analysis, Part 2'. (Eds C.A. Black et al.) pp. 1287–323. (American Society of Agronomy: Madison, WI.)

Colombo, B., and Giazzi, G. (1982). Total automatic nitrogen determination. Am. Lab. (Fairfield, CT) July, 38–45.

412

Cooper, J.E. (1986). Determination of total nitrogen in oil shale. Anal. Chem. 58, 1571-2.

Cortez, J., and Schnitzer, M. (1979). Purines and pyrimidines in soils and humic substances. Soil Sci. Soc. Am. J. 43, 958-61.

Dalal, R.C., and Henry, R.J. (1986). Simultaneous determination of moisture, organic carbon, and total nitrogen by near infrared reflectance spectrophotometry. Soil Sci. Soc. Am. J. 50, 120-3.

Dalal, R.C., Sahrawat, K.L., and Myers, R.J.K. (1984). Inclusion of nitrate and nitrite in the Kjeldahl nitrogen determination of soils and plant materials using sodium thiosulphate. Commun. Soil Sci. Plant Anal. 15, 1453-61.

Dick, W.A., and Tabatabai, M.A. (1979). Ion chromatographic determination of sulfate and nitrate in soils. Soil Sci. Soc. Am. Proc. 43, 899-904.

Douglas, L.A., and Bremner, J.M. (1970). Extraction and colorimetric determination of urea in soils. Soil Sci. Soc. Am. Proc. 34, 859-62.

Douglas, L.A., Sochtig, H., and Flaig, W. (1978). Colorimetric determination of urea in soil extracts using an automated system. Soil Sci. Soc. Am. J. 42, 291-2.

Dyck, A.W.J., and McKibbin, R.R. (1935). The non-protein nature of a fraction of soil organic nitrogen. Can. J. Res. 13, 264-8.

Elliott, L.F., Schuman, G.E., and Viets, F.G., Jr. (1971). Volatilization of nitrogen-containing compounds from beef cattle areas. Soil Sci. Soc. Am. Proc. 35, 752-5.

Fontijn, A., Sabadell, A.J., and Ronco, R.J. (1970). Homogeneous chemiluminescent measurement of nitric oxide with ozone: Implications for continuous selective monitoring of gaseous air pollutants. Anal. Chem. 42, 575-9.

Galbally, I.E., and Roy, C.R. (1978). Loss of fixed nitrogen from soils by nitric oxide exhalation. Nature (London) 275, 734-5.

Gallaher, R.N., Weldon, C.O., and Boswell, F.C. (1976). A semiautomated procedure for total nitrogen in plant and soil samples. Soil Sci. Soc. Am. J. 40, 887-9.

Henriksen, H., and Selmer-Olsen, A.A. (1970). Automatic methods for determining nitrate and nitrite in water and soil extracts. Analyst (London) 95, 514-8.

Jackson, W.A., Frost, C.E., and Hildreth, D.M. (1975). Versatile multi-range analytical manifold for automated analysis of nitrate-nitrogen. Soil Sci. Soc. Am. Proc. 39, 592-3.

Karlson, U., and Frankenberger, W.T. (1986). Determination of selenate by single-column ion chromatography. J. Chromatogr. 368, 153-61.

Keeney, D.R., and Bremner, J.M. (1967a). Use of the Coleman Model 29A Analyzer for total nitrogen analysis of soils. Soil Sci. 104, 358-63.

Keeney, D.R., and Bremner, J.M. (1967b). Determination and isotope-ratio analysis of different forms of nitrogen in soils: 7. Urea. Soil Sci. Soc. Am. Proc. 31, 317-21.

Keeney, D.R., and Nelson, D.W. (1982). Nitrogen - Inorganic forms. In 'Methods of Soil Analysis, Part 2, 2nd edition'. (Eds A.L. Page et al.) pp. 643-98. (American Society of Agronomy: Madison, WI.)

Kempers, A.J. (1974). Determination of sub-microquantities of ammonium and nitrates in soils with phenol, sodium nitroprusside and hypochlorite. Geoderma 12, 201-6.

Kirsten, W.J., and Hesselius, G.U. (1983). Rapid, automatic, high capacity Dumas determination of nitrogen. Biochem. J. 28, 529-47.

Kirsten, W.J., and Jansson, K.H. (1986). Rapid and automatic determination of nitrogen using flash combustion of large samples. Anal. Chem. 58, 2109-12.

Larson, S.L., and Peterson, H.P. (1979). Semiautomated determination of nitrogen in nitrate-containing fertilizers. Anal. Chem. 51, 2414-5.

Liao, C.F.H. (1981). Devarda's alloy method for total nitrogen determination. Soil Sci. Soc. Am. J. 45, 852-5.

Markus, D.K., McKinnon, J.P., and Buccafuri, A.F. (1985). Automated analysis of nitrite, nitrate, and ammonium nitrogen in soils. Soil Sci. Soc. Am. J. 49, 1208-15.

Moraghan, J.T., Rego, T.J., and Sahrawat, K.L. (1983). Effect of water pretreatment on total nitrogen analysis of soils by the Kjeldahl method. Soil Sci. Soc. Am. J. 47, 213-7.

Mulvaney, R.L., and Bremner, J.M. (1979). A modified diacetyl monoxime method for colorimetric determination of urea in soil extracts. Commun. Soil Sci. Plant Anal. 10, 1163-70.

Nelson, D.W., and Bremner, J.M. (1966). An evaluation of Mogilevkina's method of determining fixed ammonium in soils. Soil Sci. Soc. Am. Proc. 30, 409-10.

Nelson, D.W., and Sommers, L.E. (1980). Total nitrogen analysis for soil and plant tissues. J. Assoc. Off. Anal. Chem. 63, 770-8.

Nieto, K.F., and Frankenberger, W.T. (1985a). Single column ion chromatography: I. Analysis of inorganic anions in soils. Soil Sci. Soc. Am. J. 49, 587-92.

Nieto, K.F., and Frankenberger, W.T. (1985b). Single column ion chromatography: II. Analysis of ammonium, alkali metals, and alkaline earth cations in soils. Soil Sci. Soc. Am. J. 49, 592-6.

413

Noel, R.J. (1976). Collaborative study of an automated method for determination of crude protein in animal feeds. J. Assoc. Off. Anal. Chem. 59, 141–7.

Norman, R.J., Edberg, J.C., and Stucki, J.W. (1985). Determination of nitrate in soil extracts by dual–wavelength ultraviolet spectrophotometry. Soil Sci. Soc. Am. J. 49, 1182–5.

Norman, R.J., and Stucki, J.W. (1981). The determination of nitrate and nitrite in soil extracts by ultraviolet spectrophotometry. Soil Sci. Soc. Am. J. 45, 347–53.

Oien, A., and Selmer–Olsen, A.R. (1969). Nitrate determination in soil extracts with the nitrate electrode. Analyst 94, 888–94.

Orion Research, Inc. (1986). Nitrate measurement in environmental samples. Orion Applic. Proc. No. 115. (Orion Research, Inc.: Boston, MA).

Pruden, G., Kalembasa, S.J., and Jenkinson, D.S. (1985). Reduction of nitrate prior to Kjeldahl digestion. J. Sci. Food Agric. 36, 71–3.

Puttanna, K., and Prakạsa Rao, E.V.S. (1986). Modified method of nitrite determination in soils by sulphanilic acid/N–(1–naphthyl)ethylenediamine. Z. Pflanzenernaehr. Bodenk. 149, 517–21.

Rice, C.W., Smith, M.S., and Crutchfield, J.M. (1984). Inorganic N analysis of soil extracts by automated and distillation procedures. Commun. Soil Sci. Plant Anal. 15, 663–72.

Schuman, T.E., Stanley, M.A., and Knudsen, D. (1973). Automated total nitrogen analysis of soil and plant samples. Soil Sci. Soc. Am. Proc. 37, 480–1.

Sheldrick, B.H. (1986). Test of the Leco CHN–600 determinator for soil carbon and nitrogen analysis. Can. J. Soil Sci. 66, 543–5.

Siegel, R.S. (1980). Determination of nitrate and exchangeable ammonium in soil extracts by an ammonia electrode. Soil Sci. Soc. Am. J. 44, 943–7.

Silva, J.A., and Bremner, J.M. (1966). Determination and isotope–ratio analysis of different forms of nitrogen in soils: 5. Fixed ammonium. Soil Sci. Soc. Am. Proc. 30, 587–94.

Small, H., Stevens, T.S., and Bauman, W.C. (1975). Novel ion exchange chromatographic method using conductimetric detection. Anal. Chem. 47, 1801–9.

Smith, K.A. (1983). Gas chromatographic analysis of the soil atmosphere. In 'Soil Analysis'. (Ed. K.A. Smith.) pp. 407–54. (Marcel Dekker: New York.)

Sowden, F.J., Chen, Y., and Schnitzer, M. (1977). The nitrogen distribution in soils formed under widely differing climatic conditions. Geochim. Cosmochim. Acta 41, 1524–6.

Stevenson, F.J. (1982). Nitrogen – Organic forms. In 'Methods of Soil Analysis, Part 2, 2nd edition'. (Eds A.L. Page et al.) pp. 625–41. (American Society of Agronomy: Madison, WI.)

Stewart, B.A., Porter, L.K., and Beard, W.E. (1964). Determination of total nitrogen and carbon in soils by a commercial Dumas apparatus. Soil Sci. Soc. Am. Proc. 28, 366–8.

Stewart, B.A., Porter, L.K., and Clark, F.E. (1963). The reliability of a micro–Dumas procedure for determining total nitrogen in soil. Soil Sci. Soc. Am. Proc. 27, 377–80.

Tabatabai, M.A. (1974). Determination of nitrite in soil extracts and water samples by a nitrogen oxide electrode. Commun. Soil Sci. Plant Anal. 5, 569–78.

Thomas, R.F., and Booth, R.L. (1973). Selective electrode measurement of ammonia in water and wastes. Environ. Sci. Technol. 7, 523–6.

Walkley, A. (1935). An examination of methods for determining organic carbon and nitrogen in soils. J. Agric. Sci. 25, 598–609.

Wall, L.L., Sr., Gehrke, C.W., Neuner, T.E., Cathey, R.D., and Rexroad, P.R. (1975). Total protein nitrogen: Evaluation and comparison of four different methods. J. Assoc. Off. Anal. Chem. 58, 811–7.

Wang, L., and Oien, A. (1986). Determination of Kjeldahl nitrogen and exchangeable ammonium in soil by the indophenol method. Acta Agric. Scand. 36, 60–70.

Wong, H.K.T., and Kemp, A.L.W. (1977). The determination of total nitrogen in sediments using an induction furnace. Soil Sci. 124, 1–4.

PART V

CLOSING ADDRESS

TOWARDS THE YEAR 2000 : DIRECTIONS FOR FUTURE NITROGEN RESEARCH

E.A. Paul

ABSTRACT

Advances in knowledge concerning N in agroecosystems require a clear definition of achievable objectives and usable techniques. The impacts of computerization, molecular genetics and an increasing contribution by ecologists will play a major role in such advances. Broad goals could include complete utilization by plants of all mineral N and elimination of N losses to the environment. This will require plants with increased competitive capacity grown under management techniques that synchronize microbial release and plant uptake of soil N. I predict that the plant host, rather than the Rhizobium, will probably be found to be the limiting factor in many leguminous associations. This will require a greater research emphasis on the host. Techniques must be found for using legumes both as N sources and as weed competitors in alternative input agriculture ecosystems. Molecular genetics will affect N research in a number of ways. More basic information regarding plant— microbial interactions, microbial competition and nutrient cycling must be obtained before genetically engineered organisms can be released into the environment. It also will provide useful research tools such as genetic probes for specific microbial DNA in soil. Interdisciplinary projects will be required to ensure productive applications to field problems.

GOALS AND OBJECTIVES OF FUTURE RESEARCH

Advances in a field as important and at the same time as complex as that of N in agricultural ecosystems will depend on the recognition of research imperatives that can be successfully investigated, on the availability of useful concepts and on the applicability of imaginative techniques. The broad goals can be defined to include: (1) no losses of or pollution by mineral N, (2) increased plant competitiveness for N, (3) increased symbiotic fixation,

(4) synchronized release and plant uptake of soil or fertilizer N, (5) alteration of site characteristics in agroforestry, (6) companion or alternative N sources for high value crops that are grown in agricultural systems with lower utilization of pesticides.

Improved technology such as the measurement of the size of the microbial biomass (Smith and Paul 1987), automated rapid instrumentation for ^{15}N analysis (Marshall and Whiteway 1985) and the rapid identification of specific microorganisms in soil (Jansson et al. 1987) will have a major impact. Other advances have been made in measurement of gaseous N losses, the ecology and physiology of plant microbial interactions and the estimation of soil active N (Paul et al. 1985). Also, it is generally considered that advances in molecular genetics will be most readily applied to the plant–soil system through the activity of altered microorganisms (Halvorson et al. 1985). Soil microbiology and the study of N transformations therefore are now in a state of anticipation and excitement akin to that of a hundred years ago when the basic principles concerning N in agricultural ecosystems were being established.

It has been stated that the lack of basic understanding of the biological processes involved will be a limiting factor on future breakthroughs (Board on Agriculture 1985). It also is argued that the return from basic research will be greater than that from applied. Estimation of the annual rates of return to basic and applied research is very important but somewhat difficult to achieve. Using regression analysis that compared the ratio of an index of total output produced relative to inputs, (Evenson 1978) showed that in the 1927–1950 period basic science oriented research gave a higher rate of annual return [110%] relative to applied technology oriented research [95%] (Table 1). The opposite was true in 1948–1971 when technically oriented agricultural research gave a 2–3 fold greater return [106%] as compared to basic science oriented research [45%]. Fox's similar analyses in 1986 showed that crop research in the 1944–1983 period was equally affected by technical commodity-specific research and disciplinary biological research with a 180% return in both cases. It is of interest that during these years returns from animal science research were lower than those of plant sciences in both the basic and applied areas.

The justification for both applied and basic research in the plant sciences is very strong and consideration of the specific thrust of the research is probably more important than whether it is basic or applied.

METHODOLOGY IN NITROGEN RESEARCH

Multi-sample digestion and diffusion techniques have greatly increased the speed of handling samples for both carbon and nitrogen (Turner and Bergersen 1980). This, combined with automated methodology for conversion of ^{15}N NH_4^+ to N_2, has lead to reasonably priced ^{15}N analysis. The recent combination of an automated Dumas combustion furnace with gas chromatography

Table 1: Estimated annual rates of return to basic and applied research in U.S. agriculture.

Years	Type of research	Evenson (1978)	Fox (1986)
1858–1930	All agriculture	65	
1927–1950	Technical agricultural research	95	
	Science oriented research	110	
1948–1971	Technical agricultural research	106	
	Science oriented research	45	
1944–1983	Technical crop research		180
	Science oriented research		180
	Livestock research		130

and computer controlled mass spectrometer has made possible the direct analysis of both total N and ^{15}N contents of plant materials or soil samples (Marshall and Whiteway 1985). Elimination of Kjeldahl digestion, distillation or diffusion, concentration, and conversion to N_2, results in total N and ^{15}N analyses on a solid or liquid sample in five minutes with the capability of running 50 samples without operator attendance. The memory effects resulting from the retention of $^{15}N_2$ in the columns, the need for fine grinding and the six decimal place weighing of plant samples presents some difficulties; however, the ability to run a hundred samples in a working day on an instrument that provides ^{15}N analysis with a coefficient of variation (c.v.) of 0.2% and total N analysis with a c.v. of 1–2% will provide a major stimulus for N research (Table 2).

Table 2: Automated simultaneous analysis of solid or liquid samples for total nitrogen and ^{15}N (or total carbon and ^{13}C).

Sample requirements	* 10–100 μg N
	* solid samples ground to <250 mesh
	* liquid samples sorbed on matrix such as Chromosorb
Throughput	* cycle time 300–400 sec, 100+ samples/day
Reproducibility	* ^{15}N, c.v.<0.2% ($\pm$0.001 at natural abundance)
	* total N, c.v. 1–2%
Unsolved problems	* typically 0.5–1% memory between samples
	* variation in isotope ratio with sample size – "Pressure effect"

Soil organic matter research is central to the understanding of the mineralization—immobilization reactions involved in N availability. We now realize that even in most N_2 fixing systems a significant portion of the plant's N comes from soil organic matter. On a global basis, the size of the mineralization—immobilization process is estimated as being 30 times that of symbiotic N_2 fixation (Table 3). Measurement techniques to further delineate microbial biomass and active N pools and to measure their dynamics therefore must continue to be zealously pursued.

Table 3: Global terrestrial N fluxes on an annual basis (Paul et al. 1985).

Source	N flux
	$g \times 10^{12}$
Soil N content	105,000
Plant uptake	1,400[1]
Soil N mineralized	3,500[2]
Symbiotic N_2 fixation	120
Associative and free—living fixation	50
Fertilizer N applied	65
Fertilizer Ń utilized	26
Combustion atmospheric inputs	22
Denitrification	135
Erosion—Leaching	85

1. 2% of 70×10^{15} g C photosynthesized; 2. 40% efficiency of uptake of mineralized N.

The ability to measure the flux of N through the microbial biomass and the soil active N fraction should lead to techniques for the management of the significant stores of N within these pools. Direct measurement of the C and N after lysing the cells holds promise but must be carefully calibrated. Incubation after fumigation releases more biomass constitutents but also requires careful calibration and more careful interpretation regarding the use of controls. An internal standard can be used to measure the effect of the fumigation—incubation on the release of nonbiomass C. Also it has been found that the ratio of CO_2—C produced after fumigation (C_f) relative to the CO_2—C evolved from a nonfumigated control (C_c) can be correlated to the results obtained with internal standards and direct counts (Horton et al. 1987). This allows the recalculation of a corrected control for published data and allows the application of a correction where ^{14}C internal standards are not available. Estimates of microbial N do not have as large a problem with the background controls as do those of microbial C; however, these estimates are very sensitive to the immobilization of mineral N during the incubation subsequent to fumigation. Voroney and Paul (1984) have shown that the flush of C relative to the flush of N (C_f to N_f) can be used to correct for this immobilization. This equation however also must be corrected for the effects of fumigation on the flush of carbon C_f (Horton et al. 1987).

The uptake of N by a large biomass and its conversion to a slowly available soil organic matter fraction explains: (1) very poor competitive uptake of N by plants such as forest trees and some grasses, (2) why nitrification inhibitors which keep the N in ammonium form, preferentially absorbed by microorganisms, have not been successful, and (3) the overestimation of soil N uptake in tracer experiments. The management of a microbial biomass such that it incorporates N at times of excess and releases it at times of maximum plant growth has merit from a theoretical basis but is not easy to accomplish under practical conditions. The tillage of agricultural soils, the incorporation of green manures at appropriate times, and management of forest soils by fire, are examples of successful management techniques. In addition, selection of plants with greater competitive capacity for NH_4^+ should have merit. The mycorrhiza and soil fauna also may play a role under such conditions.

The stress on the microbial partner in a large amount of the present N_2 fixation research is providing an excellent background of knowledge on microbial biochemistry and genetics and on plant-microbial interactions (Hodgson and Stacey 1987). It should also provide the necessary basis for breakthroughs in the molecular genetics of soil microorganisms. I may be proven wrong but I predict that, in most agricultural systems, plant controls on N_2 fixation will be the major limiting factor in increasing N_2 fixation. A greater stress on the plant will probably be necessary in future research imperatives.

Molecular biological techniques are aiding the study of recognition signals in plant microbial interactions which are mutually controlled. Aspects of symbiotic fixation such as competition between rhizobia within soil, and expression of the Nif genes are dependent on microbial activity. Areas such as root colonization, O_2 availability to the N_2 fixing system, and N uptake mechanisms are probably mutually controlled. The effects of C availability and allocation, stress avoidance (drought, frost, etc.), the effects of combined N and plant competitiveness in N nutrition reside more with the host than with the bacterium.

The effects of mycorrhiza and Rhizobia on plant morphology and photosynthetic efficiency are just starting to be understood with mycorrhiza appearing to have more effects than do Rhizobia (Harris et al. 1985). Mycorrhiza have a major effect through their role of supplying P and possibly via increased competitive uptake systems for soil NH_4^+ in N deficient systems. Mycorrhiza, however, can also act as plant parasites where symbiotic systems are not well established. The role of the associations must be investigated with greater imagination than has been possible in the past.

The diversity available in legume germplasm has been exploited only to a limited extent. Commercial agriculture is based on only a few species and interesting plants such as Sesbania which nodulate on both the stem and root and have N_2 fixing systems with higher O_2 and NO_3^- tolerance have been described but have not received the attention they deserve.

Future advances in N_2 fixation research will therefore be as dependent upon plant physiological studies and plant breeding as on the activities of the bacterial geneticist and agronomist.

Intensive crop management techniques will also play a major role in future research. Successful examples from tropical countries show development of double and triple cropping where only one crop used to be grown. In temperate climates, funds are now being allocated to develop lower input, more cost-effective cropping systems. One successful example involves seeding barley into a clover residue. This was interseeded with soybeans before harvest of the barley. The soybeans were then interseeded with winter wheat. The continuous cover eliminated most weed problems and chemical N input requirements were at a minimum, as was soil erosion.

MOLECULAR BIOLOGY AND NITROGEN CYCLING

The effects of molecular genetics on N cycling will in many cases be indirect. An example of this indirect effect will probably be found in the increased infusion of funds for basic research in soil microbiology and plant microbial interactions. Conversely, if soil microbiology and N cycling work does not show the significance of its contribution and the willingness to adapt, there could be a reallocation of manpower and resources to molecular biology. Table 4 lists some of the possible applications of genetic engineering to N cycling. The list is by no means exhaustive nor is it meant to indicate that practical applications are possible in these areas by the year 2000.

The application of molecular biology will have to take into account the already great genetic diversity and high microbial competition in soils; however, possibilities such as enhanced cellulose decomposition by associative N_2 fixers should not be overlooked. The introduction of a pathogen or pathogens that eliminate nitrifying organisms is probably technically possible but would no doubt have to be preceded by lengthy discussions that would involve social as well as scientific and agronomic considerations.

Some areas require new approaches. As previously mentioned, we may require more stress on the plant both as a controller of microbial activity and as a means of adjusting our management systems. Nitrogen fixation capacity in non-leguminous plants is often stated as an objective of molecular genetics. It may however be easier to leave the legume-rhizobium association intact and adapt the legumes to a wider array of crop characteristics. Can we produce a legume that has the yield potential, nutritional characteristics and climatic adaptation of maize while still maintaining high N_2 fixing rates? This would mean the production of a legume with corn kernels rather than the more often discussed production of N_2-fixing corn.

Table 4: The application of molecular genetics to advances in N cycling in agro-ecosystems.

A. Contribution to basic knowledge
    1) Genetic probes for specific microbial DNA in soils
    2) Concepts in microbial competition and plant-microbial interactions
    3) Application of more ecological theory to soil microbiology and nutrient cycling
    4) Enhanced knowledge of soil microorganisms such as Rhizobium genetics and physiology
    5) Genetic exchange among microorganisms

B. Microbial catalyses
    1) Sewage sludge degradation
    2) Industrial fermentation of natural residues
    3) Cellulose-lignin decomposing organisms such as associative N_2 fixing microorganisms

C. N_2 fixation
    1) Inoculation and infection
    2) Increased fixation capacity
    3) Fixation in presence of soil N
    4) Resistance to abiotic stress: salinity, drought, acidity, aluminium toxicity
    5) Rhizobium in non-legumes
    6) Adaptation of Frankia to crop plants
    7) Improved Azolla-Anaebaena associations

D. Improved nitrogen nutrition
    1) Inhibition or removal of nitrifiers
    2) Controls on mineralization-immobilization
    3) Controls on denitrification

E. Plant effects
    1) N_2 fixation in leaves
    2) Increased photosynthesis and C allocation
    3) Altered rooting characteristics
    4) Increased plant competitiveness for N
    5) Altered plant constituents affecting decomposition and allelopathy
    6) Incorporation of greater yield and altered starch contents in N_2 fixing legumes

---

ACHIEVING OUR GOALS

The complexity of plant microbial interactions and soil organic matter transformations will continue to hinder activity and provide excuses for a lack of major breakthroughs. One must question whether our present scientific system is capable of rising to the challenge of effectively moving N agro—ecosystem research into the 21st century. We have access to an extensive number of journals that accept both original publications and reviews. Do we have too many journals? One usually has to read two to five associated articles to determine what is happening in a particular laboratory. Some of this comes from our need to slowly adopt techniques while we apply them to the practical questions. The methodology for biomass measurements is a case in point. It took years of study in a number of laboratories to develop methodologies applicable to most soils. In the meantime, a large number of papers have been published utilizing incomplete methodology. Fortunately corrections to some of the original work can be made. However, some of the studies will probably have to be repeated as we try to get more meaningful numbers rather than comparative data.

A number of good review articles and books on soil organic matter are now available (Stevenson 1982; Aiken et al. 1985) and the re—entry of chemists into the field is producing a re—evaluation of much of the data. However I believe that not as many new concepts as are necessary to develop this important field are now appearing. We tend to accept publication of localized data. This can be argued as being necessary because of our soil type and climatic variability. One must however question whether this allows the investigators to rest on their laurels and not strive as much as they should for new concepts that create truth and knowledge.

Our present scientific establishment requires us to spend a great deal of time chasing a limited resource base and then to publish localized results rather than striving to improve the larger picture or to develop the basic concepts that are required for future breakthroughs in our science.

The need for more rapid advances in our field may require the establishment of fairly large research centers that concentrate interdisciplinary studies applicable to more than one site. Such centers would include studies of both the basic and applied sciences. Molecular biology together with automated instrumentation should supply a large number of the needed research tools. These tools together with the expanded application of ecological principles and mathematical modelling applied to systems agriculture should provide the background for the research necessary to achieve the objectives set out in the earlier part of this paper. If in the next 13 years we can set the stage for such reorganization and recognition of our science, we can ask the younger people entering the 21st century to rise to the challenge of increasing the efficiency and utilization of nutrients while decreasing the pollution potential of our environment.

A new factor that has entered our scientific scene to an extent that is greater than that previously encountered is the role of society. Legislation such as that proposed in Iowa where all fertilizer and pesticide sales would be taxed to pay for activities in pollution abatement and research into lower input agriculture will, if enacted, have a profound effect on N research in that State. This political activity is not surprising if one considers that it is claimed that one third of Iowa wells now have NO_3-N levels greater than those set for toxicity of NO_3^--N.

The role of public opinion and ethics in the release of genetically engineered organisms is also greatly affecting our research and management potential in this field. Our science, while continuing to be exciting and well supported, will have to take into consideration many more economic, environmental and political impacts than we have in the past.

ACKNOWLEDGEMENT

This paper is published as a portion of the work conducted under NSF Grant BSR 8306181.

REFERENCES

Aiken, G.R., McKnight, D.R., Wershaw, R.L., and MacCarthy, P. (1985). 'Humic Substances in Soil Sediment and Water'. (J. Wiley and Sons: New York.)

Board on Agriculture. (1985). 'New Directions for Bioscience Research in Agriculture'. (National Academy Press: Washington, DC.)

Evenson, R.E. (1978). A century of productivity change in U.S. agriculture: An analysis of the role of invention, research and extension. Economic Growth Center, Yale University, New Haven, CT. Disc. paper 186.

Fox G. (1986). Underinvestment, myopia and commodity basis: A test of three propositions of inefficiency in the U.S. agricultural research system. Dept. Agr. Ec. and Business, Univ. of Guelph, Canada.

Halvorsen, H.O., Pramer, D., and Royal, M. (1985). Engineered organisms in the environment. Scientific Issues, Am. Soc. Microbiol. Washington, DC., U.S.A.

Harris, D., Pacovsky, R.S., and Paul, E.A. (1985). Carbon economy of soybean — Rhizobium-Glomus associations. New Phytol. 101, 427–40.

Hodgson, A.L.M., and Stacey, G. (1987). Potential for Rhizobium improvement. C.R.C. Reviews in Biotechnology Vol. 4. 1–73.

Horton, K.A., Norton, J.M., Horwath, W., and Paul, E.A. (1987). Determination of soil microbial biomass C using the chloroform fumigation incubation method: the importance of the non-fumigated control and its ecological significance. Bull. of Ecol. Soc. Amer. 68, 326.

Jansson, J.K., Holben, W.E., Dwyer, D.F., Chelm, B.K., and Tiedje, J.M. (1987). Genetic and molecular methods for detection of specific organisms in the soil environment. Microbiol. Abstr. p. 124.

Marshall, R.B., and Whiteway, J.N. (1985). Automation of an interface between a nitrogen analyser and an isotope ratio mass spectrometer. Analyst 110, 867–71.

Paul, E.A., Bauer, W.D., and Tinker, P.B. (1985). Rhizosphere dynamics. In 'Crop Productivity Research Imperatives Revisited'. (Eds M. Gibbs and C. Carlson), Boyne Highlands, October 13–18, pp. 152–76.

Smith, J.L., and Paul, E.A. (1987). The role of soil type and vegetation on microbial biomass and activity. Proc. 3rd Symp. Microb. Ecol. In Press.

Stevenson F.J. (1982). 'Humus Chemistry'. (J. Wiley and Sons: New York.)

Turner, G.L., and Bergersen, F.J. (1980). Evaluating methods for the determination of ^{15}N in nitrogen fixation studies. In 'Current Perspectives in Nitrogen Fixation'. (Eds A.H. Gibson and W.E. Newton), p. 482. (Aust. Acad. Sci.: Canberra.)

Voroney, R.P., and Paul, E.A. (1984). Determination of K_C and K_N in situ for calibration of the chloroform fumigation incubation method. Soil Biol. Biochem. 16, 9–14.

INDEX

430

431

ADDENDUM

^{15}N METHODOLOGY IN THE FIELD

P.G. Saffigna

ABSTRACT

Techniques for recovering ^{15}N-labelled soil and plant material under field conditions are described with examples from a range of crops, soils, and environments. The issues of ^{15}N application methodology, border areas, plot size, number of plants per plot, and choice of confined or unconfined plots are discussed in relation to ^{15}N recovery techniques. The innovative use of commercially-produced mechanical excavation equipment and the use of hydraulically-operated truck and tractor-mounted units that sample to a depth of 5 m are considered.

INTRODUCTION

Use of the stable isotope of nitrogen, ^{15}N, as a tracer of N cycling in soil-plant systems under field conditions has increased dramatically in recent years, particularly so since the review by Allison (1966).

It is widely recognized that the use of ^{15}N increases the sensitivity of tracing an N input in the soil-plant system. However, many authors (Martin et al. 1963; Hauck and Bremner 1976; Buresh et al. 1982) have reviewed the difficulties of obtaining quantitative recovery of ^{15}N. Errors in the determination of ^{15}N enrichment and total N content will usually be less than those introduced by the sampling of soil. Over 25 years ago, Martin and Skyring (1962) pointed out the need for careful soil sampling to avoid erroneously attributing deficits in N-balance sheets to 'losses'. Their statement — "There seems to be a tendency in the literature to regard the use of labelled N as a panacea in the study of N changes in soils, especially in the detection of N losses. This is not so, because the use of ^{15}N does not remove the necessity of accounting for all forms of N present in the system" —

– is equally as true now as it was then.

The recovery of ^{15}N–labelled soil and plant material under field conditions poses major problems in the correct interpretation of results. The problem is that the ^{15}N moiety is distributed non–uniformly, particularly in the soil but also to a lesser extent in the plant. The quantity of ^{15}N recovered is calculated from the product of the mass of any component, its N content and its ^{15}N enrichment. Thus, the final value reflects errors in the estimation of all three parameters. It is essential to obtain samples that are representative of the average ^{15}N distribution for each component.

Questions frequently asked by the researcher planning ^{15}N studies in the field include:
   (1) sources of error
   (2) confined or unconfined microplots
   (3) installation of confined microplots
   (4) plot size, shape and number of plants
   (5) size of border areas
   (6) method of ^{15}N application
   (7) sampling techniques
Reports in the literature frequently give limited information on these topics or the rationale for choosing one approach over another.

Legg and Meisinger (1982) reviewed field studies between 1967 and 1980 and emphasized ^{15}N studies with small grains, corn, rice, grassland and forest ecosystems. In this paper I do not cover all the above issues in relation to field ^{15}N studies since 1980. Rather, I consider some of the key issues pertaining to ^{15}N recovery techniques that demonstrate significant advances in field ^{15}N methodology.

SOURCES OF ERROR

Error theory

Errors in calculating ^{15}N recoveries can arise from random sources or from systematic sources leading to bias. Vallis (1969) reviewed sampling theory in relation to measuring N changes in soil–plant systems, and pointed out the important distinction between accuracy and precision in calculating N recovery. Accuracy refers to the size of the deviations from the true mean, whereas precision refers to the size of the deviations from the mean obtained by repeated application of the sampling method, i.e. the measured value. Bias is the difference between the true mean and the measured value. The bias may be constant or it may be proportional to the true mean. Incomplete N recovery by Kjeldahl analysis would lead to a biased (high) estimate of the N deficit.

Researchers often believe that errors in N balance are cumulative. This is certainly true for the bias component of error. However, it may not be true for the random component because N pools are not necessarily independent.

For example, fertilizer N taken up by the crop is not available for leaching. Legg and Meisinger (1982) demonstrate the impact that this inverse relationship has on improving the accuracy of the final N balance by using an error structure proposed by Snedecor and Cochran (1967):

$$Var(N_t) = Var(N_1) + Var(N_2) + Var(N_3) \ldots Var(N_i)$$
$$+ 2COV(N_1,N_2) + 2COV(N_1,N_3) + 2COV(N_2,N_3)$$
$$+ \ldots 2COV(N_{i-1},N_i) \tag{1}$$

where the variance (Var) of the final total N (N_t) is equal to the sum of the variances of the various N pools plus twice the covariances (COV) of all possible two-way combinations of these pools.

For a simple two-component system such as soil and plant, the relationship is given by:

$$Var(N_t) = Var(N_p) + Var(N_s) + 2COV(N_p,N_s) \tag{2}$$

where N_p = total N in plant, and N_s = total N in soil. Because the covariance term will be negative where N_p and N_s are negatively correlated, the variance of N_t will be less than the variance of either N_p or N_s.

This is apparent from the data of Carter et al. (1967), cited by Legg and Meisinger (1982). These workers compared the recoveries of ^{15}N fertilizer added to 60 cm steel cylinders driven into the soil. Sudangrass was grown for eight weeks and the ^{15}N content of the crop and the soil was measured. Var (N_p) was 31.7 (C.V. 12.9%), and Var (N_s) was 35.3 (C.V. 10.7%) for the 0–75 cm depth including roots. Since the covariance equals the product of the correlation coefficient between N_p and N_s and the standard deviations of N_p and N_s, then COV(N_p,N_s) was −32.5 (r = −0.97). Thus, from Equation 2, the variance of N_t = 31.7 + 35.3 + (2 x −32.5) = 2. Hence, the final error in N_t is much less than the sum of the random errors in each of the soil and plant components.

The correct interpretation of ^{15}N recovery is as greatly affected by the error estimate as by the mean value. Thus, a ^{15}N recovery of 85% with a standard error of ± 15% would not lead the researcher to conclude there was a N deficit of 15%. However, such a conclusion would be reasonable for a recovery of 85% ± 2%.

Soil mixing

Although "thorough" soil mixing is widely practised, few reports have quantified its effect on the uniformity of ^{15}N distribution. Yet, because only a small proportion of the soil is analysed it is necessary to ensure that the sample is representative of the average ^{15}N content. Considerable differences may exist between the ^{15}N enrichment of soil and roots. For example, Saffigna (1977) reported an average ^{15}N enrichment of 0.12 atom %

excess ^{15}N for soil total N, 1.25 atom % excess ^{15}N for soil nitrate N, and 5.65 atom % excess ^{15}N for root N in samples of a red-brown earth soil from pots that had received ^{15}N aqua ammonia and been cropped to wheat for 100 days; a nearly fiftyfold difference between the various components.

To assess the effectiveness of the "halving and quartering" mixing procedure, 12 replicate pots containing 2 kg of either a black earth or a red-brown earth receiving either ^{15}N aqua ammonia (11.46 atom % excess ^{15}N) or ^{15}N urea (5.06 atom % excess ^{15}N) were cropped to wheat for 100 days in the glasshouse. Plant tops were removed and the soil (plus roots) analyzed for soil total ^{15}N and soil mineral ^{15}N.

The variability of N and ^{15}N before mixing the soil was represented by the coefficient of variation (C.V.) calculated from the mean soil N and ^{15}N concentration of four replicate pots that were analyzed separately. The variability after mixing was determined in a similar way from the four equal portions of soil obtained after mixing and quartering the bulked soil from the remaining eight replicate pots.

The large variation in the native and fertilizer N components of the nitrate N was substantially reduced by mixing the soil (Table 1). The C.V.'s decreased from an average of 25.2% to 5.0% for the native N and from 41.5% to 3.3% for the fertilizer N component of the nitrate N. Mixing the soil did not appreciably change the variation in the native N component of the total N. Although the average C.V. decreased from 0.5% to 0.3%, both increases and decreases were recorded for individual treatments. These small C.V.'s for native total N concentration probably were due to the dominance of the organic N in the total N component. Mineral N was < 1% of the native total N at sampling. Considerable uniformity in the native organic N was expected as the soil had been thoroughly mixed before potting.

The C.V. of 0.3% for native total N concentration between subsamples from the mixed bulk of soil was similar to the 0.19% for triplicate total N determinations. This suggested that the subsampling error (by difference) was less than the analytical error and demonstrated the effectiveness of the mixing and subsampling techniques because only 0.2% of each sample was analyzed. This contrasts with the conclusion of Bartholomew et al. (1965) that subsampling errors were greater than errors in Kjeldahl total N determinations.

The C.V. in the fertilizer N component of the total N was reduced from an average of 9.7% to 2.3% by mixing the soil. The greater variation with aqua ammonia (C.V. 13.5%) than with urea (C.V. 5.9%) before mixing was in keeping with the inherent variability associated with the ammonia injection technique. The greater variability of the fertilizer N component than that of the native N was expected because more of the fertilizer N was active in N transformations and plant uptake than the relatively inert native organic N. Errors in mass spectrometric analysis would account for only a small fraction

436

Table 1: Variation in native (unlabelled) and fertilizer N concentration before and after mixing soil sampled from pots containing a red-brown earth or a black earth cropped to wheat after application of ^{15}N-labelled aqua ammonia or urea.

Treatment	Native N concentration		Fertilizer N concentration	
	Before mixing	After mixing	Before mixing	After mixing
coefficient of variation ($\%$)[1]				
Nitrate nitrogen[2]				
Red-brown earth:				
Aqua ammonia	18.6	6.8	35.0	3.0
Urea	10.9	7.3	5.5	5.4
Black earth:				
Aqua ammonia	35.3	2.9	57.1	2.6
Urea	35.9	2.8	68.4	2.1
Mean	25.2	5.0	41.5	3.3
Total nitrogen[3]				
Red-brown earth:				
Aqua ammonia	0.38	0.60	12.3	3.0
Urea	0.23	0.33	6.8	2.3
Black earth:				
Aqua ammonia	0.80	0.21	14.6	1.5
Urea	0.58	0.07	5.0	2.5
Mean	0.50	0.30	9.7	2.3

1. Calculated from the four replicate values. These values were derived from either three (native total N), two (fertilizer total N) or one determination (native and fertilizer nitrate N). 2. Nitrate N concentrations 2 to 16 ppm and ^{15}N enrichment 0.5 to 2.5 atom % excess [15]N. 3. Total N concentrations 840 to 1008 ppm and ^{15}N enrichments 0.06 to 0.30 atom % excess [15]N.

of the difference. The data in Table 1 show that the soil mixing technique effectively provided a homogeneous distribution of native and fertilizer N.

Sampling errors

Although sampling errors are usually greater than analytical errors, most workers direct more effort to reducing the latter than to improving sampling technique. Principles of soil sampling and subsampling were discussed by Cline (1944) and Petersen and Calvin (1965), and similar principles apply to plant sampling and subsampling (Jones and Steyn 1973; Barker 1974).

Vallis et al. (1973) demonstrated the large variability in recovery of ^{15}N in soil associated with taking 4.6 cm diameter soil cores in unconfined field microplots of pasture in subtropical Australia. The C.V. between cores was 56.3% for weight of ^{15}N per gram of soil, but only 6.2% for bulk density; the variability was attributed to uneven distribution of ^{15}N in the microplots, not imprecision in core—sampling techniques. Although representative sampling is essential, only rarely do workers give estimates of sampling errors. Vallis (1969) reported a C.V. of 1.9% for total N determinations on 10 g subsamples of 200 g samples of soil taken from 1600 g bulk samples previously ground to < 2 mm. However, the average C.V. between duplicate 10 g subsamples from each of the 200 g samples was 1.6%. Thus, in contrast to the findings of Bartholomew (1964) and Bartholomew et al. (1965), the subsampling error (by difference) was, in fact, less than for the Kjeldahl method. This was probably because all of the soil was ground to < 2 mm before sampling and because of the large subsample (10%) used for total N analysis.

Pruden et al. (1985) from ^{15}N field studies with wheat in England reported that between—plot C.V.'s for soil and grain % N (3.7%, 2.7%) and atom % excess ^{15}N (8.3%, 6.6%) were much higher than corresponding between—sample C.V.'s for % N (1.1%, 1.2%) and for atom % excess ^{15}N (1.1%, 0.75%) respectively. Thus, field heterogeneity contributed much more to overall sampling error than did analytical errors in the laboratory.

From a study of the variance components for sampling and analytical procedures in the design of ^{15}N experiments, Bartholomew et al. (1965) reported that the relative magnitude of the variance components were in the order: field sites > greenhouse pots > samples > subsamples > Kjeldahl determinations > mass spectrometric determinations. However, Martin and Skyring (1962) concluded that sampling errors did not influence the final N balance in most experiments. *My recommendation is that researchers should devote more effort to obtaining representative soil and plant samples in the field and less effort to obtaining the fifth decimal place on the mass spectrometer. After all, a highly accurate ^{15}N enrichment is of no advantage if the sample is unrepresentative of the field value.*

MICROPLOT TECHNIQUES

The question of whether or not to confine (enclose) microplots receiving ^{15}N under field conditions is faced by many researchers. The high cost of enriched ^{15}N ($100 U.S. per gram) usually restricts individual plot size to less than 10 m^2, with most workers using plots of 1 m^2 or less; hence, the term "microplot". Even with the less expensive depleted ^{15}N, plot sizes are small.

Unconfined microplots

Advantages and disadvantages

Their major advantage is that they closely simulate actual field conditions in respect of plant root exploration, leaching and lateral movement of fertilizer (Vallis et al. 1973; Van Cleemput et al. 1981; Woodcock et al. 1982; Khanif et al. 1983, 1984; Powlson et al. 1986a, b; Zaharah and Sharifuddin 1986; Zapata and Van Cleemput 1986a, b; Sudin and Bachik 1987). Vallis et al. (1973) proposed that unconfined microplots were adequate for fertilizer placement studies (e.g. Bartholomew 1971) and for estimating uptake of soil N and N_2 fixation by legumes in mixed pastures (Vallis et al. 1967; Henzell et al. 1968; Vallis 1969).

One major disadvantage is the sampling problem associated with unconstrained dispersal of ^{15}N in the soil (Legg and Meisinger 1982). Vallis (1969) pointed out the statistical difficulties associated with sampling concentric annuli of different radii around a circular area treated with ^{15}N under field conditions. Since the ^{15}N enrichment decreases with distance from the central application zone, this presents special difficulties for the measurement of ^{15}N recovery by core methods. As the cores are from annuli of different areas, this must be taken into account by a weighting factor when calculating total ^{15}N recovery. Unless the number of cores in each annulus is proportional to its area, cores from different annuli cannot be bulked for total N and ^{15}N analyses; hence there is a considerable practical disadvantage of an excessive number of analyses. With proportional sampling, there are an excessive number of cores required for the outer annuli where there will only be a small amount of [15]N. In contrast to this concentric ring analysis, Powlson et al. (1986a) working with wheat in England treated a 2 m x 2 m area with ^{15}N and then sampled a central 1.07 m x 1.07 m area for plant N uptake and took two 30 cm diameter soil cores for soil ^{15}N from this central area. This method gave reproducible results with the standard error of mean [15]N% recovery in soil ranging from 1.1 to 1.9.

Another disadvantage of the unconfined microplot is the larger amount of ^{15}N that must be applied to provide an adequate border and prevent edge effects. Olson (1980) reported that maize plants 36 cm from the edge of the ^{15}N-treated area contained significantly less fertilizer N than plants either 107 cm or 154–178 cm from the edge.

The validity of using ^{15}N microplots to extrapolate to the whole field was assessed by Khanif et al. (1984) with maize and barley in a sandy soil in Belgium. A plot 4 m x 4 m received ^{15}N, and at harvest plants were removed from a central 3 m x 3 m area for barley and 3.2 m x 3 m for maize. The barley plot was subdivided into nine subplots, and the maize into six subplots. An equal number of plants were taken at random from outside the ^{15}N plot. The mean N uptake for barley and maize in the whole field was not significantly different from the ^{15}N plot. Similarly, the variances for N uptake in the ^{15}N plot and the whole field were not significantly different. Thus, the results from the ^{15}N plots could be extrapolated to the whole field. Of course differences in yield between microplot and whole field may occur due to differences in sampling, e.g. mechanical vs manual harvesting.

Size, shape and plant numbers

Unconfined microplots vary greatly in size but are usually larger than confined microplots. Examples of plot sizes are 1 m x 1 m for wheat in subtropical Australia (Saffigna et al. 1985), 2 m x 2 m for wheat in England (Powlson et al. 1986a, b), 3.56 m x 3.56 m for maize in the USA (Olson 1980), 4 m x 4 m for maize in Belgium (Khanif et al. 1984), and also in Belgium, 3 m x 1.8 m plots for cropping sequences of bean, beet and wheat (Zapata and Van Cleemput 1986a, b). Usually 50 or more plants are grown per plot.

In contrast, with oil palm and rubber trees in Malaysia only one plant is included per microplot. With 2-year old oil palm ^{15}N fertilizer was applied in a circular area up to c. 2 m from the trunk (Zaharah and Sharifuddin 1986), while with 6-year old rubber trees (Sudin and Bachik 1987) the ^{15}N fertilizer was given in a 30 cm wide annulus extending from 1.2 m to 1.5 m from the trunk of the tree. These experiments represent significant advances in the development of ^{15}N methodology for field conditions.

Border areas

The possibilities of lateral movement of ^{15}N beyond the application zone, uptake of ^{15}N by plants outside the plot, and uptake of unlabelled N from outside the ^{15}N plot by the experimental plants have resulted in the use of border areas.

The border area in some instances is not treated with ^{15}N but is sampled to assess the total ^{15}N uptake by plants outside the plot. Vallis et al. (1973) found little ^{15}N in soil and plant more than 15 cm outside circular (30 cm diameter) and square (30 cm side) subtropical pasture microplots, but 10-20% of applied ^{15}N was present within 15 cm of the edge of the treated area. Van Cleemput et al. (1981) reported that wheat plants of the first row 7.5 cm outside 2 m x 2 m plots assimilated 1.4% of the applied ^{15}N while plants in the second row 22.5 cm outside the plot assimilated 0.4%. No ^{15}N was detected in the soil outside the treated area.

More commonly ^{15}N is applied to a border region outside the defined sampling plot, the area covered depending on the root distribution characteristics of the test plant. To reduce cost the aim is to have the minimum size border that is acceptable; a 50 cm border around a 1 m x 1 m plot has an area of 3 m^2 and requires three times as much ^{15}N as the plot itself.

For a 2 m x 2 m plot with 12 rows of wheat Powlson et al. (1986a) advocated sampling the central 1.07 m x 1.07 m area of 6 rows. There was considerable uptake of ^{14}N from outside the treated plot by the 2 outer 'border' rows of wheat but very little by the 3rd row from the plot edge; the border area allocated was nearly 3 times the size of the harvested area. With 4 m x 4 m plots treated with ^{15}N, Khanif et al. (1984) sampled the central 3 m x 3 m for barley and 3.2 m x 3 m for maize, allowing border areas equivalent to 78% and 67% of the area sampled. Olson (1980) reported that in a 3.56 m x 3.56 m plot with 5 rows of maize on 71 cm spacings, plants closest to the edges contained significantly less fertilizer N than those from the central 3 rows. He recommended a border zone of 107 cm for initial ^{15}N studies and 142 cm for residual ^{15}N studies, giving a border area 5 times larger than the plant sampling area.

In summary, most workers use square or rectangular unconfined microplots between 1 m^2 and 16 m^2 in surface area. Usually there are more than 50 plants per plot except with trees where there is one plant. Border areas treated with ^{15}N are essential to ensure correct estimation of fertilizer N uptake, their size depends on the root distribution of the test plant and is often 2–3 times larger than the area of the harvested plot.

Confined microplots – advantages and disadvantages

The major advantages of confined microplots are that they:
(1) restrict lateral dispersion of ^{15}N beneath the soil (thus simplifying soil sampling)
(2) prevent loss of ^{15}N from the surface soil due to runoff (particularly important if ^{15}N-labelled crop residue is the N source under test)
(3) appreciably reduce sampling problems because all the soil is usually removed (for the surface layers at least).

In tropical areas with intense rainfall, the ponded runoff water can be analysed separately for ^{15}N or stored prior to returning to the plot. Note that containment of surface water may result in increased denitrification or leaching loss over the situation where runoff occurs. Alternatively, an automatic collection system can be devised by drilling a hole in the cylinder wall that protrudes above the soil surface and siphoning off excess water through a plastic tube to a reservoir buried below ground-level outside the microplot.

The major disadvantages of confined microplots are:
(1) they may impede drainage on soils where water may drain downslope on top of a clay B horizon or where water flows along non-vertical planar voids such as slickensides in vertisols.
(2) they will excise part of the confined plant's root system and also cause an input of severed root material from surrounding plants when installed over existing plants (e.g. in permanent pasture).
(3) plant root distribution may be constrained by microplot walls so reducing the availability of water and nutrients from the soil.
(4) their installation may disturb the soil inside the microplot,

particularly if considerable force is used to drive in the cylinder.

Increased bulk density may be immediately apparent because the soil surface of the microplot will be lower than the surrounding soil. If so, a replacement plot should be installed. Disturbance to the drainage characteristics of the confined soil is more difficult to assess at the time of installation. The disturbance would be less than with repacked microplots. Ritchie et al. (1972) pointed out that soil enclosure may reduce lateral drainage and prolong waterlogging after irrigation or substantial rainfall in vertisols.

To evaluate the extent of disruption to NO_3 movement in laterally confined vertisols, Saffigna et al. (1984a) compared solute movement under field conditions in soil cores with that in identical areas of adjacent unconfined soil. The cores were obtained using PVC cylinders (11 cm diameter and 60 cm long) which were housed in a steel casing and forced into the soil by hydraulic ram. To reduce costs Br was used as a tracer of $^{15}NO_3$ movement since Saffigna et al. (1981) found that Br and $^{15}NO_3$ were excellent companion tracers under these conditions. After 130 mm rainfall over two months there was a trend towards less leaching in the 'confined' than the 'unconfined' soil.

White (1984) intalled gypsum blocks at 7 cm depth inside 40 microplots formed by forcing steel cylinders (35 cm deep, 75 cm diameter, 3 mm wall thickness), 30 cm into a vertisol in subtropical Australia. Another eight gypsum blocks were similarly placed outside the cylinders. Soil-water potential was measured daily for 19 days after 100 mm irrigation. At all times, the soil was wetter outside the cylinders than inside, indicating better drainage inside the cylinders than outside [this was contrary to expectations based on Ritchie et al. (1972)]. White (1984) hypothesized that some drainage down the inside walls of the cylinder may have occurred because water was never observed ponding in the rings. He concluded that confining this size microplot did not impede drainage in the soil. However, with 11 cm diameter and 25 cm deep microplots constructed from PVC pipe installed in a vertisol, Cogle (1986) did find ponded water in some microplots after heavy rainfall. This may not necessarily have been due to poor drainage inside the microplot, but because runoff was prevented.

Lysimeters represent the next level of confinement for ^{15}N studies with the base being sealed to facilitate collection of drainage, and thus provide an estimate of N leaching (Chichester and Smith 1978). However, the soil within them cannot be easily sampled and they are very expensive to establish and maintain.

In summary, there are advantages and disadvantages in the use of either confined or unconfined microplots. Because of the high cost of ^{15}N and the difficulties associated with soil sampling in unconfined microplots many researchers have opted for the smaller confined microplots. However, in the *past 5 years there has been a swing to the use of unconfined microplots, especially for small crops such as wheat and barley and for very large plants such as rubber trees and oil palm where a ^{15}N balance is not required. With intermediate size plants such as sugar cane and tobacco workers prefer confined microplots for ^{15}N balance studies. The disruption to drainage in confined microplots will be less the larger the surface area confined and the shallower the confining walls. I recommend that researchers use unconfined microplots or if confined microplots are necessary they should be as large and as shallow as possible.*

INSTALLATION OF CONFINED MICROPLOTS

Steel or galvanized iron frames forced into soil

This method has been used successfully with tropical pastures (Vallis et al. 1986), sorghum and wheat (White et al. 1986), maize (Chapman et al. 1987; Xu et al. 1987), tobacco (P. Saffigna, K. Ferguson and J. Littlemore, personal communication) and sugarcane (Saffigna and Wood 1986). Most workers use cylinders rather than rectangular frames. With large microplots, considerable force may be necessary to install the microplot to the required depth. White (1984) experienced great difficulty in installing microplots into a vertisol in subtropical Australia. They were steel cylinders 75 cm in diameter, 35 cm tall, with 3 mm wall thickness. The bottom edge was bevelled on the outside and lightly oiled to facilitate insertion without soil compaction. To install each cylinder a 1000 kg weight was lowered onto it, pushing it 10 cm into the ground, then it was pushed with a crane and finally pressed in with the weight of a truck. Each cylinder was installed to a depth of 30 cm, with 5 cm protruding to collect ponded water if necessary. There was no visible soil compaction.

Installation of 100 cm diameter steel cylinders, 60 cm tall, with 3 mm wall thickness to a depth of 55 cm in a loam cropped to sugar cane was achieved using a pile-driving system with a 200 kg mass raised by a tractor dropped from a height of 2 m onto the cylinder (Saffigna and Wood 1986). It took 30 minutes to install a cylinder. Although the reinforced lip of the cylinder was damaged, there was no evidence of soil compaction. Yields of cane inside and outside the microplot were similar.

In other studies (P.G. Saffigna and W.M. Strong, personal communication) no difficulty was experienced in installing 50 cm diameter, 30 cm tall, thin-walled (1 mm), galvanized iron cylinders 25 cm into a vertisol prepared for cropping to wheat or sorghum. The cylinders could be forced in 15 cm by striking a wooden frame placed on top of the cylinder by hand with a sledge-hammer. A hydraulic ram mounted on a truck (Berndt et al. 1976) was used to force the cylinder to the full 25 cm depth. A similar device mounted on a high-clearance tractor was successfully used by Chapman et al. (1987) to install 25 cm diameter, 32 cm tall galvanized iron cylinders (wall thickness 1 mm) 25 cm into an alfisol in northern Australia. Steel cylinders 30 cm diameter, 45 cm long and 2 mm wall thickness, installed 25 cm into the soil, were used for ^{15}N experiments with wheat and pasture in the Great Plains of North America (R.J.K. Myers, personal communication). Xu et al. (1987) in China used a metal frame 50 cm x 40 cm and 40 cm tall forced into soil.

Generally, the smaller the surface area enclosed, the shorter and thinner the frame, the easier it is to install microplots.

Plastic (PVC) cylinders forced into soil

Modern plastics provide an alternative to metal cylinders for microplots. Although plastic cylinders are easy to make because standard lengths of small diameter (11 cm, 15 cm) PVC drainage pipe can be readily purchased and cut to size, there are few reports of researchers driving PVC cylinders into soil to form microplots. One limitation is that large—diameter (>50 cm) PVC pipe is not readily available and another is that PVC is relatively fragile compared to steel and tends to crack when struck heavily during installation. Saffigna et al. (1985) installed 25 cm diameter, 30 cm tall, PVC cylinders 25 cm into a sandy loam and a sand in Western Australia by placing a block of wood on top of the cylinder and striking it with a sledge—hammer. The lower, outside edge was bevelled to facilitate insertion. Although mostly successful, some PVC cylinders cracked when the leading edge came in contact with a stone. The same technique was used to install 11 cm diameter, 25 cm tall PVC tubes into a vertisol (Cogle 1986).

Confining intact soil cores in plastic (PVC) cylinders

Berndt et al. (1976) developed a method for rapidly confining "undisturbed" vertisol soil cores 11 cm in diameter and 60 cm long into open—ended plastic cylinders. The cylinder was housed in a heavy steel casing which was driven into the soil using a hydraulic ram mounted on a 4-Wheel Drive truck. Saffigna et al. (1984b) adapted this technique for extensive ^{15}N microplot studies with wheat in Australia by placing the PVC-encased soil cores back into the hole in the soil. The base was unsealed to allow free drainage in contact with the soil profile. The main advantage of this method is the rapidity of microplot installation (10 min./microplot) and retrieval for soil sampling (5 min./microplot). The main disadvantage is the small surface area, so, although crops such as wheat can be grown successfully, this system would not be suitable for sorghum. Also, there may be compaction, although Berndt et al. (1976) reported that this was minimal. Although developed primarily for vertisols, this methodology has been successfully used for ^{15}N studies on a sandy soil and an alfisol at Katherine in the Northern Territory (J. Dimes, personal communication).

Excavating and back—filling

To avoid soil compaction during installation of microplots, some workers have excavated trenches around a monolith then placed constraining walls against the faces of the intact monolith to form a confined microplot. Myers and Hibberd (1985) produced microplots of c. 0.5 m^2 by cutting trenches either 5 or 15 cm wide using "ditch diggers" and then installing galvanised iron walls to a depth of either 50 or 100 cm with 5 cm above the soil surface. Sorghum was grown successfully in these vertisols. Moraghan et al. (1984a) enclosed microplots in a vertisol in India to a depth of 30 cm with steel walls. Each microplot consisted of seven 45 cm-spaced rows each 2.03 m in length and was cropped to sorghum. In further work on an alfisol, Moraghan et al. (1984b) enclosed 1 m x 2 m microplots to a depth of 30 cm with a metal wall, then to a depth of 135 cm by three layers of heavy plastic sheeting. The sheeting was placed in position during the preceding dry season after trenches were excavated to a depth of 150 cm around the defined microplot. The metal wall, which protruded 2.5 cm above the soil surface, was placed in the soil prior to excavation to improve the stability of the isolated soil block during excavation and prevented loss of ^{15}N due to runoff.

In Indonesia, Sisworo et al. (1987) confined microplots 1 m x 1 m by excavating manually with a hoe to 30 cm, then placing plastic sheet in the trench to act as a confining wall. ^{15}N-labelled fertilizer or crop residues were added to microplots which were cropped to rice, soybean, cowpea and maize in a multiple-cropping system. A novel procedure used in Nigeria is the burial of palm thatch to a depth of 10 cm to provide confining walls for 5 m x 3 m microplots cropped to 84 maize plants; the palm thatch protruded 40 cm above the soil surface to prevent runoff (Van der Kruijs, personal communication).

To summarise the characteristics of confined microplots, most researchers have used circular metal microplots of 25 to 100 cm diameter for a wide range of crops. The cylinders have usually been 50 cm or less in diameter. Microplots with surface areas > 0.2 m^2 have usually been rectangular and mostly formed by excavating a trench, placing in separate metal walls, and then backfilling.

^{15}N RECOVERY IN MICROPLOTS—CONFINED VERSUS UNCONFINED

Although various advantages (and disadvantages) can be cited to help choose between confined or unconfined microplots it is easy to lose sight of the key issue — which system gives an answer most representative of the field. A systematic comparison of results from confined and unconfined microplots is rarely reported. My own work (P.G. Saffigna and W.M. Strong) provides one such comparison on a vertisol in subtropical Australia with ^{15}N uptake by wheat grown in confined microplots (50 cm diameter cylinders 25 cm into the soil) or in adjacent 1 m x 1 m unconfined microplots. Urea, labelled with ^{15}N, was applied in bands 25 cm apart. Comparisons of ^{15}N recovery by wheat tops at anthesis were made for 2 depths of urea placement and in the presence or absence of a nitrification inhibitor. The % ^{15}N recovery from the unconfined and the confined microplots was similar for all treatments, indicating that in this study the confining system caused no artifacts in plant N uptake.

APPLICATION OF ^{15}N IN THE FIELD

The method of applying ^{15}N to microplots influences the procedure for soil sampling. For example, if ^{15}N fertilizer is banded then soil sampling with cores must take account of the likely non–uniform distribution of ^{15}N in the soil. Application technique will also obviously vary depending on the nature of the ^{15}N carrier, whether as a solution (e.g. soluble fertilizers) or as a solid (e.g. urea supergranules, soil, plant material).

^{15}N Fertilizers

^{15}N fertilizer in solution has been applied uniformly over microplots using a watering can with rubber trees (Sudin and Bachik 1987), oil palm (Zaharah and Sharifuddin 1986) and pastures (R.J.K. Myers, personal communication), and in bands with pipettes (White et al. 1986) or an automatically refilling syringe connected to a reservoir (Saffigna et al. 1985). The hand–operated mechanical device described by Woodcock et al. (1982) is a significant advance in application methodology where many large plots are to receive ^{15}N. It applies ^{15}N solution in rows 5 cm apart on areas of 2 m x 1 m or multiples of this. The apparatus weighs 62 kg, is operated by 2 people and in a single traverse applies a volume of liquid equivalent to a rainfall of 0.25 mm. Follow–up watering with 2 L of distilled water left only 1.5% of applied ^{15}N on the plants (Powlson et al. 1986a). The device has been used successfully in England for several years in experiments with pastures and wheat (Powlson et al. 1986a, b).

Solid fertilizers may be broadcast then incorporated (Olson 1980) or banded (Moraghan et al 1984a, b). Precision placement in waterlogged soil is necessary with ^{15}N supergranules in rice.

^{15}N Plant and soil residues

The release of N from plant residues may be assessed by applying ^{15}N–labelled plant material to soil. The plant material can be either chopped (or finely ground) and should be thoroughly mixed to ensure a uniform distribution of ^{15}N label. White et al. (1986) cut ^{15}N–labelled leaves and stems of mature wheat and sorghum into 2.5 cm lengths before applying to the surface or incorporating into a vertisol that was cropped to wheat. A nylon mesh with 1.5 cm x 1.5 cm holes was placed over the confined microplot to minimize loss of the applied residues. Chapman et al. (1987) partitioned ^{15}N–labelled Leucaena plants into stems, petioles and leaves before cutting the components into small pieces and applying to confined microplots. This approach was chosen because the ^{15}N label was non–uniformly distributed in some plants and because the N release characteristics of different plant parts was being evaluated. Finely ground ^{15}N–labelled sugar cane trash has been used for incorporation with or application to the soil surface (A.W. Wood and P.G. Saffigna, <u>personal communication</u>).

There are few reports of applying ^{15}N soil to microplots since most workers conduct residual ^{15}N studies directly on <u>in situ</u> microplots. One technique used was to recover ^{15}N-labelled soil from old experimental plots which was mixed and used to replace the excavated top soil in a new set of microplots (I. Vallis, <u>personal communication</u>).

In summary, a variety of techniques are used to apply ^{15}N to microplots and I believe researchers should provide more details on their application methods when publishing data. The development of an automatic device to readily apply ^{15}N solution uniformly to large plots is a major advance from the group at Rothamsted.

SAMPLING

The sampling of soil and plants is one of the most labour intensive phases of ^{15}N experimentation, and requires close attention to correct technique so that valid ^{15}N recoveries can be calculated.

Plants

Usually all above-ground herbage is harvested from the microplot, plant tops may be cut at ground level or the plants pulled loose from the soil, then partitioned into appropriate components such as grain, straw and chaff with cereals (Moraghan <u>et al</u>. 1984a, b; Powlson <u>et al</u>. 1986 a,b; White <u>et al</u>. 1986), tops, stalks, leaves and juice with sugar cane (Saffigna and Wood, 1986) or tops (leaves plus head) and roots for sugar beet (Zapata and van Cleemput 1986b).

With trees, sampling is a major task that requires the tree to be felled and sectioned. Zaharah and Sharifuddin (1986) working with 2-year old oil palm sampled the leaf, unopened spear, stump, cabbage, male inflorescence, petiole and rachis for ^{15}N analysis. The subsample of the rachis was obtained from the sawdust produced during sectioning of the tree. A similar approach was used by Sudin and Bachik (1987) who destructively sampled 6-year old rubber trees into leaves, trunks, branches and roots.

Soil

Few workers remove all the soil from a microplot for subsampling because of the size of the task, e.g. the upper 15 cm of a 2 m x 2 m soil plot contains 1000 kg of moist soil. In some circumstances cheap labour may be available for soil excavation but in most situations the adaption of commercially-produced mechanical equipment (such as trench diggers) is essential along with the construction of simple sampling tools. Having excavated the soil and measured its mass, the subsampling problem is enormous because only one millionth of the soil mass may be analysed for ^{15}N.

The most common approach to soil sampling is partial removal by coring or augering. The mass is estimated from the bulk density — a parameter often not accurately known.

The importance of representative subsampling and knowledge of the correct mass of soil is well illustrated by the findings of Carter et al. (1967). They used two methods to determine the recovery of ^{15}N fertilizer added to 60 cm diameter confined microplots which were cropped for 2 months, viz. (a) compositing seven 1.9 cm diameter cores or (b) completely removing all the soil, mixing and subsampling. The total ^{15}N recovery estimated by soil coring was too high (113%) and very variable (86 to 137%). In contrast, the complete removal method averaged 100% and ranged from 98% to 101%. The poor results obtained with soil cores were attributed to difficulties in obtaining a representative sample when roots were present and to inaccuracy in calculating total soil mass.

Complete removal of soil

With confined 75 cm diameter microplots, White et al. (1986) excavated 0-10, 10-20 and 20-30 cm depths and mixed the soil within each layer in a concrete mixer before subsampling. Below 30 cm, six cores each 3 cm diameter were bulked for the 30-60 and 60-90 cm depths. Most of the ^{15}N was in the 0-10 cm depth and only a trace below 30 cm. Myers and Hibberd (1985) used a similar technique of complete excavation of 0-15 cm depth and coring for the 15-30 and 30-45 cm depths. The bulk of the ^{15}N residual in soil was in the upper 30 cm with usually < 5% below that depth. Although large ^{15}N deficits occurred the authors did not attribute this to leaching beyond 45 cm. Saffigna and Wood (1986) completely excavated the 0-10 cm and 10-20 cm depths of soil in 100 cm diameter confined microplots and cored for depths below this.

Saffigna et al. (1982) removed all soil from confined microplots of 11 cm diameter PVC cylinders 60 cm long. Both sides of the PVC tubes were cut with a circular saw to expose the intact soil column which was sectioned using thin wire or a spatula.

For large unconfined 1 m x 1 m microplots Saffigna et al. (1985) completely excavated soil to a depth of 30 cm from the central 60 cm x 50 cm. This was done by forcing in a 40 cm tall, 60 cm x 50 cm galvanized iron frame to a depth of 10 cm and then removing the soil inside (and outside) to a depth of 10 cm. Intersecting trenches 5 cm wide were excavated around the frame to a depth of 50 cm using a commercially available trenching machine to isolate the central block. This was then removed as 10-20 and 20-30 cm layers by special sampling tools that took a 10 cm deep semicircular wedge of soil 20 cm in diameter. This technique is satisfactory on most vertisols because of the plastic nature of the soil. Samples from 30-60, 60-90 and 90-120 cm depths were taken as 5 cm diameter soil cores. Most of the ^{15}N in the soil was in the upper 30 cm with an average of only 2-3% below 30 cm.

Soil coring

Many implements have been used to obtain soil cores. They include manually operated screw and Jarret augers, mechanical devices for inserting and extracting tubes, petrol-driven post-hole screw augers, hydraulically driven rams mounted on high clearance tractors (2 m above soil surface) or 4WD trucks that rapidly insert and withdraw steel tubes to a depth of 5 m, and petrol-driven jack hammers. Mostly cores are 2 to 5 cm in diameter.

Larger diameter cores are sometimes taken from the surface soil layers which contain most of the residual ^{15}N and where distribution is often less uniform due to banding of fertilizers. Powlson et al. (1986a) used a petrol-driven post-hole auger and took two 30 cm diameter cores to 23 cm depth from the central 1.07 m x 1.07 m area of 2 m x 2 m unconfined microplots. The auger passed through a steel annulus on which the soil was collected. For the 23-50 and 50-70 cm depths a 15 cm diameter auger was used. Most of the ^{15}N was in the 0-23 cm depth, < 5% in the 23-50 cm depth, and c. 0.5% in the 50-70 cm depth.

To reduce sampling errors associated with banding of ^{15}N fertilizer, Moraghan et al. (1986a) sampled a soil block 25 cm wide, 30 cm deep and extending the width of 1 m wide x 2 m long unconfined microplots in 0-15 and 15-30 cm depth increments. Five soil cores, 5 cm in diameter and in depth increments 0-15, 15-30, 30-60 and 60-90 cm, were composited to assess ^{15}N at depth. Most of the ^{15}N was in the top 30 cm with a maximum of 2.5% of the applied ^{15}N being in the 60-90 cm depth.

In summary, substantial errors can occur in ^{15}N balance sheets due to poor sampling of soil (in particular) and plants. I recommend that researchers pay more attention to improving sampling techniques and report them more fully when publishing results. Local ingenuity has resulted in a wide range of devices and techniques for sampling soil. Although soil cores are extensively used they are not satisfactory where ^{15}N is non-uniformly distributed. Complete excavation of soil is preferred but often is logistically not feasible. Since most of the residual ^{15}N is usually present in the upper 20 cm of soil I recommend that researchers fully excavate to 20 cm and take soil cores below that depth.

REFERENCES

Allison, F.E. (1966). The fate of nitrogen applied to soils. Adv. Agron. 18, 219-58.

Barker, A.V. (1974). Nitrate determinations in soil, water and plants. Massachusetts Agric. Exp. Sta. Res. Bull. 611.

Bartholomew, W.V. (1964). Guides in extending the use of tracer nitrogen in soils and fertilizer research. In 'Soil and Fertilizer Nitrogen Research, a Projection into the Future — a Symposium' pp. 81-96 (Tennessee Valley Authority, Wilson Dam: Alabama.)

Bartholomew, W.V. (1971). ^{15}N research on the availability and use of crop nitrogen. In 'Nitrogen-15 in Soil-Plant Studies'. pp. 1-20. (International Atomic Energy Agency: Vienna.)

Bartholomew, W.V., Nelson, L.R., and Volk, R.J. (1965). Consideration of variance components for sampling and analytical procedures in the design for ^{15}N experiments. Agron. Abstr. p. 82.

Berndt, R.D., Strong, W.M., and Craswell, E.T. (1976). A device for procuring undisturbed cores of cracking clay soils. Queensland J. Agric. Anim. Sci. 33, 115-20.

Buresh, R.J., Austin, E.R, and Craswell, E.T. (1982). Analytical methods in ^{15}N research. Fert. Res. 3, 37-62.

Carter, J.N., Bennett, O.L., and Pearson, R.W. (1967). Recovery of fertilizer nitrogen under field conditions using nitrogen-15. Soil Sci. Soc. Am. Proc. 31, 50-6.

Chapman, A.L., Xu, Z.H., Saffigna, P.G., Myers, R.J.K., and McCown, R.L. (1987). Nitrogen cycling in Leucaena alley cropping in the semi-arid tropics. In 'Poster Abstracts of Advances in Nitrogen Cycling in Agricultural Ecosystems, 11-15 May 1987' (Ed. G.T. Adams). pp. 105-6 (CSIRO: Brisbane).

Chichester, F.W. and Smith, S.J. (1978). Disposition of ^{15}N-labelled fertilizer nitrate applied during corn culture in field lysimeters. J. Environ. Qual. 7, 227-33.

Cline, M.G. (1944). Principles of soil sampling. Soil Sci. 58, 275-88.

Cogle, A.L. (1986). Carbon and nitrogen transformations during the decomposition of crop residues. PhD Thesis. Griffith University, Brisbane, Qld., Australia.

Craswell, E.T. (1979). Isotopic studies on the nitrogen balance in a cracking clay. IV. Fate of the three nitrogen fertilizers in fallow soil in the field. Aust. J. Soil Res. 14, 317-23.

Hauck, R.D., and Bremner, J.M. (1976). Use of tracers for soil and fertilizer nitrogen research. Adv. Agron. 28, 219-66.

Henzell, E.F., Martin, A.E., Ross, P.J., and Haydock, K.P. (1968). Isotopic studies on the uptake of nitrogen by pasture plants. IV. Uptake of nitrogen from labelled plant material by Rhodes grass and Siratro. Aust. J. Agric. Res. 19, 65-77.

Jones, J.B., and Steyn, W.J.A. (1973). Sampling, handling and analysing plant tissue samples. In 'Soil testing and Plant analysis'. (Eds. L.M. Walsh and J.D. Beaton). pp. 249-70. (Soil Sci. Soc. Am.: Madison, Wisconsin).

Khanif, Y.M., Van Cleemput, O., and Baert, L. (1983). Fate of field-applied labelled fertilizer nitrate on sandy soils. Plant Soil 74, 473-6.

Khanif, Y.M., Van Cleemput, O., and Baert, L. (1984). Field study of the fate of labelled fertilizer nitrate applied to barley and maize in sandy soils. Fert. Res. 5, 289-94.

Legg, J.O., and Meisinger, J.J. (1982). Soil nitrogen budgets. In 'Nitrogen in Agricultural Soils'. Agronomy 22. (Ed. F.J. Stevenson). pp. 503-66. (Am. Soc. Agron. Madison: Wisconsin.)

Martin, A.E., Henzell, E.F., Ross, P.J., and Haydock, K.P. (1963). Isotopic studies on the uptake of nitrogen by pasture grasses. I. Recovery of fertilizer nitrogen from the soil: plant system using Rhodes grass in pots. Aust. J. Soil Res. 1, 169-84.

Martin, A.E., and Skyring, G.W. (1962). Losses of nitrogen from the soil-plant system. In 'Review of Nitrogen in the Tropics with Particular Reference to Pastures: a Symposium'. Bull. No. 46. pp. 19-34. (Commonw. Bur. Past. Fld. Crops: Berkshire, England).

Moraghan, J.T., Rego, J.T., and Buresh, R.J. (1984b). Labelled nitrogen fertilizer research with urea in the semi-arid tropics. III. Field studies on alfisol. Plant Soil 82, 193-203.

Moraghan, J.T., Rego, J.T., Buresh, R.J., Vlek, P.L.G., Burford, J.R., Singh, S., and Sahrawat, K.L. (1984a). Labelled nitrogen fertilizer research with urea in the semi-arid tropics. II. Field studies on a vertisol. Plant Soil 80, 21-33.

Myers, R.J.K. and Hibberd, D.E. (1985). Fate of fertilizer nitrogen applied to cracking clay soils in central Queensland under dryland and irrigated conditions. In 'Root Zone Limitation to Crop Production on Clay Soils'. (Eds. W.A. Muirhead and E. Humphreys). pp. 283-90. (CSIRO: Melbourne.)

Olson, R.V. (1980). Plot size requirements for measuring residual fertilizer nitrogen and nitrogen uptake by corn. Soil Sci. Soc. Am. J. 44, 428-9.

Peterson, R.G., and Calvin, L.D. (1965). Sampling. In "Methods of Soil Analysis. Physical and Mineralogical Properties. (Eds. C.A. Black et al.). Agron. No. 9, Part 1, pp. 54-72. (Am. Soc. Agron.: Madison).

Powlson, D.S., Pruden, G., Johnston, A.E., and Jenkinson, D.S. (1986a). The nitrogen cycle in the Broadbalk Wheat Experiment: recovery and losses of ^{15}N-labelled fertilizer applied in spring and inputs of nitrogen from the atmosphere. J. Agric. Sci. Camb. 107, 591-609.

Powlson, D.S., Hart, P.B.S., Pruden, G., and Jenkinson, D.S. (1986b). Recovery of ^{15}N-labelled fertilizer applied in autumn to winter wheat at four sites in eastern England. J. Agric. Sci. Camb. 107, 611-20.

Pruden, G., Powlson, D.S., and Jenkinson, D.S. (1985). The measurement of ^{15}N in soil and plant material. Fert. Res. 6, 205-18.

Ritchie, J.T., Kissel, D.E., and Burnett, G. (1972). Water movement in undisturbed swelling clay soil. Soil Sci. Soc. Am. Proc. 36, 874-9.

Saffigna, P.G. (1977). Fertilizer nitrogen balance and transformations in Queensland wheat soils using ^{15}N. M. Agr. Sc. Thesis, University of Queensland, Brisbane, Australia.

Saffigna, P.G., Cogle, A.L., McMahon, G., and Prove, B. (1984a). Evaluation of the usefulness of in situ field cores for measuring solute movement in a vertisol. In 'The Properties and Utilization of Cracking Clay Soils'. (Eds. J.W. McGarity, E.H. Hoult, and H.B. So). pp. 181–3. Rev. Rur. Sci. 5 (Univ. New England: Armidale).

Saffigna, P.G., Cogle, A.L., Strong, W.M., and Waring, S.A. (1982). The effect of carbonaceous residue on ^{15}N fertilizer nitrogen transformations in the field. In 'The Cyling of Carbon, Nitrogen, Sulfur and Phosphorus in Terrestrial and Aquatic ecosystems'. (Eds. I.E. Galbally and J.R. Freney). pp. 83–7. (Aust. Acad. Sci.: Canberra).

Saffigna, P.G., Cogle, A.L., Strong, W.M., and Waring, S.A. (1984b). The effect of early application of ^{15}N fertilizer on recovery of nitrogen by wheat grown in a vertisol. In 'The Properties and Utilization of Cracking Clay Soils'. (Eds. J.W. McGarity, E.H. Hoult and H.B. So). pp. 227–30. Rev. Rur. Sci. 5 (Univ. New England: Armidale).

Saffigna, P.G., Cogle, A.L., and Turk, M.J. (1981). Leaching of bromide and ^{15}N-labelled nitrate in a cracking clay. In "Salinity and Water Quality". (Eds. A.J. Rixon and R.J. Smith). pp. 87–92. (Darling Downs Inst. Adv. Educ.: Toowoomba, Australia).

Saffigna, P.G., Strong, W.M., and Mason, M.G. (1985). Nitrogen fertilizer availability to wheat underfield conditions in Queensland and Western Australia measured with ^{15}N. Proc. 3rd Aust. Agron. Conf., 30 Jan.–1 Feb. 1985, Univ. Tasmania, Hobart, pp. 249. (Aust. Soc. Agron.: Parkville, Australia).

Saffigna, P.G., and Wood, A.W. (1986). Nitrogen studies with sugar cane trash in the Herbert Valley. Aust. Soc. Sugar Cane Technologists, Ann. Conf., Townsville, Australia, p. 103.

Sisworo, W.H., Rasyio, H., Mardjo, M., and Myers, R.J.K. (1987). The relative roles of N fixation, fertilizer, crop residues and the soil in supplying N in multiple cropping systems in Lampung Province, Indonesia. In 'Poster Abstracts of Advances in Nitrogen Cycling in Agricultural Ecosystems, 11–15 May 1987'. (Ed. G.T. Adams). pp. 156–8. (CSIRO: Brisbane.)

Snedecor, G.W., and Cochran, W.G. (1967). Statistical Methods. 6th ed. (Iowa State University Press: Amers.)

Sudin, M.N., and Bachik, A.T. (1987). Study of uptake and distribution of nitrogen in HEVEA using ^{15}N techniques. In 'Urea–Tech 87'. Intern. Symp. Urea Techn. Utiliz., p. 35. (Malaysian Soc. Soil Sci.: Kuala Lumpur).

Vallis, I. (1969). The measurement of gains and losses of nitrogen in grazed pastures. PhD Thesis, University of Queensland, Brisbane, Australia.

Vallis, I., Catchpoole, V.R., Ludlow, M.M., and McGill, W.B. (1986). Nitrogen and carbon dynamics in a green panic sward. CSIRO Div. Trop. Crops Past. Ann. Rep. 1985–86, pp. 51–2.

Vallis, I., Haydock, K.P., Ross, P.J., and Henzell, E.F. (1967). Isotopic studies on the uptake of small additions of ^{15}N-labelled fertilizer by Rhodes grass and Townsville stylo. Aust. J. Agric. Res. 18, 865–77.

Vallis, I., Henzell, E.F., Martin, A.E. and Ross, P.J. (1973). Isotopic studies on the uptake of nitrogen by pasture plants. V. ^{15}N balance experiments in field microplots. Aust. J. Agric. Res. 24, 693–702.

Van Cleemput, O., Hofman, G., and Baert, L. (1981). Fertilizer nitrogen balance study on sandy loam with winter wheat. Fert. Res. 2, 119–26. (Commonw. Agric. Bur. Intern.: Wallingford, U.K.)

White, P.J. (1984). The effect of stubble management practices on nitrogen availability to field-grown wheat. PhD Thesis, Griffith University, Brisbane, Australia.

White, P.J., Vallis, I., and Saffigna, P.G. (1986). The effect of stubble management on the availability of ^{15}N-labelled residual fertilizer nitrogen and crop stubble nitrogen in an irrigated black earth. Aust. J. Exp. Agric. 26, 99–106.

Woodcock, T.M., Pruden, G., Powlson, D.S., and Jenkinson, D.S. (1982). Apparatus for applying ^{15}N-labelled fertilizer uniformly to field micro-plots. J. Agric. Engng. Res. 27, 369–72.

Xu, Z.H., Cao, Z.H., and Li, C.K. (1987). Studies on the effect of urea supergranules on maize and the fate of nitrogen in calcareous sandy soil. In 'Poster Abstracts of Advances in Nitrogen Cycling in Agricultural Ecosystems, 11–15 May 1987'. (Ed. G.T. Adams). pp. 173–5. (CSIRO: Brisbane).

Zaharah, A.R., and Sharifuddin, H.A.H. (1986). Measurement of nitrogen and phosphorous uptake by oil palm trees using isotope techniques. In Workshop Report. (Eds. J. Sharifuddin and Y. Othman). pp. 118–25. (Soil Sci. Dept., Fac. Agric., Universiti Pertanian Malaysia: Selangor.)

Zapata, F., and Van Cleemput, O. (1986a). Fertilizer nitrogen recovery and biological nitrogen fixation in fababean – sugar beet and spring wheat – fababean cropping sequences. Fert. Res. 8, 263–8.

Zapata, F., and Van Cleemput, O. (1986b). Recovery of ^{15}N-labelled fertilizer by sugar beet-spring wheat and winter rye – sugar beet cropping sequences. Fert. Res. 8, 269–78.